Study Guide for Atkins and Jones's

CHEMISTRY

Molecules, Matter, and Change

THIRD EDITION

DAVID BECKER

Oakland Community College
Oakland University

W. H. FREEMAN AND COMPANY

NEW YORK

ISBN: 0-7167-3032-4

Printed in the United States of America

First printing 1997, VB

CONTENTS

PREFACE

This *Study Guide* is designed to accompany the third edition of *Chemistry: Molecules, Matter, and Change* by Peter Atkins and Loretta Jones. The overwhelmingly positive response to the first and second editions, especially by chemistry students, has been truly heartening. I have, therefore, followed the format and style of the first two editions. I have also tried to anticipate and elaborate on the fundamental and critical information that is required by a student in learning the textbook material. Rather than being comprehensive, I have chosen to cover in depth what I consider to be the most important topics so that the *Study Guide* will be helpful not only to the student who requires basics to build on, but also to the student who desires some elaboration of topics. (Even so, approximately 95 percent of the textbook material is covered in some form or another.) Because I believe that it is largely through working problems that a student develops a full appreciation of chemical principles, the focus of the *Study Guide* is decidedly on problem solving. With the addition of questions to the Self-Test exercises, the third edition now provides the student with more than 1075 such exercises; the Self-Test exercises are similar to those I use on my own exams, thus they should provide excellent practice for the type of material a student might encounter on an exam.

The symbols, concepts, and style of the *Study Guide* are identical to those in the Atkins/Jones text, so students should have little difficulty moving from one to the other. An unusually large number of graphs, figures, and chemical structures enriches the *Study Guide* material.

Each *Study Guide* chapter parallels a chapter in *Chemistry: Molecules, Matter, and Change,* 3e. The key ideas of each textbook section are stated and sometimes expanded on in a parallel **Key Concepts** section. The Key Concepts introduce no new material, but restate textbook material in different words. Each Key Concept is accompanied by at least one worked out (in detail!) example problem and a parallel exercise that has an answer but no detailed solution. A **Pitfalls** section sometimes points out common errors or misconceptions; at other times it offers insights on concepts or problem-solving techniques. A **Key Words** section is included for each major chapter division. The descriptive chemistry that appears in Chapters 1–18 and 22 of the text is reinforced at the end of the *Study Guide* chapters under the headings **Descriptive Chemistry to Remember** and **Chemical Equations to Know**. Also included is a **Mathematical Equations to Know and Understand** section. Finally, a large variety of multiple-choice questions appear as **Self-Test** exercises. These questions vary in difficulty from remedial to very challenging, although basic single-concept problems have been emphasized.

The usefulness of this *Study Guide* depends on its accuracy and clarity. If you discover any errors or have an idea regarding how to say something more clearly than it is presented, please take the time to drop me a note with your ideas. I would especially like to expand the Pitfalls sections—if you encounter a problem area in the course that would benefit from a Pitfall comment, I would be particularly interested in hearing from you. If your idea does result in a Pitfall in future editions of the *Study Guide,* you will be credited with the idea (if you wish). I can be reached at: David Becker, Department of Chemistry, Oakland University, Rochester, MI 48309; FAX: 810 370-2321; email: dbecker@ouchem.chem.oakland.edu

TO THE STUDENT

You have registered for a General Chemistry course with a goal in mind, and possibly a long-term career plan that requires a knowledge of chemistry. If you have been permitted by your college or university to sign up for a chemistry course, you can succeed in it; but it will take time, discipline, and perseverance. As a rule of thumb, a four-credit chemistry course requires a minimum of 10 to 15 hours of home study per week. It would be wise for you to set up a weekly schedule that includes specific times for home study and stick to it! (A blank weekly calendar follows this section.)

It would be foolhardy to try to write a rigid prescription for studying chemistry; every one of us is different. Nevertheless, the following steps have proved successful for many students (including many of my own). I suspect that any successful plan of study will not deviate too much from these steps. Do the following for each chapter in your textbook:

1. Read the chapter once lightly. Try to get an overview of the material without taking detailed notes or underlining. Take a break.

2. Reread the chapter (or required sections) very carefully. Underline or take notes. Work out derivations. Try to work all example problems and exercises within the chapter. When you finish each textbook section, read the relevant Key Concepts in the *Study Guide* and try to work out the *Study Guide* examples and exercises that accompany the Key Concepts. Write the definitions of all the Key Words in the *Study Guide* (writing the definitions of all the boldfaced words in the text narrative would be even better.)

3. Try to do the assigned homework problems. It is most important to write all the answers to the problems in an organized way. At this stage you will probably not be able to do all the assigned problems; don't spend more than 15 minutes on any one problem.

4. Attend the lecture. Take notes and listen. But don't let taking notes deter you from listening to your instructor. Don't be afraid to ask questions in the lecture. (I have found in my own lectures that when one student asks a question, many other students have the same question in mind. There is no such thing as a "dumb question.")

5. Complete the homework problems and exercises from the text and *Study Guide* that you could not do before the lecture. If you are unable to do any problems at this stage with the textbook and *Study Guide* alone, get help from your instructor, college tutors, or fellow students. In many colleges, computer-aided instruction (CAI) is available. (You don't have to know how to use a computer to use CAI; it's a marvelous opportunity for self-paced learning.)

6. At this point you will understand most of the material in the chapter. Reread the chapter carefully to try to put all the material together into a unified whole.

7. Do the Self-Test exercises in the *Study Guide* as a test of your knowledge. If you can't work any of the exercises, get help.

8. Before an exam, go over your text underlinings, your notes, and any *Study Guide* material that your instructor has assigned (such as the Descriptive Chemistry to Remember, Chemical Equations to Know, and Mathematical Equations to Know and Understand).

9. Try to pace your studying so that you have a few hours to relax and get a good night's sleep before the exam. Last minute cramming is notoriously unsuccessful in chemistry.

It is most important to set up a disciplined study schedule that you will follow each week. After consideration of your schedule of classes and job and family obligations, block out, on the following calendar, at least 10 hours a week (more if your instructor suggests) when you can faithfully study chemistry. If possible, reserve one day (or half a day) a week for recreation and relaxation; try to avoid doing any schoolwork at this time.

	Mon	Tues	Wed	Thurs	Fri	Sat	Sun
7 A.M.							
8							
9							
10							
11							
Noon							
1 P.M.							
2							
3							
4							
5							
6							
7							
8							
9							
10							
11							

ACKNOWLEDGMENTS

Any author who writes a scholarly text quickly discovers that, to be successful, any such effort must be a group effort. Many people contributed to the three editions of this *Study Guide* and helped to improve it in countless ways.

Matthew E. Hackett of Dinosaur Hill Nature Preserve in Rochester, Michigan, checked all of the problems for accuracy and clarity. Erica Seifert and Matthew Fitzpatrick of W. H. Freeman could not have been more helpful with editorial support; the unsung heroine of the editing process, copy editor Jodi Simpson, helped immensely to make material clear, complete, and readable.

George W. Eastland, Jr., of Saginaw Valley State University, University Center, Michigan, suggested a number of insightful improvements to the second edition; Margot E. Getman and Georgia Lee Hadler of Scientific American Books provided outstanding editorial support with remarkable good humor. The copy editor for the second edition, Yvonne Howell, also made substantial and thoughtful contributions to the text.

A large number of individuals carefully reviewed the first edition and suggested many important changes. Among them are Judith R. Fish, Oakland University, Rochester, Michigan; John Forsberg, Saint Louis University; Forrest C. Hentz, Jr., North Carolina State University; Jeffrey A. Hurlburt, Metropolitan State College, Denver, Colorado; Earl S. Huyser, University of Kansas; Wayne Lowder, Department of Energy, New York; Richard A. Potts, University of Michigan, Dearborn Campus; Don Roach, Miami Dade Community College; and Robert L. Stern, Oakland University. Peter W. Atkins of Oxford University offered invaluable ideas and moral support at many stages of the writing. Gary Carlson, who originally conceived of this *Study Guide* and guided it through its development and culmination, deserves special mention.

My wife, Rita, and sons, Shanan, Seth, and Daniel, again deserve thanks for their patience and support.

CHAPTER 1

MATTER

Experience indicates that all matter is made from approximately 100 basic building blocks called **elements**. These elements can combine to form the millions of substances known to science.

THE ELEMENTS

1.1   ATOMS

KEY CONCEPT   An element is composed of one kind of atom

Imagine cutting a piece of copper in half, and then cutting it in half again and again until we get the smallest piece of matter that is still copper. This smallest piece is an atom of copper. Any further division of the atom into smaller pieces would result in matter that is no longer copper. Because it is a common observation that any sample of a particular element has the same characteristics as any other sample of the same element, it must follow that any one atom of an element must be, in its most important attributes, the same as any other atom of that element.

EXAMPLE   Applying the atomic hypothesis

A sample of a type of atom is found to be magnetic; that is, the atoms respond to the presence of a nearby magnet. Another sample of atoms, under identical conditions, is not magnetic. Are these samples made up of atoms of the same element?

SOLUTION No. According to Dalton's hypothesis, all the atoms of a given element are identical. Because the two samples respond differently to a magnetic field, the atoms are not identical and, therefore, not of the same element.

EXERCISE Atoms are often considered to be spheres; the radius of an atom can then be thought of as the distance from the center of the atom to its surface. Are atoms of the same element likely to have the same radii or different radii?

[*Answer:* The same]

PITFALL   Are atoms of a single element completely identical?

Atoms of any one element are all very similar; so similar, in fact, that we are confident in saying that an element is composed of only one kind of atom. However, atoms of an element do not have to be identical in every respect. Just as we humans are animals of one kind, we are not all identical!

1.2   NAMES OF THE ELEMENTS

KEY CONCEPT   Names and symbols of the elements

Chemists (and chemistry students) must be able to write the **symbols** of the elements in their day-to-day work. When writing the symbols that have two letters, remember that the first letter of an elemental symbol is always uppercase and the second lowercase. Thus, for mercury, Hg is the only correct symbol. In some cases, the use of only uppercase letters can lead to an erroneous representation, because a symbol improperly written in uppercase can be confused with the symbols for two elements.

EXAMPLE   Writing elemental symbols properly

Find four examples in which writing an elemental symbol in uppercase letters could lead to confusion.

SOLUTION Chemists expect an elemental symbol to be an uppercase letter if the symbol has only one letter and to be an uppercase followed by a lowercase letter if the symbol has two letters. Thus, two uppercase letters look like adjacent symbols for two elements. To answer the question, we must look at the periodic table for symbols that have two letters, each of which corresponds to the symbol for an element.

Symbol for One Element	Possible Confusing Misrepresentation
Sc scandium	SC sulfur, carbon
Hf hafnium	HF hydrogen, fluorine
Ni nickel	NI nitrogen, iodine
Pu plutonium	PU phosphorus, uranium

EXERCISE Formulate a situation in which writing the symbol of an element in lowercase letters may cause confusion.

[*Answer:* NO intended (combination of nitrogen and oxygen), but No written (element nobelium).]

1.3   THE NUCLEAR ATOM

KEY CONCEPT A   The nuclear atom: Electrons, protons, and neutrons

The three **subatomic particles** in the atom are the electron, the proton, and the neutron. The **electron**, discovered by J. J. Thomson, is a negatively charged particle that is a substituent of all matter. Because all atoms have zero electrical charge, some positively charged particle must be present in matter to offset the negatively charged electron. The positive particle is called a **proton**. The electron has a charge of -1 and the proton $+1$; thus, for the atom to be electrically neutral, the number of protons in an atom must be the same as the number of electrons. Rutherford established the location of the electrons and protons in the atom with a scattering experiment. The experiment consists of firing α particles at gold atoms and measuring how the particles deflect as they interact with the atoms. The results Rutherford observed were

1. Most of the α particles went through the gold foil undeflected.
2. A few of the α particles were deflected a bit.
3. Very few α particles (about 1 in 20,000) were strongly deflected; some even bounced straight back.

We explain these results by assuming that a gold atom consists of an extremely small, heavy, positively charged nucleus surrounded by tenuous, light electrons. The accompanying figure shows the experiment schematically: The small black dots represent the massive, positively charged gold nuclei. α particles (which are positively charged) are repulsed by the positive charge of the nucleus.

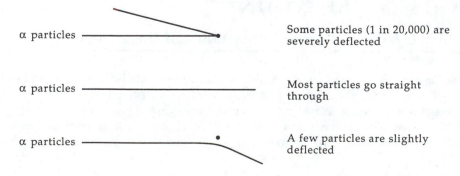

α particles ——————— Some particles (1 in 20,000) are severely deflected

α particles ——————— Most particles go straight through

α particles ——————— A few particles are slightly deflected

The **neutron,** which has zero charge and a mass close to that of the proton, completes the model of the **nuclear atom.** In this model, the positively charged protons and the electrically neutral neutrons, which are both massive, form the tiny nucleus. The nucleus is positively charged because the proton carries a positive charge. The electron, which is a light particle, moves about the nucleus in such a way that most of the volume of the atom is empty space; the negative charges of the electrons offset the positive charges of the protons.

EXAMPLE   The nuclear atom

The text states that the nucleus of a gold atom is a dense sphere about 10^{-14} m in diameter and that the diameter of the whole atom is about 10^{-9} m in diameter (100,000 times greater). Use the formula for the volume of a sphere, $V = \frac{4}{3}\pi r^3$, to calculate the percentage of the volume of the gold atom taken up by the nucleus. The r in the formula is the radius of the sphere.

SOLUTION We expect the nucleus to have a much smaller volume than the whole atom because its diameter is so much smaller than that of the atom. To get the percentage of the volume of the atom taken up by the nucleus, we use the following formula:

$$\text{Percentage volume occupied by nucleus} = \frac{\text{volume of nucleus}}{\text{volume of atom}} \times 100$$

For the nucleus, $r = 10^{-14}/2$ m $= 5 \times 10^{-15}$ m, and

$$V = \tfrac{4}{3}\pi r^3 = \tfrac{4}{3}\pi(5 \times 10^{-15}\ \text{m})^3 = \tfrac{4}{3}\pi(125 \times 10^{-45}\ \text{m}^3) = 5 \times 10^{-43}\ \text{m}^3$$

For the whole atom, $r = 10^{-9}/2$ m $= 5 \times 10^{-10}$ m, and

$$V = \tfrac{4}{3}\pi r^3 = \tfrac{4}{3}\pi(5 \times 10^{-10}\ \text{m})^3 = \tfrac{4}{3}\pi(125 \times 10^{-30}\ \text{m}^3) = 5 \times 10^{-28}\ \text{m}^3$$

Now, using the volumes just calculated, we can answer the question:

$$\text{Percentage volume occupied by nucleus} = \frac{5 \times 10^{-43}\ \text{m}^3}{5 \times 10^{-28}\ \text{m}^3} \times 100 = 1 \times 10^{-13}\%$$

This is 0.000 000 000 000 1% of the volume of the atom and confirms that only a tiny bit of the volume of the atom is taken up by the nucleus.

EXERCISE A gold atom has 79 protons and 79 electrons. Assume it has 118 neutrons. Use the data in Table 1.1 of the text to calculate the percentage of the mass of the atom that is in the nucleus. Is it correct to say that most of the mass of the atom is in the nucleus?

[*Answer:* 99.98%; yes]

KEY CONCEPT B   Atomic number

The number of protons in an atom is called the **atomic number** of the atom. Because all atoms are electrically neutral (have zero electrical charge), the number of electrons in the atom equals the number of protons, and the atomic number equals the number of electrons for a neutral atom. The atomic number specifies the element that any atom corresponds to. For instance, every oxygen atom has an atomic number of 8, and any atom with an atomic number of 8 must be an oxygen atom. In the periodic table, the atomic number is the integer that accompanies each elemental symbol: 1 for hydrogen, 2 for helium, 92 for uranium. The atomic number is often symbolized by Z.

EXAMPLE 1   The protons and electrons in an atom

How many protons and how many electrons are present in a single atom of cadmium (Cd)?

SOLUTION Both the number of protons and the number of electrons in an atom are given by the atomic number. Cadmium is a transition metal in the fifth period; its atomic number is 48. Cadmium, therefore, has 48 protons and 48 electrons.

EXAMPLE 2 Using the atomic number

A 15-g crystal of sodium chloride contains 1.5×10^{23} atoms of sodium and an equal number of chlorine atoms. What is the mass of the electrons in the sample? Is it a large or small percentage of the total mass of the crystal?

SOLUTION The atomic numbers of sodium and chlorine are 11 and 17, respectively; so each sodium atom must contribute 11 electrons to the crystal, and each chlorine must contribute 17 electrons. Thus, we can calculate the total number of electrons in the crystal:

$$\text{Electrons from Na} = 1.5 \times 10^{23} \text{ atom Na} \times \frac{11 \text{ electrons}}{1 \text{ atom Na}} = 1.7 \times 10^{24} \text{ electrons}$$

$$\text{Electrons from Cl} = 1.5 \times 10^{23} \text{ atom Cl} \times \frac{17 \text{ electrons}}{1 \text{ atom Cl}} = 2.6 \times 10^{24} \text{ electrons}$$

$$\text{Total number of electrons} = 1.7 \times 10^{24} \text{ electrons} + 2.6 \times 10^{24} \text{ electrons} = 4.3 \times 10^{24} \text{ electrons}$$

The mass of each electron is 9.11×10^{-28} g (Table 1.1 of the text), so

$$\text{Total mass of electrons} = 4.3 \times 10^{24} \text{ electrons} \times \frac{9.11 \times 10^{-28}}{1 \text{ electron}} = 3.9 \times 10^{-3} \text{ g}$$

The percentage of the mass of the crystal due to the electrons is

$$\text{Percentage} = \frac{\text{mass electrons} \times 100}{\text{total mass}} = \frac{3.9 \times 10^{-3} \text{ g}}{15.0 \text{ g}} \times 100 = 2.6 \times 10^{-2}\%$$

The electrons, as expected, constitute a small percentage of the mass of the crystal.

EXERCISE A cup containing 250 g of water contains approximately 8.4×10^{24} atoms of oxygen and 1.7×10^{25} atoms of hydrogen. What is the mass of the electrons in the water? What percentage of the mass of the water is contributed by the electrons?

[*Answer:* 7.7×10^{-2} g; 0.031%]

1.4   ISOTOPES

KEY CONCEPT   Isotopes

The development of an instrument called a mass spectrometer enabled scientists to obtain very precise measurements of the masses of atoms. In a mass spectrometer, an atom is converted to a positive ion by removal of an electron. (A chemical species with a positive or negative charge is called an ion.) The ions are then separated according to their masses. The precise measurement of masses resulted in a completely unexpected discovery. For almost every element, a variety of atoms exist with slightly different masses. For instance, three types of neon atoms can be characterized, as shown in the following table:

Element	Number of Protons	Mass (g)	Mass Relative to Hydrogen
neon-20	10	3.32×10^{-23}	20
neon-21	10	3.49×10^{-23}	21
neon-22	10	3.65×10^{-23}	22

Each of the three atoms shown has 10 protons, so each is a form of neon. Therefore, all are chemically equivalent and all are represented by the symbol Ne on the periodic table. Each has a slightly different mass, however. When the same element exists with different masses, the different forms are called **isotopes** of the element. In this case, three isotopes of neon exist: neon-20, neon-21, and neon-22.

EXAMPLE

How much of the total mass in a neon-20 atom can be accounted for by the 10 protons and 10 electrons in the atom? Is there a significant mass left unaccounted for?

SOLUTION We know the total mass of a neon-20 atom and the individual masses of an electron and a proton (Table 1.1 of the text). Because we also know the number of electrons and protons in the atom, we can calculate how much mass they account for by multiplying the number of each by its mass. The mass unaccounted for is obtained by subtracting the total mass of the protons and electrons from the mass of the atom:

$$\text{Total mass of protons} = 10 \text{ protons} \times \frac{1.67 \times 10^{-24} \text{ g}}{1 \text{ proton}} = 1.67 \times 10^{-23} \text{ g}$$

$$\text{Total mass of electrons} = 10 \text{ electrons} \times \frac{9.11 \times 10^{-28} \text{ g}}{1 \text{ electron}} = 9.11 \times 10^{-27} \text{ g}$$

$$\text{Total mass accounted for} = 1.67 \times 10^{-23} \text{ g} + 9.11 \times 10^{-27} \text{ g} = 1.67 \times 10^{-23} \text{ g}$$

The total mass unaccounted for equals the mass of the atom minus the mass of the electrons and protons:

$$\text{Unaccounted mass} = (\text{mass of neon-20}) - (\text{mass of protons} + \text{electrons})$$

$$= 3.32 \times 10^{-23} \text{ g} - 1.67 \times 10^{-23} \text{ g} = 1.65 \times 10^{-23} \text{ g}$$

A significant amount of mass, 49.7% of the mass of the atom, is left unaccounted for:

$$\frac{1.65 \times 10^{-23} \text{ g}}{3.32 \times 10^{-23} \text{ g}} \times 100 = 49.7\%$$

(Owing to effects to be described in Chapter 22, this calculation is an approximation.)

EXERCISE Assume that the difference in the masses of neon-20 and neon-21 is due to a single extra subatomic particle in the neon-21. What is the mass of the particle? Could it be any of the particles shown in Table 1.1 of the text?

[*Answer:* 1.7×10^{-24} g; neutron (could not be proton because proton would change identity of element)]

1.5 WHERE DID THE ELEMENTS COME FROM?

KEY CONCEPT   The elements are formed in the stars

In stars, a process called nuclear fusion creates larger elements from smaller ones. In our own Sun, for instance, hydrogen atoms are combined to form helium atoms (this process results in the release of enormous amounts of heat and light). Ordinary stars, such as the Sun, can produce the lighter elements, but the extraordinary heat and pressure that occur in novas and supernovas are required for formation of the heavier elements. It is astounding to realize that there is a high probability that one of the carbon atoms in your body originated, billions of years ago, in the far reaches of the universe.

EXAMPLE   The products of fusion

One of the fusion reactions that can occur in a star is the combination of ^{12}C and ^{4}He to produce a single new element. What is the identity of the new element that is produced?

SOLUTION Because a single new element is produced, all of the protons in the original carbon and helium atoms must be used to form the nucleus of the new element. Carbon has six protons and helium has two protons, so the new element must have eight protons. Thus, the new element must be oxygen. The process can be represented by the following nuclear reaction. Note that, in this process, the six neutrons in ^{12}C and the two neutrons in ^4He also become part of the ^{16}O nucleus.

$$^{12}C + {}^4He \longrightarrow {}^{16}O$$

EXERCISE In another fusion process that occurs in some stars, two ^{12}C atoms combine to form ^1H plus one other element. What is the other element?

[*Answer:* ^{23}Na]

1.6 THE PERIODIC TABLE

KEY CONCEPT   Groups and periods in the periodic table

The periodic table is a map of the elements, arranged, among other things (see Chapter 7), according to certain of their properties. It is divided into major subdivisions called groups and periods. A **period** is a horizontal row of elements; there are seven periods in the periodic table. Some of the elements in the sixth and seventh periods are placed, for convenience, under the main part of the table instead of in the main part where they actually belong. For instance, tracing across Period 6, we start at cesium (Cs, $Z = 55$), move across to barium (Ba, $Z = 56$), down to lanthanum (La, $Z = 57$), across to Ytterbium (Yb, $Z = 70$), up to lutetium (Lu, $Z = 71$), and finally, across to the end of the period at xenon (Xe, $Z = 86$). The **groups** (also called **families**) are the vertical columns in the periodic table. The elements in a group frequently show similarities in chemical and physical properties. In addition, it is common to observe a smooth change in properties as we scan down a group. This smooth change is evident in the following examples (two entries are left blank intentionally; hydrogen is not included because it is a nonmetal).

Group 1	Melting Point (K)	Group 18	Boiling Point (K)
Li	454	He	4
Na	371	Ne	27
K	?	Ar	87
Rb	337	Kr	?
Cs	301	Xe	165
Fr	300	Rn	211

Note the smooth change in melting points as we go down Group 1 and in boiling points as we go down Group 18. However, the smoothness is *not inviolate;* it is merely a tendency, not a certainty.

EXAMPLE   Using the gradation in properties of a group to estimate a property

Estimate the melting point of potassium (K).

SOLUTION To make the estimate, we depend on the regularity of changes in properties as we go down the elements in a group (and hope we have not encountered an exception). From the table, the melting point of potassium must lie between 371 K and 337 K. Without further information, the most reasonable approach would be to assume that the melting point is approximately halfway between 371 K and 337 K, or 354 K [(371 + 337)/2 = 354]. The actual melting point of potassium is 366 K. The estimate is remarkably good.

EXERCISE Estimate the boiling point of krypton (Kr).

[*Answer:* 126 K estimate, 116 K actual]

1.7 METALS, NONMETALS, AND METALLOIDS

KEY CONCEPT   Location of elements in the periodic table

All of the elements in the periodic table can be classified as either a metal, a nonmetal, or a metalloid. Common knowledge tells us much about the differences between a metal and nonmetal. The following table illustrates a few of these differences.

Metals	Nonmetals
good electrical conductors	poor conductors or nonconductors
good heat conductors	poor heat conductors
lustrous (shiny)	dull in appearance
malleable, ductile	brittle
mostly solids	many gases

As an example, let us compare a copper wire (a metal) with graphite (carbon, a nonmetal that is the main component of pencil lead). Copper possesses all of the properties of a metal, including being malleable (able to be shaped by hammering or bending) and ductile (able to be drawn into wire or hammered thin). Graphite possesses most of the typical properties of a nonmetal. Nonmetals are in the upper right of the periodic table; metals are lower and to the left. The few metalloids (Si, Ge, As, Sb, Te, Po) have properties intermediate between those of a metal and those of a nonmetal and are located in between the metals and metalloids in the periodic table. (See Fig. 1.18 of the text.)

EXAMPLE   The location of metals and nonmetals in the periodic table

Use the periodic table to decide whether the elements oxygen (O) and iron (Fe) are metals or nonmetals. List three properties of each to verify your conclusion.

SOLUTION The position of an element in the periodic table indicates whether it is a metal or a nonmetal. Oxygen is high and toward the right, so we conclude it is a nonmetal. Iron is lower and in the center, but far to the left of the nonmetals, so we conclude it is a metal. A list of properties verifies this conclusion.

Oxygen	Iron
colorless	shiny
electrically nonconducting	good electrical conductor
gas	workable solid (can be bent, hammered)

EXERCISE Give the number of a group that has a nonmetal at the top of the group and a metal at the bottom.

[*Answer:* Group 14 carbon (nonmetal) and lead (metal)]

KEY WORDS   Define or explain each term in a written sentence or two.

atom	group	metalloid	nucleon
atomic number	isotope	neutron	nucleus
electron	mass number	noble gas	periodic table
element	metal	nonmetal	proton

COMPOUNDS

1.8 WHAT ARE COMPOUNDS?

KEY CONCEPT   Ionic and molecular compounds

Most compounds can be classified as either molecular or ionic. A molecule is a unit of matter consisting of two or more atoms bonded together. A **molecular compound** is a compound made up of molecules. For instance, a water molecule consists of two hydrogen atoms bonded to one oxygen atom; it is a distinct, definite group of bonded atoms. A collection of water molecules (a glass of water, for instance) is like a bag of marbles; each water molecule is a distinct entity, just as each marble is. An **ion** is a charged chemical species. An **ionic compound** is a compound made up of a collection of oppositely charged ions positioned in a huge lattice. No ions are present in a molecular compound; molecules are not the principal structural unit in an ionic compound.

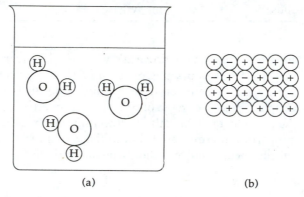

(a)   (b)

In the figure, (a) shows three discrete water molecules in a beaker of water; (b) shows a cross section of an ionic crystal, with a 1:1 ratio of positive to negative ions. No distinct molecules are present in the ionic crystal lattice.

EXAMPLE   Understanding the nature of a molecule

A picture of a model of ethanol is shown in Fig. 1.20 of the text. How many atoms of each kind are bound together to make up one ethanol molecule? Describe, in writing, which atoms are bonded to each other. What is the charge on this molecule?

SOLUTION The picture of ethanol in Fig. 1.20 shows two black spheres (representing two carbon atoms), six white spheres (representing six hydrogen atoms), and one red sphere (representing one oxygen atom). Thus, there are two carbons, six hydrogens, and one oxygen in ethanol. In the molecule, the two carbons are bonded to each other, and the single oxygen is bonded to one of the carbons; five hydrogens are each bonded to carbon, and one hydrogen is bonded to oxygen. None of the hydrogens is bonded to another hydrogen. The molecule is made up of neutral atoms bound together, so it has zero charge.

EXERCISE Imagine a compound with the same number of carbon atoms (2), hydrogen atoms (6), and oxygen atoms (1) as ethanol, but in which the atoms are bonded together differently than in ethanol. Is this new compound also ethanol?

[*Answer:* No. It is not only the number of atoms but also the way in which they bond together that determines the identity of the compound.]

1.9   MOLECULAR COMPOUNDS

KEY CONCEPT   Molecules

A **molecular formula** symbolizes how many of each kind of atoms join together to make a molecule.

A water molecule is made up of two hydrogen atoms and one oxygen joined together; hence, the molecular formula of water is H_2O. The subscript 2 indicates two hydrogen atoms. When only one atom is present in a molecule, the subscript 1 is omitted from the formula; so water is written H_2O rather than H_2O_1. We should note that, although a molecular formula gives the number of atoms of each kind that join together to form a molecule, it does not tell us how the atoms are linked.

EXAMPLE Interpreting a molecular formula

How many atoms of each kind are present in one molecule of sucrose, $C_{12}H_{22}O_{11}$?

SOLUTION The subscripts after each element symbol tell us how many atoms of that element are present in the formula. In this case, the subscript 12 following carbon tells us there are 12 carbon atoms; the subscript 22 after hydrogen indicates 22 hydrogen atoms; and the 11 after oxygen, 11 oxygen atoms.

EXERCISE How many atoms of each kind are present in one molecule of the Freon $C_2Cl_3F_3$?

[*Answer:* 2 carbon, 3 chlorine, and 3 fluorine atoms]

PITFALL   Similar symbols in an element and a compound

Even though they may look the same, the H_2 symbol for the hydrogen molecule and the H_2 in H_2O do not mean exactly the same thing. In the hydrogen molecule, the two hydrogen atoms are joined to each other. In water, two hydrogen atoms are present, but they are not joined to each other. Each is linked to the centrally located oxygen atom. For any molecule, we must remember that the formula tells us which atoms are present but not how they are linked, and the same symbols in two different molecules may not represent exactly the same thing.

KEY WORDS   Define or explain each term in a written sentence or two.

chemical formula
molecular formula
structural formula

1.10   IONIC COMPOUNDS AND IONS

KEY CONCEPT A   Cations

A **cation** is an ion with a positive charge. Before proceeding, let's make certain the origin of the positive charge is clear by examining the familiar sodium ion (Na^+). A sodium ion is formed by removing an electron from a sodium atom. If we account for the charges on the sodium atom before and after removal of the electron, we obtain the following:

Sodium Atom (Na)			Sodium Ion (Na^+)	
11 protons	+11		11 protons	+11
11 electrons	−11	remove 1 electron →	10 electrons	−10
net charge on atom	0		net charge on ion	+1

All positive ions, whether monatomic or not, have a positive charge for the same reason: They have an excess of protons over electrons. Table 1.4 of the text lists some important cations; Figure 1.24 indicates the charge on some others. From the information in these items, we note the following:

1. For Groups 1 and 2, the charge of a cation is the same as the group number: for example, K^+ in Group 1 and Mg^{2+} in Group 2.
2. For Groups 13 and 14, the maximum charge of a cation is the group number minus 10: for example, Tl^{3+} in Group 13 and Pb^{4+} in Group 14. Also, Al^{3+} and Ga^{3+} (group number − 10) are the only common ions that Al and Ga form.

3. Some atoms, especially the transition metals, are able to form differently charged cations: for example, copper forms Cu^+ and Cu^{2+}.

> ### EXAMPLE   The charge on a cation
>
> What is the charge of the ion formed by radium, element number 88? State how many electrons the ion has.
>
> **SOLUTION** Radium is a Group 2 element (check your periodic table.) We know that the Group 2 elements form ions with charge $+2$ only, so the ion is Ra^{2+}. The ion is formed by loss of two electrons from the element, so it has 86 electrons ($88 - 2 = 86$).
>
> **EXERCISE** Write the symbols of the two ions formed by tin and state how many electrons each has.
>
> [*Answer:* Sn^{2+}, 48 electrons; Sn^{4+}, 46 electrons]

KEY CONCEPT B   Anions

An **anion** is a negatively charged ion. An anion is formed when an atom gains one or more electrons. Let's reconcile the charge on the chloride ion (Cl^-) as we did earlier for the sodium ion:

Chlorine Atom (Cl)		Chloride Ion (Cl⁻)	
17 protons	$+17$	17 protons	$+17$
17 electrons	-17 $\xrightarrow{\text{gain 1 electron}}$	18 electrons	-18
net charge on atom	0	net charge on ion	-1

All anions have a negative charge because they have an excess of electrons over protons. Table 1.5 and Figure 1.27 of the text list some important anions; we note two facts from these tables:

1. Most of the simple anion names end with the suffix *-ide*. For some of the more common ions (carbide, nitride, phosphide, oxide, sulfide, fluoride, chloride, bromide, and iodide), this suffix is added to the first part of the name (the stem) of the element. For example, bromide ion is *brom* (first part of bromine) + *ide*.
2. The charge number of many of the simple anions (carbide, nitride, phosphide, oxide, sulfide, fluoride, chloride, bromide, and iodide) is equal to (group number) $- 18$. For example, the charge on phosphide, a Group 15 ion, is $15 - 18 = -3$.

> ### EXAMPLE   The charge on an anion
>
> What is the charge of the anion formed by selenium, number 34? How many electrons does the ion have?
>
> **SOLUTION** We use the periodic table as a guide to determine the charge of the ion. Selenium is a Group 16 atom. Thus, using the rule that the charge on a simple anion equals (group number) $- 18$, we expect the charge on the selenium anion to be $16 - 18 = -2$. The ion is formed by addition of two electrons to a selenium ion, so it has 36 electrons ($34 + 2 = 36$).
>
> **EXERCISE** Refer only to the periodic table to predict the charge on the simple anion formed by nitrogen. Name the ion, calculate the number of electrons on it, and write its symbol.
>
> [*Answer:* -3; nitride; 10 electrons; N^{3-}]

KEY CONCEPT C   Polyatomic ions

Let's envision a group of atoms joined together to form a unit, but with extra electrons or too few electrons. Such a unit, which would have a negative charge (extra electrons) or a positive charge (too few electrons), is called a **polyatomic ion**. A large number of common polyatomic anions contain one or more oxygen atoms and are, therefore, called **oxoanions**. Table 1.5 of the text shows some oxoanions.

EXAMPLE **The origin of the charge on a polyatomic ion**

What is the origin of the charge on the ammonium ion, NH_4^+?

SOLUTION The charge on an ion originates with an imbalance between the number of electrons and the number of protons in the ion. Nitrogen (atomic number 7) contributes 7 protons to the ion and each of the four hydrogen atoms (atomic number 1) contributes 1 proton; so 11 protons are present. For an overall charge of $+1$, 10 electrons must be present.

EXERCISE Account for the charge on the sulfate ion, SO_4^{2-}.

[*Answer:* 48 protons + 50 electrons result in -2 charge.]

PITFALL   The charge on polyatomic ions

Some confusion occasionally arises regarding the charge of the various oxoanions described in the text. In an oxoanion such as SO_4^{2-}, the subscript 4 has nothing directly to do with the charge on the anion; it refers to the number of oxygen atoms in the anion. The $2-$ superscript gives the correct charge. In a similar vein, PO_4^{3-} has a -3 charge, NO_3^- a -1 charge, and ClO_2^- a -1 charge.

KEY CONCEPT D   The formula unit

When ions come together to form an ionic compound, molecules do not form. In an ionic crystal, the positive and negative ions form a huge, repeating lattice in which there are no distinct molecules. The **formula unit** is the formula that expresses the relative number of positive ions to negative ions in the lattice, usually in a form that gives the simplest ratio of positive to negative ions. We can derive the formula unit by using the fact that all salts are electrically neutral. We assure the electrical neutrality by writing a formula unit with zero charge, one in which the sum of the charges of all the ions equals zero. Consider magnesium chloride as an example. This ionic salt is made up of Mg^{2+} ions combined with Cl^- ions. To construct a formula unit with zero charge, we must combine two Cl^- ions with one Mg^{2+} ion to give the formula unit $MgCl_2$. The two chlorides contribute a charge of -2 and the magnesium a charge of $+2$, so the sum of charges is $-2 + 2 = 0$, as required by electrical neutrality.

EXAMPLE   **Predicting a formula unit**

What is the correct formula for the formula unit of the salt composed of Na^+ (sodium ion) and CO_3^{2-} (carbonate ion)?

SOLUTION The criterion for writing a correct formula unit is that the charges of all the ions must sum to zero. The sodium ion has a $+1$ charge and the carbonate ion a -2 charge. It takes two sodium ions, which together have a total charge of $+2$ to cancel the -2 charge on the carbonate ion. The correct formula is Na_2CO_3, indicating that two Na^+ ions combine with one CO_3^{2-} ion to form the Na_2CO_3 formula unit.

EXERCISE What is for formula unit for the salt formed from the combination of Mg^{2+} ion with PO_4^{3-} ion?

[*Answer:* $Mg_3(PO_4)_2$]

PITFALL   Symbols for elements and compounds

In a formula unit such as $MgCl_2$, the Cl_2 has a different meaning than the Cl_2 used as the symbol for the diatomic chlorine molecule: The Cl_2 in $MgCl_2$ indicates two Cl^- ions, which are part of the formula unit. In the diatomic Cl_2 molecule, there are no Cl^- ions; two chlorine atoms join together to form a neutral molecule.

KEY WORDS Define or explain each term in a written sentence or two.

anion   formula unit   molecule
cation   inorganic compounds   organic compounds
chemical formula   ion   oxoanion
compound   law of constant composition   polyatomic ion
empirical formula   molecular formula   structural formula

MIXTURES

1.11 TYPES OF MIXTURES

KEY CONCEPT Solutions

A **solution** is a **homogeneous** mixture. It is homogeneous because a visual inspection of a solution does not reveal its individual components. You can try this yourself: Thoroughly dissolve $\frac{1}{4}$ teaspoon of table salt in a cup of water to make an NaCl solution and examine it carefully with a magnifying lens. The presence of the salt is not detectable. In a solution, the **solvent** is the component in which the **solute** is dissolved. Usually much more solvent is present than solute. Aqueous solutions are *always clear,* but they can also be colored. To understand what the **solubility** of a solute is, imagine trying to dissolve 5 pounds (lb) of sugar in a cup of water. After you dissolve a certain amount of sugar, the solution will contain as much sugar as it can possibly hold; it will be impossible to dissolve any more. The maximum amount that dissolves is called the solubility of the sugar; in our example, the solubility is approximately 1 lb/cup. It is possible for a solution to momentarily contain too much of a solid solute, in which case the extra solute will spontaneously form a solid. When the extra solute forms a solid slowly, large crystals tend to form. This process is called **crystallization.** When the extra solute forms a solid rapidly, a fine powder tends to form. The formation of the solid is called **precipitation.**

EXAMPLE Solutions

When we mix a small amount of sodium metal with excess liquid ammonia (which is colorless), the sodium dissolves, resulting in a homogeneous, clear, blue mixture. Is this a solution? If it is, identify the solute and solvent and state whether it is aqueous or nonaqueous.

SOLUTION A homogeneous mixture is a solution; hence the mixture described in the problem is a solution. (Solutions may be colored.) The component in excess, in this case ammonia, is the solvent; and the dissolved substance, in this case sodium, is the solute. Because the solvent is not water, the solution is nonaqueous.

EXERCISE When we mix a colorless aqueous solutions in a test tube with a light blue aqueous solution in a beaker, the solution in the beaker becomes cloudy. What has happened?

[*Answer:* A chemical reaction has occurred and a precipitate has formed.]

1.12 SEPARATION TECHNIQUES

KEY CONCEPT Physical properties and the separation of mixtures

When we take two or more substances and mix them together (without permitting a chemical reaction to occur), the result is a **mixture.** A mixture has a combination of physical properties that originate with the different components in the mixture. It can be separated into its various components by physical methods, which are methods that take advantage of the differences in certain physical properties of the components. A compound cannot be separated into elements by physical methods because separation by a physical method requires the presence of at least two pure substances with two different sets of physical properties. A compound has only a single set of physical properties.

EXAMPLE   Using physical properties

Develop a method to separate the components in a mixture of sugar and sand.

SOLUTION We must find a physical property of sugar and of sand that can be exploited to separate the two. For the separation to be successful, however, not only must the property be different for the two substances, but a technique must be available to exploit the difference. Because sugar is very soluble in water and sand is insoluble, we could accomplish a separation by adding the mixture of sugar and sand to a beaker of water and stirring thoroughly. The sugar dissolves, but the sand does not. If we now pour the contents of the beaker into a piece of filter paper (such as a coffee filter), the sand would stay behind in the filter paper and the dissolved sugar would pass through the filter paper with the water. The water can now be evaporated or *carefully* boiled off to leave the pure sugar.

EXERCISE You purchase a pound of table salt (NaCl, sodium chloride) from one company and mix it with a pound of table salt purchased from another company. Would it be possible to separate the table salt from the two different sources by a physical method?

[*Answer:* No; only one substance is present.]

KEY WORDS   Define or explain each term in a written sentence or two.

aqueous solution	distillation	physical property
chemical property	gas-liquid chromatography	precipitation
chromatogram	heterogeneous mixture	solid solution
chromatography	mixtures	
crystallization	nonaqueous solution	

THE NOMENCLATURE OF COMPOUNDS

1.13   NAMES OF CATIONS

KEY CONCEPT   Cations with different charges

The name of a cation depends on whether the atom is able to form ions of different charges; if not, the name of the cation is the same as the name of the element. The most commonly encountered cations of this type are the Group 1 and Group 2 cations and Zn^{2+}, Ag^+, Cd^{2+}, Al^{3+}, and Ga^{3+}. Some examples are

$$Li^+ \text{ lithium ion} \qquad Ca^{2+} \text{ calcium ion} \qquad Cd^{2+} \text{ cadmium ion}$$

However, if the atom forms ions of different charges, we add to the name of the element the charge on the ion, expressed as roman numerals in parentheses. Some examples are

$$Cu^+ \text{ copper(I) ion} \qquad Fe^{2+} \text{ iron(II) ion} \qquad Sn^{2+} \text{ tin(II) ion}$$

$$Cu^{2+} \text{ copper(II) ion} \qquad Fe^{3+} \text{ iron(III) ion} \qquad Sn^{4+} \text{ tin(IV) ion}$$

An older system for naming cations, which is still occasionally encountered, uses the stem of the name of the element with the suffix *-ous* for the lower charged ion or *-ic* for the higher charged ion. If the name of the element originates with a Latin name, the stem is also from the Latin. This two-suffix system can handle only the two most common ions formed by an element.

EXAMPLE Naming cations

Without consulting any references except a periodic table, name the following ions: Rb^+, Mg^{2+}, Au^{3+}, Cu^{2+}, Tl^+, Al^{3+}.

SOLUTION The name of a cation is the same as the name of the element; if the element can form differently charged ions, then the Stock number (in parentheses) must be added to the name. Metals in Groups 1 and 2 and zinc and aluminum form ions of only one charge, so no Stock number is required. Au, Cu, and Tl form differently charged ions (Fig. 1.24 of the text), thus they require a Stock number.

Rb^+	rubidium ion	Cu^{2+}	copper(II) ion
Mg^{2+}	magnesium ion	Tl^+	thallium(I) ion
Au^{3+}	gold(III) ion	Al^{3+}	aluminum ion

EXERCISE Without consulting any references, name the following ions: K^+, Ba^{2+}, Fe^{3+}, Pb^{2+}, Tl^{3+}, Zn^{2+}.

[*Answer:* K^+, potassium ion; Ba^{2+}, barium ion; Fe^{3+}, iron(III) ion; Pb^{2+}, lead(II) ion; Tl^{3+}, thallium(III) ion; Zn^{2+}, zinc ion]

PITFALL The ions formed by mercury

The mercury(I) ion exists not as a single Hg^+ ion but as a dimer. It may be thought of as two Hg^+ ions joined together. Its correct formula is Hg_2^{2+}; it should not be confused with the mercury(II) ion, Hg^{2+}:

$$\text{mercury(I)} \quad Hg_2^{2+} \qquad \text{mercury(II)} \quad Hg^{2+}$$

1.14 NAMES OF ANIONS

KEY CONCEPT Monatomic anions and oxoanions

Monatomic anions are named by adding the suffix *-ide* to the stem name of the element. For instance,

$$Br^- \text{ bromide} \qquad S^{2-} \text{ sulfide} \qquad N^{3-} \text{ nitride}$$

As described in the text, oxoanions are named according to the oxygen content of the ion:

CO_3^{2-} carbonate	SO_4^{2-} sulfate	NO_3^- nitrate	ClO^- hypochlorite
	SO_3^{2-} sulfite	NO_2^- nitrite	ClO_2^- chlorate
			ClO_3^- chlorate
			ClO_4^- perchlorate

Many other variations are shown in Table 1.5 of the text. The names and formulas of these oxoanions must be learned, because much of the chemistry that follows depends on recognizing them.

EXAMPLE Naming anions

Without consulting any references, name the following ions: O^{2-}, F^-, PO_4^{3-}, CN^-, HSO_4^{2-}, NO_2^-.

SOLUTION Monatomic anions (anions containing only one atom) are named by using the stem name of the element and the suffix *-ide*. For example, O^{2-} is named by combining *ox* (for oxygen) with *ide* to get oxide. For more complicated ions, including oxoanions (anions that contain oxygen), memorizing the names in Table 1.5 is probably the best way to learn their names.

O^{2-}	oxide ion	CN^-	cyanide ion
F^-	fluoride ion	HSO_4^{2-}	hydrogen sulfate or bisulfate ion
PO_4^{3-}	phosphate ion	NO_2^-	nitrite ion

EXERCISE Name these ions: SO_3^{2-}, NO_3^-, OH^-, CrO_4^{2-}, $Cr_2O_7^{2-}$, and HCO_3^- (two names). For the ions containing chromium, consult an outside reference.

[*Answer:* SO_3^{2-} sulfite ion; NO_3^- nitrate ion; OH^- hydroxide ion; CrO_4^{2-} chromate ion; $Cr_2O_7^{2-}$ dichromate ion; HCO_3^- hydrogen carbonate ion, bicarbonate ion]

PITFALL   How is fluoride spelled?

Note the correct spelling of "fluoride." It is not "flouride." The "u" comes before the "o"!

1.15   NAMES OF IONIC COMPOUNDS

KEY CONCEPT   Ionic compounds have a first name and a last name

In an ionic compound, the first name is the name of the positive ion; the last name, that of the negative ion. The name of the positive ion includes the charge in roman numerals in parentheses, if appropriate. To name an ionic compound, we must

1. Identify what ions are present.
2. Determine the correct name for each of the two ions.
3. Name the compound as the positive ion name followed by the negative ion name.

In certain ionic compounds, individual water molecules are incorporated into the crystal lattice. This water can be driven off by heating the crystal. When the water is present, the compound is called a **hydrate;** when the water is absent, the compound is said to be **anhydrous.** The formula of the hydrate shows the presence of water in the crystal by use of a centered dot followed by the number of water molecules per formula unit: for instance, $MgCO_3 \cdot 3H_2O$ and $Na_3PO_4 \cdot 12H_2O$. This water can be driven off by heating the crystal, as the following equation indicates:

$$Hydrate \xrightarrow{heat} anhydrous\ form\ +\ water$$
$$MgCO_3 \cdot 3H_2O \longrightarrow MgCO_3 + 3H_2O$$

The Greek prefixes shown in Table 1.6 of the text are used to indicate the number of waters of hydration in the formula unit. So the fourth rule for naming ionic compounds is

4. If the compound is a hydrate, indicate the number of waters of hydration by using a Greek prefix plus the term "hydrate" as the last part of the name.

EXAMPLE 1   Naming ionic compounds

Name the compounds Cu_2SO_4, $Na_2MnO_4 \cdot 10H_2O$, and $AuBr_3$.

SOLUTION Cu_2SO_4: We recognize the SO_4^{2-} as the sulfate ion. To balance the -2 charge number on SO_4^{2-}, each of the two copper ions must have a $+1$ charge: Each is a Cu^+ ion. Because copper forms different charged ions, this ion is the copper(I) ion (rather than "copper ion") and Cu_2SO_4 is copper(I) sulfate.

$Na_2MnO_4 \cdot 10H_2O$: The ions present are sodium ion and manganate ion. Sodium forms Na^+ and does not form different charged ions, so its name is "sodium." The 10 waters of hydration are indicated by using the term "decahydrate" at the end of the name, so $Na_2MnO_4 \cdot 10H_2O$ is sodium manganate decahydrate.

$AuBr_3$: The anion in this formula is a bromide, Br^-, ion with charge -1. The three bromides contribute a total charge of -3 to the formula unit, so the gold is present as a Au^{3+} ion. Because gold forms different charged ions, the name of Au^{3+} is gold(III) ion and $AuBr_3$ is gold(III) bromide.

EXERCISE Name $Mg(NO_3)_2$, $Pb(N_3)_2$, and $LiClO_4 \cdot 3H_2O$. (N_3^- is the azide ion.)

[*Answer:* magnesium nitrate, lead(II) azide, lithium perchlorate trihydrate]

EXAMPLE 2   Writing the formulas of ionic compounds from their names

Write the formulas of cobalt(II) chlorate hexahydrate, mercury(I) bromide, and strontium peroxide. (Peroxide ion is O_2^{2-}.)

SOLUTION Cobalt(II) chlorate hexahydrate: The cobalt(II) ion is Co^{2+} and the chlorate ion is ClO_3^-. For a formula unit with zero charge, we must combine two chlorate ions (total charge -2) with one cobalt(II) ion (charge $+2$). Hexahydrate means "six waters" of hydration, so the complete formula of cobalt(II) chlorate hexahydrate is $Co(ClO_3)_2 \cdot 6H_2O$.

Mercury(I) bromide: The mercury(I) ion is Hg_2^{2+}. It has a total charge of $+2$. Each bromide has a -1 charge. For a formula unit with zero charge, two bromide ions (total charge -2) are combined with one mercury(I) dimer (total charge $+2$). The formula of mercury(I) bromide is Hg_2Br_2.

Strontium peroxide: The strontium ion is Sr^{2+} and the peroxide ion is O_2^{2-}. For a neutral formula unit, one strontium ion should be combined with one peroxide. The formula of strontium peroxide is SrO_2.

EXERCISE Write the formulas of tin(IV) chloride tetrahydrate, ammonium hydrogen carbonate, and barium fluoride.

[*Answer:* $SnCl_4 \cdot 4H_2O$, NH_4HCO_3, BaF_2]

1.16   NAMES OF MOLECULAR COMPOUNDS

KEY CONCEPT A   Naming simple binary compounds (nonacids)

A **binary compound** is one that contains only two elements (although it may contain more than two atoms). Examples are HCl, H_2O, and PCl_5. Table 1.7 of the text shows some common binary compounds that are usually referred to by traditional, nonsystematic names that must be learned. Binary compounds that are not acids are named as follows: The first part of the name is the name of the first element in the formula with a Greek prefix (Table 1.6) that gives the number of atoms in the formula. The second part of the name is the stem of the name of the second element in the formula, with the suffix *-ide*. For the second element, as for the first, a Greek prefix is used to indicate the number of atoms in the formula. The Greek prefix *mono-* is usually not used; an important counterexample to this rule is the name *carbon monoxide*. Some examples are OF_2, oxygen difluoride; N_2O_5, dinitrogen pentoxide; P_4O_{10}, tetraphosphorus decaoxide.

EXAMPLE   Naming simple binary compounds

Without looking at any references, write the names of the following compounds: N_2O_4, NH_3, PCl_5, CO_2.

SOLUTION NH_3 has a traditional name that should be memorized. The remaining three compounds are named by using Greek prefixes to indicate the number of atoms of each element present. The first element name is simply the name of the element, the second element name uses the element stem followed by *-ide*.

N_2O_4	dinitrogen tetroxide (not tetraoxide, which is too difficult to say)
PCl_5	phosphorus pentachloride (not monophosphorus pentachloride)
NH_3	ammonia (not nitrogen trihydride)
CO_2	carbon dioxide (not monocarbon dioxide)

EXERCISE Without looking at any references, write the names of the following compounds: S_2Cl_2, CH_4, XeF_4, NO_2.

[*Answer:* S_2Cl_2, disulfur dichloride; CH_4, methane; XeF_4, xenon tetrafluoride; NO_2, nitrogen dioxide]

KEY CONCEPT B   Naming simple binary compounds (acids)

An acid is a compound that releases H^+ when it is placed in water. The formulas for inorganic acids usually have "H" as the first letter in the formula, as in the acids HBr and HNO_3. The formula

H_2O is an exception because water is not considered an acid. CH_4 is not an acid even though it contains hydrogen; its hydrogens are not released as H^+ when it is dissolved in water.

Binary acids can exist as molecular gases when not dissolved in water. In this form, the first part of the name is "hydrogen" and the second part is the stem of the second element with the suffix *-ide*. When dissolved in water, the acid properties become apparent and the name changes. The first name is the prefix *hydro-* followed by the stem of the name of the second element in the formula, and then the suffix *-ide*. The second part of the name is "acid." Examples are shown in the table.

Molecular Formula	Name of Molecular Gas	Name of Dissolved Acid
HF	hydrogen fluoride	hydrofluoric acid
HCl	hydrogen chloride	hydrochloric acid
H_2S	hydrogen sulfide	hydrosulfuric acid

EXAMPLE   Naming simple binary acids

Without looking at any references, write the names of the following compounds: HBr(g), HI(aq), HF(aq), HCl(g).

SOLUTION The symbol (g) indicates that the compound is in the gaseous state, whereas the symbol (aq) indicates that the compound is dissolved in water. Following the rules cited, we have HBr(g), hydrogen bromide; HI(aq), hydroiodic acid; HF(aq), hydrofluoric acid; HCl(g), hydrogen chloride.

EXERCISE Without looking at any references, write the names of the following compounds: $H_2S(g)$, HBr(aq), HCl(aq), HF(g).

[*Answer:* $H_2S(g)$, hydrogen sulfide; HBr(aq), hydrobromic acid; HCl(aq), hydrochloric acid; HF(g), hydrogen fluoride]

KEY CONCEPT C   Naming oxoacids

The oxoacids are all molecular compounds; their names are based on the name of the parent anion. An anion ending in *-ate* gives an acid ending in *-ic*, and an anion ending in *-ite* gives an acid ending in *-ous. Per* and *hypo* are used for the acid name if one is part of the anion name:

Molecular Formula	Name of Compound	Anion Name and Formula
H_2SO_4	sulfuric acid	sulfate ion (SO_4^{2-})
H_2SO_3	sulfurous acid	sulfite ion (SO_3^{2-})
H_3PO_4	phosphoric acid	phosphate ion (PO_4^{3-})
H_3PO_3	phosphorous acid	phosphite ion (PO_3^{2-})
HNO_3	nitric acid	nitrate ion (NO_3^-)
HNO_2	nitrous acid	nitrite ion (NO_2^-)
HClO	hypochlorous acid	hypochlorite ion (ClO^-)

EXAMPLE   Naming oxoacids

Without looking at any references, write the names of the following compounds: H_2SO_4, HNO_3, $HClO_2$.

SOLUTION Oxoacids are generally found dissolved in water and are named as acids. Table 1.5 of the text and the preceding table give the names of the common oxoacids. They should be learned. The compounds in this example are H_2SO_4, sulfuric acid; HNO_3, nitric acid; $HClO_2$, chlorous acid.

EXERCISE Without looking at any references, write the names of the following compounds: H_2SO_3, H_2CO_3, $HClO_4$.

[*Answer:* H_2SO_3, sulfurous acid; H_2CO_3, carbonic acid; $HClO_4$, perchloric acid]

PITFALL   Sulfuric and hydrosulfuric acids

Be certain to distinguish sulfuric acid (H_2SO_4) from hydrosulfuric acid (H_2S).

KEY WORDS   Define or explain each term in a written sentence or two.

hydrates   Stock number
oxoacids   systematic name

DESCRIPTIVE CHEMISTRY TO REMEMBER

• The **halogens** have the following colors: Fluorine is a very pale yellow gas. Chlorine is a yellow-green gas. Bromine is a red-brown liquid. Iodine is a purple-black solid.
• **Dueterium** is an isotope of hydrogen with 1 neutron and 1 proton in the nucleus.
• **Heavy water** contains two deuterium atoms (D_2O) instead of two hydrogen atoms (H_2O).
• The **alkali metals** are all soft, silvery metals that melt at low temperatures. They all react with water.
• The **coinage metals** are copper (Cu), silver (Ag), and gold (Au).
• **Sulfur dioxide** is a colorless, pungent, poisonous gas.

SELF-TEST EXERCISES

The elements

For Exercises 1–4, associate the symbol of the element with the name.

1. Sodium
(a) So   (b) S   (c) nA   (d) SO   (e) Na

2. Phosphorus
(a) Ph   (b) ph   (c) P   (d) Po   (e) Pb

3. Kr
(a) kryptonite   (b) kallium   (c) krypton   (d) korium   (e) cobalt

4. Au
(a) gold   (b) silver   (c) actimium   (d) arsenic   (e) aluminum

5. The charge to mass ratio of the electron is -1.76×10^8 C·g^{-1} and its mass is 9.11×10^{-28} g. What is the charge of the electron in coulombs (C)?
(a) -6.23×10^{18} C   (b) -5.18×10^{-36} C   (c) -1.60×10^{-19} C
(d) -1 C   (e) -1.93×10^{35} C

6. The nuclear atom model states that atoms consist of a
(a) very large, heavy, positive nucleus surrounded by light, negative electrons.
(b) very large, light, negative nucleus surrounded by heavy, positive protons.
(c) very small, heavy, positive nucleus surrounded by light, negative electrons.
(d) jelly like positive substance made of protons in which negative electrons are embedded.
(e) small, light, negative nucleus surrounded by heavy, positive protons.

7. Use a periodic table to find the atomic number of potassium.
(a) 39   (b) 39.1   (c) 20   (d) 58   (e) 19

8. What element has an atomic number of 32?
(a) sulfur   (b) oxygen   (c) argon   (d) rubidium   (e) germanium

9. How many electrons are present in a zinc atom?
(a) 65   (b) 35   (c) 30   (d) 50   (e) 95

10. How many protons are present in an atom of xenon (Xe)?
(a) 131   (b) 77   (c) 54   (d) 8   (e) 32

11. Which of the following elements has only 28 electrons?
(a) Fe   (b) Ar   (c) Si   (d) Ba   (e) Ni

12. What is the mass number of an atom of bromine with 39 neutrons?
(a) 78   (b) 70   (c) 39   (d) 35   (e) 74

13. An atom has mass number 106 and 56 neutrons. What element is it?
(a) Unh   (b) Ba   (c) Sn   (d) U   (e) Rn

14. How many protons, electrons, and neutrons (p, e, n) are there in an atom of ^{44}Ca?
(a) (20, 20, 44)   (b) (20, 20, 24)   (c) (10, 24, 24)
(d) (24, 24, 20)   (e) (24, 24, 44)

15. Which of the following represents an isotope of Ni (p = 28, A = 58) in which p is the number of protons and A the mass number?
(a) p = 29, A = 58   (b) p = 28, A = 59   (c) p = 29, A = 57
(d) p = 27, A = 60   (e) p = 59, A = 82

16. What is the mass number of an atom with 25 electrons, 25 protons, and 30 neutrons?
(a) 25   (b) 50   (c) 55   (d) 80   (e) 35

17. How many neutrons are present in an atom of ^{56}Fe?
(a) 30   (b) 56   (c) 86   (d) 26   (e) 52

18. One of the processes that occurs in stars is the combination of neon-20 with helium-4 to form only one new element. What is the element?

$$^{20}Ne + {}^{4}He \longrightarrow ?$$

(a) ^{24}Na   (b) ^{24}Ne   (c) ^{24}Al   (d) ^{24}Mg   (e) ^{24}Si

19. One of the processes that occurs in stars is the combination of two neon-20 atoms to give magnesium-24 and one other element. What is the element?

$$^{20}Ne + {}^{20}Ne \longrightarrow {}^{24}Mg + ?$$

(a) ^{16}N   (b) ^{16}C   (c) ^{16}O   (d) ^{24}Ne   (e) ^{18}N

20. Which is an alakli metal?
(a) Na   (b) Ca   (c) Al   (d) Pb   (e) Fe

21. Which is a transition metal?
(a) Na   (b) Ba   (c) Fe   (d) Ra   (e) Pb

22. All of the following are typical properties of nonmetals except one. Which one?
(a) poor electrical conductor   (b) dull appearance   (c) brittle
(d) poor heat conductor   (e) ductile

23. Which of the following elements is most likely to have properties similar to Ca?
(a) S   (b) Sr   (c) Sc   (d) K   (e) Ra

24. Which of the following elements is a metal?
(a) S   (b) F   (c) P   (d) I   (e) Rb

25. Which of the following elements is a nonmetal?
(a) Ba   (b) S   (c) K   (d) Hg   (e) U

26. Which of the following elements does not conduct electricity?
(a) Fe   (b) Cu   (c) Au   (d) P   (e) Ag

27. The term *ductile* refers to which of the following properties of an element?
(a) can be hammered into thin sheets   (b) can be drawn into a wire
(c) conducts electricity   (d) is lustrous
(e) is a gas at room temperature

Compounds

28. How many hydrogen atoms are there in one molecule of pentane (C_5H_{12})?
(a) 17   (b) 12   (c) 5   (d) 2   (e) 7

29. What is the total number of atoms in one molecule of cholesterol, $C_{27}H_{46}O$?
(a) 27   (b) 47   (c) 46   (d) 74   (e) 1242

30. A sample of calcium carbonate ($CaCO_3$) from Kentucky is found to have the following composition: C, 12.00%; Ca, 40.04%; O, 47.96%. Predict the percentage oxygen in a sample of calcium carbonate from Australia (in the southern hemisphere).
(a) 51.00%   (b) 43.96%   (c) 47.96%   (d) 32.00%
(e) need more information to tell

31. The overall charge on each of the compounds water (H_2O), sodium chloride (NaCl), and methane (CH_4) is the same. What is it?
(a) −2   (b) −1   (c) 0   (d) +1   (e) +2

32. The structural formulas of two molecules, both containing hydrogen and oxygen, follow. Are the compounds the same or different?

$$H-O-H \qquad H-O-O-H$$

(a) different   (b) same   (c) need more information to tell

33. The structural formulas of two molecules, both with formula C_3H_8O follow. Are the compounds the same or different?

(a) same   (b) different   (c) need more information to tell

34. Use a periodic table to predict the charge on the monatomic ion formed by Se.
(a) +3   (b) +1   (c) −4   (d) −1   (e) −2

35. Use a periodic table to predict the charge on the monatomic ion formed by Sr.
(a) +1   (b) +2   (c) −3   (d) −1   (e) −2

36. Use a periodic table to predict the charge on the monatomic ion formed by Ga.
(a) +1   (b) +3   (c) −5   (d) −3   (e) −2

37. Only one of the elements shown will not form a monatomic ion. Which one?
(a) K   (b) Kr   (c) Br   (d) O   (e) Al

38. How many electrons are there in the Ba^{2+} ion?
(a) 54   (b) 58   (c) 53   (d) 56   (e) 55

39. How many electrons are there in the S^{2-} ion?
(a) 16   (b) 17   (c) 15   (d) 18   (e) 14

40. How many electrons are there in the NO_2^- ion?
(a) 21   (b) 22   (c) 24   (d) 23   (e) 25

41. What is the formula of the compound formed by Ga^{3+} and O^{2-} ions?
(a) Ga_3O_2 (b) Ga_2O_3 (c) Ga_3O (d) GaO (e) GaO_2

42. What is the formula of the compound formed by Ba and S_8?
(a) BaS   (b) Ba_8S_8 (c) Ba_2S (d) BaS_8 (e) BaS_2

43. Without looking at any references, name the ClO_3^- ion.
(a) perchlorate ion   (b) chlorine trioxide ion
(c) chlorate ion   (d) hypochlorous ion

44. Which of the ions shown is the hydroxide ion?
(a) NH_2^- (b) O^{2-} (c) OH^- (d) H^- (e) HS^-

Mixtures

45. A mixture of sodium chloride (NaCl) and sucrose ($C_{12}H_{22}O_{11}$) from Kentucky is found to contain 42.22% sodium chloride and 57.78% sucrose. Predict the percentage sucrose in a mixture of sodium chloride and sucrose from Australia (in the southern hemisphere).
(a) 50.00%   (b) 57.78%   (c) 33.33%   (d) 8.00%
(e) need more information to tell

46. Which of the following mixtures is a homogeneous mixture?
(a) concrete   (b) ketchup   (c) mud   (d) milk   (e) seawater

47. What separation technique takes advantage of the difference in solubility of the components in a mixture?
(a) chromatography   (b) filtration   (c) distillation

The nomenclature of compounds

48. What is the name of the Fe^{2+} cation?
(a) iron(III)   (b) iron(II)   (c) ferric   (d) iron   (e) fluorine

49. What is the systematic name of $FeCO_3$?
(a) iron(III) carbonate   (b) iron(II) carbonate
(c) iron carbonate   (d) iron(VI) carbonate

50. What is the name of $Ba(OH)_2$?
(a) barium oxyhydride   (b) barium(II) hydroxide   (c) barium hydroxide
(d) barium dihydroxide   (e) barium 2-hydroxide

51. What is the formula of potassium dichromate?
(a) $KCrO_4$ (b) K_2CrO_4 (c) KCr_2O_7 (d) $K_2Cr_2O_7$ (e) KCr_2

52. What is the formula of iridium(IV) chloride?
(a) Ir_4Cl (b) $IrCl_4$ (c) $IrCl_2$ (d) $IrCl$ (e) Ir_2Cl_3

53. What is the formula of ammonia? (Do not refer to any references.)
(a) H_2O_2 (b) N_2H_4 (c) PH_3 (d) NH_3 (e) NH_4

54. What is the formula of dinitrogen pentoxide?
(a) N_2O_4 (b) NO (c) $(NO_5)_2$ (d) N_2O (e) N_2O_5

Descriptive chemistry

When answering these questions, use no references except a periodic table.

55. One of the halogens is a purple-black solid. Which one?
(a) F   (b) Cl   (c) Br   (d) I   (e) At

56. Which of the following alkali metals reacts most violently with water?
(a) Cs   (b) Rb   (c) K   (d) Na   (e) Li

57. Which of the following is a soft, silvery metal?
(a) Na   (b) Fe   (c) Cr   (d) Au   (e) Ti

58. Which of the following elements is extremely unreactive?
(a) O   (b) P   (c) Ar   (d) Fe   (e) Ca

59. One of the metals shown will not react with water to produce H_2 gas, even when it is red hot. Which one?
(a) Li   (b) Be   (c) Ca   (d) Mg   (e) Cs

CHAPTER 2

MEASUREMENTS
AND MOLES

In many instances, chemical observations are quantitative; in particular, when the quantitative measurement involves the amount of a substance, the unit of measure is called a mole. In this chapter, we explore how to properly express numerical values that have been experimentally determined.

MEASUREMENTS AND UNITS

2.1 THE METRIC SYSTEM

KEY CONCEPT **Most measured values are expressed in a specific unit**

To facilitate the sharing of information among scientists, a system of units called the **International System of Units** (SI units) has been developed. With such a system, a measured value reported by a scientist in one laboratory can immediately be understood and used by a scientist in a distant laboratory because the units used for the measurement will be familiar to both. The SI base unit of mass is the **kilogram** (kg), which is defined by a standard mass maintained in France. The **gram** (g) and **milligram** (mg) are other mass units that are related to the kilogram in a simple way and are more convenient to use in certain situations. The SI base unit of length is the **meter** (m). A convenient unit of length that is simply related to the meter is the **centimeter** (cm). The SI unit of time, the **second** (s), is the familiar unit of time used in everyday activities.

> **EXAMPLE Choosing the correct unit**
>
> The speed of an object is expressed as distance divided by time. What are the correct units for speed in the SI system; include the abbreviation of the units in your answer.
>
> **SOLUTION** The unit of distance in the SI system is the meter (m) and the unit of time is the second (s). The units for speed, distance divided by time, must be meters divided by seconds; the abbreviation is m/s, or equivalently, $m \cdot s^{-1}$.
>
> **EXERCISE** Two astronauts, Smith and Gold, have masses of 75 kg and 85 kg, respectively. Which astronaut weighs more on the Earth's surface and which weighs more on the space shuttle?
>
> [*Answer:* Gold weighs more on Earth; both weigh the same on the shuttle (zero weight).]

2.2 PREFIXES FOR UNITS

KEY CONCEPT A **A prefix determines the size of a metric system unit**

How tall are you? What do you think is the distance from Detroit to Los Angeles? The typical answers to these questions might be, 6 ft tall (for a typical American male) and about 2000 mi. We should note that two different sized units are used for these answers. A smaller unit (feet) is used for a small distance and a larger unit (miles) for a large distance. In the metric system, larger and smaller units are formed by adding a prefix to a base unit. For example, using the prefixes from Table 2.1 in the text and the base unit of gram, we obtain the following:

Prefix	Abbreviation	Multiplier	Relationship for Gram
kilo	k	$1000 \ (10^3)$	$1 \ \text{kg} = 1000 \ \text{g}$
centi	c	$0.01 \ (10^{-2})$	$1 \ \text{cg} = 0.01 \ \text{g}$
milli	m	$0.001 \ (10^{-3})$	$1 \ \text{mg} = 0.001 \ \text{g}$
micro	μ*	$0.000 \ 001 \ (10^{-6})$	$1 \ \mu\text{g} = 10^{-6} \ \text{g}$
nano	n	10^{-9}	$1 \ \text{ng} = 10^{-9} \ \text{g}$
pico	p	10^{-12}	$1 \ \text{pg} = 10^{-12} \ \text{g}$

*This is a lowercase Greek "mu." You can make this symbol by adding length to the descenders on a "u."

EXAMPLE   Understanding prefixes

Which is longer, a pen 62,000 μm long or a pencil 5.5 cm long?

SOLUTION To make a comparison between the two lengths, we must express both in the same units. The simplest choice for the common unit is the meter, since both lengths can be directly related to the meter. Because 1 μm = 0.000 001 m, 62,000 μm is the same as 62,000 $\times$ 0.000 001 m; this is 0.062 m. Similarly, because 1 cm = 0.01 m, 5.5 cm equals 5.5 $\times$ 0.01 m, or 0.055 m. The pen is longer.

EXERCISE The newspapers often report that college professors make "kilobuck" salaries. If your Shakespeare professor earns about 44 kilobucks per year, how much does he earn (in dollars per year)?

[*Answer:* \$44,000 yr^{-1}]

KEY CONCEPT B   Powers of 10 and scientific notation: Numbers > 10

Scientists must often deal with very large numbers (such as 300,000,000 $\text{m} \cdot \text{s}^{-1}$, the speed of light) or very small numbers (such as 0.000 000 000 05 m, the radius of a hydrogen atom). Writing numbers in **scientific notation** enables us to write very large numbers or very small numbers in a particularly convenient way. To write a number in scientific notation, we must use powers of 10, so we start with a review of this topic. When any number is raised to some power, the number is multiplied by itself the number of times given by the power:

$$2^3 = 2 \times 2 \times 2 = 8$$

$$2.4^2 = 2.4 \times 2.4 = 5.76$$

$$10^5 = 10 \times 10 \times 10 \times 10 \times 10 = 100,000$$

If we want to write a large number in scientific notation, we first factor the large number into two parts: a number between one and 10 and a second number, such as 100 or 1000 or 10,000 or any similar multiple of 10. For the speed of light, for instance, this gives

$$300,000,000 \ \text{m} \cdot \text{s}^{-1} = 3.00 \times 100,000,000 \ \text{m} \cdot \text{s}^{-1}$$

Number between 1 and 10   Multiple of 10

The multiple of 10 can be written as a power of 10 and substituted into the number:

$$100,000,000 = 10 \times 10 \times 10 \times 10 \times 10 \times 10 \times 10 \times 10 = 10^8$$

$$300,000,000 \ \text{m/s} = 3.00 \times 10^8 \ \text{m} \cdot \text{s}^{-1}$$

In this case, 3.00×10^8 is the speed of light, in meters per second, written in scientific notation. The reason for writing 3.00 instead of 3 as the number between 1 and 10 will be discussed later. To enter this number into a scientific calculator, use the following steps, which work for the common calculators that use arithmetic notation:

Step	Display Shows
1. enter 3	3.
2. press EE or EXP key	3. 00
3. enter 8	3. 08

The 3. 08 now displayed is your calculator's way of indicating the number 3×10^8.

EXAMPLE   Writing numbers larger than 10 in scientific notation

The national debt is approximately \$4,600,000,000,000. Express this in scientific notation.

SOLUTION We first factor \$4,600,000,000,000 into a number between 1 and 10 and a multiple of 10. Usually, in writing the number between 1 and 10, we drop all extra zeros:

$$\$4,600,000,000,000 = \$4.6 \times 1,000,000,000,000$$

We then decide how to write the multiple of 10 as a power of 10:

$$1,000,000,000,000 = 10 \times 10 \times 10 \times 10 \times 10 \times 10 \times 10 \times 10 \times 10 \times 10 \times 10 \times 10 = 10^{12}$$

Putting together the two results, we get

$$\$4,600,000,000,000 = \$4.6 \times 10^{12}$$

EXERCISE Approximately 12,500,000 students are enrolled in colleges and universities in the United States. Write this number in scientific notation.

[*Answer:* 1.25×10^7]

KEY CONCEPT C   Powers of 10 and scientific notation: Numbers < 1

To write small numbers in scientific notation, we must deal with negative exponents. The meaning of a negative exponent is defined by the relationship

$$Y^{-n} = \frac{1}{Y^n} = \frac{1}{Y \times Y \times Y \times Y} \quad (n \text{ times})$$

For powers of 10, for example,

$$10^{-5} = \frac{1}{10^{+5}} = \frac{1}{10 \times 10 \times 10 \times 10 \times 10} = \frac{1}{100,000} = 0.000\ 01$$

To write a number less than 1 in scientific notation, we first factor the number into a number between 1 and 10 and a multiple of 10, and then substitute in a power of 10 for the multiple of 10. For instance,

$$0.000\ 567 = 5.67 \times 0.0001$$

$$0.0001 = \frac{1}{10000} = 10^{-4}$$

$$0.000\ 567 = 5.67 \times 10^{-4}$$

To enter this number into your calculator requires the following steps, which will work for most scientific calculators:

Step	Display Shows
1. enter 5.67	5.67
2. press EE or EXP	5.67 00
3. enter 4	5.67 04
4. press the $+/-$ key or change-sign key	5.67 $-$ 04

The display 5.67 $-$ 04 is your calculator's way of indicating the number 5.67×10^{-4}.

EXAMPLE Using scientific notation

Rhode Island, the smallest state, makes up 0.033% of the total area of the United States. Write this number in scientific notation.

SOLUTION Because 0.0033 is less than 1, we recognize that the power of 10 that results must be negative if we write the number in standard scientific notation. We first factor the number into a number between 1 and 10 and a multiple of 10:

$$0.0033 = 3.3 \times 0.001$$

We then convert 0.001 into a power of 10:

$$0.001 = \frac{1}{1000} = 10^{-3}$$

Finally, we substitute 10^{-3} for 0.001 in the original factored number:

$$0.0033 = 3.3 \times 10^{-3}$$

EXERCISE An average chemistry class of 25 students contains approximately 0.000 000 61% of the world's population. Express this number in scientific notation.

[*Answer:* 6.1×10^{-7}]

PITFALL Entering negative exponents on a calculator

On many calculators, you cannot enter a negative exponent by using the minus operation key, $-$. Only the change-sign key, $+/-$, or its equivalent works.

2.3 DERIVED UNITS

KEY CONCEPT A Derived units: Volume

The sciences use a vast number of **derived units,** or units that are a combination of two or more base units. For instance, consider the unit associated with volume. If we calculate the volume of any object, for example, a rectangular solid, a sphere, a pyramid, or even an irregularly shaped object (such as a banana), the answer will always be in units of (length)^3. The **liter** (L) is a common unit of volume. It is not, specifically, a (length)^3 but is related to the cubic centimeter through the relationship that 1 L $= 1000$ cm^3 (exactly).

EXAMPLE Units of volume

A quart is equal to 0.946 L (Table 2.2 in the text). How many milliliters are there in a quart?

SOLUTION From Table 2.1 of the text, we conclude that 1 L = 1000 mL, or equivalently 0.001 L = 1 mL. Using the method of ratios or unit analysis, we conclude that 0.946 L = 946 mL; so 1 qt = 946 mL:

$$0.946 \; \cancel{L} \times \frac{1000 \; \text{mL}}{1 \; \cancel{L}} = 946 \; \text{mL}$$

EXERCISE How many milliliters are there in a cup? (4 cup = 1 qt)

[*Answer:* 236]

KEY CONCEPT B   Derived units: Density

The **density** (*d*) of a substance is a measure of how much mass can be contained in a given volume of it. A substance that has a large mass packed in a small volume has a high density; a substance that packs a small mass into a large volume has a low density. For example, a brick has a relatively large mass in a relatively small volume and, therefore, has a high density; a feather, on the other hand, uses a large volume for a relatively small mass and, therefore, has a low density. To calculate the density of a sample, we divide the mass of the sample by its volume:

$$d = \frac{\text{mass}}{\text{volume}} = \frac{m}{V}$$

The units of density must agree with the definition of density as a mass divided by a volume; units such as gram per milliliter ($\text{g} \cdot \text{mL}^{-1}$), gram per cubic centimeter ($\text{g} \cdot \text{cm}^{-3}$), and gram per liter ($\text{g} \cdot \text{L}^{-1}$) are common. The density depends on both temperature and pressure.

EXAMPLE 1   Calculating the density of a sample

What is the density of a brick with a mass of 1600 g and a volume of 1100 cm^3?

SOLUTION To calculate the density, we use the defining equation

$$d = \frac{m}{V}$$

Inserting $m = 1600$ g and $V = 1100 \; \text{cm}^3$ into the equation gives us

$$d = \frac{1600 \; \text{g}}{1100 \; \text{cm}^3} = 1.455 \; \text{g} \cdot \text{cm}^{-3}$$

EXERCISE A helium balloon with a volume of 2.0 L contains 0.36 g of helium. What is the density of helium?

[*Answer:* $0.18 \; \text{g} \cdot \text{L}^{-1}$]

EXAMPLE 2   Calculating the volume of a sample of known density and mass

One ounce of gold cost approximately \$382 in 1996. What is the volume of 1.00 oz (28.4 g) of gold? Its density is 19.3 $\text{g} \cdot \text{cm}^{-3}$.

SOLUTION The formula for calculating density is

$$d = \frac{m}{V}$$

Because the mass and density are given and the volume requested, we isolate the volume on one side of the equation by multiplying both sides of the equation by *V* and dividing both sides by *d*:

$$V = \frac{m}{d}$$

Substituting in $m = 28.4$ g and $d = 19.3 \; \text{g} \cdot \text{cm}^{-3}$ gives us

$$V = \frac{28.4 \ \cancel{g}}{19.3 \ \cancel{g} \cdot cm^{-3}} = 1.47 \ cm^3$$

Thus 1.00 oz of gold occupies 1.47 cm³ (about ½ teaspoon).

EXERCISE Calculate the volume of 1.00 oz of nitrogen, which has a density of 1.25 g·L⁻¹.

[*Answer:* 22.7 L]

EXAMPLE 3 Calculating the mass of a sample of known density and volume

What is the mass of 10.0 L of oxygen gas, which has a density of 1.43 g·L⁻¹?

SOLUTION The formula for calculating density is

$$d = \frac{m}{V}$$

Because the volume and density are given and the mass requested, we isolate the mass on one side by multiplying both sides of the equation by V:

$$m = d \times V$$

Substituting d = 1.43 g·L⁻¹ and V = 10.0 L gives

$$m = (1.43 \ g \cdot \cancel{L}^{-1})(10.0 \ \cancel{L}) = 14.3 \ g$$

EXERCISE What is the mass of an ice cube 3.0 cm on a side? The density of ice is 0.92 g·cm⁻³.

[*Answer:* 25 g]

KEY CONCEPT C   Extensive and intensive properties

An **extensive property** of a sample is a property that depends on how much of the sample is present, that is, the extent of the sample. Imagine that we have three samples of water and that we tabulate the volume, mass, and density of each sample.

Sample Number	Relative Amount Present	Mass (g)	Volume (mL)	Density (g·mL⁻¹)
I	little (about 20 drops)	1.0	1.0 mL	1.0
II	more (1 cup)	237	237 mL	1.0
III	most (1 quart)	946	946 mL	1.0

Clearly, both the volume and the mass of the water depend on how much water is present; the volume and mass are, therefore, extensive properties. The density, however, is an **intensive property**, a property that is independent of the size of the sample. The density of water is 1.0 g·mL⁻¹ regardless of the size of the sample. Many intensive properties are defined by dividing one extensive property by another, as we define density.

EXAMPLE Extensive and intensive properties

Examine a sample consisting of a teaspoon of NaCl (sodium chloride or table salt). Assume that your sample has a mass of 11 g and a volume of 5 mL. Give three intensive properties and two extensive properties of the sample.

SOLUTION An intensive property is one that does not depend on the size of the sample; an extensive property is one that does. To answer the question, we imagine what would happen to various properties if we enlarge or shrink the size of the sample (either will do). We can easily obtain a larger sample of table salt, for example, by pouring 1 cup of NaCl (approximately 237 mL, or 513 g) out of its container. We compare this sample with the smaller sample and list some properties that are different in the two samples and properties that are not.

Property	Sample 1	Sample 2	Intensive or Extensive
mass	11 g	513 g	extensive
volume	5 mL	237 mL	extensive
color	white	white	intensive
density	2.2 g·mL^{-1}	2.2 g·mL^{-1}	intensive
taste	salty	salty	intensive

EXERCISE Is the electrical conductivity of copper wire (the ease of electrical conduction per meter of wire) an extensive or intensive property?

[*Answer:* Intensive]

2.4   UNIT CONVERSIONS

KEY CONCEPT A   Unit analysis and conversion factors

A working scientist frequently has to change a measured or calculated number from one set of units to another. As an example, if we are asked to fill a dime wrapper for a bank, the only information given on the wrapper is that it is meant to hold $5; we must decide how many dimes are equivalent to $5. This procedure is actually a unit conversion, from a certain number of dollars to a certain number of dimes:

$$5 \text{ dollar} \longrightarrow ? \text{ dime}$$

Most of us can solve this problem easily because we are so familiar with our money system, but many chemical problems are similar. A powerful calculational technique called **unit analysis** can also be used to solve such a problem; with this method we use the formulation

(Answer with new units) = (given number with old units) × (conversion factor)

For the problem of converting 5 dollars to the equivalent number of dimes, we have

? dime = 5 dollar × (conversion factor)

In this problem, the old units are "dollar" and the new units "dime." The next step is to find the proper **conversion factor,** a factor that expresses a known relationship between the old units and the new units. For our example, we use the fact that 1 dollar = 10 dime to derive two conversion factors

$$\frac{1 \text{ dollar}}{10 \text{ dime}} \quad \text{and} \quad \frac{10 \text{ dime}}{1 \text{ dollar}}$$

The first conversion factor gives the number of dollars per dime ($\frac{1}{10}$), the second, the number of dimes per dollar (10). It is most important to remember that conversion factors always come in pairs, as just indicated, with one factor the inverse of the other. The correct conversion factor for any calculation will result in cancellation of the old units; the remaining units are the correct new units for the desired answer:

$$50 \text{ dime} = 5 \text{ \sout{dollar}} \times \frac{10 \text{ dime}}{1 \text{ \sout{dollar}}}$$

When we use the correct conversion factor, the dollar units on top of the fraction cancel the dollar units on the bottom, leaving the expected units of dime for the answer. If we try the second (incorrect) conversion factor of the pair, we get

$$0.50 \frac{(\text{dollar})^2}{\text{dime}} = 5 \text{ dollar} \times \frac{1 \text{ dollar}}{10 \text{ dime}} \quad \text{(incorrect answer, wrong units)}$$

The second conversion factor does not permit cancellation of units and thus is the wrong one to use. We use five steps in a unit conversion problem:

1. Analyze the problem. Decide which number with old units is to be converted into a new number with new units. Write your analysis in the following form:

    Given number with old units $\longrightarrow$ answer with new units

2. Find a known relationship between the old and the new units.
3. Set up the two conversion factors that result from the relationship between the units.
4. Set up the formulation

    (Answer with new units) = (given numbers with old units) $\times$ (conversion factor)

5. Do the indicated calculation, taking care that units cancel properly.

EXAMPLE   Unit conversions

What is the mass, in kilograms, of a 16-lb sledge hammer?

SOLUTION We follow steps 1 through 5.

1. Analyze the problem. Decide which number with old units is to be converted into a new number with new units. Write your analysis in the form "given number with old units $\rightarrow$ answer with new units."

    $$16 \text{ lb} \longrightarrow ? \text{ kg}$$

2. Find a known relationship between the old and the new units:

    $$2.205 \text{ lb} = 1 \text{ kg} \quad \text{(refer to inside back cover of the text)}$$

3. Set up the two conversion factors that result from the relationship between the units:

    $$\frac{2.205 \text{ lb}}{1 \text{ kg}} \quad \text{and} \quad \frac{1 \text{ kg}}{2.205 \text{ lb}}$$

4. Set up the formulation (given number with old units) $\times$ (conversion factor) = (answer with new units):

    $$16 \text{ lb} \times \frac{1 \text{ kg}}{2.205 \text{ lb}} = ?$$

5. Do the indicated calculation, taking care that units cancel properly:

    $$16 \cancel{\text{ lb}} \times \frac{1 \text{ kg}}{2.205 \cancel{\text{ lb}}} = 7.3 \text{ kg}$$

EXERCISE How many minutes are there in 245 seconds?

[*Answer:* 4.08 min]

KEY CONCEPT B   Converting units in a denominator

Unit conversions can be used to convert a unit in the denominator of a number. For example, if we know the density of a substance in grams per milliliter and want to calculate the density in grams per liter, we must conver milliliters (mL) in the denominator of the original unit to liters (L). The principle involved is the same as in other unit conversions: The conversion factor must be such that the old units cancel and leave the correct new units.

EXAMPLE   Converting units in the denominator of a complex unit

The density of $CHCl_3$ (chloroform) is 1480 $\text{g} \cdot \text{L}^{-1}$. What is the density in grams per milliliter?

SOLUTION We apply the five steps used earlier.

1. Analyze the problem. Decide which number with old units is to be converted into a new number with new units. Write your analysis in the form "given number with old units" → "answer with new units." In this case, we must change the unit L in $g \cdot L^{-1}$ to mL:

$$1480 \ g \cdot L^{-1} \longrightarrow ? \ g \cdot mL^{-1}$$

2. Find a known relationship between the old and the new units. Table 2.1 of the text gives us the clue to the relationship between liters and milliliters:

$$1000 \ mL = 1 \ L$$

3. Set up the two conversion factors that result from the relationship between the units:

$$\frac{1000 \ mL}{1 \ L} \quad \text{and} \quad \frac{1 \ L}{1000 \ mL}$$

4. Set up the formulation (given number with old units) × (conversion factor) = (answer with new units):

$$1480 \frac{g}{L} \times \frac{1 \ L}{1000 \ mL} = ?$$

5. Do the indicated calculation, taking care that units cancel properly:

$$1480 \frac{g}{\cancel{L}} \times \frac{1 \ \cancel{L}}{1000 \ mL} = 1.480 \frac{g}{mL}$$

EXERCISE A leaky faucet drips water at the rate of $0.055 \ mL \cdot s^{-1}$. How many milliliters per hour (h = hour) does this correspond to?

[*Answer:* $2.0 \times 10^2 \ mL \cdot h^{-1}$]

PITFALL   Choosing the correct conversion factor

A common mistake made in unit conversion calculations is choosing the wrong conversion factor from the pair of possible factors. To avoid this error, make a quick, rough mental calculation of the answer you expect and compare your estimate with the answer you get from a detailed calculation. At the very least, estimate whether the answer should be larger or smaller than the given number. For example, in a problem that asks for the number of centimeters in a mile, we should note that a centimeter is much smaller than a mile, so there should be a lot of centimeters in 1 mile. If we get an answer such as 161 cm or 1.609×10^{-5} cm (0.000 016 09 cm) for the number of centimeters in 1 mile, we would immediately recognize that something was wrong and look for an error somewhere in the problem solution. The moral: *Look carefully at your answer!*

2.5   TEMPERATURE

KEY CONCEPT   The three common temperature scales

Three temperature scales are in use throughout the world today: the Celsius (°C) scale, the Fahrenheit (°F) scale, and the Kelvin (K) scale. To convert from one of these scales to an other, we use the following relationships:

From °C to °F: $°F = 32 + \frac{9}{5} (°C)$

From °F to °C: $°C = \frac{5}{9} \times (°F - 32)$

From °C to K: $K = 273.15 + °C = 237 + °C$ (for everyday use, this form has enough precision)

From K to °C: $°C = K - 273.15 = K - 273$

It is helpful to recognize that the two equations that relate degrees Celsius to degrees Fahrenheit are equivalent. It is necessary to learn only one of them because it is straightforward to algebraically convert to the other.

EXAMPLE   Converting between Celsius and Fahrenheit temperatures

The highly **volatile** (low boiling point) liquid ether boils at 34.5°C. What is this temperature on the Kelvin and Fahrenheit scales?

SOLUTION To convert 34.5°C to the Kelvin scale, we use the relationship in which the Kelvin temperature is expressed on one side of the equation by itself (because this is the unknown) and everything else, including degrees Celsius, on the other side: K = 273.15 + °C. Because 34.5°C is expressed to a tenth of a degree, we use 273.15 in the conversion formula rather than 273:

$$K = 273.15 + 34.5 = 307.65 = 307.6$$

The answer is rounded to a tenth of a degree to agree with the tenth of a degree precision in the number 34.5. For conversion to the Fahrenheit scale, we use the equation in which degrees Fahrenheit (the unknown) is on one side and everything else, including degrees Celsius (the known), on the other side:

$$°F = 32 + \frac{9}{5}(°C)$$

Because °C = 34.5,

$$°F = 32 + 62.1 = 94.1°$$

Because the number 32 is an exact number, we express the answer to a tenth of a degree precision, the original precision in the number 34.5.

EXERCISE The coldest (official) temperature ever recorded in the United States was −79.8°F at Prospect Creek Camp, Alaska, on January 23, 1971. What is this temperature in degrees Celsius and in kelvins?

[*Answer:* −62.1°C, 211.1 K]

PITFALL   Negative temperatures in temperature conversions

As we have seen, negative temperatures are possible on the Celsius and Fahrenheit temperature scales. In doing a conversion with a negative temperature, care must be taken to properly subtract and add signed numbers. Consider the conversion of −22°F to degrees Celsius. The equation to use is

$$°C = \frac{5}{9} \times (°F - 32)$$

Substituting in °F gives

$$°C = \frac{5}{9} \times (-22 - 32)$$

The subtraction in the parentheses involves signed numbers. Here we note that −22 − 32 = −54. The final answer for the Celsius temperature is $\frac{5}{9} \times (-54) = -30°C$.

2.6   THE UNCERTAINTY OF MEASUREMENTS

KEY CONCEPT A   The uncertainty in a measurement

Every measured number must be somewhat uncertain. For instance, the diameter of a nickel, measured with a vernier calipers, is 2.115 cm. When we use the calipers, we must estimate the last digit (the 5); thus, there is some uncertainty in this digit. For most of the equipment we encounter in a laboratory, similar estimates are required, and there is an uncertainty of ± 1 in the last digit read. For our measurement, reporting the diameter is 2.115 cm means that the actual diameter is between 2.114 cm (2.115 − 0.001, corresponding to −1 in the last digit) and 2.116 cm (2.115 + 0.001, corresponding to +1 in the last digit). In summary:

If we report a measurement as 2.115,	$\longrightarrow$	we assume there is an uncertainty of ± 1 in the last digit,	$\longrightarrow$	which means the actual value is between 2.114 and 2.116

EXAMPLE   Uncertainty in measured values

The mass of a beaker is reported as 22.56 g. Interpret this number with regard to its uncertainty.

SOLUTION Unless a statement is made to the contrary, it is safe to assume that the last digit in any measured number is reliable only to ± 1. For the value 22.56 g, the 6 at the end of the number could be a 7 (6 + 1) or a 5 (6 − 1). Thus, the actual value of the mass is between 22.55 g and 22.57 g. That is, it could be 22.550 g, 22.551 g, 22.552 g, 22.553 g, . . ., 22.567 g, 22.568 g, 22.569 g, or 22.570 g.

EXERCISE The volume of an industrial reaction vessel is determined to be 1346 L. What is the uncertainty associated with this value?

[*Answer:* It is uncertain by ± 1 L, so the volume is between 1345 L and 1347 L.]

KEY CONCEPT B   Significant figures

A subtle but direct interplay exists between the uncertainty of a number and the number of significant figures in the number. The word "significant" in the term gives a clue to its meaning: **significant figures** (written sf) are the meaningful digits in a reported number. The last digit in the number, which is known only to ± 1, is counted as a significant digit. When we write a number, we must be certain to use the proper number of significant figures, neither too few nor too many. As an example, assume we determine the mass of a sample on a balance that is calibrated to 0.01 g, and get 4.33 g. If we write 4.3 g (instead of the correct 4.33 g), we have used too few significant figures because our balance is capable of better precision. On the other hand, if we write 4.331 g (instead of the correct 4.33 g), we have used too many significant figures because our balance is not calibrated to 0.001 g and is incapable of the precision expressed in 4.331 g. When no zeros are present, the number of significant figures is the same as the total number of digits in the number. For our example,

4.3 g	2 sf	too few for the balance used
4.33 g	3 sf	correct number for the balance used
4.331 g	4 sf	too many for the balance used

Whether a zero is a significant figure or not depends on its role in a number. If a zero sets the position of the decimal point or is written by itself in front of a decimal (such as in 0.866), it is not significant. All other zeros are significant. The following examples illustrate the rules for handling zeros and the alternate technique of writing a number in standard scientific notation in order to count the significant figures in the number. The digits in boldface are significant:

3.045	$\mathbf{3.045} \times 10^0$	4 sf
0.0**33**	$\mathbf{3.3} \times 10^{-2}$	2 sf
83.670	$\mathbf{8.3470} \times 10^1$	5 sf
0.00**802**	$\mathbf{8.02} \times 10^{-3}$	3 sf

Numbers such as 360 present one last ambiguity regarding the handling of zeros. The zero in 360 may simply set the position of the decimal point; in this case, the 6 in the number is known to ± 1, and the number contains two significant figures. Or the zero may be a measured number, one that has been determined to ± 1; in this case, three significant numbers are present. We shall resolve this problem as it was resolved in the text: All zeros at the end of numbers are significant unless otherwise stated. If a number such as 360 has only two significant figures, it will always be written in scientific notation.

EXAMPLE Counting significant figures

How many significant figures are there in each of the following numbers? 33.4, 0.6600, 0.044×10^{-2}

SOLUTION We have two ways to solve this problem. We may use the rules for determining the significance of zeros, or we may write the numbers in standard scientific notation and count digits. The correct answers are

33.4	3.34×10^1	3 sf
0.6600	6.600×10^{-1}	4 sf
0.044×10^{-2}	4.4×10^{-4}	2 sf

Note, in the last answer, that a number must be written in *standard* scientific notation (as a number between 1 and 10 times a power of 10) to directly give the correct number of significant figures; 0.044×10^{-2} is not in standard scientific notation because 0.044 is not between 1 and 10.

EXERCISE How many significant figures are there in each of the following numbers? 350, 0.00431, 3.005

[*Answer:* 3; 3, and 4 sf]

KEY CONCEPT C   Significant figures in addition and subtraction

Measured properties are frequently used to calculate other properties. How does the result of a calculation reflect the precision of the data used in the calculation? As we expect, the number of significant figures in the result depends on the numbers of significant figures in the data used in the calculation. For **addition** and **subtraction**, we express only the place values that are known for every number used in the calculation. If a place value is unknown for any number in the calculation, that place value cannot be known for the answer. For example, let us find the sum $2.233 + 10.11 + 5.6$. This sum is indicated in the following standard manner except that unknown place values in the addends are denoted by question marks:

$$
\begin{array}{l}
2.233 \\
10.11? \\
\underline{5.6??} \\
17.943 \quad \text{(sum before correct rounding off)} \\
17.9 \quad\quad \text{(correct number of significant figures)}
\end{array}
$$

The question marks indicate that the thousandths place value is unknown for the number 10.11 and that the thousandths and hundredths are unknown for 5.6. Therefore, the hundredths and thousandths place values cannot be known in the answer, and the last two digits (43) must be rounded off. The correct answer is 17.9. Notice in this case, that the answer has three significant figures, more than the two significant figures in 5.6 and less than the four significant figures in either 2.233 or 10.11.

EXAMPLE Significant figures in a subtraction problem

Express $863 - 20$ to the correct number of significant figures, assuming that 20 has one significant figure.

SOLUTION On the left, we write the subtraction in the normal way. However, the number 20 has only one significant figure, so the zero in 20 represents an unknown place value. The subtraction is rewritten on the right, with a question mark in the units place value for 20 to emphasize that it is unknown.

$$
\begin{array}{ll}
\quad 863 & \quad 863 \\
\underline{-\ 20} & \underline{-\ 2?} \\
\quad 843 \quad \text{(incorrect)} & \quad 84? \quad \text{(which rounds to } 8.4 \times 10^2)
\end{array}
$$

Because the units place value is unknown for the number 20, we cannot express the units place value in the answer; so the 3 must be rounded off. This leaves 840 (or better, 8.4×10^2) as the correct answer.

EXERCISE Express 863 − 20 (with two significant figures in 20) to the correct number of significant figures.

[*Answer:* 843]

KEY CONCEPT D Significant figures in multiplication and division

The correct number of significant figures in the result of a **multiplication** or **division** problem is the same as the smallest number of significant figures present in the data used in the calculation. In the following calculation, the number of significant figures is indicated for each number used in the calculation. The smallest number of significant figures used is two (for 4.2), so the answer must be expressed in two significant figures.

$$\frac{\overset{(3\text{ sf})}{6.78} \times \overset{(2\text{ sf})}{4.2} \times \overset{(4\text{ sf})}{5.115}}{\underset{(4\text{ sf})}{10.22} \times \underset{(3\text{ sf})}{5.58}} = 2.5541096 \quad \text{(calculator answer)}$$

$$= 2.6 \quad \text{(only two significant figures allowed)}$$

EXAMPLE Significant figures in a combined problem

Give the answer using the correct number of significant figures: $(61.82 \times 0.0212)/1.5 = ?$

SOLUTION The number of significant figures allowed in the answer is equal to the smallest number of significant figures in the numbers used in the calculation. Thus, we count the significant figures in each of the numbers: 61.82 (4 sf), 0.0212 (3 sf), and 1.5 (2 sf). The smallest number of significant figures used in the calculation is two, for the number 1.5. Therefore, the answer should be expressed to two significant figures.

$$\frac{61.82 \times 0.0212}{1.5} = 0.873723 \quad \text{(calculator answer)}$$

$$= 0.87 \quad \text{(two significant figures allowed)}$$

EXERCISE What is $(3.50 \times 1.234)/113.3$?

[*Answer:* 0.0381]

2.7 ACCURACY AND PRECISION

KEY CONCEPT A Precision

The **precision** of a number is a measure of its uncertainty. A number with a large relative uncertainty is said to be imprecise; one with a small relative uncertainty is said to be precise—that is, it has been "well characterized." A number with high precision contains a large number of significant figures, one with low precision a small number of significant figures. Assume we measure the volume of a sample of water with two different instruments. One instrument, calibrated in 1-mL increments, gives us 21 mL for the volume; the other, calibrated in 0.01-mL increments, gives us 20.88 mL. The uncertainty in the first measurement is ±1 mL whereas the uncertainty in the second is ±0.01 mL. The second measurement has a smaller uncertainty and is therefore more precise.

EXAMPLE Deciding which of two numbers is more precise

The diameter of an iron nail is measured with vernier calipers and with a micrometer to be 2.2 mm and 2.167 mm, respectively. Which measurement is more precise? Do the measurements agree or disagree?

SOLUTION In general, the number with higher precision has more significant figures. On this basis, 2.167 mm is a more precise measurement than 2.2 mm. In agreement with this idea, we note that the value 2.2 mm has an uncertainty of ± 0.1 mm and the value 2.167 mm an uncertainty of ± 0.001, indicating that the value 2.167 mm is much more precise than the value 2.2 mm. Regarding the agreement of the two values, we note that a reported value of 2.2 mm means that the correct value lies between 2.1 mm and 2.3 mm. The more precise measurement is also between 2.1 mm and 2.3 mm, so the two measurements agree nicely.

EXERCISE Which of the two measured values for the length of a pencil, 2×10^1 cm and 15 cm, is more precise? Do the measurements agree?

[*Answer:* 15 cm; yes]

KEY CONCEPT B   Accuracy

The **accuracy** of a measured number is determined solely by how close the measured value is to the correct value of the property. Accuracy and precision are different qualities of a measured value and should not be confused. The precision of a number relates to its uncertainty and is given by the number of significant figures, whereas the accuracy is concerned with how correct the measurement is. It is worth noting that a working scientist generally cannot determine the accuracy of a measurement at the time it is taken because the correct value is not usually known in advance of the experiment. In student laboratories, the correct answer is often known by an instructor, who can inform a student of the accuracy of the day's laboratory work.

EXAMPLE   Determining accuracy

The actual mass of a piece of brass is known to be 16.3355 g. Of the three determinations—16.3336 g, 16.3387 g, and 16.3379 g—which is the least accurate?

SOLUTION The determination that is furthest from the correct value is the least accurate determination. We analyze the results by setting up a table showing how far each is from the correct mass.

Determination (g)	Deviation from Correct Value (g)
16.3336	0.0019 too low
16.3387	0.0032 too high
16.3379	0.0024 too high

The second determination, 16.3387 g, is the furthest from the correct value and is, therefore, the least accurate.

EXERCISE The distance between two lakes is 3.22 km. Of the three independent measurements for this distance—1.89 mile, 3.331 km, and 3160 m—which is the most accurate?

[*Answer:* 3160 m]

KEY WORDS   Define or explain each term in a written sentence or two.

accuracy	derived units	significant figures
base units	extensive property	Système International
conversion factor	intensive property	
density	precision	

CHEMICAL AMOUNTS

2.8   THE MOLE

KEY CONCEPT   The mole concept

The **mole** (mol) is formally defined as the number of atoms in exactly 12 g of carbon-12. This definition is rarely used, except to arrive at the conclusion that 1 mol of anything is 6.022 $\times$

10^{23} of those things. The mole is often called the "chemist's dozen," and the idea behind it is exactly like that behind the dozen. A dozen is a certain number of objects (12) and a mole is a certain number of objects (6.022×10^{23}). The number 6.022×10^{23} mol^{-1} is called **Avogadro's constant.** The following comparisons illustrate the idea behind the mole:

$$1 \text{ dozen eggs} = 12 \text{ eggs}$$

$$1 \text{ gross pencils} = 144 \text{ pencils}$$

$$1 \text{ ream paper} = 500 \text{ sheets paper}$$

$$1 \text{ mol atoms} = 6.022 \times 10^{23} \text{ atoms}$$

The mole is a frequently used quantity, so it will be to your advantage to learn and understand how to use it. Avogadro's constant, $N_A = 6.022 \times 10^{23}$ mol^{-1}, should be memorized.

EXAMPLE   Using the mole

How many bromine atoms are there in 2.44 mol Br?

SOLUTION The relation between the number of bromine atoms and the moles of bromine atoms is given by the definition of the mole:

$$1 \text{ mol Br} = 6.022 \times 10^{23} \text{ Br atom}$$

From this relationship, we can write two conversion factors:

$$\frac{1 \text{ mol Br}}{6.022 \times 10^{23} \text{ Br atoms}} \quad \text{and} \quad \frac{6.022 \times 10^{23} \text{ Br atoms}}{1 \text{ mol Br}}$$

The question requires that 2.44 mol of Br be converted to number of Br atoms. We use the second conversion factor so that the units "mol Br" cancel properly, leaving us with the desired units "Br atom":

$$2.44 \text{ mol Br} \times \frac{6.022 \times 10^{23} \text{ Br atoms}}{1 \text{ mol Br}} = 14.7 \times 10^{23} \text{ Br atoms} = 1.47 \times 10^{24} \text{ Br atoms}$$

One thing to look for in a problem such as this is the reasonableness of the answer. Whenever we consider the number of atoms in a sample, we are likely to obtain a very, very large number: If we get a small number of atoms or a fraction of one atom, we can be certain that we have solved the problem incorrectly.

EXERCISE How many moles of mercury are present in 8.66×10^{22} Hg atoms?

[*Answer:* 0.144 mol Hg]

2.9   MOLAR MASS

KEY CONCEPT·   The molar mass of an element

The **molar mass** of an element is the mass of a sample that contains 1 mol (6.022×10^{23}) of atoms of the element; it is included in the periodic table for each element. Because many elements exist in nature as a mix of different isotopes, the molar mass of an element is a weighted average of the masses of the isotopes. The following table illustrates the relationships that spring from the definition of the molar mass.

Element	Molar Mass (g·mol^{-1})	Interpretation of Molar Mass
Fe	55.85	55.85 g Fe = 1 mol Fe = 6.022×10^{23} Fe atoms
P	30.97	30.97 g P = 1 mol P = 6.022×10^{23} P atoms
Au	197.0	197.0 g Au = 1 mol Au = 6.022×10^{23} Au atoms

The relationships expressed in this table allow for a number of types of calculations, as the following examples illustrate. These are some of the most important calculations you will do in this course and merit close attention.

EXAMPLE 1 Calculating the moles in a given mass

How many moles of iron are equivalent to 122 g Fe?

SOLUTION The molar mass of Fe is 55.85 $g \cdot mol^{-1}$. This tells us that 55.85 g Fe is equivalent to 1 mol Fe:

$$1 \text{ mol Fe} = 55.85 \text{ g Fe}$$

This relationship can be used to determine the moles of iron in any mass of iron. For our problem,

$$122 \text{ g Fe} \times \frac{1 \text{ mol Fe}}{55.85 \text{ g}} = 2.18 \text{ mol Fe}$$

EXERCISE How many moles of argon atoms are there in 35.8 g Ar?

[*Answer:* 0.896 mol Ar]

EXAMPLE 2 Calculating the mass in a certain number of moles

How many kilograms of cadmium are present in 1.00×10^2 mol Cd?

SOLUTION Cadmium's molar mass is 112.4 $g \cdot mol^{-1}$:

$$1 \text{ mol Cd} = 112.4 \text{ g}$$

Using unit analysis so that mol Cd cancels and g Cd remain as the units in the answer gives

$$1.00 \times 10^2 \text{ mol Cd} \times \frac{112.4 \text{ g}}{1 \text{ mol Cd}} = 1.12 \times 10^4 \text{ g}$$

The answer is converted to kilograms (1 kg = 1000 g):

$$1.12 \times 10^4 \text{ g} \times \frac{1 \text{ kg}}{1000 \text{ g}} = 11.2 \text{ kg Cd}$$

EXERCISE How many kilograms of iridium are present in 325 mol Ir?

[*Answer:* 62.5 kg]

EXAMPLE 3 Calculating the number of atoms in a given mass

How many atoms of nickel are present in 25.0 g Ni?

SOLUTION We have no direct connection between the number of atoms in a sample and the number of grams of sample. However, the moles of nickel can be calculated from the grams, and the number of nickel atoms from the moles:

$$g \text{ Ni} \longrightarrow \text{mol Ni} \longrightarrow \text{number of nickel atoms}$$

We start by calculating the moles of nickel, using the molar mass of 58.71 $g \cdot mol^{-1}$:

$$25.0 \text{ g Ni} \times \frac{1 \text{ mol Ni}}{58.71 \text{ g Ni}} = 0.426 \text{ mol Ni}$$

The number of atoms of nickel can be calculated from the fact that 1 mol Ni = 6.022×10^{23} atoms Ni:

$$0.426 \text{ mol Ni} \times \frac{6.022 \times 10^{23} \text{ atoms Ni}}{1 \text{ mol Ni}} = 2.57 \times 10^{23} \text{ atoms Ni}$$

EXERCISE How many neon atoms are present in 58.2 g Ne?

[*Answer:* 1.74×10^{24} atoms Ne]

2.10   MEASURING OUT COMPOUNDS

KEY CONCEPT A   The molar mass of compounds

The molar mass of a compound is the mass in 1 mol of a compound. It is equal to the sum of the molar masses of the atoms in the compound. (For elements that exist as molecules, the molar mass is the sum of the molar masses in a molecule of the element.) The molar mass is calculated by adding the molar masses of all the atoms in the molecule. For instance, C_2H_6O (ethanol) contains 2 carbon atoms, 6 hydrogen atoms, and 1 oxygen atom Its molar mass is the sum of the molar masses of these atoms:

$$2 \text{ C}: \quad 2 \times 12.01 \quad \text{g} \cdot \text{mol}^{-1} = 24.02 \quad \text{g} \cdot \text{mol}^{-1}$$
$$6 \text{ H}: \quad 6 \times \quad 1.008 \text{ g} \cdot \text{mol}^{-1} = \quad 6.048 \text{ g} \cdot \text{mol}^{-1}$$
$$1 \text{ O}: \quad 1 \times 16.00 \quad \text{g} \cdot \text{mol}^{-1} = \underline{16.00 \quad \text{g} \cdot \text{mol}^{-1}}$$
$$46.068 \text{ g} \cdot \text{mol}^{-1} = 46.07 \text{ g} \cdot \text{mol}^{-1}$$

EXAMPLE   Calculating and interpreting molar masses

Alanine, an important biological compound, has the molecular formula $C_3H_7NO_2$. What is the molar mass of alanine? How many molecules are present in a molar mass of alanine?

SOLUTION   To calculate the molar mass of a compound, we must add together the molar mass of all of the atoms in the compound. Alanine contains 3 carbon atoms, 7 hydrogen atoms, 1 nitrogen atom, and 2 oxygen atoms. Thus, we obtain

$$3 \text{ C}: \quad 3 \times 12.01 \quad \text{g} \cdot \text{mol}^{-1} = 36.03 \quad \text{g} \cdot \text{mol}^{-1}$$
$$7 \text{ H}: \quad 7 \times \quad 1.008 \text{ g} \cdot \text{mol}^{-1} = \quad 7.056 \text{ g} \cdot \text{mol}^{-1}$$
$$1 \text{ N}: \quad 1 \times 14.01 \quad \text{g} \cdot \text{mol}^{-1} = 14.01 \quad \text{g} \cdot \text{mol}^{-1}$$
$$2 \text{ O}: \quad 2 \times 16.00 \quad \text{g} \cdot \text{mol}^{-1} = \underline{32.00 \quad \text{g} \cdot \text{mol}^{-1}}$$
$$89.10 \quad \text{g} \cdot \text{mol}^{-1}$$

The molar mass of a compound is the mass of 1 mol. Thus, 89.10 g of alanine contains 6.022×10^{23} molecules of alanine.

EXERCISE   Calculate the molar mass of each of the following and state the number of molecules of each in one molar mass: (a) HNO_3 (nitric acid); (b) O_2 (oxygen).

[*Answer:* (a) 63.01 g·mol^{-1}, 6.022×10^{23} HNO_3 molecules; (b) 32.00 g·mol^{-1}, 6.022×10^{23} O_2 molecules]

KEY CONCEPT B   Molar mass of an ionic compound

The molar mass of an **ionic compound** is the sum of the molar masses of the atoms in the formula unit. It is the mass of compound that contains 1 mol of formula units. For instance, for the molar mass of calcium sulfate ($CaSO_4$), we get

$$1 \text{ Ca}: \quad 1 \times 40.08 \text{ g} \cdot \text{mol}^{-1} = \quad 40.08 \text{ g} \cdot \text{mol}^{-1}$$
$$1 \text{ S}: \quad 1 \times 32.06 \text{ g} \cdot \text{mol}^{-1} = \quad 32.06 \text{ g} \cdot \text{mol}^{-1}$$
$$4 \text{ O}: \quad 4 \times 16.00 \text{ g} \cdot \text{mol}^{-1} = \underline{\quad 64.00 \text{ g} \cdot \text{mol}^{-1}}$$
$$136.14 \text{ g} \cdot \text{mol}^{-1}$$

The following table gives the interpretation of the molar mass for ionic compounds.

Compound	Molar Mass (g·mol^{-1})	Interpretation of Molar Mass
$BaCO_3$	197.35	197.35 g $BaCO_3$ = 1 mol $BaCO_3$ = 6.022×10^{23} $BaCO_3$ formula units
$Ca(NO_3)_2$	164.09	164.09 g $Ca(NO_3)_2$ = 1 mol $Ca(NO_3)_2$ = 6.022×10^{23} $Ca(NO_3)_2$ formula units
$Fe_2(Cr_2O_7)_3$	759.66	759.66 g $Fe_2(Cr_2O_7)_3$ = $Fe_2(Cr_2O_7)_3$ = 6.022×10^{23} $Fe_2(Cr_2O_7)_3$ formula units

EXAMPLE   Calculating the molar mass of an ionic compound

What is the molar mass of $Sr(HSO_4)_2$ (strontium hydrogen sulfate)?

SOLUTION The molar mass of a compound is calculated by summing the masses of all of the atoms in the compound. We must remember that the subscript "2" following the HSO_4^- ion, in parentheses, indicates there are two hydrogen sulfate ions; thus, the hydrogen sulfate ion contributes two hydrogen atoms, two sulfur atoms, and eight hydrogen atoms to the formula unit.

$$
\begin{array}{llr}
1\ Sr: & 1 \times 87.62 = & 87.62 \\
2\ H: & 2 \times 1.01 = & 2.02 \\
2\ S: & 2 \times 32.06 = & 64.12 \\
8\ O: & 8 \times 16.00 = & \underline{128.00} \\
& & 281.76\ g \cdot mol^{-1}\ Sr(HSO_4)_2
\end{array}
$$

EXERCISE What is the molar mass of ammonium chromate, $(NH_4)_2CrO_4$?

[*Answer:* $152.07\ g \cdot mol^{-1}$]

KEY WORDS   Define or explain each term in a written sentence or two.

Avogadro's constant
molar mass
mole

DETERMINATION OF CHEMICAL FORMULAS

2.11   MASS PERCENTAGE COMPOSITION

KEY CONCEPT   Mass percentage

Before discussing mass percentage, we shall briefly review how percentages work. Let's assume we have a combination of three components A, B, and C. We can represent this combination with a pie chart:

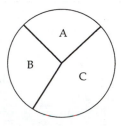

By definition, the percentage of any component is given by

$$
Percentage = \frac{piece\ of\ pie}{whole\ pie} \times 100
$$

For any component, A, for instance,

$$
Percentage\ A = \frac{amount\ of\ A}{amount\ of\ A\ +\ amount\ of\ B\ +\ amount\ of\ C} \times 100
$$

Chemists usually base percentage calculations on the mass of each component. For example, a typical piece of 14-carat gold jewelry with a mass of 3.672 g might contain 2.140 g gold (Au), 0.661 g copper (Cu), and 0.871 g silver (Ag). The following pie chart represents the composition of this item of jewelry:

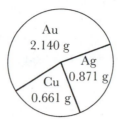

The mass percentage of each component, which means the percentage of each based on the mass present, is calculated by the formulation given earlier:

$$\text{Mass percentage gold} = \frac{\text{mass gold}}{\text{total mass}} \times 100 = \frac{2.140 \text{ g}}{3.672 \text{ g}} \times 100 = 58.28\%$$

$$\text{Mass percentage copper} = \frac{\text{mass copper}}{\text{total mass}} \times 100 = \frac{0.661 \text{ g}}{3.672 \text{ g}} \times 100 = 18.0\%$$

$$\text{Mass percentage silver} = \frac{\text{mass silver}}{\text{total mass}} \times 100 = \frac{0.871 \text{ g}}{3.672 \text{ g}} \times 100 = 23.7\%$$

EXAMPLE   Calculating the percentage composition of compound

The compound isoamyl acetate gives bananas their familiar smell. A sample of isoamyl acetate is analyzed and found to contain 0.8408 g C, 0.1411 g H, and 0.3199 g O (and nothing else). What is the percentage composition of this compound?

SOLUTION The mass percentage is calculated by taking the "piece over the whole" and multiplying by 100. More specifically, we take the mass of each component, divide by the total mass of the sample, and multiply the result by 100. The total mass of the sample is 0.8408 g + 0.1411 g + 0.3200 g = 1.3019 g. Thus, the mass percentage of each element is

$$\text{Mass percentage C} = \frac{\text{mass C}}{\text{mass sample}} \times 100 = \frac{0.8408 \text{ g}}{1.3019 \text{ g}} \times 100 = 64.58\%$$

$$\text{Mass percentage H} = \frac{\text{mass H}}{\text{mass sample}} \times 100 = \frac{0.1411 \text{ g}}{1.3019 \text{ g}} \times 100 = 10.84\%$$

$$\text{Mass percentage O} = \frac{\text{mass O}}{\text{mass sample}} \times 100 = \frac{0.3200 \text{ g}}{1.3019 \text{ g}} \times 100 = 24.58\%$$

EXERCISE What is the percentage composition of octane (a component of gasoline) if a sample of octane is found to contain 0.9609 g C and 0.1814 g H, and nothing else?

[*Answer:* 84.12% C; 15.89% H]

2.12   DETERMINING EMPIRICAL FORMULAS

KEY CONCEPT   Empirical formulas

The **empirical formula** of a compound is a chemical formula that gives the simplest whole-number ratio of atoms in the compound. If we know the molecular formula or formula unit for a compound, we can determine the empirical formula by taking the relative numbers of atoms in the compound and reducing the relative numbers to the simplest whole-number ratio. Different compounds can have the same empirical formula; for example, benzene (C_6H_6), acetylene (C_2H_2), and cyclobutadiene (C_4H_4) all have the empirical formula CH. Also, the correct molecular formula or formula unit for a compound may be the same as the empirical formula, as for water (H_2O) and calcium sulfate ($CaSO_4$).

EXAMPLE 1 Understanding empirical formulas

Give the empirical formulas for the following compounds: (a) anise alcohol, $C_8H_{10}O_2$; (b) aluminum chloride, $AlCl_3$.

SOLUTION (a) Here we are given the whole-number ratio of atoms in the compound, 8:10:2. To determine the empirical formula, we need the simplest whole-number ratio; we get it by dividing each of the numbers in the ratio 8:10:2 by the smallest of the numbers, 2, which gives 4:5:1. The empirical formula is therefore C_4H_5O. (b) The ratio of atoms in the formula unit (1:3) is already the simplest whole-number ratio, so the empirical formula is the same as the formula unit, $AlCl_3$.

EXERCISE Give the empirical formula of the following compounds: (a) ethylene glycol, $C_2H_6O_2$; (b) propylene glycol, $C_3H_8O_2$.

[*Answer:* (a) CH_3O; (b) $C_3H_8O_2$]

EXAMPLE 2 Determining an empirical formula from percentage composition

Tetroquinone has the following percentage composition: C, 41.87%; H, 2.340%; O, 55.78%. What is the empirical formula of tetroquinone?

SOLUTION The percentage composition gives the relative mass of each element present; we must convert to the relative moles of each element present. First, assume we are dealing with exactly 100 g of compound. (Any amount can be used, but using 100 g makes the calculation simpler because the mass of each element then equals the percentage of each.)

$$g\ C = 0.4187 \times 100\ g = 41.87\ g$$

$$g\ H = 0.02340 \times 100\ g = 2.340\ g$$

$$g\ O = 0.5578 \times 100\ g = 55.78\ g$$

We next convert the mass of each element to moles of each, using the molar mass of each element:

$$mol\ C = 41.87\ g \times \frac{1\ mol\ C}{12.01\ g} = 3.486\ mol\ C$$

$$mol\ H = 2.340\ g \times \frac{1\ mol\ H}{1.008\ g} = 2.321\ mol\ H$$

$$mol\ O = 55.78\ g \times \frac{1\ mol\ O}{16.00\ g} = 3.486\ mol\ O$$

The numbers 3.486, 2.321, and 3.486 give the ratio of moles of atoms. The empirical formula is the simplest whole-number ratio of moles of atoms, so the ratio 3.486:2.321:3.486 must be converted to a ratio of simple whole numbers. To do this, we divide by the smallest of the numbers in the ratio:

$$C:\quad \frac{3.486}{2.321} = 1.502$$

$$H:\quad \frac{2.321}{2.321} = 1.000$$

$$O:\quad \frac{3.486}{2.321} = 1.502$$

Multiplying each number in the ratio by 2 results in the desired simplest whole-number ratio of $3:2:2$.

$$C: \quad 2 \times 1.502 = 3.004$$

$$H: \quad 2 \times 1.000 = 2.000$$

$$O: \quad 2 \times 1.502 = 3.004$$

The empirical formula of the compound is $C_3H_2O_3$.

EXERCISE The percentage composition of dibromoanthracene is C, 50.04%; H, 2.400%; Br, 47.56%. What is the empirical formula of dibromoanthracene?

[*Answer:* C_7H_4Br]

2.13 DETERMINING MOLECULAR FORMULAS

KEY CONCEPT   Empirical formulas to molecular formulas

The molecular formula of a compound can be determined if its molar mass and empirical formula mass are both known. For instance, consider compounds with the empirical formula CH_2O. The simplest compound with the empirical formula CH_2O has the molecular formula CH_2O; the molar mass equals the empirical formula mass, 30.03 g/mol. The next compound with the same empirical formula contains the equivalent of not one, but two empirical formula units. By this we do not mean that two empirical units are bound together, but rather that the compound has twice the number of each atom as the empirical formula unit has. The compound, therefore, has 2 carbon atoms, 4 hydrogen atoms, and 2 oxygen atoms bound together. Its molecular formula is $C_2H_4O_2$, because the equivalent of two empirical formula units is present, its molar mass (60.06 g/mol) must be twice the empirical formula mass (30.03 g/mol). This result can be extended to compounds with three, four, or more empirical formula units. It becomes evident that since the molecular formula contains an integral number of empirical formula units, the molecular mass must be an integral multiple of the empirical formula mass. The table that follows summarizes the results for the empirical formula CH_2O up to the molecule $C_5H_{10}O_5$.

Molecular Formula	Molar Mass (g·mol^{-1})	Molar Mass/Empirical Formula Mass	Number of Empirical Formulas in the Molecular Formula
CH_2O	30.03	1	1
$C_2H_4O_2$	60.06	2	2
$C_3H_6O_3$	90.09	3	3
$C_4H_8O_4$	120.12	4	4
$C_5H_{10}O_5$	150.15	5	5

We see that by dividing the molar mass by the empirical formula mass, we find the number of empirical formulas that are contained in the molecular formula. Thus, if we know the empirical formula and the molar mass of a compound, we can determine the molecular formula.

EXAMPLE   Determining the molecular formula from the empirical formula and the molar mass

A compound with the empirical formula C_5H_5O is found to have a molar mass of 324.36 g·mol^{-1}. What is the molecular formula of the compound?

SOLUTION We must determine how many empirical formula units are contained in the molecular formula. The empirical formula mass of C_5H_5O is 81.09 g·mol^{-1}. We divide the molar mass by the empirical formula mass to find the number of empirical units in the molecular formula:

$$\frac{324.36 \text{ g·mol}^{-1}}{81.09 \text{ g·mol}^{-1}} = 4.000 = 4 \quad \text{(remember, the answer must be an integer)}$$

The molecular formula contains four of the C_5H_5O empirical formula units, so the molecular formula must be $C_{20}H_{20}O_4$. A check shows the molar mass of $C_{20}H_{20}O_4$ to be 324.36 $g \cdot mol^{-1}$.

EXERCISE A compound with the empirical formula C_2H_2O has a molar mass of 210.18 $g \cdot mol^{-1}$. What is the molecular formula of the compound?

[*Answer:* $C_{10}H_{10}O_5$]

KEY WORDS   Define or explain each term in a written sentence or two.

empirical formula   mass percentage composition
formula unit   molecular formula

SOLUTIONS IN CHEMISTRY

2.14   MOLARITY

KEY CONCEPT A   Molar concentration

Solutions have a variable composition. To completely characterize a solution, we must know its composition. The most widely used expression of solution composition is the **molarity,** which is the moles of solute per 1 liter of solution. To calculate the molarity, we determine the moles of solute in a solution and divide by the liters of solution:

$$\text{Molarity} = \frac{\text{moles of solute}}{\text{liters of solution}}$$

It is important to note that the factor in the denominator is the volume of *solution,* not the volume of *solvent.* The volume of the solution includes the volume of solvent plus any changes made in the volume by the addition of the solute to the solvent.

EXAMPLE   Determining the molarity of solute in a solution

Adding 53.5 g NaCl to 482 mL H_2O results in 500 mL of solution. What is the molarity of the NaCl solution?

SOLUTION We use the definition

$$\text{Molarity} = \frac{\text{moles of solute}}{\text{liters of solution}}$$

For NaCl:

$$\text{Molarity of NaCl} = \frac{\text{mol NaCl}}{\text{liters of solution}}$$

We calculate the moles of NaCl, using the molar mass of NaCl, which is 58.5 $g \cdot mol^{-1}$:

$$\text{Mol NaCl} = 53.5 \text{ g} \times \frac{1 \text{ mol NaCl}}{58.5 \text{ g}} = 0.915 \text{ mol}$$

The liters of solution are calculated from the 500 mL of solution given in the problem:

$$500 \text{ mL} \times \frac{1 \text{ L}}{1000 \text{ mL}} = 0.500 \text{ L}$$

Substituting these values into the defining equation gives

$$\text{Molarity of NaCl} = \frac{\text{mol NaCl}}{\text{liters of solution}}$$

$$= \frac{0.915 \text{ mol}}{0.500 \text{ L}}$$

$$= 1.83 \text{ mol} \cdot L^{-1}$$

$$= 1.83 \text{ M}$$

Note that the amount of water (482 mL) is irrelevant to solving the problem. The volume of solvent is not important in calculating molarity; the volume of solution is.

EXERCISE Adding 50.0 g $MgSO_4$ to 200 mL of water results in 205 mL of solution. What is the molarity of the $MgSO_4$ in the solution?

[*Answer:* 2.03 M]

KEY CONCEPT B   Moles of solute in a volume of solution

The molarity is widespread in use because it provides a convenient way to measure out a given number of moles of solute. The moles of solute in a given volume of solution is given by the equation

$$\text{Moles of solute} = \text{molarity} \times \text{volume (in L)}$$

$$\text{mol} \quad = \quad M \quad \times \quad V$$

where it is understood that the volume is in liters. Because chemists must frequently obtain a particular number of moles of a substance, solutions of known molarity are convenient, indeed. In addition, if a chemist is required to prepare a specific volume of a solution with a specific molarity, this equation specifies the number of moles (and, therefore, the number of grams) of solute to be used. We finally note that if the volume is expressed in milliliters, $M \times V$ gives the millimoles of solute. This is so because molarity equals millimole per milliliter (as well as mole per liter).

$$\text{mmol} = M \times V \quad \text{(volume in mL)}$$

EXAMPLE   Calculating the milliliters of solution that contain a specific number of moles

A chemist is required to obtain 0.433 mol of Na_2SO_4 from a 2.00 M solution of Na_2SO_4. How many milliliters of solution are required?

SOLUTION For a solution of known molarity, the relationship mol $= M \times V$ relates the moles of solute to the liters of solution. We first substitute the stated molarity and moles of solute into this equation and then solve for the liters of solution:

$$0.433 \text{ mol Na}_2\text{SO}_4 = 2.00 \frac{\text{mol Na}_2\text{SO}_4}{\text{L}} \times V$$

$$V = \frac{0.433 \text{ mol Na}_2\text{SO}_4}{2.00 \text{ mol Na}_2\text{SO}_4/\text{L}} = 0.217 \text{ L}$$

Converting to milliliters gives

$$0.217 \text{ L} \times \frac{1000 \text{ mL}}{\text{L}} = 217 \text{ mL}$$

We could also find the milliliters of solution needed by treating the molarity as a conversion factor. In this example, the molarity of 2.00 M gives

$$2.00 \text{ mol Na}_2\text{SO}_4 = 1 \text{ L solution}$$

Because we started with 0.433 mol Na_2SO_4, the conversion problem becomes

$$0.433 \text{ mol Na}_2\text{SO}_4 \times \frac{1 \text{ L solution}}{2.00 \text{ mol Na}_2\text{SO}_4} = 0.217 \text{ L} = 217 \text{ mL}$$

As we expect, the answer is the same as we obtained earlier. The answer to a problem does not depend on which technique we use to solve it.

EXERCISE Write a recipe for preparing 250 mL of a 3.00 M solution of NaOH.

[*Answer:* Add enough water to 30.0 g NaOH to make a final volume of 250 mL of solution.]

2.15 DILUTION

KEY CONCEPT   Dilutions

To **dilute** a solution, we add solvent to the solution, which results in a new solution with a relatively smaller amount of solute per liter of solution. If we denote the molarity and volume of the original concentrated solution as M_{conc}, V_{conc}, and the molarity and concentration after dilution as M_{dil}, V_{dil}, then

$$M_{conc} \times V_{conc} = M_{dil} \times V_{dil}$$

EXAMPLE   Making a dilute solution

A laboratory technician is asked to make up 2.00 L of a 0.200 M solution of NaBr from a bottle of 1.78 M NaBr. Calculate the volume of 1.78 M NaBr needed and write the steps to be taken to make the solution.

SOLUTION The equation $M_{conc} \times V_{conc} = M_{dil} \times V_{dil}$ is used in calculations involving the dilution of a solution when the composition of the solution is expressed as molarity. For this example we have

$$M_{conc} = 1.78 \text{ M} \qquad V_{conc} = ?$$

$$M_{dil} = 0.200 \text{ M} \qquad V_{dil} = 2.00 \text{ L}$$

Substituting into the equation gives

$$1.78 \text{ M} \times V_{conc} = 0.200 \text{ M} \times 2.00 \text{ L}$$

$$V_{conc} = \frac{0.200 \text{ M} \times 2.00 \text{ L}}{1.78 \text{ M}}$$

$$= 0.225 \text{ L} = 225 \text{ mL}$$

To prepare the solution, the laboratory technician should measure 225 mL (0.225 L) of the 1.78 M NaBr into a 2.00-L volumetric flask and add enough water (approximately 1775 mL) to make 2.00 L of solution.

EXERCISE Describe how you would make 200 mL of a 0.100 M solution of HCl from a 6.0 M stock solution.

[*Answer:* Take 3.3 mL of the 6.0 M HCl and add enough water to make 200 mL of solution.]

KEY WORDS   Define or explain each term in a written sentence or two.

dilute
molar concentration
molarity

DESCRIPTIVE CHEMISTRY TO REMEMBER

- Nothing can be cooled below **absolute zero** (0 K).
- One mole of objects contains an **Avogradro's constant**, 6.022×10^{23} mol^{-1}, of the objects.
- Water can be decomposed to **oxygen** and **hydrogen** by electrolysis.
- Magnesium (Mg) burns in nitrogen (N_2) to form **magnesium nitride** (Mg_3N_2).
- One of the first known compounds of xenon (XeF_4) was prepared in the 1960s.
- **Water** freezes at 0°C and boils at 100°C.

MATHEMATICAL EQUATIONS TO KNOW AND UNDERSTAND

$$°F = 32 + \left(\frac{9}{5}\right)°C$$

conversion from °C to °F

$$°C = \left(\frac{5}{9}\right)(°F - 32)$$

conversion from °F to °C

$$K = 273.15 + °C$$

relation between kelvins and °C

$$\text{Mass \% A} = \frac{\text{mass of A in a sample}}{\text{total mass of sample}} \times 100$$

mass percentage of component in a mixture

$$\text{Molarity (M)} = \frac{\text{moles solute}}{\text{volume solution (in liters)}}$$

definition of molarity

$$n = M \times V\text{(in liters)}$$

moles solute (n) from molarity

$$M_{conc}V_{conc} = M_{dil}V_{dil}$$

dilution of a solution

SELF-TEST EXERCISES

Measurements and units

1. Which of the following describes the quantity of matter in a sample?
(a) weight   (b) volume   (c) mass   (d) temperature   (e) length

2. How many kilograms are in 675 g?
(a) 6.75   (b) 67.5   (c) 0.00675   (d) 0.675   (e) 675,000

3. How many milligrams are in 1.25 g?
(a) 1250   (b) 0.00125   (c) 0.125   (d) 125   (e) 12.5

4. What unit of length is most convenient for measuring the height of a human?
(a) millimeter   (b) centimeter   (c) meter
(d) kilometer   (e) megameter

5. The prefix "nano" corresponds to what multiplier?
(a) 10^{-3} (b) 10^3 (c) 10^{-2} (d) 10^{-6} (e) 10^{-9}

6. What is the volume (in cubic meters) of a rectangular solid 1.2 m long, 158 mm deep, and 62 cm wide?
(a) 12,000 m^3 (b) 1.2 m^3 (c) 0.12 m^3 (d) 1200 m^3 (e) 12 m^3

7. How many milliliters are there in 0.35 L?
(a) 0.00035 mL   (b) 35 mL   (c) 0.0035 mL
(d) 0.035 mL   (e) 350 mL

8. A sample of metal has a mass of 63.22 g and a volume of 8.89 cm^3. What is its density?

(a) 562 $g \cdot cm^{-3}$ (b) 14.1 $g \cdot cm^{-3}$ (c) 0.00178 $g \cdot cm^{-3}$
(d) 7.11 $g \cdot cm^{-3}$ (e) 0.141 $g \cdot cm^{-3}$

9. What is the volume of a 15.65-g piece of wood with a density of 0.857 $g \cdot cm^{-3}$?
(a) 7.89 cm^3 (b) 18.3 cm^3 (c) 16.5 cm^3
(d) 12.5 cm^3 (e) 54.6 cm^3

10. The density of aluminum is 2.70 $g \cdot cm^{-3}$. What is the mass of a 252-cm^3 sample of aluminum?
(a) 10.7 g   (b) 680 g   (c) 392 g   (d) 1.47 g   (e) 93.3 g

11. 15.2 oz =
(a) 2.32 g   (b) 1.87 g   (c) 0.536 g   (d) 477 g   (e) 431 g

12. 52 km =
(a) 62 mile   (b) 84 mile   (c) 32,000 mile
(d) 32 mile   (e) 84,000 mile

13. Convert 8.2 gal to microliters
(a) 3.2×10^{-7} μL   (b) 3.2×10^{-8} μL   (c) 3.1×10^{-6} μL
(d) 3.1×10^{7} μL   (e) 3.1×10^{6} μL

14. Convert 0.00421 mi to centimeters.
(a) 6.77 cm   (b) 678 cm   (c) 382 cm   (d) 1.48 cm   (e) 148 cm

15. How many cubic kilometers of water are there in a lake that contains 2.0 mi^3 of water?
(a) 8.3 km^3 (b) 66 km^3 (c) 3.2 km^3 (d) 33 km^3 (e) 5.0 km^3

16. What is the density of iron (7.86 $g \cdot cm^{-3}$) in grams per tablespoon ($g \cdot tbl^{-1}$), where 1 tbl = 4.93 mL?
(a) 0.626 $g \cdot tbl^{-1}$ (b) 24.1 $g \cdot tbl^{-1}$ (c) 1.60 $g \cdot tbl^{-1}$
(d) 0.0258 $g \cdot tbl^{-1}$ (e) 38.7 $g \cdot tbl^{-1}$

17. Convert 30 $m \cdot hr^{-1}$ to feet per second.
(a) 20 $ft \cdot s^{-1}$ (b) 57 $ft \cdot s^{-1}$ (c) 44 $ft \cdot s^{-1}$
(d) 30 $ft \cdot s^{-1}$ (e) 81 $ft \cdot s^{-1}$

18. Convert 36.0°F to degrees Celsius.
(a) −12°C   (b) 2.2°C   (c) 7.2°C   (d) 61°C   (e) 309°C

19. Convert −8°C to degrees Fahrenheit.
(a) −22°F   (b) 28°F   (c) −18°F   (d) 18°F   (e) −13°F

20. Convert 31°C to kelvins.
(a) 304 K   (b) 88 K   (c) −0.55 K   (d) 63 K   (e) 242 K

21. How many significant figures are in the number 9.225?
(a) 4   (b) 3   (c) 2   (d) 1   (e) 0

22. How many significant figures are in the number 1680?
(a) 4   (b) 3   (c) 2   (d) 1   (e) 0

23. How many significant figures are in the number 0.0552?
(a) 3   (b) 4   (c) 0   (d) 5   (e) 2

24. What is 0.00950 in standard scientific notation?
(a) 950   (b) 0.950×10^{-2} (c) 9.50×10^{3}
(d) 9.5×10^{-3} (e) 9.50×10^{-3}

25. What is 6.200×10^2 in decimal notation?
(a) 62   (b) 62,000   (c) 620   (d) 6200   (e) 620.0

26. $2.235 + .01 =$
(a) 2   (b) 2.245   (c) 2.250   (d) 2.25   (e) 2.24

27. $2.235 \times .01 =$
(a) 0.02   (b) 0.022   (c) 0.0223   (d) 0.0224   (e) 0.02235

28. $(10.24 - 0.11) \times 6.55 =$
(a) 66.3515   (b) 67   (c) 66.3   (d) 66.4   (e) 66

29. What is the total mass (in grams) carried by 9 trucks, each carrying an average mass of 2643 g?
(a) 2.379×10^4 (b) 2×10^4 (c) 3×10^4
(d) 23,787   (e) 2.38×10^4

30. Which of the following values for the mass of a dime is most precise?
(a) 2 g   (b) 2.3 g   (c) 2.28 g   (d) 2.278 g   (e) 2.2777 g

31. A digital watch shows the time to be 8:05.22 A.M. when it is actually 8:10.22 A.M. An analog watch gives the time as 8:09 A.M. Which of the following is correct?
(a) analog more accurate and precise
(b) analog more precise, digital more accurate
(c) analog more accurate, digital more precise
(d) digital more accurate and precise

32. Which of the following describes a systematic error in an experiment?
(a) Temperature fluctuations cause random differences in a balance reading.
(b) Vibrations in a building cause a meter to jiggle around the correct reading.
(c) A miscalibrated thermometer consistently reads 2°C low.
(d) Human judgment in estimating a measurement causes some uncertainty in the last digit.

Chemical amounts

33. How many silicon atoms are present in 0.778 mol of silicon?
(a) 4.79×10^{23} (b) 7.74×10^{23} (c) 6.022×10^{23}
(d) 8.77×10^{24} (e) 2.22×10^{24}

34. A sample of uranium contains 4.59×10^{24} atoms. How many moles of uranium are present?
(a) 0.131 mol   (b) 7.62 mol   (c) 0.762 mol
(d) 1.31 mol   (e) 276 mol

35. A sample of water contains 4.45×10^{21} molecules of water. How many millimoles of water are present in the sample?
(a) 7.39×10^{-3} (b) 7.39   (c) 0.135   (d) 7.39×10^3 (e) 135

36. What is the molar mass of cholesterol, $C_{27}H_{46}O$, in $g \cdot mol^{-1}$?
(a) 216   (b) 244   (c) 387   (d) 410   (e) 444

37. What is the molar mass of $(NH_4)_2SO_4$?

(a) $114.1 \text{ g} \cdot \text{mol}^{-1}$ (b) $146.2 \text{ g} \cdot \text{mol}^{-1}$ (c) $210.2 \text{ g} \cdot \text{mol}^{-1}$

(d) $132.1 \text{ g} \cdot \text{mol}^{-1}$ (e) $70.0 \text{ g} \cdot \text{mol}^{-1}$

38. What is the molar mass of the acid that gives vinegar its sour taste, acetic acid, $HC_2H_3O_2$?

(a) $32.0 \text{ g} \cdot \text{mol}^{-1}$ (b) $44.2 \text{ g} \cdot \text{mol}^{-1}$ (c) $60.1 \text{ g} \cdot \text{mol}^{-1}$

(d) $102.4 \text{ g} \cdot \text{mol}^{-1}$ (e) $98.1 \text{ g} \cdot \text{mol}^{-1}$

39. How many moles of $C_4H_8Br_2$ (dibromobutane) are present in 56.6 g of $C_4H_8Br_2$?

(a) 3.81   (b) 0.262   (c) 1.22×10^4 (d) 8.18×10^{-5} (e) 0.555

40. How many moles of $C_8H_{18}O$ (octanol) are present in 175 g of $C_8H_{18}O$?

(a) 1.34   (b) 1.43   (c) 0.744   (d) 0.697   (e) 22.7

41. How many grams of H_2SO_3 (sulfurous acid) are present in 5.22 mol H_2SO_3?

(a) 566   (b) 0.0636   (c) 15.7   (d) 2.34×10^{-3} (e) 428

42. How many grams of PF_5 (phosphorus pentafluoride) are there in 7.88×10^{-3} mol PF_5?

(a) 1.23   (b) 0.993   (c) 1.01   (d) 6.26×10^{-5} (e) 1.60×10^4

Determination of chemical formulas

43. Fructose (also known as fruit sugar) occurs in a large number of fruits. It is the sweetest of the sugars. A 1.0766-g sample of fructose is found to contain 0.4307 g C, 0.0722 g H, and 0.5737 g O. What is the percentage of carbon (C) in fructose?

(a) 25.00%   (b) 40.01%   (c) 66.68%   (d) 43.07%   (e) 46.55%

44. Glycerol is an important component of body fat and cell membranes. It is an adulterant in some wines since it imparts a sweet taste, but it is toxic in its pure form and wines containing glycerol cannot be sold legally in the US. A 0.8753 g sample of glycerol is found to contain 0.4562 g O. What is the percentage of oxygen (O) in glycerol?

(a) 10.89%   (b) 45.62%   (c) 47.88%   (d) 52.12%   (e) 61.11%

45. The sugar L-threose has the mass percentage composition 40.00% C, 6.71% H, and 53.29% O. What is the empirical formula of L-threose?

(a) CH_2O (b) $C_4H_6O_5$ (c) C_4H_2O (d) CHO (e) $C_4H_7O_5$

46. The compound purine is the structural antecedent of a large portion of the biological compounds RNA, and DNA. It has the mass percentage composition 50.00% C, 3.36% H, and 46.65% N. What is the empirical formula of purine?

(a) CH_2N (b) $C_5H_3N_5$ (c) $C_5H_4N_4$ (d) $C_6H_4N_3$ (e) C_4H_8N

47. What is the empirical formula of the compound formed by the reaction of 1.000 g O_2 with 1.000 g S to form 2.000 g product?

(a) S_2O_3 (b) SO_3 (c) SO_2 (d) SO (e) S_2O

48. What is the molecular formula of the compound with the empirical formula C_2H_3 and molar mass $135.2 \text{ g} \cdot \text{mol}^{-1}$?

(a) C_4H_6 (b) C_8H_{12} (c) $C_{10}H_{15}$ (d) C_2H_3 (e) CH

49. What is the molecular formula of a compound with the empirical formula C_5H_8O and molar mass $336.5 \text{ g} \cdot \text{mol}^{-1}$?

(a) C_5H_8O (b) $C_{20}H_{32}O_4$ (c) $C_{50}H_{80}O_5$

(d) $C_{21}H_{36}O_3$ (e) $C_{19}H_{28}O_5$

Solutions in chemistry

50. A solution is prepared by adding 230 mL H_2O to 0.645 mol HNO_3, resulting in 250 mL of solution. What is the molar concentration of HNO_3?

(a) 0.161 M   (b) 2.80 M   (c) 2.58 M   (d) 1.34 M   (e) 0.645 M

51. 194 mL H_2O is added to 37.0 g $BaCl_2$, resulting in 200 mL of solution. What is the molar concentration of $BaCl_2$?

(a) 0.178 M   (b) 0.466 M   (c) 0.918 M   (d) 0.883 M   (e) 1.11 M

52. How many moles of NaCl are required to prepare 500 mL of a 0.250 M NaCl solution?

(a) 0.100   (b) 1.00   (c) 0.500   (d) 0.750   (e) 0.125

53. How many grams of $NaNO_3$ are required to prepare 125 mL of 3.00 M $NaNO_3$?

(a) 46.1   (b) 31.9   (c) 227   (d) 3.54   (e) 28.3

54. How many moles of $Zn(NO_3)_2$ are present in 115 mL of a 0.65 M $Zn(NO_3)_2$ solution?

(a) 0.25   (b) 13   (c) 0.075   (d) 5.7   (e) 0.18

55. How many grams of Na_2SO_4 are present in 250 mL of a 1.72 M solution of Na_2SO_4?

(a) 61.1   (b) 20.6   (c) 244   (d) 35.5   (e) 977

56. 25.0 mL of a 18 M H_2SO_4 solution are diluted to 500 mL with water. What is the molar concentration of the H_2SO_4?

(a) 0.45 M   (b) 0.050 M   (c) 0.23 M   (d) 0.36 M   (e) 0.90 M

57. 432 mL H_2O is added to 50.0 mL of a 3.00 M $NiCl_2$ solution, resulting in 500 mL of solution. What is the molar concentration of the $NiCl_2$?

(a) 0.347 M   (b) 0.311 M   (c) 0.300 M   (d) 2.59 M   (e) 3.33 M

Descriptive chemistry

58. What is the temperature below which nothing can be cooled?

(a) 0°F   (b) 0°C   (c) 77 K   (d) 0 K   (e) −253°C

59. Which of the following metals forms a nitride when burned in nitrogen?

(a) Au   (b) Mg   (c) Pb   (d) Ca   (e) Fe

60. Which of the following is one of the first known compounds of a noble gas?

(a) KrF_6 (b) KrF_4 (c) HeXe   (d) Ar_2O (e) XeF_4

CHAPTER 3

CHEMICAL REACTIONS: MODIFYING MATTER

In a chemical reaction, one or more substances is changed into an entirely new substance or substances. For instance, plants, using sunlight and photosynthesis, convert the two substances carbon dioxide and water into the new substance glucose (and related compounds). Although there are millions of known chemical reactions, this chapter will explore only a few.

CHEMICAL EQUATIONS AND CHEMICAL REACTIONS

3.1 SYMBOLIZING CHEMICAL REACTIONS

KEY CONCEPT Chemical equations show a rearrangement of atoms

During a chemical reaction, a specific substance (or substances) changes to a different substance (or substances). The substances we start with in the reaction are called **reactants,** and the new substances formed by the chemical reaction are called **products.** The chemical change from reactants to products is symbolized by a **chemical equation,** in which the formulas for the reactants are written to the left of an arrow and the products to the right of the arrow. The arrow symbolizes a chemical change. A **stoichiometric coefficient** in front of each substance is needed to ensure that the number of atoms of each element is the same for the reactants as for the products. The physical state of the reactants and products is often indicated by placing a symbol next to each substance:

(s) solid   (l) liquid   (g) gas   (aq) aqueous (substance dissolved in water)

EXAMPLE   Writing a chemical equation

Write the chemical equation that symbolizes the reaction in which solid ammonium nitrate is heated to 200°C to form the gas dinitrogen oxide and water vapor.

SOLUTION Using techniques from Chapter 1, we get the following formulas for the substances involved: ammonium nitrate, NH_4NO_3; dinitrogen oxide, N_2O; and water, H_2O. We now assemble these into the required chemical equation. The statement in the problem makes clear that the reactant, or substance we start with, is NH_4NO_3 and that the products, or substances formed, are N_2O and H_2O. We indicate the physical state of each substance with the proper symbol and write 200°C over the arrow to indicate the temperature needed to make the reaction proceed:

$$NH_4NO_3(s) \xrightarrow{200°C} N_2O(g) + 2H_2O(g)$$

The 2 is needed in front of H_2O to assure that the number of atoms of each element is the same to the left and right of the arrow.

EXERCISE When solid magnesium is heated to 600°C in the presence of oxygen, magnesium oxide is formed. Write the chemical equation for this reaction.

[*Answer:* $2Mg(s) + O_2(g) \xrightarrow{600°C} 2MgO(s)$]

PITFALL   Magnesium and manganese

Be certain to distinguish magnesium (Mg, atomic number 12) from manganese (Mn, atomic number 25).

3.2   BALANCING CHEMICAL EQUATIONS

KEY CONCEPT   All reactant atoms must be present in the products

All chemical equations must be balanced. We **balance a chemical equation** by adjusting the stoichiometric coefficients in the equation until the number of atoms of each element on the left-hand side of the arrow is equal to the number of atoms of the same element on the right-hand side. The simplest equations are balanced by inspection: In these, only one or two coefficients have to be adjusted, and the equation is balanced in one or two steps by a commonsense approach. The reaction between H_2 and O_2 to form water is a good example:

$$H_2(g) + O_2(g) \longrightarrow H_2O(g) \quad \triangle$$

Inspection of the unbalanced equation (symbolized by $\triangle$) reveals that there are two oxygen atoms on the left and only one on the right, so a 2 must be placed in front of $H_2O(g)$.

$$H_2(g) + O_2(g) \longrightarrow 2H_2O(g) \quad \triangle$$

With this change, there are now four hydrogen atoms on the right, which are balanced by placing a 2 in front of $H_2(g)$.

$$2H_2(g) + O_2(g) \longrightarrow 2H_2O(g)$$

Atom	Left	Right
H	4	4
O	2	2

The equation is now balanced. We indicate the balance in the small table, which gives the number of each atom to the left and the right of the arrow. A second level of difficulty occurs when more than one or two steps are involved or more complicated compounds are involved. A few rules will suffice to balance most such equations. (The rules here are slightly different from the ones in the text: Use the ones you are more comfortable with.)

Step 1. Balance any element that appears in only one compound on each side of the arrow.
Step 2. By inspection, balance any element that appears in more than one compound on either the right or left.
Step 3. Balance any element that appears in elemental form on the right or left.
Step 4. Clear any fractions present, if desired.
Hint. Balance polyatomic ions that appear on both sides of the equation as units rather than as separate atoms.

EXAMPLE   Balancing chemical equations

The compound $C_2H_8N_2$ (dimethylhydrazine), related to hydrazine, is used as a rocket fuel in the following reaction. Balance this equation.

$$C_2H_8N_2(l) + N_2O_4(g) \longrightarrow N_2(g) + H_2O(g) + CO_2(g) \quad \triangle$$

Step 1. Balance any element that appears in only one compound on each side of the arrow. Carbon and hydrogen are balanced in this step. There are two C atoms on the left, so a 2 is placed in front of $CO_2(g)$. There are eight hydrogen atoms on the left, which require $4H_2O(g)$ molecules on the right:

$$C_2H_8N_2(l) + N_2O_4(g) \longrightarrow N_2(g) + 4H_2O(g) + 2CO_2(g) \quad \triangle$$

Step 2. By inspection, balance any element that appears in more than one compound on either the right or the left. Only oxygen is balanced here, because nitrogen is balanced in step 3. There are four oxygen atoms on the left and eight on the right, so we place a 2 in front of $N_2O_4(g)$ to get eight oxygen atoms on the left:

$$C_2H_8N_2(l) + 2N_2O_4(g) \longrightarrow N_2(g) + 4H_2O(g) + 2CO_2(g) \quad \triangle$$

Step 3. Balance any element that appears in elemental form on the right or left. Here, we balance the nitrogen atoms by placing a 3 in front of $N_2(g)$ on the right, so that there are six nitrogen atoms on each side of the arrow:

$$C_2H_8N_2(l) + 2N_2O_4(g) \longrightarrow 3N_2(g) + 4H_2O(g) + 2CO_2(g)$$

Atom	Left	Right
C	2	2
H	8	8
N	6	6
O	8	8

Step 4 is not used in this example.

EXERCISE Balance the following equation and include the table showing the number of atoms on the left and the right. Balance SO_4^{2-} as a unit, rather than as separate sulfur and oxygen atoms:

$$Fe_2(SO_4)_3(aq) + BaCl_2(aq) \longrightarrow BaSO_4(s) + FeCl_3(aq)$$

Answer: $Fe_2(SO_4)_3(aq) + 3BaCl_2(aq) \longrightarrow 3BaSO_4(s) + 2FeCl_3(aq)$

Atom or Ion	Left	Right
Fe	2	2
SO_4^{2-}	3	3
Ba	3	3
Cl	6	6

PITFALL Changing formulas to balance an equation

Once the formulas of the products and reactants are written correctly, they *cannot* be changed to balance the equation. Changing the formulas would change the meaning of the equation.

PITFALL Incorrect formulas in an equation

Occasionally, with a complicated equation, you may find that balancing one atom unbalances another, which then has to be rebalanced, and that rebalancing this atom causes another (or the original atom) to go out of balance. If you find you are going in circles with this process, so that no matter what you do, one element always remains unbalanced, you probably have written a formula incorrectly or have forgotten to include a required formula. If balancing one element always unbalances another, with no end in sight, check your formulas!

KEY WORDS Define or explain each term in a written sentence or two.

balanced equation   law of conservation of mass   reagent
chemical equation   product   stoichiometric coefficient
chemical reaction   reactant

PRECIPITATION REACTIONS

3.3   AQUEOUS SOLUTIONS

KEY CONCEPT   **Many compounds break up into ions when dissolved in H_2O**

When NaCl is dissolved in water, it breaks up into separate Na^+ and Cl^- ions that are free to move, independent of each other, from place to place in the solution. This process is called **ionization.** If magnesium bromide ($MgBr_2$) is dissolved, it ionizes into one Mg^{2+} ion and two Br^- ions, reflecting the composition of the compound. Some molecular compounds also form ions when dissolved in water; for instance, HNO_3 forms hydrated H^+ and NO_3^- ions. These three compounds are all called **strong electrolytes** because they are highly soluble and ionize completely, thereby producing many ions in solution. If only partial ionization occurs, the compound is called a **weak electrolyte,** even if it is very soluble. Because electrolytes form charged species, electrolyte solutions are electrical conductors; they conduct well or poorly, depending on the amount of ionization that occurs. The more ions that form, the better the conductivity.

EXAMPLE   Predicting the number of ions formed

How many ions form when one formula unit of $FeCl_3$ is dissolved in water? Will the resulting solution be conducting or nonconducting?

SOLUTION Iron(III) chloride ($FeCl_3$) is made up of one Fe^{3+} ion and three Cl^- ions. Thus, when one formula unit of $FeCl_3$ dissolves in water, a total of four ions will be liberated. Because ions form, and because $FeCl_3$ is very soluble, the solution that results will be a good conductor.

EXERCISE Acetic acid is a weak electrolyte that undergoes about 4% ionization into hydrated H^+ and CH_3COO^- ions? How many ions will form if 100 acetic acid molecules dissolve in water?

[Answer: 8 ions]

3.4   REACTIONS BETWEEN STRONG ELECTROLYTE SOLUTIONS

KEY CONCEPT   **Precipitates are insoluble substances that form in reactions**

A **precipitation reaction** is one in which an insoluble solid forms as a product after two solutions are mixed. By definition, an insoluble solid does not dissolve. As it forms, it appears in the reaction mixture as a cloudy, opaque material. A precipitation may be represented as

Solution	+	Solution	$\longrightarrow$	Precipitate
Transparent, but possibly colored		Transparent, but possibly colored		Opaque, either white or colored

In the following safe home experiment, you can see a precipitation reaction. Thoroughly dissolve $\frac{1}{2}$ teaspoon of Epsom Salts in $\frac{1}{2}$ cup of water; in a separate cup, thoroughly dissolve $\frac{1}{2}$ teaspoon of washing soda (not baking soda) in $\frac{1}{2}$ cup of water. You should now have two clear, colorless solutions. Mix the two solutions in a clear glass: A white precipitate forms. The equation for this reaction is

$$MgSO_4(aq) + Na_2CO_3(aq) \longrightarrow MgCO_3(s) + Na_2SO_4(aq)$$

Epsom Salts	Washing soda	White precipitate	Soluble salt

Both $MgSO_4$ and Na_2CO_3 are soluble and dissolve easily in water, giving colorless, transparent solutions. When they are mixed, a reaction occurs. The Mg^{2+} ion replaces the two Na^+ ions in Na_2CO_3 to form insoluble $MgCO_3$. The two Na^+ ions replace Mg^{2+} in $MgSO_4$ to form the soluble salt Na_2SO_4.

EXAMPLE   Writing a precipitation equation

Write the equation for the precipitation reaction between NaF and Ca(NO$_3$)$_2$ to form products. Be certain to label all the substances appropriately with (aq) or (s). CaF$_2$ is insoluble.

SOLUTION We first deduce what the products are and which is a precipitate:

$$\text{NaF} + \text{Ca(NO}_3)_2 \longrightarrow ?$$

Na$^+$ replaces Ca^{2+} in Ca(NO$_3$)$_2$ and, in doing so, unites with NO$_3^-$ to form NaNO$_3$; in addition, Ca^{2+} replaces Na$^+$ in NaF and, in doing so, unites with F$^-$ to form CaF$_2$. So the unbalanced equation is

$$\text{NaF} + \text{Ca(NO}_3)_2 \longrightarrow \text{NaNO}_3 + \text{CaF}_2 \quad \triangle$$

Table 3.1 of the text states that salts of the Group 1 elements are soluble, so NaNO$_3$ does not form a precipitate. The problem states that CaF$_2$ is insoluble; therefore, it is the precipitate. (As expected, Table 3.1 also confirms that NaF and Ca(NO$_3$)$_2$ are soluble.) To balance the equation, we note that the left side has one F$^-$ ion and that the right has two, so a 2 is placed in front of NaF to balance fluorine. In addition, there are two NO$_3^-$ ions on the left but only one on the right, so a 2 is needed in front of NaNO$_3$. The balanced equation is

$$2\text{NaF} + \text{Ca(NO}_3)_2 \longrightarrow 2\text{NaNO}_3 + \text{CaF}_2$$

Ion or Atom	Left	Right
Na	2	2
F	2	2
Ca	1	1
NO$_3^-$	2	2

Because CaF$_2$ is insoluble, it forms a solid precipitate and is labeled (s). The final equation is

$$2\text{NaF(aq)} + \text{Ca(NO}_3)_2\text{(aq)} \longrightarrow 2\text{NaNO}_3\text{(aq)} + \text{CaF}_2\text{(s)}$$

EXERCISE Write the equation for the precipitation reaction between (NH$_4$)$_2$S and AgNO$_3$. Construct a table showing that the equation is balanced.

Answer: (NH$_4$)$_2$S(aq) + 2AgNO$_3$(aq) $\longrightarrow$ Ag$_2$S(s) + 2NH$_4$NO$_3$(aq)

Ion or Atom	Left	Right
NH$_4^+$	2	2
S	1	1
Ag	2	2
NO$_3^-$	2	2

PITFALL   Writing the correct formulas in a precipitation reaction

When writing the products for a precipitation reaction, avoid the temptation to automatically carry forward the subscript of an ion in a reactant as the ion becomes part of the product. For example, in the reaction

$$\text{FeCl}_3\text{(aq)} + 3\text{NaOH(aq)} \longrightarrow \text{Fe(OH)}_3\text{(s)} + 3\text{NaCl(aq)} \quad \text{(correct)}$$

we take care not to automatically keep the subscript 3 on Cl$^-$ and the (implied) 1 on OH$^-$ and write

$$\text{FeCl}_3\text{(aq)} + 3\text{NaOH(aq)} \longrightarrow \text{FeOH(s)} + \text{NaCl}_3\text{(aq)} \quad \text{(incorrect)}$$

When writing the formulas of the products of any reaction, one must purposefully check the formulas of the products to write them correctly.

3.5   IONIC AND NET IONIC EQUATIONS

KEY CONCEPT   Net ionic equations show an overall chemical change

In the precipitation reactions that we have written, there are a total of four different ions in the reactants. Before we mix the reactant solutions, these ions exist as separated ions in aqueous solution. After they are mixed, two of the ions combine to form the solid precipitate, and the other two simply stay in aqueous solution. Chemically, nothing has happened to the two ions that stay in solution; they start out as aqueous ions and end up as aqueous ions. On the other hand, the ions that form the precipitate have undergone a chemical change, from separate aqueous ions to a solid bonded precipitate. We often choose to write a **net ionic equation** in which we show only the overall chemical change and ignore the ions that stay in solution. For example, consider the reaction between silver nitrate ($AgNO_3$) and potassium sulfate (K_2SO_4):

$$2AgNO_3(aq) + K_2SO_4(aq) \longrightarrow Ag_2SO_4(s) + 2KNO_3(aq)$$

At the start of the reaction, we have Ag^+, NO_3^-, K^+, and SO_4^{2-} ions in solution; the abbreviation (aq) tells us that the salts are dissolved. After the reaction, Ag^+ and SO_4^{2-} have reacted to form insoluble $Ag_2SO_4(s)$, whereas NO_3^- and K^+ stay in solution and do not react. The net ionic equation shows only the species that react and is, therefore,

$$2Ag^+(aq) + SO_4^{2-}(aq) \longrightarrow Ag_2SO_4(s)$$

K^+ and NO_3^- are called **spectator ions** because they do not participate in the reaction but stay in solution and "watch" as the reaction occurs.

EXAMPLE   Writing a net ionic equation

Write the net ionic equation for the precipitation reaction between silver fluoride (AgF) and sodium sulfate (Na_2SO_4). Ag_2SO_4 is insoluble.

SOLUTION Because the way we write a net ionic equation depends on the products formed, the first step is to determine the formulas of the products. In this case, Ag^+ replaces Na^+ in Na_2SO_4, resulting in Ag_2SO_4, and Na^+ replaces Ag^+ in AgF, resulting in NaF:

$$AgF(aq) + Na_2SO_4(aq) \longrightarrow Ag_2SO_4 + NaF \quad \triangle$$

Table 3.1 of the text informs us that NaF is soluble, so NaF exists as $Na^+(aq)$ ions and $F^-(aq)$ ions on both sides of the equation, that is, as spectator ions. The net reaction is the reaction of $Ag^+(aq)$ and $SO_4^{2-}(aq)$ to form $Ag_2SO_4(s)$:

$$Ag^+(aq) + SO_4^{2-}(aq) \longrightarrow Ag_2SO_4(s) \quad \triangle$$

To balance the equation, two $Ag^+(aq)$ are needed:

$$2Ag^+(aq) + SO_4^{2-}(aq) \longrightarrow Ag_2SO_4(s)$$

Ion or Atom	Left	Right
Ag^+	2	2
SO_4^{2-}	1	1

EXERCISE Write the net ionic equation for the precipitation reaction between $Pb(NO_3)_2$ and K_2CrO_4.

[*Answer:* $Pb^{2+}(aq) + CrO_4^{2-}(aq) \longrightarrow PbCrO_4(s)$]

PITFALL   Spectator ions are present!

From looking at a net ionic equation, it is easy to imagine that the spectator ions are not present. This is not the case; the spectator ions are present. We just choose to ignore them because they are not important to the overall chemical process that occurs.

PITFALL   Equations for nonexistent reactions

It is possible to write an apparent equation for a nonexistent process. For example, we can write the presumed equation with $NaNO_3$ and K_2SO_4 as reactants:

$$2NaNO_3(aq) + K_2SO_4(aq) \longrightarrow Na_2SO_4(aq) + 2KNO_3(aq)$$

As Table 3.1 of the text confirms, the reactants exist as $Na^+(aq)$, $NO_3^-(aq)$, $K^+(aq)$, and $SO_4^{2-}(aq)$ in solution. However, the proposed products, Na_2SO_4 and KNO_3, are both soluble and also exist as ions in solution. Thus, no precipitate forms. We conclude from this observation that when $NaNO_3(aq)$ and $K_2SO_4(aq)$ are mixed, *no chemical process occurs*. The only process accomplished is the mixing of $NaNO_3(aq)$ with $K_2SO_4(aq)$. This situation is often represented by N.R., which stands for "no reaction," on the right side of the arrow.

$$NaNO_3(aq) + K_2SO_4(aq) \longrightarrow N.R.$$

3.6   PUTTING PRECIPITATION TO WORK

KEY CONCEPT   Solubilities help predict precipitation reactions

Let's assume we want to form a precipitate of magnesium carbonate ($MgCO_3$). What reagents should we choose as reactants? We need two soluble reagents, one that will provide magnesium ions (Mg^{2+}) and one that will provide carbonate ions (CO_3^{2-}); when these reagents are dissolved in water and mixed, they will form the insoluble product $MgCO_3$. For obtaining Mg^{2+}, we note that Table 3.1 of the text indicates that $MgCl_2$ is soluble; it is also true that most nitrates are soluble, so $Mg(NO_3)_2$ could also be used to provide Mg^{2+} ions. To get CO_3^{2-} ions, the information in Table 3.1 indicates that Group 1 carbonates are soluble, as is ammonium carbonate [$(NH_4)_2CO_3$]. So, for instance, by mixing a solution of magnesium nitrate with a solution of sodium carbonate, we will get the desired product, $MgCO_3$.

$$Mg(NO_3)_2(aq) + Na_2CO_3(aq) \longrightarrow MgCO_3(s) + 2NaNO_3(aq)$$

The net ionic equation for this reaction is

$$Mg^{2+}(aq) + CO_3^{2-}(aq) \longrightarrow MgCO_3(s)$$

EXAMPLE   Generating the desired precipitate

Assume that you wish to precipitate $Ca_3(PO_4)_2$ to demonstrate the properties of phosphate rock (which is the name given to calcium phosphate as a mineral). What two reagents could be used to form $Ca_3(PO_4)_2$? Write the net ionic equation for the reaction involved.

SOLUTION We need a source of Ca^{2+} ions and of PO_4^{3-} ions in order to form $Ca_3(PO_4)_2$. Thus, we must find a soluble compound that contains Ca^{2+} and a soluble compound that contains PO_4^{3-}. For Ca^{2+}, Table 3.1 of the text indicates that the calcium halides $CaCl_2$, $CaBr_2$, and CaI_2 are all soluble; in addition, almost all metal nitrates are soluble, so $Ca(NO_3)_2$ could also be used. Not many phosphates are soluble, but Table 3.1 indicates that Group 1 phosphates and ammonium phosphate are soluble. Thus, if we mixed, for instance, solutions of calcium iodide (CaI_2) and potassium phosphate (K_3PO_4), we would get the desired product,

$$2K_3PO_4(aq) + 3CaI_2(aq) \longrightarrow Ca_3(PO_4)_2(s) + 6KI(aq)$$

The net ionic equation is

$$2PO_4^{3-}(aq) + 3Ca^{2+}(aq) \longrightarrow Ca_3(PO_4)_2(s)$$

EXERCISE What two compounds could be used as reactants to generate the precipitate CdS? Write the net ionic equation involved.

[*Answer:* $Cd(NO_3)_2$ and Na_2S; $Cd^{2+}(aq) + S^{2-}(aq) \longrightarrow CdS(s)$]

KEY WORDS   Define or explain each term in a written sentence or two.

complete ionic equation   nonelectrolyte   spectator ion
hydrated   precipitate   strong electrolyte
insoluble substance   precipitation reaction   weak electrolyte
net ionic equation   soluble substance

THE REACTIONS OF ACIDS AND BASES

3.7   ACIDS AND BASES IN AQUEOUS SOLUTIONS

KEY CONCEPT   Recognizing acids and bases

Acids and bases are discussed together because acids react with bases in specific and characteristic ways. An **acid** is a substance that contains hydrogen and releases H^+ ions when dissolved in water. HCl, HNO_3, and H_2SO_4 are acids. The released H^+ immediately attaches to a water molecule; so, for instance, when HCl is dissolved in water, the following process occurs:

$$HCl(aq) + H_2O(l) \longrightarrow H_3O^+(aq) + Cl^-(aq)$$

A **base** is a substance that produces OH^- (hydroxide ions) in water. For instance, the ionic compounds $NaOH$ and $Ca(OH)_2$ are bases; so is the molecular compound ammonia, NH_3, which does not contain OH^- ions but produces them when dissolved in water.

$$NaOH(aq) \longrightarrow Na^+(aq) + OH^-(aq)$$

$$NH_3(aq) + H_2O(l) \longrightarrow NH_4^+(aq) + OH^-(aq)$$

Chemists frequently (but not always) write the formulas of acids with the ionizable hydrogens (the ones that are released as H^+) as the first symbol in the formula. For instance, H_2SO_4 has two ionizable hydrogens, and $HC_2H_3O_2$ has one.

EXAMPLE   Identifying and understanding acids and bases

Identify each of the following compounds as an acid, a base, or neither: H_2O, HBr, CH_3OH, $H_2C_2O_4$, KOH, NH_3, and C_3H_8. State the effect of aqueous solutions of each acid and base on litmus.

SOLUTION The formulas for acids frequently have hydrogen as the first letter in the formula. (One notable exception is water, which has H as the first letter in its formula but is not considered to be an acid or base.) Both HBr and $H_2C_2O_4$ have H as the first letter and both are acids (hydrobromic acid and oxalic acid, respectively). Oxalic acid is a carboxylic acid with two —COOH groups. Its formula may be written as HOOCCOOH. C_3H_8 (propane) contains eight hydrogen atoms, but they do not form hydrogen ions in water; so propane is not an acid. Because acids turn litmus red, aqueous solutions of HBr and $H_2C_2O_4$ will turn litmus red. Bases produce OH^- when the base is dissolved in water, either because the base contains hydroxide ion or because the base reacts with water to produce hydroxide. The bases that contain hydroxide are invariably ionic compounds, so KOH (potassium hydroxide) is recognized as a base. NH_3 (ammonia) is a base because it reacts with water to form hydroxide ion

$$NH_3(g) + H_2O(l) \longrightarrow NH_4^+(aq) + OH^-(aq)$$

CH_3OH (methanol) contains an OH group, but the group is not a hydroxide ion and does not ionize in water, so CH_3OH is not a base; it is an alcohol. Because bases turn litmus blue, an aqueous solution of KOH or NH_3 will turn litmus blue.

EXERCISE Identify each of the following as either an acid, a base, or neither: HI, C_2H_5OH, $Ca(OH)_2$, $HC_3H_5O_2$, and C_5H_{12}. State the effect of aqueous solutions of each acid and base on litmus.

[*Answer:* Acids: HI, $HC_3H_5O_2$; aqueous solutions turn litmus red.
Base: $Ca(OH)_2$; aqueous solution turns litmus blue.
Neither: C_2H_5OH and C_5H_{12}.]

3.8 STRONG AND WEAK ACIDS AND BASES

KEY CONCEPT A   The relative strengths of acids

An acid is classified as strong or weak, depending on the percentage of the acid that ionizes when it is dissolved in water.

1. A **strong acid** is 100% ionized in aqueous solution at moderate concentrations. For example, when HCl is dissolved in water, all of the HCl molecules ionize into H_3O^+ and Cl^- ions.
2. A **weak acid** is only partially ionized in aqueous solution at moderate concentrations. For example, when CH_3COOH is dissolved in water, most of the CH_3COOH molecules do not ionize. Only a few percent ionize into H_3O^+ and CH_3COO^- ions.

EXAMPLE   Classifying acids as weak or strong

When an unknown acid HX is dissolved in water, it is determined that for every 100 HX molecules that dissolve, 100 H_3O^+ ions and 100 X^- ions are formed. For a second unknown acid HY, for every 100 HY molecules that dissolve, 3 H_3O^+ and 3 Y^- form. What percentage of each acid dissolves? Classify each acid as weak or strong.

SOLUTION For acid HX, every molecule that dissolves also ionizes. It is 100% ionized and is, therefore, a strong acid; it is "completely" ionized. For acid HY, only 3 out of every 100 molecules ionize. It is 3% ionized and is, therefore, a weak acid; it is "incompletely" ionized.

EXERCISE In the preceding example, how many ions are formed from the 100 molecules HX and how many from the 100 molecules HY?

[*Answer:* HX: 200 ions, 100 H_3O^+ and 100 X^-
HY: 6 ions, 3 H_3O^+ and 3 Y^-]

KEY CONCEPT B   The relative strengths of bases

A base is classified as weak or strong, depending on the percentage of the base that ionizes to release (or form) OH^- ions when it is dissolved in water.

1. A strong base is 100% ionized in aqueous solution at moderate concentrations, forming large amounts of OH^-. There are two kinds of strong bases. The Group 1 hydroxides and alkaline earth metal hydroxides [$Ca(OH)_2$, $Sr(OH)_2$, $Ba(OH)_2$] are all strong bases. They contain OH^- ions and ionize completely in aqueous solution. Many metal oxides (such as CaO) are strong bases. They do not contain OH^- ions, but when dissolved in water, the oxide ion strongly undergoes a reaction to form large amounts of OH^-.

$$BaO(s) + H_2O(l) \longrightarrow Ba^{2+}(aq) + 2OH^-(aq) \quad \text{(strong reaction)}$$

2. A weak base is one that is incompletely ionized in aqueous solution. The common weak bases are ammonia and amines. These compounds do not contain any OH^- but react weakly with water to form small amounts of OH^-.

$$CH_3NH_2(aq) + H_2O(l) \longrightarrow CH_3NH_3^+(aq) + OH^-(aq) \quad \text{(weak reaction)}$$

EXAMPLE   Formation of OH^- ions in solution

Why is potassium hydroxide, KOH, considered to be a strong base? Dimethylamine [$(CH_3)_2NH$] is considered to be a weak base, even though it contains no OH^- ion. Explain why. In both cases, write the equations involved.

SOLUTION Potassium hydroxide is an ionic compound that completely ionizes in aqueous solution, forming large amounts of OH^- ion. Because of the formation of large amounts of OH^-, it is considered a strong base.

$$KOH(s) \xrightarrow{H_2O} K^+(aq) + OH^-(aq) \quad \text{(strong reaction)}$$

Dimethylamine contains no OH^-, but it reacts weakly with water to form small amounts of OH^-. Because only a small amount of OH^- is formed, it is considered to be a weak base.

$$(CH_3)_2NH(aq) + H_2O(l) \longrightarrow (CH_3)_2NH_2^+(aq) + OH^-(aq) \quad \text{(weak reaction)}$$

EXERCISE Based on the examples given so far, classify LiOH and $(CH_3)_3N$ (trimethylamine) as strong or weak bases.

[*Answer:* LiOH, strong base; $(CH_3)_3N$, weak base]

PITFALL   Formulas of amines

Amines are analogues of ammonia. In ammonia, only three hydrogen atoms are attached to the central nitrogen. Similarly, in the methyl amines, only three groups (a combination of H and/or CH_3) are attached to the central nitrogen. So, the formulas of the methyl amines are

$(CH_3)NH_2$ methylamine

$(CH_3)_2NH$ dimethylamine

$(CH_3)_3N$ trimethylamine

3.9   ACIDIC AND BASIC CHARACTER IN THE PERIODIC TABLE

KEY CONCEPT   Oxides of the elements are often acidic or basic

The oxides of a number of elements have acidic or basic character. Nonmetal oxides, for the most part, are acidic. They demonstrate their acidic character in two ways; first, they form acid solutions when dissolved in water and, second, they react with bases (usually to form a salt and water). For instance, carbon dioxide forms carbonic acid when dissolved in water, and it reacts with calcium hydroxide to form water and the salt calcium carbonate.

$$CO_2(g) + H_2O(l) \longrightarrow H_2CO_3(aq)$$
$$CO_2(g) + Ca(OH)_2(aq) \longrightarrow CaCO_3(aq) + H_2O(l)$$

Metal oxides, for the most part, are basic. They demonstrate their basic character in two ways; first, they form basic solutions when dissolved in water and, second, they react with acids (usually to form a salt and water). For instance, barium oxide forms barium hydroxide when dissolved in water and reacts with hydrochloric acid to form water and the salt barium chloride.

$$BaO(s) + H_2O(l) \longrightarrow Ba(OH)_2(aq)$$
$$BaO(s) + 2HCl(aq) \longrightarrow BaCl_2(aq) + H_2O(l)$$

The oxides of a few elements that lie near the boundary between the metals and nonmetals, as might be expected, share some of the properties of the metal oxides and of the nonmetal oxides. Their oxides show both acidic and basic character; the character that is manifested (either acid or base) depends on what is done to the oxide. Such oxides are called **amphoteric.**

EXAMPLE   Classification of oxides

Classify both N_2O_5 and K_2O as acidic, basic, or amphoteric. Write one reaction for each that demonstrates its character.

SOLUTION Dinitrogen pentoxide (N_2O_5) is a nonmetal oxide, so it is likely to be an acidic oxide. One reaction of acidic oxides is a reaction with water to form an acid.

$$N_2O_5(l) + H_2O(l) \longrightarrow 2HNO_3(aq)$$

Potassium oxide (K_2O) is a Group 1 metal oxide, so it most certainly is a basic oxide. One reaction of basic oxides is a reaction with water to form a base.

$$K_2O(s) + H_2O(l) \longrightarrow 2KOH(aq)$$

EXERCISE Classify SnO_2 as acidic, basic, or amphoteric.

[*Answer:* Amphoteric]

3.10 NEUTRALIZATION

KEY CONCEPT Acid-base neutralization

When an acid reacts with a base, the acid and base are said to **neutralize** each other because the acid and base are both destroyed in an acid-base reaction. In neutralization reactions, a salt and water are often (but not always) formed. For example,

$$HBr(aq) + KOH(aq) \longrightarrow KBr(aq) + H_2O(l)$$
$$\text{Acid} \qquad \text{Base} \qquad \text{Salt} \qquad \text{Water}$$

Any ionic compound that does not contain OH^- ion or O^{2-} ion is called a **salt.** In an acid-base reaction, the anion of the salt formed originates with the acid and the cation originates with the base.

EXAMPLE 1 Writing a neutralization reaction

Write the neutralization reaction between nitric acid and calcium hydroxide. What is the name of the salt formed?

SOLUTION The reaction of an acid with a base results in the formation of a salt and water. The salt formed is calcium nitrate.

$$2HNO_3(aq) + Ca(OH)_2(aq) \longrightarrow Ca(NO_3)_2(aq) + 2H_2O(l)$$
$$\text{Acid} \qquad\qquad \text{Base} \qquad\qquad \text{Salt} \qquad\qquad \text{Water}$$

EXERCISE Write the neutralization reaction between rubidium hydroxide and sulfuric acid.

[*Answer:* $2RbOH(aq) + H_2SO_4(aq) \longrightarrow Rb_2SO_4(aq) + 2H_2O(l)$]

3.11 THE FORMATION OF GASES

KEY CONCEPT Acids can be used to form gases

Acids can, in certain reactions, be used to produce a gas. The gas may be formed directly by transfer of an H^+ ion to the appropriate anion; for instance, in the first reaction that follows, H^+ transfers to CN^- to form the highly toxic gas HCN. In other cases, transfer of H^+ results in formation of a product that immediately decomposes into a gas and, quite often, water. Many carbonates and bicarbonates undergo this reaction to form the gas CO_2 and H_2O.

$$NaCN(s) + HCl(aq) \longrightarrow NaCl(aq) + HCN(g)$$

$$MgCO_3(s) + HCl(aq) \longrightarrow H_2CO_3(aq) + MgCl_2(aq) \longrightarrow CO_2(g) + H_2O(l) + MgCl_2(aq)$$

EXAMPLE Using an acid to form a gas

Suggest the reagents that may be suitable for making the gas H_2S. Write the proposed equation.

SOLUTION In order to form H_2S, we need one reagent to supply H^+ and a second to supply S^{2-}. When these combine, the desired gas will result. HCl is a common source of H^+ and Na_2S a good source of S^{2-}.

$$Na_2S(aq) + 2HCl(aq) \longrightarrow H_2S(g) + 2NaCl(aq)$$

EXERCISE Using the reaction given above to form CO_2 as an example, suggest a reaction for forming the gas SO_2.

[*Answer:* $2HCl(aq) + K_2SO_3(s) \longrightarrow 2KCl(aq) + H_2SO_3(aq) \longrightarrow SO_2(g) + H_2O(l) + 2KCl(aq)$]

KEY WORDS   Define or explain each term in a written sentence or two.

acid	base	salt	weak base
acidic hydrogen	basic oxide	strong acid	
acidic oxide	carboxyl group	strong base	
amphoteric	neutralization reaction	weak acid	

REDOX REACTIONS

Oxidation-reduction reactions abound in our biochemistry and in our everyday life. They are often called **redox reactions.** Whenever oxidation occurs in a reaction, reduction must also occur; and whenever reduction occurs, so must oxidation.

3.12   OXIDATION AND REDUCTION

KEY CONCEPT   Oxidation-reduction as electron transfer

Oxidation and reduction reactions involve the transfer of electrons from one substance to another. We start with the definitions that a substance is **oxidized** when it **loses electrons** and is **reduced** when it **gains electrons.** (The mnemonic device "LEO the lion goes GER" can help you remember this: *Lose Electrons Oxidation and Gain Electrons Reduction.*) We should note that with this definition of oxidation and reduction, the electron transfer is assumed to be from one atom to another. Whenever oxidation occurs in a reaction, reduction *must* occur at the same time. It is impossible to have an oxidation without a reduction, or a reduction without an oxidation. A reaction in which oxidation and reduction occur is called a **redox** (reduction-oxidation) reaction.

EXAMPLE   Electron loss and gain in a redox reaction

Construct a table showing what is oxidized and what is reduced for the reaction

$$2AgNO_3(aq) + Cu(s) \longrightarrow Cu(NO_3)_2 + 2Ag(s)$$

SOLUTION We first analyze the charges of the reactants and products to see what electron transfers have occurred. For reactants, we have Ag^+ ion, NO_3^- ion, and a neutral Cu atom. The products contain the Cu^{2+} ion, the NO_3^- ion, and a neutral Ag atom. Thus, Ag^+ has gained an electron to form Ag and in doing so has been reduced (GER). Cu has lost two electrons to form Cu^{2+} and has therefore been oxidized (LEO). The NO_3^- ion is a spectator ion because it is in the same form on the right side of the equation as on the left. It has been neither oxidized nor reduced. The summary table is

Atom	Starting Charge	Ending Charge	Electron Change (per atom)	Redox Change
Cu	0	+2	loss of 2 electrons	oxidized
Ag	+1	0	gain of 1 electron	reduced

EXERCISE Construct a table showing what is oxidized and what is reduced in the reaction $Mg(s) + Cl_2(g) \longrightarrow MgCl_2(s)$.

Atom	Starting Charge	Ending Charge	Electron Change (per atom)	Redox Change
Mg	0	+2	loss of 2 electrons	oxidized
Cl	0	−1	gain of 1 electron	reduced

3.13 KEEPING TRACK OF ELECTRONS: OXIDATION NUMBERS

KEY CONCEPT A   Oxidation numbers

An **oxidation number** is a number assigned to an atom according to the effective charge of the atom. The assignment is based on the set of rules given in Toolbox 3.2 of the text. Work through the rules in the order given and stop when you have arrived at an oxidation number for an atom. Two inferences can be drawn from these rules: (a) atoms in their elemental form (such as O_2, N_2, Na, and P_4) have zero oxidation number, and (b) the oxidation state of a monatomic ion (such as Na^+, Cl^-, S^{2-}, and Al^{3+}) is the same as the charge of the ion. Rules higher in the toolbox take precedence over rules that are lower. The rules will give the right answer for the oxidation numbers *most* of the time. (Some complicated compounds need a few additional rules, which we do not cover in this course.)

EXAMPLE   Determining oxidation numbers

Determine the oxidation numbers of (a) Na in Na(s); (b) O in OF_2; (c) C in $C_2O_4^{2-}$.

SOLUTION (a) Na in Na(s) is in the elemental form and, according to rule 1, is assigned oxidation number 0. (b) For OF_2, rule 2 tells us that the oxidation numbers of one O and two F must sum to zero. Because the rules also state that fluorine must have oxidation number −1 in its compounds, the two fluorines contribute −2 ($2 \times -1 = -2$) to the sum of oxidation numbers in OF_2. In order that the sum of oxidation numbers equal zero, the oxidation number of O must be +2 (a rare oxidation state for O). (c) The sum of the oxidation numbers in $C_2O_4^{2-}$ must sum to −2, the charge on the ion. Each O has a −2 oxidation number (toolbox rules), so the four O contribute −8 toward the −2 charge. This means that the two C must contribute +6, since $(+6) + (-8) = -2$. If two C contribute +6, each must contribute +3, and the oxidation number of C is +3.

EXERCISE Determine the oxidation number of (a) Xe in XeO_6^{4-}; (b) H in H_2; (c) Mg in $MgCl_2$.

[*Answer:* (a) +8; (b) 0; (c) +2]

KEY CONCEPT B   Oxidation and reduction as a change in oxidation number

As we might expect from the term *oxidation number,* the higher the oxidation number of an atom, the more oxidized it is. For example, SO_2 is oxidized when it reacts with O_2 to form SO_3. The oxidation number of S is +4 in SO_2 and +6 in SO_3, showing that the oxidation number is increased when oxidation occurs. From the point of view of electron transfer, sulfur most lose two electrons to go from a +4 oxidation number to a +6 oxidation number, confirming that oxidation viewed as the loss of electrons corresponds to oxidation viewed as an increase in oxidation number. The following table summarizes the three points of view discussed in the text.

Process	Oxygen Content	Electron Transfer	Oxidation Number
oxidation	increase in O content	loss of electrons	increase in oxidation number
reduction	decrease in O content	gain of electrons	decrease in oxidation number

EXAMPLE   Using oxidation-number change to determine oxidation and reduction

Use the change in oxidation number to determine what is oxidized and what is reduced in the reaction

$$Cu(s) + 4HNO_3(aq, conc.) \longrightarrow Cu(NO_3)_2(aq) + 2NO_2(g) + 2H_2O(l)$$

SOLUTION When using the change in oxidation number to characterize a redox reaction, you must first determine the oxidation number of all the elements in the reaction. The table shows the changes.

Element	Left Side of Equation		Right Side of Equation		Process
	Species	Oxidation Number	Species	Oxidation Number	
Cu	Cu(s)	0	Cu^{2+}	+2	oxidation
H	H^+	+1	H_2O	+1	no change
N	NO_3^-	+5	NO_3^-	+5	no change
N	NO_3^-	+5	NO_2	+4	reduction
O	NO_3^-	−2	NO_3^-, NO_2, H_2O	−2	no change

Note that the oxidation number of Cu increases from 0 to +2, so Cu has been oxidized. The N in some of the NO_3^- has undergone a decrease in oxidation number, so it has been reduced.

EXERCISE Use the change in oxidation number to determine what is oxidized and what is reduced in the reaction that occurs in a lead storage battery:

$$2H_2SO_4(aq) + Pb(s) + PbO_2(s) \longrightarrow 2PbSO_4(s) + 2H_2O(l)$$

[*Answer:* Pb(s) is oxidized; Pb in $PbO_2(s)$ is reduced.]

PITFALL   Oxidation numbers are not always actual charges

Even though the sum of all of the oxidation numbers of the atoms in a compound or molecular ion must equal the total charge on the compound or ion, it is not necessarily true that the individual oxidation number on an atom equals the actual charge on the atom. In some cases, the oxidation number equals the charge; in many cases, it doesn't.

3.14   OXIDIZING AND REDUCING AGENTS

KEY CONCEPT   The oxidizing and reducing agents in a redox reaction

Consider the following redox reaction between copper and oxygen to form copper(II) oxide. The copper atoms each lose two electrons and donate them to two oxygen atoms. The copper is oxidized (LEO) and the oxygen reduced (GER).

$$2Cu(s) + O_2(g) \longrightarrow 2CuO(s)$$

If we desired to couch our description of the reaction by looking for agents of change, we could say that the copper, by donating electrons to oxygen has forced the oxygen to be reduced. The copper, therefore, is the **reducing agent.** In a similar fashion, the oxygen, by taking electrons from the copper, has forced the copper to be oxidized. The oxygen is, as might be expected, the **oxidizing agent.** The relationships found in this reaction are always true for redox reactions:

The substance oxidized is the reducing agent.

The substance reduced is the oxidizing agent.

EXAMPLE 1 The substances oxidized and reduced: reducing and oxidizing agents

Identify the substance oxidized, the substance reduced, the oxidizing agent, and the reducing agent in the following equation:

$$Ca(s) + Cl_2(g) \longrightarrow CaCl_2(s)$$

SOLUTION The atom or substance that loses electrons is oxidized, and the atom or substance that gains electrons is reduced. In this reaction, the charge on calcium changes from 0 to +2, so calcium loses electrons and is oxidized. The charge on chlorine changes from 0 to −2, so chlorine gains electrons and is reduced. The electrons gained by chlorine come from calcium, so calcium is the reducing agent. The (same) electrons lost by calcium are accepted by chlorine, so chlorine is the oxidizing agent. As in all redox reactions, the oxidizing agent is reduced and the reducing agent is oxidized:

Reduced: Cl   *Oxidized:* Ca

Oxidizing agent: Cl   *Reducing agent:* Ca

EXERCISE Identify the substance oxidized, the substance reduced, the oxidizing agent, and the reducing agent in the following equation:

$$CuO(s) + H_2(g) \longrightarrow Cu(s) + H_2O(g)$$

[*Answer:* Reduced, CuO (or Cu in CuO); oxidized, $H_2(g)$; oxidizing agent, CuO (or Cu in CuO); reducing agent, $H_2(g)$]

EXAMPLE 2 Acids as oxidizing agents

Write the net ionic equation for the reaction between HCl and aluminum. State what is oxidized, what is reduced, and what the oxidizing and reducing agents are.

SOLUTION Acids act as oxidizing agents in two different ways: through the oxidizing ability of H^+ and through the oxidizing ability of the acid anion. In this case, H^+ can oxidize aluminum; the reaction is

$$6H^+(aq) + 2Al(s) \longrightarrow 3H_2(g) + 2Al^{3+}(aq)$$

H^+ gains electrons and is reduced; it is the oxidizing agent. Al loses electrons and is oxidized; it is the reducing agent.

EXERCISE Write the net ionic equation for the reaction between iron and HCl. Assume conditions are such that iron forms Fe^{2+}. State what is oxidized, what is reduced, and what the oxidizing and reducing agents are.

[*Answer:* $2H^+(aq) + Fe(s) \longrightarrow H_2(g) + Fe^{2+}(aq)$: Fe is oxidized and is the reducing agent; H^+ is reduced and is the oxidizing agent.]

3.15 BALANCING SIMPLE REDOX REACTIONS

KEY CONCEPT Charge is conserved in a chemical reaction

We have already learned that an equation, in order to be mass balanced, must have the same number of atoms of each kind on both sides. However, this is not enough to assure that an equation is actually balanced. To assure that no electrons have magically appeared or disappeared in a reaction, an equation must have the same charge on each side of the arrow. This is called **charge balance** and indicates that charge is conserved just as mass is.

EXAMPLE Charge balance in an equation

Balance the equation, $Zn(s) + Cu^+(aq) \longrightarrow Cu(s) + Zn^{2+}(aq)$ ⚠

SOLUTION For an equation to be balanced, it must be mass balanced and charge balanced. The equation given is already mass balanced; there is one atom of Zn on each side and one atom of Cu on each side. However, it is not charge balanced. The charge on the left is +1 and on the right +2. To get a +2 on the left, we must place a two in front of Cu^+.

$$Zn(s) + 2Cu^+(aq) \longrightarrow Cu(s) + Zn^{2+}(aq) \quad \triangle$$

The equation is now charge balanced, with a +2 on the left ($2 \times +1 = +2$) and a +2 on the right. It is no longer mass balanced, however. To achieve mass balance, we must place a 2 in front of the Cu(s) on the right. The equation is now completely balanced.

$$Zn(s) + 2Cu^+(aq) \longrightarrow 2Cu(s) + Zn^{2+}(aq)$$

Atom	Left	Right
Zn	1	2
Cu	2	2
charge	+2	+2

EXERCISE Balance the equation $Mg(s) + Cr^{3+}(aq) \longrightarrow Mg^{2+}(aq) + Cr(s)$ $\quad \triangle$

Answer: $3Mg(s) + 2Cr^{3+}(aq) \longrightarrow 3Mg^{2+}(aq) + 2Cr(s)$

Atom	Left	Right
Mg	3	3
Cr	2	2
charge	+6	+6

PITFALL   Charge balancing a chemical equation

To be balanced, an equation must have the same total charge on the left as on the right. Balance does not mean that the charge on the right cancels the charge on the left but that charge on the left equals that on the right. For example, the following equation is mass balanced because the number of atoms of each element is the same on both sides of the equation. However, it is not charge balanced because the total charge on the left is +1 and the total charge on the right is +2:

$$Ag^+(aq) + Cu(s) \longrightarrow Cu^{2+}(aq) + Ag(s) \quad \triangle$$

Total charge +1 $\neq$ Total charge +2

To balance this equation, we multiply $Ag^+(aq)$ by 2 and $Ag(s)$ by 2:

$$Ag^+(aq) + Cu(s) \longrightarrow Cu^{2+}(aq) + Ag(s)$$

Total charge +2 $=$ Total charge +2

Atom	Left	Right
Ag	2	2
Cu	1	1
charge	+2	+2

3.16   CLASSIFYING REACTIONS

KEY CONCEPT   The three types of reactions

We have discussed three types of reactions so far.

1. Precipitation reaction: Two soluble electrolytes react in aqueous solution to form an insoluble substance called a **precipitate.**
2. Neutralization reaction: An acid and a base react, through H^+ **transfer,** to form a salt and water.
3. Redox reaction: A reducing agent is oxided and an oxidizing agent is reduced through **electron transfer.**

EXAMPLE   Identifying the type of reaction

Balance and identify the type of reaction that is given by the equation,

$$Zn(s) + HCl(aq) \longrightarrow ZnCl_2(aq) + H_2(g) \quad \triangle$$

SOLUTION In this reaction, Zn changes oxidation state from 0 to +2 and H from +1 to 0. The oxidation number of Cl remains −1. The reaction is a redox reaction.

$$Zn(s) + 2HCl(aq) \longrightarrow ZnCl_2(aq) + H_2(g)$$

Atom	Left	Right
Zn	1	1
H	2	2
Cl	2	2
charge	0	0

EXERCISE Balance and identify the type of reaction that is given by the equation,

$$ZnO(s) + HNO_3(aq) \longrightarrow Zn(NO_3)_2(aq) + H_2O(l) \quad \triangle$$

Answer: Neutralization; $ZnO(s) + 2HNO_3(aq) \longrightarrow Zn(NO_3)_2(aq) + H_2O(l)$

Atom	Left	Right
Zn	1	2
NO_3^-	2	2
H	2	2
O	1	1
charge	0	0

KEY WORDS   Define or explain each term in a written sentence or two.

charge balance

oxidation

oxidation number

oxidation state

oxidizing agent

redox reaction

reducing agent

reduction

DESCRIPTIVE CHEMISTRY TO REMEMBER

- **Methane** (CH_4) is the major component of natural gas.
- **Acetic acid** (CH_3COOH) is a weak electrolyte.
- A **precipitation reaction** is a reaction in which an insoluble substance (a precipitate) is formed.
- Most **Group 1** compounds, **ammonium** (NH_4^+) compounds, and **nitrates** are soluble.
- Shellfish form their shells from **calcium carbonate** ($CaCO_3$).
- The three common lab acids, HCl, HNO_3, and H_2SO_4 are all **strong acids.**
- The **carboxyl group** (—COOH) contains an acidic hydrogen.
- **Calcium oxide** (CaO) is called lime; **calcium hydroxide** [$Ca(OH)_2$] is called slaked lime.
- **Aluminum oxide** (Al_2O_3) is an amphoteric oxide.
- The **nitrate** ion (NO_3^-) and **permanganate** ion (MnO_4^-) are powerful oxidizing agents.
- **Hydrogen sulfide** (H_2S) is an extremely poisonous gas with the smell of rotting eggs.

CHEMICAL EQUATIONS TO KNOW

- **Sodium** metal reacts violently with water to form sodium hydroxide and hydrogen gas:

$$2Na(s) + 2H_2O(l) \longrightarrow 2NaOH(aq) + H_2(g)$$

- In a reaction typical of all carbonates, **calcium carbonate** ($CaCO_3$) decomposes to carbon dioxide and an oxide when heated:

$$CaCO_3(s) \xrightarrow{800°} CaO(s) + CO_2(g)$$

- In a reaction typical of organic compounds, **methane** (CH_4) burns in a plentiful supply of air to form water and carbon dioxide. **Butane** (C_4H_{10}) undergoes the same reaction:

$$CH_4(g) + 2O_2(g) \longrightarrow CO_2(g) + 2H_2O(g)$$
$$2C_4H_{10}(g) + 13O_2(g) \longrightarrow 8CO_2(g) + 10H_2O(g)$$

- In a precipitation reaction, aqueous **silver nitrate** reacts with aqueous **sodium chloride** to give a silver chloride precipitate:

$$AgNO_3(aq) + NaCl(aq) \longrightarrow NaNO_3(aq) + AgCl(s)$$

- The bright, yellow precipitate **lead(II) chromate** is formed when a solution of lead(II) nitrate is mixed with a solution of potassium chromate:

$$Pb(NO_3)_2(aq) + K_2CrO_4(aq) \longrightarrow 2KNO_3(aq) + PbCrO_4(s)$$

- **Ammonia** reacts with water to form ammonium ions and hydroxide ions:

$$NH_3(aq) + H_2O(l) \longrightarrow NH_4^+(aq) + OH^-(aq)$$

- **Oxide ion** strongly reacts with water to form hydroxide ion:

$$O^{2-}(aq) + H_2O(l) \longrightarrow 2OH^-(aq)$$

- In a reaction typical of nonmetal oxides, CO_2, SO_2, and P_4O_{10} react with water to form **acids** (carbonic acid, sulfurous acid, and phosphoric acid, respectively):

$$CO_2(g) + H_2O(l) \longrightarrow H_2CO_3(aq)$$
$$SO_2(g) + H_2O(l) \longrightarrow H_2SO_3(aq)$$
$$P_4O_{10}(s) + 6H_2O(l) \longrightarrow 4H_3PO_4(aq)$$

- In an **acid-base reaction,** magnesium oxide (base) reacts with hydrochloric acid to form the salt magnesium chloride and water:

$$MgO(s) + 2HCl(aq) \longrightarrow MgCl_2(aq) + H_2O(l)$$

- In an **acid-base reaction,** sodium hydroxide (base) reacts with carbon dioxide (acid) to form the salt sodium hydrogen carbonate (but no water):

$$NaOH(s) + CO_2(g) \longrightarrow NaHCO_3(s)$$

- In an **acid-base reaction,** nitric acid reacts with magnesium hydroxide (base) to form the salt magnesium nitrate and water:

$$2HNO_3(aq) + Mg(OH)_2(s) \longrightarrow Mg(NO_3)_2(aq) + 2H_2O(l)$$

- In a **redox reaction,** zinc dissolves in hydrochloric acid to form hydrogen gas and aqueous zinc chloride:

$$Zn(s) + 2HCl(aq) \longrightarrow ZnCl_2(aq) + H_2(g)$$

- In a **redox reaction,** iron(III) oxide is reduced to iron metal. At the same time, the reducing agent hydrogen (gas) is oxidized:

$$Fe_2O_3(s) + 3H_2(g) \xrightarrow{\Delta} 2Fe(l) + 3H_2O(g)$$

SELF-TEST EXERCISES

Chemical equations and chemical reactions

1. All the equations given below, except one, are balanced. Which one is not?

(a) $6Li(s) + N_2(g) \longrightarrow 2Li_3N(s)$

(b) $2SeO_3(s) \longrightarrow 2SeO_2(s) + O_2(g)$

(c) $P_4O_{10}(s) + 5H_2O(l) \longrightarrow 4H_3PO_4(aq)$

(d) $2KNO_3(s) \longrightarrow 2KNO_2(s) + O_2(g)$

(e) $Cu(s) + 2H_2SO_4(aq) \longrightarrow CuSO_4(aq) + SO_2(g) + 2H_2O(l)$

2. True or false? To balance the following equation, it is proper to change the formula CaO to CaO$_2$.

$$Ca(s) + O_2(g) \longrightarrow CaO(s)$$

(a) true   (b) false

3. What coefficient appears in front of O$_2$(g) when the following equation is balanced?

$$P_4(s) + O_2(g) \longrightarrow P_4O_{10}(s) \quad \triangle$$

(a) $\frac{5}{2}$ (b) 4   (c) 5   (d) 10   (e) 8

4. What coefficient appears in front of H$_2$O when the following equation is balanced?

$$CuSO_4 \cdot 5H_2O(s) \longrightarrow CuSO_4(s) + H_2O(g) \quad \triangle$$

(a) 5   (b) 1   (c) 10   (d) 3   (e) 4

5. What coefficient appears in front of H$_2$O when the following equation is balanced?

$$Ca(OH)_2(aq) + HCl(aq) \longrightarrow CaCl_2(aq) + H_2O(l) \quad \triangle$$

(a) 5   (b) 3   (c) 6   (d) 2   (e) 1

6. What coefficient appears in front of Se when the following equation is balanced?

$$Se_2Br_2(l) + H_2O(l) \longrightarrow HBr(g) + H_2SeO_3(aq) + Se(s) \quad \triangle$$

(a) 5   (b) 4   (c) 3   (d) 1   (e) 2

7. What coefficient appears in front of NaOH when the following equation is balanced?

$$MnO_2(s) + NaOH(aq) + O_2(g) \longrightarrow Na_2MnO_4(aq) + H_2O(l) \quad \triangle$$

(a) 2   (b) 4   (c) 1   (d) 6   (e) 3

8. What coefficient appears in front of O$_2$ when the following equation is balanced?

$$NH_3(g) + O_2(g) \longrightarrow NO(g) + H_2O(g) \quad \triangle$$

(a) 4   (b) 5   (c) 2   (d) 1   (e) 3

9. What coefficient appears in front of SiH$_4$ when the following equation is balanced?

$$HCl(g) + Mg_2Si(s) \longrightarrow MgCl_2(s) + SiH_4(g) \quad \triangle$$

(a) 2   (b) 1   (c) 4   (d) 3   (e) 5

Precipitation reactions

10. Which of the following is a nonelectrolyte?
(a) NaCl   (b) O$_2$ (c) HNO$_3$ (d) Na$_2$SO$_4$ (e) CH$_3$COOH

11. What makes a solution of an electrolyte conducting?
(a) formation of ions   (b) formation of a precipitate
(c) presence of O atoms   (d) formation of a gas
(e) hydration of dissolved species

12. Which of the following is a precipitation reaction?
(a) $2Ca(s) + O_2(g) \longrightarrow 2CaO(s)$
(b) $2HNO_3(aq) + CaO(s) \longrightarrow Ca(NO_3)_2(aq) + H_2O(l)$
(c) $MgCO_3(s) \longrightarrow MgO(s) + CO_2(g)$
(d) $SiO_2(s) + 4HF(aq) \longrightarrow SiF_4(g) + 2H_2O(l)$
(e) $K_2SO_4(aq) + Ba(NO_3)_2(aq) \longrightarrow BaSO_4(s) + 2KNO_3(aq)$

13. What are the spectator ions in the following equation?

$$SrCl_2(aq) + K_2SO_4(aq) \longrightarrow SrSO_4(s) + 2KCl(aq)$$

(a) Sr^{2+} only   (b) K^+ only   (c) Sr^+, SO_4^{2-}
(d) Sr^{2+}, K^+ (e) K^+, Cl^-

14. What are the spectator ions in the following equation?

$$Na_2S(aq) + Ba(NO_3)_2(aq) \longrightarrow BaS(s) + 2NaNO_3(aq)$$

(a) Na^+, NO_3^- (b) Na^+ only   (c) Na^+, Ba^{2+}
(d) S^{2-}, NO_3^- (e) Ba^{2+}, S^{2-}

15. What is the net ionic equation for the reaction of $Pb(NO_3)_2$ with H_2SO_4?
(a) $Pb(NO_3)_2(aq) + SO_4^{2-}(aq) \longrightarrow PbSO_4(s) + 2NO_3^-(aq)$
(b) $Pb(NO_3)_2(aq) + H_2SO_4(aq) \longrightarrow PbSO_4(s) + 2HNO_3(aq)$
(c) $Pb^{2+}(aq) + SO_4^{2-}(aq) \longrightarrow PbSO_4(s)$
(d) $Pb^{2+}(aq) + H_2SO_4(aq) \longrightarrow PbSO_4(s) + 2H^+(aq)$

16. What is the net ionic equation for the reaction of $Cd(NO_3)_2$ with Na_2S?
(a) $Cd(NO_3)_2(aq) + S^{2-}(aq) \longrightarrow CdS(s) + 2NO_3^-(aq)$
(b) $Cd^{2+}(aq) + Na_2S(aq) \longrightarrow CdS(s) + 2Na^+(aq)$
(c) $Na^+(aq) + NO_3^-(aq) \longrightarrow NaNO_3(aq)$
(d) $Cd^{2+}(aq) + S^{2-}(aq) \longrightarrow CdS(s)$
(e) $Cd(NO_3)_2(aq) + Na_2S(aq) \longrightarrow 2NaNO_3(aq) + CdS(s)$

17. What two reagents would you use to prepare the solid $BaCO_3$? (Check solubilities before you answer!)
(a) $Ba(NO_3)_2$, Na_2CO_3 (b) $Ba_3(PO_3)_2$, Na_2CO_3 (c) $Ba(NO_3)_2$, $ZnCO_3$
(d) $Ba_3(PO_3)_2$, $ZnCO_3$ (e) $BaSO_4$, Ag_2CO_3

18. Which of the following will form a precipitate when it is the product of a reaction?
(a) $(NH_4)_2CO_3$ (b) Na_3PO_4 (c) MnC_2O_4
(d) FeI_2 (e) $Sn(NO_3)_4$

The reactions of acids and bases

19. Which of the following is an acid?
(a) $NaOH$ (b) CH_4 (c) H_2O (d) NH_3 (e) $HC_2H_3O_2$

20. All of the following but one are bases. Which one is not a base?
(a) $NaOH$ (b) C_2H_5OH (c) NH_3 (d) CaO (e) $Mg(OH)_2$

21. An aqueous solution of $HCHO_2$ contains approximately 96 molecules of unionized $HCHO_2$ for every 4 ionized (into H^+ and CHO_2^-) molecules. How would you classify $HCHO_2$?
(a) weak acid   (b) strong acid   (c) strong base
(d) weak base   (e) salt

22. What is O^{2-} converted to when it is dissolved in water?
(a) H_2O (b) H_2O_2 (c) H_3O^+ (d) OH^- (e) O_2

23. Which of the following is a weak base?
(a) $NaOH$ (b) CH_3NH_2 (c) CH_4 (d) $Ca(OH)_2$ (e) $HC_2H_3O_2$

24. One of the following acids is a weak acid. Which one?

(a) HNO_3 (b) H_2SO_4 (c) HCl   (d) C_6H_5COOH (e) $HClO_4$

25. Which of the following oxides is likely to be an acidic oxide?

(a) Al_2O_3 (b) CaO   (c) Na_2O (d) BaO   (e) N_2O_3

26. What compound is formed when NO_2 is placed in water?

(a) HNO_2 (b) HNO_3 (c) NCl_3 (d) CH_3NH_2 (e) $NaNO_3$

27. Which of the reactions shown is a neutralization reaction?

(a) $CH_4(g) + 2O_2(g) \longrightarrow CO_2(g) + 2H_2O(l)$

(b) $Mg(s) + H_2O(g) \longrightarrow MgO(s) + H_2(g)$

(c) $KOH(aq) + H_3PO_4(aq) \longrightarrow KH_2PO_4(aq) + H_2O(l)$

(d) $MgCO_3(s) \longrightarrow MgO(s) + CO_2(g)$

(e) $CaO(s) + H_2O(l) \longrightarrow Ca(OH)_2(s)$

28. What salt is formed when HBr is neutralized with $Ca(OH)_2$?

(a) $CaBr_2$ (b) CaO   (c) CaH_2 (d) $CaCl_2$ (e) $Ca(BrO_3)_2$

29. Which of the following compounds forms the gas CO_2 when reacted with HCl?

(a) $Ca(OH)_2$ (b) $FeCl_3$ (c) $MgCO_3$ (d) Na_2SO_4 (e) Al_2O_3

Redox reactions

30. What is oxidized in the reaction $MgO(s) + H_2(g) \longrightarrow H_2O(l) + Mg(s)$?

(a) H_2 (b) MgO   (c) H_2O (d) Mg

31. What is reduced in the reaction $C(s) + 2N_2O(g) \longrightarrow CO_2(g) + 2N_2(g)$?

(a) C   (b) N_2O (c) CO_2 (d) N_2

32. What is oxidized in the reaction $Br_2(l) + CaI_2(aq) \longrightarrow CaBr_2(aq) + I_2(s)$?

(a) Ca^{2+} (b) Br_2 (c) I_2 (d) I^- (e) Br^-

33. What is the oxidation number of N in N_2O_5?

(a) 0   (b) +10   (c) +2   (d) +5   (e) −3

34. What is the oxidation number of Cr in CrO_3^{3-}?

(a) +6   (b) +7   (c) −6   (d) +9   (e) +3

35. What is the oxidation number of Fe in Fe_2S_3?

(a) +2   (b) +3   (c) +6   (d) 0   (e) +4

36. Use oxidation numbers to determine what atom is oxidized in the following reaction.

$$8H_2S(g) + 8I_2(s) \longrightarrow 16HI(g) + S_8(s)$$

(a) H   (b) S   (c) I

37. What is the oxidizing agent in the following reaction?

$$Br_2(l) + 5Cl_2(g) + 12KOH(aq) \longrightarrow 2KBrO_3(aq) + 10KCl(aq) + 6H_2O(l)$$

(a) Cl_2 (b) K^+ (c) OH^- (d) Br_2 (e) H_2

38. Use oxidation numbers to determine the reducing agent in the following reaction.

$$14H^+(aq) + Cr_2O_7^{2-}(aq) + 6Fe^{2+}(aq) \longrightarrow 2Cr^{3+}(aq) + 7H_2O(l) + 6Fe^{3+}(aq)$$

(a) H^+ (b) $Cr_2O_7^{2-}$ (c) Fe^{2+}

39. What is the oxidizing agent in the reaction $2Mg(l) + TiCl_4(g) \longrightarrow 2MgCl_2(s) + Ti(s)$?

(a) $MgCl_2$ (b) Mg   (c) $TiCl_4$ (d) Ti

40. What coefficient appears in front of Cu when the equation shown is balanced?

$$Cu^+(aq) + Mg(s) \longrightarrow Mg^{2+}(aq) + Cu(s) \quad \triangle$$

(a) 2   (b) 1   (c) 3   (d) 6   (e) 5

41. What coefficient appears in front of Br^- when the equation shown is balanced?

$$Cl_2(g) + Br^-(aq) \longrightarrow Br_2(g) + Cl^-(aq) \quad \triangle$$

(a) 4   (b) 3   (c) 1   (d) 2   (e) 5

42. What coefficient appears in front of Fe^{2+} when the equation shown is balanced?

$$Cu(aq) + Fe^{3+}(aq) \longrightarrow Cu^{2+}(aq) + Fe^{2+}(aq) \quad \triangle$$

(a) 3   (b) 2   (c) 1   (d) 4   (e) 6

Descriptive chemistry

43. What is the major component of natural gas?

(a) CH_4 (b) C_2H_2 (c) CO_2 (d) CH_3COOH (e) H_2S

44. Which of the following is insoluble? Do not refer to any sources.

(a) $Sn(NO_3)_2$ (b) $PbCrO_4$ (c) $FeCl_3$ (d) $(NH_4)_2SO_4$ (e) KF

45. Which of the following anions is a strong oxidizing agent?

(a) SO_4^{2-} (b) S^{2-} (c) CO_3^{2-} (d) MnO_4^- (e) F^-

46. Which of the following is a poisonous gas with the smell of rotting eggs?

(a) CO_2 (b) C_2H_2 (c) H_2S (d) H_2 (e) N_2

47. Which of the following is a bright yellow solid?

(a) NaCl   (b) $PbCrO_4$ (c) AgCl   (d) O_2 (e) NH_3

48. What compound imparts the sour taste to vinegar?

(a) NaCl   (b) CH_3COOH (c) CO_2 (d) O_2 (e) HCl

49. What is the color of the dye litmus in an acid solution?

(a) red   (b) blue   (c) colorless   (d) orange   (e) green

50. What is the formula of trimethylamine, a compound found in rotting fish?

(a) CH_3NH_2 (b) NH_3 (c) CH_4 (d) $(CH_3)_3N$ (e) H_2S

51. What is the formula of ethyne (also called acetylene), which is used in welding?

(a) CH_4 (b) P_4O_{10} (c) NH_3 (d) CH_3COOH (e) C_2H_2

52. What is the primary product of photosynthesis?

(a) H_2O (b) $C_6H_{12}O_6$ (c) CO_2 (d) C_6H_5COOH (e) $HClO_3$

CHAPTER 4

REACTION STOICHIOMETRY: CHEMISTRY'S ACCOUNTING

In any chemical reaction, it is possible to calculate the amount of reactants and/or products involved in the reaction from the balanced chemical equation. Either the mass of the substances involved and/or the moles can be determined.

HOW TO USE REACTION STOICHIOMETRY

4.1 MOLE-TO-MOLE PREDICTIONS

KEY CONCEPT A Stoichiometric coefficients as moles of a substance

The stoichiometric coefficients in a chemical equation can be interpreted as the number of moles of reactants and products that react:

$$N_2(g) + 3H_2(g) \longrightarrow 2NH_3(g)$$
$$\text{1 mol } N_2 \quad \text{3 mol } H_2 \quad \text{2 mol } NH_3$$

$$2KClO_3(s) \longrightarrow 2KCl(s) + 3O_2(g)$$
$$\text{2 mol } KClO_3 \quad \text{2 mol } KCl \quad \text{3 mol } O_2$$

EXAMPLE   Interpreting equations in terms of moles

Write an interpretation of the following equation in terms of the moles of reactants and products:

$$Mg_3N_2(s) + 6H_2O(l) \longrightarrow 3Mg(OH)_2(s) + 2NH_3(g)$$

SOLUTION The coefficient in front of a product or reactant represents the number of moles of each reactant participating in the reaction and the number of moles of each product that results. For this reaction, 1 mol Mg_3N_2 reacts with 6 mol H_2O to form 3 mol $Mg(OH)_2$ and 2 mol NH_3. More specifically,

$$Mg_3N_2(s) + 6H_2O(l) \longrightarrow 3Mg(OH)_2(s) + 2NH_3(g)$$
$$\text{1 mol } Mg_3N_2 \quad \text{6 mol } H_2O \quad \text{3 mol } Mg(OH)_2 \quad \text{2 mol } NH_3$$

EXERCISE Write a sentence that interprets the following equation in terms of the moles of reactants and products:

$$C_2H_6(g) + \tfrac{7}{2}O_2(g) \longrightarrow 2CO_2(g) + 3H_2O(g)$$

[*Answer:* 1 mol C_2H_6 reacts with 3.5 mol O_2 to form 2 mol CO_2 and 3 mol H_2O.]

KEY CONCEPT B   Mole calcuations

The interpretation of stoichiometric coefficients as moles of reactants and products enables us to do calculations involving these quantities. We approach calculations by setting up the proper conversion factors, based on the coefficients, and using unit conversion. Let us consider the following equation,

$$P_4(s) + 5O_2(g) \longrightarrow P_4O_{10}(s)$$

and the meaning of the coefficients and the conversion factors that arise. The coefficients tell us that 1 mol P_4 reacts with 5 mol O_2 to yield 1 mol P_4O_{10}. This means that 1 mol P_4 requires 5 mol O_2 to completely react and that 5 mol O_2 requires 1 mol P_4 to completely react. In addition, when the 1 mol P_4 and 5 mol O_2 do react, 1 mol P_4O_{10} should result. From these coefficients, we get the following conversion factors:

Because 1 mol P_4 requires 5 mol O_2, we get $\dfrac{1 \text{ mol } P_4}{5 \text{ mol } O_2}$ and $\dfrac{5 \text{ mol } O_2}{1 \text{ mol } P_4}$

Because 1 mol P_4 can give 1 mol P_4O_{10}, we get $\dfrac{1 \text{ mol } P_4}{1 \text{ mol } P_4O_{10}}$ and $\dfrac{1 \text{ mol } P_4O_{10}}{1 \text{ mol } P_4}$

Because 5 mol O_2 can give 1 mol P_4O_{10}, we get $\dfrac{5 \text{ mol } O_2}{1 \text{ mol } P_4O_{10}}$ and $\dfrac{1 \text{ mol } P_4O_{10}}{5 \text{ mol } O_2}$

EXAMPLE   Calculating moles of product from moles of reactant

How many moles of P_4O_{10} can be produced from 2.27 mol of O_2?

SOLUTION Here we seek a relationship between moles of P_4O_{10} and moles of O_2:

$$\frac{5 \text{ mol } O_2}{1 \text{ mol } P_4O_{10}} \quad \text{and} \quad \frac{1 \text{ mol } P_4O_{10}}{5 \text{ mol } O_2}$$

To set up the unit conversion, we start with the given amount of O_2 and multiply by the proper conversion factor to get the correct units for the answer. We note that the number of moles of P_4O_{10} formed is less than the number of moles of O_2, so we expect the answer to be less than 2.27 mol:

$$2.27 \text{ mol } O_2 \times \frac{1 \text{ mol } P_4O_{10}}{5 \text{ mol } O_2} = 0.454 \text{ mol } P_4O_{10}$$

EXERCISE How many moles of P_4 are required to react with 2.27 mol O_2?

[*Answer:* 0.454 mol]

4.2   MASS-TO-MASS PREDICTIONS

KEY CONCEPT   The masses taking part in reactions

We discovered that it is always possible to find the moles of a substance from the mass of the substance, or the mass from the moles, using the molar mass of the substance:

$$\boxed{\text{Moles}} \xleftrightarrow{\text{molar mass}} \boxed{\text{Mass}}$$

Now, a chemical equation relates the moles of reactants and products in a reaction:

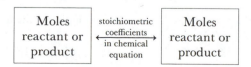

By combining these two relationships, we can in a three-step calculation relate the mass of a reactant or product in a chemical reaction to the mass of another reactant or product. In a typical problem, the mass of a reactant or product is given and we are asked to calculate the mass of another reactant or product (the unknown). The three steps are illustrated in the following example.

EXAMPLE 1   Calculating the mass of reactant needed to react with another reactant

How many grams of H_2O are needed to react with 25.0 g of P_4O_{10} in the reaction given below?

$$P_4O_{10}(s) + 6H_2O(l) \longrightarrow 4H_3PO_4(aq)$$

SOLUTION We use a three-step calculation.

Step 1. Use the molar mass of the given substance to calculate the moles of the substance. In this case, we use the molar mass of P_4O_{10}, which is $283.89 \text{ g} \cdot \text{mol}^{-1}$:

$$25.0 \text{ g } P_4O_{10} \times \frac{1 \text{ mol } P_4O_{10}}{283.89 \text{ g } P_4O_{10}} = 0.0881 \text{ mol } P_4O_{10}$$

Step 2. Use the balanced chemical equation to determine the moles of the unknown reactant or product. The equation tells us that 6 mol of H_2O are required for every mole of P_4O_{10}:

$$0.0881 \text{ mol } P_4O_{10} \times \frac{6 \text{ mol } H_2O}{1 \text{ mol } P_4O_{10}} = 0.529 \text{ mol } H_2O$$

Step 3. Use the molar mass of the unknown to determine its mass. Here, we use the molar mass of H_2O, which is $18.02 \text{ g} \cdot \text{mol}^{-1}$:

$$0.529 \text{ mol } H_2O \times \frac{18.02 \text{ g } H_2O}{1 \text{ mol } H_2O} = 9.53 \text{ g } H_2O$$

We can also do this problem in a single step by using the three conversion factors, one after the other:

$$25.0 \text{ g } P_4O_{10} \times \frac{1 \text{ mol } P_4O_{10}}{283.89 \text{ g } P_4O_{10}} \times \frac{6 \text{ mol } H_2O}{1 \text{ mol } P_4O_{10}} \times \frac{18.02 \text{ g } H_2O}{1 \text{ mol } H_2O} = 9.52 \text{ g } H_2O$$

You may choose to do unit conversions with separate steps or with a single long step; use the method you are most comfortable with. The small difference in the two answers in Step 3 is due to rounding error in the first method of calculation.

EXERCISE How many grams of H_3PO_4 would be expected to form from 100 g H_2O in the reaction just given?

[*Answer:* 363 g]

EXAMPLE 2   Calculating the mass of product from the amount of reactant

How many grams of hydrogen cyanide (HCN) can be produced from the complete reaction of 50.0 g O_2 from the reaction

$$2CH_4(g) + 3O_2(g) + 2NH_3(g) \longrightarrow 2HCN(g) + 6H_2O(g)$$

SOLUTION

Step 1. Use the molar mass of the given substance to calculate the moles of the substance:

$$50.0 \text{ g } O_2 \times \frac{1 \text{ mol } O_2}{32.0 \text{ g } O_2} = 1.56 \text{ mol } O_2$$

Step 2. Use the balanced chemical equation to determine the moles of the unknown reactant or product:

$$1.56 \text{ mol } O_2 \times \frac{2 \text{ mol HCN}}{3 \text{ mol } O_2} = 1.04 \text{ mol HCN}$$

Step 3. Use the molar mass of the unknown to determine its mass:

$$1.04 \ \text{mol HCN} \times \frac{27.03 \ \text{g}}{1 \ \text{mol HCN}} = 28.1 \ \text{g HCN}$$

Again, we can do this calculation in one long step:

$$50.0 \ \text{g O}_2 \times \frac{1 \ \text{mol O}_2}{32.0 \ \text{g}} \times \frac{2 \ \text{mol HCN}}{3 \ \text{mol O}_2} \times \frac{27.03 \ \text{g}}{1 \ \text{mol HCN}} = 28.2 \ \text{g HCN}$$

EXERCISE How many grams of HNO_3 can be produced from the complete reaction of 65.0 g NO_2?

$$3NO_2(g) + H_2O(l) \longrightarrow 2HNO_3(aq) + NO(g)$$

[Answer: 59.4 g]

4.3   THE VOLUME OF SOLUTION REQUIRED FOR REACTION

KEY CONCEPT A   The volume of solution required for reaction

We learned earlier that the quantitative relationships describing the moles of reactants and products that take part in a chemical reaction are given by the stoichiometric coefficients of the equation. In addition, we had previously developed the relationship "molarity × volume = moles." By putting these two ideas together, we can calculate the volume of a solution that must be used to react with a certain amount of a reagent.

EXAMPLE   Calculating the volume of solution required for reaction

How many milliliters of 0.112 M NaOH are required to neutralize 10.0 mL of a 0.352 M solution of H_2SO_4? Assume the reaction is

$$2NaOH(aq) + H_2SO_4(aq) \longrightarrow 2H_2O(l) + Na_2SO_4(aq)$$

SOLUTION Three steps are used to solve this problem.

Step 1. The number of moles of the reactant with known molarity and volume are calculated with the equation

$$\text{Moles} = \text{molarity} \times \text{volume}$$

We are concerned here with the H_2SO_4 because it is the reactant for which we have a known volume and known molarity. We note that 10.0 mL = 0.0100 L.

$$\text{Moles } H_2SO_4 = 0.352 \ \frac{\text{mol } H_2SO_4}{L} \times 0.0100 \ L$$

$$= 3.52 \times 10^{-3} \ \text{mol } H_2SO_4$$

Step 2. The number of moles of the second reactant are calculated by using the stoichiometric coefficients of the balanced equation. The equation tells us that 2 mol NaOH react with 1 mol H_2SO_4.

$$3.52 \times 10^{-3} \ \text{mol } H_2SO_4 \times \frac{2 \ \text{mol NaOH}}{1 \ \text{mol } H_2SO_4} = 7.04 \times 10^{-3} \ \text{mol NaOH}$$

Step 3. The volume of solution of the second reactant is calculated by the equation

$$\text{Volume} = \frac{\text{moles}}{\text{molarity}}$$

The problem states that the molarity of the NaOH is 0.112 M.

$$\text{Volume NaOH} = \frac{7.04 \times 10^{-3} \text{ mol } \cancel{\text{NaOH}}}{0.112 \text{ mol } \cancel{\text{NaOH}} \cdot \text{L}^{-1}}$$

$$= 0.0629 \text{ L}$$

$$= 62.9 \text{ mL}$$

EXERCISE How many milliliters of 0.125 M $KMnO_4$ are required to react with 25.0 mL of 0.831 M $FeCl_2$? The reaction is

$$KMnO_4(aq) + 8HCl(aq) + 5FeCl_2(aq) \longrightarrow 5FeCl_3(aq) + MnCl_2(aq) + KCl(aq) + 4H_2O(l)$$

[Answer: 33.2 mL]

KEY CONCEPT B   Acid-base titrations

A titration is an experiment in which a chemical reaction is done with carefully measured amounts of reactants in order to determine the number of moles of one of the reactants. There are many ways to do titrations, but the following procedure illustrates the general approach taken.

1. The analyte (for instance, the acid in an acid-base titration) is placed into a clean flask. If it is a solid, it is weighed carefully on an analytical balance. If it is in solution, the volume of solution is determined accurately by obtaining it with a buret or a pipet.
2. A solution of the titrant is added to the analyte. The goal is to add the volume of titrant that will exactly react with the quantity of the analyte originally present. We know when we have added enough by using an indicator that changes color when the correct amount has been added; this point is called the stoichiometric point. If too much or too little titrant is added, the titration must be redone.
3. The volume of the titrant needed to just react with the analyte is written down.
4. The data obtained are used to calculate the unknown moles of one of the reactants (either the analyte or titrant) from the known moles of the other. A titration is designed so that the moles of one of the reactants can be calculated from known factors, such as the molarity and volume of the reactant (for a solution) or its mass and molar mass (for a solid). With this information and the stoichiometric coefficients of the balanced chemical equation, the moles of the other reactant (the unknown) are calculated.

EXAMPLE   Determining the molarity of an unknown base in an acid-base titration

A student prepares a solution of NaOH by placing 4 g of it into a flask and adding enough water to make 1 L of solution: an approximate 0.1 M solution. The exact molarity of the solution is determined by a titration, in which 10.00 mL of the solution requires 11.22 mL of a 0.1004 M solution of HCl to reach the stoichiometric point. What is the exact molarity of the NaOH? The reaction is

$$NaOH(aq) + HCl(aq) \longrightarrow NaCl(aq) + H_2O(l)$$

SOLUTION The steps for this calculation are almost identical to the three steps in the previous example. You should compare the two calculations to see where they are similar and where they differ.

Step 1. The number of moles of the known reactant is determined by

$$\text{Moles} = \text{molarity} \times \text{volume}$$

The known reactant is the titrant HCl, because its molarity is given. We note that 11.22 mL = 0.01122 L.

$$\text{Moles HCl} = 0.1004 \frac{\text{mol HCl}}{\cancel{\text{L}}} \times 0.01122 \cancel{\text{L}}$$

$$= 0.001126 \text{ mol HCl}$$

$$= 1.126 \times 10^{-3} \text{ mol HCl}$$

Step 2. The stoichiometric coefficients of the balanced chemical equation are used to calculate the moles of the other reactant present:

$$1.126 \times 10^{-3} \ \text{mol HCl} \times \frac{1 \ \text{mol NaOH}}{1 \ \text{mol HCl}} = 1.126 \times 10^{-3} \ \text{mol NaOH}$$

Step 3. The molarity of the unknown solution is calculated by using the definition of molarity:

$$\text{Molarity} = \frac{\text{moles solute}}{\text{liters solution}}$$

1.126×10^{-3} mol NaOH is the number of moles that were originally present in the 10.00 mL of NaOH solution used in the titration. Because 10.00 mL = 0.01000 L,

$$\text{Molarity} = \frac{1.126 \times 10^{-3} \ \text{mol NaOH}}{0.0100 \ \text{L}}$$

$$= 0.1126 \ \text{M}$$

EXERCISE The student now uses the 0.1126 M standard NaOH solution to titrate 10.00 mL of an unknown H_2SO_4 solution. 23.65 mL of the NaOH were required. What is the molarity of the H_2SO_4? Assume the reaction is

$$2NaOH(aq) + H_2SO_4(aq) \longrightarrow Na_2SO_4(aq) + 2H_2O(l)$$

[*Answer:* 0.1331 M]

KEY WORDS   Define or explain each term in a written sentence or two.

analyte   titrant
stoichiometric point   titration
stoichiometric relation

THE LIMITS OF REACTION

4.4   REACTION YIELD

KEY CONCEPT   Theoretical and percentage yields

The **theoretical yield** is the mass of product we would expect if *all* of the limiting reactant converted to the expected product; it is, therefore, the maximum mass of product obtainable in a given reaction. At times, not all of a reactant converts to the desired product. For a variety of reasons, the actual amount of product obtained is often less than the expected theoretical yield. The **percentage yield** relates the theoretical yield to the amount of product obtained in an actual laboratory or industrial process:

$$\text{Percentage yield} = \frac{\text{actual mass of product obtained}}{\text{theoretical yield of product}} \times 100$$

EXAMPLE   Calculating the percentage yield

When 2.3 kg CS_2 reacts with excess Cl_2, 3.6 kg CCl_4 is formed. What is the percentage yield of CCl_4?

$$CS_2(l) + 2Cl_2(g) \longrightarrow CCl_4(l) + 2S(s)$$

SOLUTION We must substitute into the equation

$$\text{Percentage yield} = \frac{\text{actual mass of product obtained}}{\text{theoretical yield of product}} \times 100$$

The actual mass of the product CCl_4 is given as 3.6 kg, so we have

$$\text{Percentage yield} = \frac{3.6 \text{ kg}}{\text{theoretical yield of } CCl_4} \times 100$$

Now we need the theoretical yield, or the amount we expect, from the type of calculation illustrated earlier.

Step 1. Use the molar mass of the given substance to calculate the moles of that substance:

$$2.3 \text{ kg } CS_2 \times \frac{10^3 \text{ g } CS_2}{1 \text{ kg } CS_2} \times \frac{1 \text{ mol } CS_2}{76.13 \text{ g } CS_2} = 30 \text{ mol } CS_2$$

Step 2. Use the balanced chemical equation to determine the moles of the unknown reactant or product:

$$30 \text{ mol } CS_2 \times \frac{1 \text{ mol } CCl_4}{1 \text{ mol } CS_2} = 30 \text{ mol } CCl_4$$

Step 3. Use the molar mass of the unknown to determine its mass:

$$30 \text{ mol } CCl_4 \times \frac{154 \text{ g } CCl_4}{1 \text{ mol } CCl_4} \times \frac{1 \text{ kg}}{1000 \text{ g}} = 4.6 \text{ kg } CCl_4$$

Thus, 4.6 kg is the theoretical yield of CCl_4. It must be substituted in the defining equation for percentage yield to get the final answer:

$$\text{Percentage yield} = \frac{3.6 \text{ kg}}{4.6 \text{ kg}} \times 100 = 78\%$$

EXERCISE What is the percentage yield of Fe if the reaction of 233 g of Fe_2O_3 with excess C results in 128 g Fe?

$$3C(s) + 2Fe_2O_3(s) \longrightarrow 4Fe(s) + 3CO_2(g)$$

[*Answer*: 78.5%]

4.5 LIMITING REACTANTS

KEY CONCEPT   One reactant may be used up before the others

Part of the process of getting zinc metal from its ore involves converting zinc sulfide to its oxide:

$$2ZnS(s) + 3O_2(g) \longrightarrow 2ZnO(s) + 2SO_2(g)$$

In this reaction, the stoichiometric coefficients tell us that 3 mol O_2 are required to react with 2 mol ZnS. Now imagine a situation in which a chemist mixes 100.0 mol O_2 with 1.0 mol ZnS. Will all of the ZnS react? Will all of the O_2 react? (Answer these questions before you proceed.) It is evident that there is a great excess of oxygen. All the ZnS will react, but most of the O_2 will not. In fact, we can calculate that 1.0 mol ZnS requires 1.5 mol O_2; thus, when all of the ZnS reacts, 98.5 mol O_2 remain unreacted (100.0 mol − 1.5 mol = 98.5 mol). In this situation, ZnS is called the **limiting reactant** because it limits the extent of reaction and the amount of product formed. To analyze a limiting reactant problem, we use the three-step procedure illustrated in the following example.

EXAMPLE   Identifying the limiting reactant

48.0 g of O_2 is mixed with 85.0 g ZnS preparatory to converting the zinc sulfide to zinc oxide in the reaction $2ZnS(s) + 3O_2(g) \rightarrow 2ZnO(s) + 2SO_2(g)$. What is the limiting reactant?

SOLUTION We use a three-step procedure.

Step 1. Calculate the moles of each reactant present, using the molar mass of each. The molar mass of O_2 is 32.0 g·mol^{-1} and that of ZnS is 97.4 g·mol^{-1}:

$$48.0 \text{ g } O_2 \times \frac{1 \text{ mol } O_2}{32.0 \text{ g } O_2} = 1.50 \text{ mol } O_2$$

$$85.0 \text{ g ZnS} \times \frac{1 \text{ mol ZnS}}{97.4 \text{ g ZnS}} = 0.872 \text{ mol ZnS}$$

Step 2. Calculate the moles of any one of the products that should form from the given amount of each reactant:

$$1.50 \text{ mol } O_2 \times \frac{2 \text{ mol ZnO}}{3 \text{ mol } O_2} = 1.00 \text{ mol ZnO}$$

$$0.872 \text{ mol ZnS} \times \frac{2 \text{ mol ZnO}}{2 \text{ mol ZnS}} = 0.872 \text{ mol ZnO}$$

Step 3. The reactant that results in the formation of the smallest amount of product is the limiting reactant. The amount of product that forms is determined by the limiting reactant. Because the smaller amount of product results from the amount of ZnS present, ZnS is the limiting reactant. (In addition, 0.872 mol of ZnO should form.)

EXERCISE 110 g CO is mixed with 355 g I_2O_5 for the following reaction. What is the limiting reactant?

$$5CO(g) + I_2O_5(s) \longrightarrow I_2(s) + 5CO_2(g)$$

[*Answer:* CO]

PITFALL   The factors that determine the limiting reactant

The limiting reactant is not necessarily the reagent that is present with the smallest mass or the fewest moles. For example, in the problems just given, the following results were obtained:

Reagent	Grams Present	Moles Present	Limiting Reagent?
ZnS	85.0	0.872	yes
O_2	48.0	1.50	no
CO	110	3.73	yes
I_2O_5	355	1.10	no

In the worked problem, there are more grams of ZnS than of O_2, but ZnS is still the limiting reactant. In the exercise problem, there are more moles of CO than of I_2O_5, but CO is still the limiting reactant. The identity of the limiting reactant depends on the mass of each reactant present, the molar mass of each, and the stoichiometric coefficients in the balanced chemical equation.

4.6   COMBUSTION ANALYSIS

KEY CONCEPT   Analysis through conservation of mass

Combustion analysis is typically done on organic compounds containing carbon, hydrogen, and perhaps some other element. The goal of the analysis is to determine either the mass percentage of each element in the compound or the empirical formula of the compound. The analysis is done by combusting the compound in a large excess of O_2, so that CO_2 and H_2O are the products. Three key facts underlie the experiment:

1. All the carbon atoms present in the original sample end up in the CO_2 formed by the combustion. In terms of conversion factors, there is 1 mol C in the original sample per mol CO_2 produced:

$$\frac{1 \text{ mol C (original sample)}}{1 \text{ mol CO}_2}$$

2. All of the hydrogen atoms in the original sample end up in the H_2O formed by the combustion. Because there are 2 mol H per mol H_2O (as indicated by the formula H_2O), 2 mol H in the original sample will produce 1 mol H_2O in the products. In terms of conversion factors:

$$\frac{2 \text{ mol H (original sample)}}{1 \text{ mol H}_2O}$$

3. The oxygen atoms in the CO_2 and H_2O produced do not originate with the sample.

EXAMPLE 1   Determining an empirical formula through combustion analysis

Analysis of 1.008 g of a compound containing only carbon and hydrogen results in the formation of 3.163 g CO_2 and 1.294 g H_2O in a combustion analysis experiment. What is the empirical formula of the compound?

SOLUTION For a compound containing only carbon and hydrogen, we use three steps.

Step 1. Use the molar mass of CO_2 and H_2O to calculate the moles of each produced:

$$3.163 \text{ g CO}_2 \times \frac{1 \text{ mol CO}_2}{44.01 \text{ g CO}_2} = 0.07187 \text{ mol CO}_2$$

$$1.294 \text{ g H}_2O \times \frac{1 \text{ mol H}_2O}{18.02 \text{ g H}_2O} = 0.07181 \text{ mol H}_2O$$

Step 2. Calculate the moles of C and moles of H in the original sample:

$$0.07187 \text{ mol CO}_2 \times \frac{1 \text{ mol C (original sample)}}{1 \text{ mol CO}_2} = 0.07187 \text{ mol C (in original sample)}$$

$$0.07181 \text{ mol H}_2O \times \frac{2 \text{ mol H (original sample)}}{1 \text{ mol H}_2O} = 0.1436 \text{ mol H (in original sample)}$$

Step 3. Find the simplest whole number ratio of moles of elements in the compound. This defines the empirical formula:

$$\frac{0.07187 \text{ mol C}}{0.07187} = 1 \text{ mol C}$$

$$\frac{0.1436 \text{ mol H}}{0.07187} = 1.998 \text{ mol H} = 2 \text{ mol H} \quad \text{(an integer)}$$

The empirical formula is CH_2.

EXERCISE Combustion of 1.333 g of a sample of a compound containing only carbon and hydrogen results in the formation of 4.334 g CO_2 and 1.332 g H_2O. What is the empirical formula of the compound?

[*Answer:* C_2H_3]

EXAMPLE 2   Determining an empirical formula through combustion analysis

Combustion of 1.426 g of an unknown sample containing carbon, hydrogen, and oxygen results in formation of 1.394 g CO_2 and 0.2855 g H_2O. What is the empirical formula of the compound?

SOLUTION Because the compound contains more than just hydrogen and carbon, some extra work (Steps 2A and 2B) is required to calculate the amount of oxygen present.

Step 1. Use the molar mass of CO_2 and H_2O to calculate the moles of each produced:

$$1.394 \text{ g CO}_2 \times \frac{1 \text{ mol CO}_2}{44.01 \text{ g CO}_2} = 0.03167 \text{ mol CO}_2$$

$$0.2855 \text{ g H}_2\text{O} \times \frac{1 \text{ mol H}_2\text{O}}{18.02 \text{ g H}_2\text{O}} = 0.01584 \text{ mol H}_2\text{O}$$

Step 2. Calculate the moles of C and moles of H in the original sample:

$$0.03167 \text{ mol CO}_2 \times \frac{1 \text{ mol C (original sample)}}{1 \text{ mol CO}_2} = 0.03167 \text{ mol C (in original sample)}$$

$$0.01584 \text{ mol H}_2\text{O} \times \frac{2 \text{ mol H (original sample)}}{1 \text{ mol H}_2\text{O}} = 0.03168 \text{ mol H (in original sample)}$$

Step 2A. Using the molar mass of C and H, calculate the grams of C and H in the original sample:

$$0.03167 \text{ mol C} \times \frac{12.01 \text{ g}}{1 \text{ mol C}} = 0.3804 \text{ g C}$$

$$0.03168 \text{ mol H} \times \frac{1.008 \text{ g}}{1 \text{ mol H}} = 0.03193 \text{ g H}$$

Step 2B. Calculate the mass of oxygen and moles of oxygen atoms in the original sample. Because the sample contains carbon, hydrogen, and oxygen only, the mass of oxygen in the sample equals the total mass of the sample minus the mass of carbon and mass of oxygen:

$$\text{mass O} = \text{mass sample} - (\text{mass carbon} + \text{mass hydrogen})$$

$$= 1.426 \text{ g} - (0.3804 \text{ g} + 0.03193 \text{ g}) = 1.014 \text{ g}$$

To calculate the moles of oxygen atoms, we use the molar mass of O, which is $16.00 \text{ g} \cdot \text{mol}^{-1}$:

$$1.014 \text{ g O} \times \frac{1 \text{ mol O}}{16.00 \text{ g O}} = 0.06338 \text{ mol O}$$

Step 3. Find the simplest whole number ratio of moles of elements in the compound. We calculated the moles of C and H in Step 2 and the moles of O in Step 2B. To find the simplest whole number of moles from the values already calculated, we divide each of the calculated values by the smallest one:

$$\frac{0.03168 \text{ mol H}}{0.03167} = 1.000 \text{ mol H} = 1 \text{ mol H} \quad \text{(an integer)}$$

$$\frac{0.03167 \text{ mol C}}{0.03167} = 1 \text{ mol C}$$

$$\frac{0.06338 \text{ mol O}}{0.03167} = 2.001 \text{ mol O} = 2 \text{ mol O} \quad \text{(an integer)}$$

The empirical formula is thus CHO_2.

EXERCISE What is the empirical formula of ethylene glycol if combustion of a 1.652-g sample results in formation of 2.343 g CO_2 and 1.438 g H_2O?

[*Answer:* CH_3O]

KEY WORDS   Define or explain each term in a written sentence or two.

actual yield	limiting reactant
combustion analysis	percentage yield
competing reaction	theoretical yield

DESCRIPTIVE CHEMISTRY TO REMEMBER

- **Hydrogen gas** (H_2) and **carbon monoxide** (CO) are industrial reducing agents.
- **Pyrite** (FeS_2) decomposes when exposed to air to form sulfuric acid.
- **Octane** (C_8H_{18}) is one of the many compounds found in gasoline. When it is burned in a limited supply of oxygen, both CO_2 and CO are formed (as is H_2O).
- A promising substitute for Freons is **HFC-134a** ($C_2H_2F_4$).
- **Combustion analysis** is a procedure in which the composition of an organic compound is determined by burning the compound in an unlimited supply of oxygen and measuring the masses of carbon dioxide and water produced.
- **Phosphorus pentoxide** (P_4O_{10}) acts as a drying agent because it strongly absorbs water.
- **Sodium hydroxide** (NaOH) strongly absorbs carbon dioxide and is used to determine the amount of carbon dioxide liberated in a combustion experiment.

CHEMICAL EQUATIONS TO KNOW

- In the **Haber syntheses,** hydrogen gas and nitrogen gas are reacted to form ammonia:

$$N_2(g) + 3H_2(g) \longrightarrow 2NH_3(g)$$

- Iron is produced from hematite ore (Fe_2O_3) through **reduction** with carbon monoxide:

$$Fe_2O_3(s) + 3CO(g) \longrightarrow 2Fe(s) + 3CO_2(g)$$

- **Calcium carbide** (CaC_2) reacts with water to form calcium hydroxide and acetylene (C_2H_2):

$$CaC_2(s) + 2H_2O(l) \longrightarrow Ca(OH)_2(s) + C_2H_2(g)$$

- Aluminum metal can be reacted with chromium (III) oxide to produce **chromium metal:**

$$2Al(l) + Cr_2O_3(s) \xrightarrow{\Delta} Al_2O_3(s) + 2Cr(l)$$

- Carbon dioxide can be removed from power plant exhaust gases by reaction with an aqueous slurry of **calcium silicate:**

$$2CO_2(g) + H_2O(l) + CaSiO_3(s) \longrightarrow SiO_2(s) + Ca(HCO_3)_2(aq)$$

- Phosphorus reacts with elemental chlorine to produce **phosphorus trichloride:**

$$2P(s) + 3Cl_2(g) \longrightarrow 2PCl_3(l)$$

- A promising substitute for Freons, **HFC-134a,** is produced by the reaction of trifluoroethene with hydrogen fluoride:

$$C_2HF_3(l) + HF(g) \longrightarrow C_2H_2F_4(l)$$

- Sodium metal reacts with aluminum oxide to form **aluminum** and **sodium oxide:**

$$6Na(l) + Al_2O_3(s) \xrightarrow{\Delta} 3Na_2O(s) + 2Al(l)$$

- **Urea** can be formed by the reaction of ammonia with carbon dioxide:

$$2NH_3(g) + CO_2(g) \longrightarrow OC(NH_2)_2(s) + H_2O(l)$$

SELF-TEST EXERCISES

How to use reaction stoichiometry

1. How many moles of $SiCl_4$ can be formed from 3.0 mol Cl_2?

$$Si(s) + 2Cl_2(g) \longrightarrow SiCl_4(g)$$

(a) 3.0   (b) 6.0   (c) 1.5   (d) 2.5   (e) 0.67

2. How many moles of HCl can be formed from 2.3 mol of H_2O?

$$PCl_5(s) + 4H_2O(l) \longrightarrow 5HCl(aq) + H_3PO_4(aq)$$

(a) 1.8　　　(b) 2.3　　　(c) 0.58　　　(d) 11.5　　　(e) 2.9

3. How many grams of Fe_2O_3 can be formed from 6.50 mol of Fe?

$$4Fe(s) + 3O_2(g) \longrightarrow 2Fe_2O_3(s)$$

(a) 519　　　(b) 1.04×10^3　　　(c) 12.3　　　(d) 2.08×10^3　　　(e) 653

4. How many kilograms of $NaHCO_3$ are required to produce 51.0 mol CO_2?

$$2NaHCO_3(s) \longrightarrow Na_2CO_3(s) + CO_2(g) + H_2O(g)$$

(a) 4.28　　　(b) 8.57　　　(c) 2.14　　　(d) 195　　　(e) 77.5

5. How many grams of PbO can be formed from 25.0 g PbS?

$$2PbS(s) + 3O_2(g) \longrightarrow 2PbO(s) + 2SO_2(g)$$

(a) 26.8　　　(b) 23.3　　　(c) 11.7　　　(d) 46.6　　　(e) 13.4

6. How many grams of HNO_2 are needed to produce 125 g of NO?

$$3HNO_2(aq) \longrightarrow HNO_3(aq) + 2NO(g) + H_2O(l)$$

(a) 8.82×10^3　　　(b) 196　　　(c) 294　　　(d) 588　　　(e) 131

7. How many kilograms of NH_4F can be prepared from 1.22 kg NH_3?

$$4NH_3(g) + 3F_2(g) \longrightarrow 3NH_4F(s) + NF_3(g)$$

(a) 1.99　　　(b) 1.56　　　(c) 2.66　　　(d) 3.54　　　(e) 2.87

8. In a titration, the solution in the buret is called the

(a) analyte　　　(b) indicator　　　(c) unknown　　　(d) titrant　　　(e) dye

9. In a titration, 17.88 mL of 0.1246 M NaOH is required to titrate 10.00 mL of an unknown HCl solution. What is the molarity of the HCl solution?

(a) 0.04489 M　　　(b) 14.35 M　　　(c) 0.06969 M
(d) 4.489 M　　　(e) 0.2228 M

10. In a titration, 17.88 mL of 0.1246 M NaOH is required to titrate 10.00 mL of an unknown H_2SO_4 solution. What is the molarity of the H_2SO_4 solution?

(a) 8.977 M　　　(b) 0.1114 M　　　(c) 0.1394 M
(d) 0.4489 M　　　(e) 0.2228 M

11. When 0.2476 g of benzoic acid ($HC_7H_5O_2$, a white crystalline solid) is dissolved in water and titrated with an unknown solution of KOH, 23.55 mL of the NaOH solution is required. What is the molarity of the NaOH solution? Benzoic acid is a monoprotic acid.

(a) 0.08609 M　　　(b) 0.01051 M　　　(c) 5.831 M
(d) 0.5020 M　　　(e) 0.47749 M

12. A 25.00-mL sample of HCl requires 33.26 mL of 0.1026 M NaOH for titration. What is the molarity of the HCl?

(a) 0.2730 M　　　(b) 0.06825 M　　　(c) 0.7712 M
(d) 0.1365 M　　　(e) 0.3856 M

13. 5.00 mL of a citric acid ($H_3C_6H_5O_7$) solution require 27.35 mL of 0.1116 M NaOH for titration. What is the molarity of the citric acid solution?

$$3NaOH(aq) + H_3C_6H_5O_7(aq) \longrightarrow Na_3C_6H_5O_7(aq) + 3H_2O(l)$$

(a) 0.2035 M   (b) 0.1256 M   (c) 1.832 M

(d) 0.6105 M   (e) 0.06121 M

14. When a 0.7185-g sample of a solid acid is dissolved in about 10 mL of water and titrated with 0.1655 M NaOH, 21.26 mL of the NaOH are required. What is the molar mass of the acid? 1 mol NaOH is required to react with 1 mol of the acid.

(a) 43.40 g·mol^{-1} (b) 351.8 g·mol^{-1} (c) 36.46 g·mol^{-1}

(d) 4.897 g·mol^{-1} (e) 204.2 g·mol^{-1}

The limits of reaction

15. What is the limiting reagent when 30.0 g CO and 10.0 g H_2 are mixed and reacted in the following reaction:

$$3H_2(g) + CO(g) \longrightarrow CH_4(g) + H_2O(g)$$

(a) CO   (b) H_2 (c) neither

16. How much Fe can be produced if 100 g Fe_2O_3 and 100 g CO are mixed and reacted?

$$Fe_2O_3(s) + 3CO(g) \longrightarrow 2Fe(s) + 3CO_2(g)$$

(a) 97.9 g   (b) 35.0 g   (c) 147 g   (d) 70.0 g   (e) 200 g

17. When 50.0 g Ca are reacted with excess N_2, 41.8 g Ca_3N_2 are produced. What is the percentage yield of Ca_3N_2?

$$3Ca(s) + N_2(g) \longrightarrow Ca_3N_2(s)$$

(a) 67.9%   (b) 83.6%   (c) 22.6%   (d) 100%   (e) 33.4%

18. 19.9 g H_2O are formed when 45.0 g $Mg(OH)_2$ is reacted with excess HCl. What is the percentage yield of H_2O?

$$Mg(OH)_2(s) + 2HCl(aq) \longrightarrow 2H_2O(l) + MgCl_2(aq)$$

(a) 69.8%   (b) 71.6%   (c) 44.2%   (d) 34.9%   (e) 100%

19. A 1.532-g sample of a compound containing only carbon and hydrogen produces 4.807 g CO_2 and 1.968 g H_2O in a combustion analysis. What is the empirical formula of the compound?

(a) C_2H (b) CH   (c) CH_2 (d) C_2H_3 (e) C_5H_2

20. A 0.8135-g sample of a compound containing carbon, hydrogen, and oxygen produces 2.104 g CO_2 and 0.4306 g H_2O in a combustion analysis. What is the empirical formula of the compound?

(a) CHO   (b) C_2H_2O (c) CHO_2 (d) C_3HO_3 (e) C_4H_4O

Descriptive chemistry

21. Combustion analysis is the determination of the composition of an unknown organic compound based on its combustion to

(a) C + H_2 (b) CO + H_2 (c) CH_4

(d) CO_2 + H_2O (e) CO_2 + H_2

22. Which of the following is used as a rocket fuel?

(a) CH_4 (b) C_2H_2 (c) N_2H_4 (d) NH_3 (e) H_2O

23. What is the reducing agent in the production of iron from hematite (Fe_2O_3)?

(a) C (b) H_2 (c) CO (d) O_2 (e) N_2

24. Which of the following is the mineral pyrite, found near abandoned mines, that contributes to water pollution by its decomposition to form sulfuric acid?

(a) S (b) Fe_2O_3 (c) CH_4 (d) FeS_2 (e) $CaSiO_3$

25. Which of the following is octane, a compound found in gasoline?

(a) C_8H_{18} (b) CH_4 (c) CO_2 (d) C_2H_4 (e) N_2H_4

26. Acetylene (C_2H_2) is produced by the reaction of water with

(a) $CaCO_3$ (b) CaC_2 (c) C_2H_4 (d) NaF (e) NH_3

27. Phosphorus trichloride (PCl_3) can be produced by the reaction of phosphorus with

(a) $CaCl_2$ (b) CCl_4 (c) HCl (d) $NaCl$ (e) Cl_2

28. The formula of HFC-134a, a compound designed to replace Freons, is

(a) CF_4 (b) HF (c) C_2HF_3 (d) $C_2H_2F_4$ (e) C_6H_6

29. Which of the following compounds is used to absorb the water formed in combustion analysis?

(a) P_4O_{10} (b) $NaOH$ (c) H_2O_2 (d) $MgCl_2$ (e) $Ca(OH)_2$

CHAPTER 5

THE PROPERTIES OF GASES

Gases play an important role both in our daily lives and in the chemical sciences. Our atmosphere is made up of gases, as are eleven of the elements (H_2, F_2, Cl_2, N_2, O_2, He, Ne, Ar, Kr, Xe, Rn). The fuels methane (CH_4), propane (C_3H_8), and butane (C_4H_{10}) are also gases.

THE NATURE OF GASES

5.1 THE STATES OF MATTER

KEY CONCEPT   Solids, liquids, and gases

The three common forms of matter can be naively characterized by the ease of movement of the particles that make up the matter. Solids are rigid; the particles are held tightly to each other without the opportunity for any substantial motion. In liquids, the particles adhere to each other but slip and slide past each other, thereby giving liquids their typical fluidity. In gases, the particles making up the gas move about without, for the most part, interacting with one another. Because gases flow easily, they are also considered fluids.

> **EXAMPLE   Predicting a state of matter**
>
> A sample of matter is poured into a cylinder and the cylinder fitted with a piston. When the piston is pushed down, the matter does not give way; it cannot be compressed. Which of the three states of matter is present?
>
> **SOLUTION** Because the matter can be poured into a cylinder, it fills up the bottom of the container it is in. It is also not compressible. These properties are typical of a liquid.
>
> **EXERCISE** Diffusion is the process that occurs when a chemical moves from an area of high concentration to one of low concentration. In which form of matter is diffusion least likely to occur?
>
> [*Answer:* Solid]

5.2 THE MOLECULAR CHARACTER OF GASES

KEY CONCEPT   Gas molecules move about freely

The molecules that make up a gas are very widely separated from one another and move rapidly in a straight line until they suffer a collision with either the walls of the confining container or each other. To get an image of the scale involved, imagine just five flies buzzing about at the speed of a jet airliner in an average classroom! This is what typical gas molecules might look like. The molecules of a gas do not attract one another very strongly, so when they collide they simply rebound (like two billiard balls colliding) rather than sticking to each other. In an actual gas, some molecules move more rapidly than the others; at any instant, a few might even be motionless. The average speed of the molecules, however, is determined by the temperature of the gas. As the temperature increases, the average speed of the molecules increases.

EXAMPLE The motion of gas molecules

How likely is it that all (or most) of the molecules in a sample of gas will be moving in the same direction at the same time?

SOLUTION There is virtually no chance this could happen. The molecules in a gas move in chaotic motion, at different speeds and in different directions.

EXERCISE Imagine two samples of helium gas, one at 20°C and one at twice the temperature in degrees Celsius, 40°C. Is it necessarily true that the molecules in the warmer sample are moving twice as fast as those in the cooler sample?

[*Answer:* No. The molecules in the warmer sample move faster, but not twice as fast. (The actual relationship between temperature and speed will be discussed later).]

5.3   PRESSURE

KEY CONCEPT A   Molecules colliding with a surface exert a pressure

When a gas molecule collides with the wall of a container, it pushes on the wall as it rebounds back and changes its direction of motion. This "pushing" effect is how a molecule exerts a pressure. When many billions of molecules collide with a wall (at some instant) over a specific area, the pressure is the force exerted by the molecules divided by the area of the wall with which the molecules collide.

EXAMPLE The pressure of gas molecules

Imagine a flask that contains 0.10 mol of neon. At some point, another 0.10 mol of neon is introduced into the flask. What happens to the pressure when the additional 0.10 mol of gas is introduced? Be specific and quantitative!

SOLUTION When the second 0.10 mol of gas is introduced, the number of gas molecules is doubled, that is, 0.20 mol of Ne contains twice as many molecules as 0.10 mol. Because there are twice the number of molecules, the number of collisions with the walls of the flask will double. This will cause the pressure to double. Thus, the new pressure is twice the original pressure.

EXERCISE Now, we take the sample of gas containing 0.20 mol of neon and raise the temperature. In qualitative terms, predict what will happen to the pressure?

[*Answer:* Because the molecules move faster when the temperature is increased, they collide with the walls of the flask more frequently and more strongly. Thus, the pressure will increase.]

KEY CONCEPT B   Reading a barometer

In reading a barometer, remember that the liquid (usually mercury or water) in the closed tube tries to "seek its own level," which means the level of the liquid in the open-to-the-air container. The pressure of the atmosphere prevents the liquid from finding its level and determines the actual level of the liquid. Thus, at zero atmospheric pressure, the barometer looks like illustration (a), where the two liquid levels are the same. As the atmospheric pressure increases, it pushes the mercury column up the tube, as shown in (b) and (c). Because the pressure inside the closed barometer tube is (essentially) zero, the height of the mercury above the liquid level in the cup is a direct measure of atmospheric pressure.

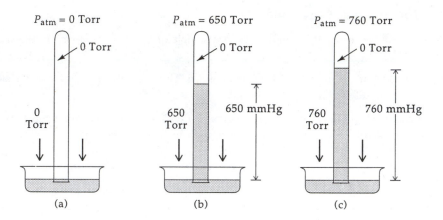

$P_{atm} = 0$ Torr $P_{atm} = 650$ Torr $P_{atm} = 760$ Torr

0 Torr   0 Torr   0 Torr

0 Torr   650 Torr   650 mmHg   760 Torr   760 mmHg

(a)   (b)   (c)

EXAMPLE   Understanding a barometer

Would the barometer in illustration (b) function if there were a small leak at its top? Explain your answer.

SOLUTION No, it would not work. Air would leak into the barometer, and the pressure inside the barometer would eventually reach atmospheric pressure. Then, the inside pressure pushing the column downward would equal the outside pressure pushing it upward, and the two liquid levels would be the same.

EXERCISE Could a highly volatile liquid be used as a barometer liquid?

[*Answer:* No. The liquid would evaporate into the empty space inside the barometer, thereby causing gas pressure to build up inside; this would result in inaccurate readings.]

5.4   UNITS OF PRESSURE

KEY CONCEPT   The common units of pressure

The two most commonly used pressure units in the sciences are the Torr and the atmosphere (atm). The pressure at sea level under normal conditions is close to 1 atm and 760 Torr. In addition, 1 atm = 760 Torr, exactly. The SI unit of pressure is the pascal (Pa); the pascal is a very small unit of pressure. 1 atm = $1.013\,25 \times 10^5$ Pa. Conversions from one unit of pressure to another is most easily done by using unit conversions.

EXAMPLE   Pressure unit conversions

It is thought that the pressure at the center of a tornado may be as low as 600 Torr. What is this pressure in atmospheres?

SOLUTION From Table 5.2 of the text, we know that 760 Torr = 1 atm exactly. With this information, the problem can be solved by the unit conversion method:

$$600 \cancel{\text{ Torr}} \times \frac{1 \text{ atm}}{760 \cancel{\text{ Torr}}} = 0.789 \text{ atm}$$

EXERCISE A typical high-pressure cooker can attain a pressure of 1.15 atm. What is this pressure in kilopascals?

[*Answer:* 117 kPa]

KEY WORDS   Define or explain each term in a written sentence or two.

atmosphere (unit)   kinetic model   solid
barometer   liquid   Torr
gas   pressure

THE GAS LAWS

5.5   BOYLE'S LAW

KEY CONCEPT   The volume of a gas is inversely proportional to pressure

It is intuitive that when we increase the pressure on a gas (as in a bicycle pump), the gas yields and undergoes a decrease in volume. Quantitatively, for instance, when the pressure is doubled, the volume is halved; when it is tripled, the volume is decreased to one-third its original value. Similar relationships hold when the pressure is decreased, except that the volume now increases. These relationships hold only when the temperature is held constant and are collectively described by **Boyle's law.** Mathematically, the relationship is,

$$P \propto \frac{1}{V}$$

EXAMPLE   Using Boyle's Law

9.0 L of a gas originally at 1.0 atm undergoes a change in pressure to 2.0 atm at constant temperature. What is the new volume of the gas?

SOLUTION Because the new pressure is larger than the original pressure, the new volume must be smaller than the original volume. If the pressure is doubled, the new volume must be one-half the original volume, or, in this case, 4.5 L.

EXERCISE 9.0 L of a gas originally at 1.0 atm undergoes a change in pressure to 0.33 atm at constant temperature. What is the new volume of the gas?

[*Answer:* 27 L]

5.6   CHARLES'S LAW

KEY CONCEPT   The volume of a gas is directly proportional to absolute temperature

When the temperature of a gas is increased at constant pressure, the gas molecules move faster and push out on the walls of the container, causing the volume to increase. The experiment must be done in a container that can expand, such as a balloon or a piston fitted with a cylinder. Quantitatively, for instance, when the absolute temperature (in kelvins) is doubled, the volume is doubled; when it is tripled, the volume is tripled. Similar relationships hold when the temperature is decreased, except that the volume now decreases. These relationships hold only when the pressure is held constant and is known as **Charles's law.** Mathematically, the relationship is

$$V \propto T$$

EXAMPLE   Using Charles's law

3.2 L of a gas originally at 150 K undergoes a change in temperature, at constant pressure, to 300 K. What is the new volume of the gas?

SOLUTION According to Charles' law, at constant pressure, the volume of a gas is proportional to the Kelvin temperature. When the Kelvin temperature doubles, as in this case, the volume also doubles. The new volume is 6.4 L.

EXERCISE 3.3 L of a gas originally at 150 K undergoes a change in temperature to 50 K at constant pressure. What is the new volume of the gas?

[*Answer:* 1.1 L]

5.7   AVOGADRO'S PRINCIPLE

KEY CONCEPT   The volume of a gas is proportional to the moles of gas present

To illustrate Avogadro's principle, we must again use a container that can expand or contract. This is how constant pressure is maintained; also, the temperature must be kept constant. Under these conditions, it is intuitive that the volume occupied by a gas must increase as the number of moles of gas increases. If we think about it, when no gas is present, the volume must be zero; as gas is added, the volume increases. Quantitatively, the volume of gas is proportional to the moles of gas present. This means that if the number of moles doubles, the volume must double; if it triples, the volume must triple. Similarly, if the number of moles is halved, the new volume will be one-half the original volume. Mathematically, the relationship is

$$V \propto n$$

EXAMPLE   Using Avogadro's principle

A chemical reaction is run, at constant pressure and temperature, inside a cylinder fitted with a piston. The reactants contain 0.20 mol of a gas in a volume of 100 mL. The products contain 2.4 mol of gas. What is the volume of the products?

SOLUTION According to Avogadro's principle, at constant pressure and temperature, the volume of a gas is proportional to the number of moles of gas present. In this case, there are 12 times more moles of gas present at the end of the reaction than at the beginning. The identity of the gas changes in going from reactants to products, but this does not affect the results because all gases have (approximately) the same molar volume at the same temperature and pressure. Because there are 12 times the moles of gas present, the new volume must be 12 times the original volume. The new volume is 1200 mL or 1.2 L.

EXERCISE When a balloon contains 1.00 mol of gas, its volume is 50.0 mL. What is the volume of the balloon if 2.00 additional moles of gas are blown into it? Assume temperature and pressure are constant.

[*Answer:* 150 mL]

5.8   USING THE GAS LAWS

KEY CONCEPT A   Using the individual gas laws

The individual gas laws can be used to calculate changes in one of the variables, P, V, T, or n when one or more of the other variables changes. There are a number of methods to do such calculations; in this study guide we use equations derived directly from the individual laws. The mathematical results are exactly the same as the examples in the text. To use the individual gas laws, we must determine which variables are to be held constant and which change, because this determines which law to use. The table that follows summarizes the derived equation for each law and which law to use under different circumstances. The subscript 1 indicates the value of a variable before some change occurs, and subscript 2 indicates the value after a change occurs.

Law	Equation	Constant Variables	Variables that Change	Derived Equation
Boyle's law	$V \propto \dfrac{1}{P}$	n, T	P, V	$P_1 V_1 = P_2 V_2$
Charles's law	$V \propto T$	n, P	T, V	$\dfrac{V_1}{T_1} = \dfrac{V_2}{T_2}$
Avogadro's principle	$V \propto n$	P, T	n, V	$\dfrac{V_1}{n_1} = \dfrac{V_2}{n_2}$
combined gas law	$\dfrac{PV}{T} = \text{constant}$	n	P, V, T	$\dfrac{P_1 V_1}{T_1} = \dfrac{P_2 V_2}{T_2}$

EXAMPLE 1   Boyle's law calculation

The pressure of a 5.2-L sample of gas at 700 Torr is increased to 780 Torr. What is the new volume of the gas? The temperature of the gas is held constant.

SOLUTION We must first ascertain that this is a Boyle's law problem. The problem states that a change in pressure results in a change in volume for a sample of gas at constant temperature. So, P and V change while n and T are held constant; thus, this is a Boyle's law problem with

$$P_1 = 700 \text{ Torr} \qquad P_2 = 780 \text{ Torr}$$
$$V_1 = 5.2 \text{ L} \qquad V_2 = ?$$

Substituting into Boyle's law gives

$$P_1 V_1 = P_2 V_2$$
$$(700 \text{ Torr})(5.2 \text{ L}) = (780 \text{ Torr})(V_2)$$

Dividing both sides of the equation by 780 Torr gives

$$V_2 = 5.2 \text{ L} \times \frac{700 \text{ Torr}}{780 \text{ Torr}}$$
$$= 4.7 \text{ L}$$

Notice that this calculation takes on the same form as those shown in the text, with the initial volume multiplied by the factor that makes the final volume smaller.

EXERCISE A 65-L sample of gas at 2.2 atm undergoes a change in pressure to 1.5 atm at constant temperature. What is the new volume of the gas?

[*Answer:* 95 L]

EXAMPLE 2   Charles's law calculation

A 200-mL sample of gas at 25°C undergoes a change in volume to 100 mL at constant pressure. What is the new temperature of the gas in kelvins?

SOLUTION Because the number of moles and the pressure of the gas are held constant while the volume and temperature are changed, this is a Charles's law problem. The temperature must be expressed in kelvins. Thus, using $K = °C + 273$, we have

$$V_1 = 200 \text{ mL} \qquad T_1 = 298 \text{ K}$$
$$V_2 = 100 \text{ mL} \qquad T_2 = ?$$

Substituting into Charles's law gives

$$\frac{V_1}{T_1} = \frac{V_2}{T_2}$$

$$\frac{200 \text{ mL}}{298 \text{ K}} = \frac{100 \text{ mL}}{T_2}$$

To solve this equation, we first multiply both sides of the equation by T_2 and by 298 K and then divide both sides by 200 mL. This gives

$$T_2 = 298 \text{ K} \times \frac{100 \text{ mL}}{200 \text{ mL}}$$

$$= 149 \text{ K}$$

EXERCISE A 3.66-L sample of gas at 350°C undergoes a change in temperature to 283°C. What is the new volume of the gas?

[*Answer:* 3.27 L]

EXAMPLE 3   Avogadro's principle calculation

In a particular gas-phase reaction, 1.25 L of gas-phase reactant converts to 3.75 L of gas-phase product. At the start of the reaction, 0.113 mol of reactant is present. How many moles of product are formed? The temperature and pressure of the reaction system are the same at the end of the reaction as at the start.

SOLUTION Because the number of moles and the volume of gas change at constant pressure and temperature, this is an Avogadro's principle problem with

$$V_1 = 1.25 \text{ mL} \qquad n_1 = 0.113 \text{ mol}$$

$$V_2 = 3.75 \text{ mL} \qquad n_2 = ?$$

Substituting into the equation for Avogadro's principle gives

$$\frac{V_1}{n_1} = \frac{V_2}{n_2}$$

$$\frac{1.25 \text{ L}}{0.113 \text{ mol}} = \frac{3.75 \text{ L}}{n_2}$$

To solve this equation, we first multiply both sides of the equation by n_2 and 0.113 mol, and then divide both sides by 1.25 L. This gives

$$n_2 = 0.113 \text{ mol} \times \frac{3.75 \text{ L}}{1.25 \text{ L}}$$

$$= 0.339 \text{ mol}$$

EXERCISE During the course of a reaction, 0.25 mol of gas-phase reactant becomes 0.50 mol of gas-phase product. The initial volume of gaseous reactant is 125 mL. What is the volume of gaseous product?

[*Answer:* 2.5×10^2 mL]

PITFALL   The correct temperature in gas law problems

In Charles's law and combined law calculations, be sure you express the temperature in kelvins.

PITFALL   Is it V_1/V_2 or V_2/V_1?

For any of the calculations involving Boyle's and Charles's laws, Avogadro's principle, or the combined gas law, it is possible to inadvertently use a term such as V_1/V_2 when you really meant to use the inverse, V_2/V_1.

If you look at the answer to all your calculations and ask yourself whether the answer seems reasonable, you can often catch and correct this type of error. For example, if a problem specifies that a sample undergoes a decrease in pressure at constant temperature, the volume must increase. If you solve the problem mathematically and end up with a smaller volume, you have probably made the kind of error mentioned above.

PITFALL   There must be consistency of units in gas law problems

In gas law calculations, the units used for volume must be the same for V_1 and V_2. Similarly, the units for pressure must be the same for P_1 and P_2.

KEY CONCEPT B   Using the combined gas law

If the number of moles of a gas is kept constant and the temperature, pressure, and volume change, the change in conditions is described by the combined gas law equation:

$$\frac{P_1 V_1}{T_1} = \frac{P_2 V_2}{T_2}$$

EXAMPLE   Combined gas law calculation

If 3.2 L of a gas at 25°C and 732 Torr undergo a change to 50°C and 943 Torr, what is the new volume of the gas?

SOLUTION   In this problem, changes in pressure and temperature cause a change in volume for a constant number of moles of a gas. When n is constant and P, V, and T change, we use the combined gas law:

$$\frac{P_1 V_1}{T_1} = \frac{P_2 V_2}{T_2}$$

$P_1 = 732$ Torr	$P_2 = 943$ Torr
$T_1 = 25 + 273.15 = 298$ K	$T_2 = 50 + 273.15 = 323$ K
$V_1 = 3.2$ L	$V_2 = ?$

Because V_2 is the unknown, we rearrange the original equation so that V_2 is on the left-hand side of the equation and everything else on the right:

$$V_2 = V_1 \times \frac{P_1}{P_2} \times \frac{T_2}{T_1}$$

Substituting the given values, we get

$$V_2 = 3.2 \text{ L} \times \frac{732 \text{ Torr}}{943 \text{ Torr}} \times \frac{323 \text{ K}}{298 \text{ K}}$$

$$= 2.7 \text{ L}$$

EXERCISE   5.5 L of a gas at 0.955 atm and 12.2°C undergo a change in volume and pressure to 5.3 L and 1.22 atm. What is the new temperature of the gas?

[*Answer:* 3.5×10^2 K]

5.9   THE IDEAL GAS LAW

KEY CONCEPT   The proper units for the ideal gas law

The ideal gas law relates the pressure P, volume V, temperature T (in kelvins), and number of moles n of an ideal gas:

$$PV = nRT$$

In a typical problem using the ideal gas law, any three of the parameters P, V, T, and n are given and the fourth must be calculated. The value of R, expressed in the proper units, is also needed. This can be looked up when needed and is usually supplied on examinations. When you use the ideal gas law equation, write the units with all of the values substituted in and cancel units correctly. If the units of the answer come out wrong, you most certainly have substituted into the equation incorrectly.

EXAMPLE   Unit consistency in the ideal gas law

Assume you are required to use the ideal gas law with the value of $R = 8.206 \times 10^{-2}$ L·atm· K^{-1}·mol^{-1}. What units must be used to express the volume of gas?

SOLUTION The units used for the volume of gas must be consistent with the units of volume expressed for R. The units of R, L·atm·K^{-1}·mol^{-1}, uses liters (L) for volume. Thus, the volume of gas must be expressed in liters.

EXERCISE Assume you are required to use the ideal gas law with the value of $R = 8.206 \times 10^{-2}$ L· atm·K^{-1}·mol^{-1}. What units must be used to express the pressure of gas?

[*Answer:* atm]

5.10   USING THE IDEAL GAS LAW TO MAKE PREDICTIONS

KEY CONCEPT   Direct substitution into the ideal gas law

Many types of predictions can be made using the ideal gas law. In doing calculations with the law, a few hints might prove helpful:

- The temperature must always be expressed in Kelvins.
- The pressure and volume can be expressed in any units as long as the value of R has units consistent with those of the pressure and volume.
- The parameter n always has units of moles. However, we can use the molar mass to convert this to grams or to convert from grams to moles if required.

In any ideal gas law calculation you should always evaluate the reasonableness of your answer by using the fact that 1 mol of an ideal gas at 25°C occupies about 24 liters (see section 5.11 for the exact value). For example, if you are asked to calculate the volume of 0.10 mol of a gas at 45°C, you might reason as follows: Since 1 mol of an ideal gas at 25°C occupies about 24 L, 0.10 mol should occupy 1/10 of that volume, or about 2.4 L. In addition, since a gas expands when its temperature is raised, a gas at 45°C should have a slightly larger volume than the same gas at 25°C. Thus, you would expect your answer to be a bit larger than 2.4 L. If it isn't, check both your substitutions into the ideal gas law and your arithmetic.

EXAMPLE 1   Using the ideal gas law

A typical child's helium balloon has a volume of about 2 L. How many moles of helium (He) will the balloon hold if it is filled to a pressure of 770 Torr on a day when the temperature is 24°C? How many grams of helium does it hold?

SOLUTION The problem gives the volume, temperature, and pressure (V, T, P) of a gas and asks for the moles (n) of gas. The ideal gas law expresses the relationship among these parameters and is used to solve the problem. Because the unknown is the number of moles n, we divide both sides of the equation by RT to get the unknown on one side of the equation by itself:

$$PV = nRT$$

$$n = \frac{PV}{RT}$$

The values given for the known parameters are $P = 770$ Torr, $V = 2$ L, and $T = 24 + 273.15 = 297$ K. Because we use $R = 0.08206$ L·atm·mol^{-1}·K^{-1}, we first convert the pressure into units of atmospheres, using the fact that 1 atm = 760 Torr:

$$770 \text{ Torr} \times \frac{1 \text{ atm}}{760 \text{ Torr}} = 1.01 \text{ atm}$$

Substituting these values into the equation gives

$$n = \frac{(1.01 \text{ atm})(2 \text{ L})}{(0.082\,06 \text{ L·atm·K}^{-1}\text{·mol}^{-1})(297 \text{ K})}$$

$$= 0.082\,842\,2 \text{ mol He}$$

$$= 0.08 \text{ mol He}$$

The number of grams of He is calculated by using the molar mass of He, which is 4.003 g·mol^{-1}:

$$\text{Grams He} = 0.08 \text{ mol He} \times \frac{4.003 \text{ g He}}{1 \text{ mol He}}$$

$$= 0.3 \text{ g He}$$

EXERCISE How many moles of an ideal gas occupy 125 mL at 35.2°C and 750 Torr?

[*Answer:* 0.004 87 mol]

EXAMPLE 2   Using the ideal gas law

What is the volume of 3.45 mol of an ideal gas at 35°C and 0.766 atm?

SOLUTION The temperature must be expressed in kelvins for an ideal gas law calculation: 35°C equals 308 K (273 + 35 = 308). Substituting into the ideal gas law gives

$$pV = nRT$$

$$(0.766 \text{ atm}) \; V = (3.45 \text{ mol})(0.082\,06 \text{ L·atm·mol}^{-1}\text{·K}^{-1})(308 \text{ K})$$

$$V = \frac{(3.45 \text{ mol})(0.082\,06 \text{ L·atm·mol}^{-1}\text{·K}^{-1})(308 \text{ K})}{0.766 \text{ atm}}$$

$$= 114 \text{ L}$$

EXERCISE What is the volume of 0.444 mol of an ideal gas at 0.95 atm and 77.3°C?

[*Answer:* 13 L]

EXAMPLE 3   Using the ideal gas law

If 0.0640 mol of an ideal gas occupies 3.68 L at 742 Torr, what is the temperature of the gas?

SOLUTION Because we use $R = 0.08206$ L·atm·mol^{-1}·K^{-1}, the pressure must be converted to atmospheres before substituting into the ideal gas law:

$$742 \text{ Torr} \times \frac{1 \text{ atm}}{760 \text{ Torr}} = 0.976 \text{ atm}$$

$$PV = nRT$$

$$(0.976 \text{ atm})(3.68 \text{ L}) = (0.0640 \text{ mol})(0.08206 \text{ L·atm·mol}^{-1}\text{·K}^{-1})T$$

$$T = \frac{(0.976 \text{ atm})(3.68 \text{ L})}{(0.082\,06 \text{ L·atm·mol}^{-1}\text{·K}^{-1})(0.0640 \text{ mol})}$$

$$= 684 \text{ K}$$

EXERCISE If 2.33 mol of an ideal gas occupies 1.21×10^4 mL at 651 Torr, what is the temperature of the gas in kelvins?

[*Answer:* 54.2 K]

PITFALL   The mass of a gas in the ideal gas law

At times, an ideal gas calculation calls for the mass of a gas. Although the ideal gas law does not deal directly with the mass of a gas, remember that the mass can always be calculated from the moles and molar mass of the gas:

$$\text{Mass} = \text{moles} \times \text{molar mass}$$

5.11   MOLAR VOLUME

KEY CONCEPT   Molar volume of gases

The **molar volume**, V_m, of a substance is the volume of 1 mol of the substance, or equivalently, the volume per mole of the substance. Because the volume of a gas depends on the temperature and pressure, the molar volume of a gas is usually defined at a reference temperature and pressure. Chemists use two different reference points. At 25.00°C and 1.000 atm pressure, $V_m = 24.47$ L. Let's make sure the meaning of this is clear; at 25.00°C and 1.000 atm pressure, 1 mol of an ideal gas occupies a volume of 24.47 L. Another common reference is point is 0°C and 1 atm pressure; this is called standard temperature and pressure (STP). At STP, $V_m = 22.41$ L.

EXAMPLE   Calculating the volume of a sample of gas using V_m

What is the volume, at 25.0°C and 1 atm pressure, of 11.0 g of CH_4 (methane, a gas).

SOLUTION We first calculate the moles of CH_4 present by using its molar mass ($16.04 \text{ g} \cdot \text{mol}^{-1}$).

$$11.0 \text{ g } CH_4 \times \frac{1 \text{ mol } CH_4}{16.04 \text{ g}} = 0.686 \text{ mol } CH_4$$

We know that 1 mol of an ideal gas occupies 24.47 L at 25.0°C and 1 atm; we can use this fact to calculate the volume of any number of moles of a gas at the same temperature and pressure:

$$0.686 \text{ mol } CH_4 \times \frac{24.47 \text{ L}}{1 \text{ mol}} = 16.8 \text{ L } CH_4$$

EXERCISE What is the volume, at 25.0°C and 1 atm pressure, of 11.0 g of CF_4?

[*Answer:* 3.06 L]

5.12   THE STOICHIOMETRY OF REACTING GASES

KEY CONCEPT   The volume of gas taking part in a reaction

A balanced chemical equation gives the relative number of moles of reactant and product. Thus, in a typical stoichiometric problem, the moles of the substances in the reaction become the focus of the calculations. When a gaseous reactant or product is involved, we are often given the volume of the gas or asked to calculate the volume. For this type of problem, the ideal gas law or the molar volume of a gas is used to determine the moles of the gas from the volume, or the volume from the moles, depending on the nature of the problem.

EXAMPLE 1   The volume of gaseous product from the mass of a second product

Magnesium reacts with cold dilute HCl to form the gas H_2:

$$Mg(s) + 2HCl(aq) \longrightarrow MgCl_2(aq) + H_2(g)$$

In a specific experiment, 12.2 g of $MgCl_2$ were formed. What volume of H_2 was also formed? Assume the temperature is 25.0°C and the pressure is 1.0 atm.

SOLUTION We first calculate the moles of $MgCl_2$ formed, using the molar mass of $MgCl_2$, 95.2 g·mol^{-1}:

$$12.2 \text{ g } \cancel{MgCl_2} \times \frac{1 \text{ mol } \cancel{MgCl_2}}{95.2 \text{ } \cancel{g}} = 0.128 \text{ mol } MgCl_2$$

The equation states that the number of moles of $MgCl_2$ formed is the same as the number of moles of H_2 formed. This information is used to calculate the moles of H_2:

$$0.128 \text{ } \cancel{\text{mol } MgCl_2} \times \frac{1 \text{ mol } H_2}{1 \text{ } \cancel{\text{mol } MgCl_2}} = 0.128 \text{ mol } H_2$$

Because the volume at 25.0°C and 1.0 atm is asked for, the easiest way to proceed is to use the molar volume of a gas at these conditions, 24.47 L.

$$0.128 \text{ mol } H_2 \times \frac{24.47 \text{ L}}{\text{mol}} = 3.13 \text{ L } H_2$$

EXERCISE How many liters of N_2 (at 25.0°C and 1 atm) are required to react with 91.5 g Li?

$$6Li(s) + N_2(g) \longrightarrow 2Li_3N(s)$$

[*Answer:* 53.8 L]

EXAMPLE 2   Calculating the volume of gas in a reaction

How many liters of HCl(g), measured at STP, can be produced from 0.70 L of $Cl_2(g)$, also measured at STP?

$$H_2(g) + Cl_2(g) \longrightarrow 2HCl(g)$$

SOLUTION The equation tells us that 1 mol H_2 reacts with 1 mol Cl_2 to produce 2 mol HCl. By using Avogadro's principle, we conclude that the volumes of reactants and products follow the same ratio as the number of moles of each. Thus, 1 L H_2 reacts with 1 L Cl_2 to produce 2 L HCl. Because 2 mol HCl are produced from 1 mol Cl_2, we get

$$0.70 \text{ } \cancel{L} \text{ } Cl_2 \times \frac{2 \text{ L HCl}}{1 \text{ } \cancel{L} \text{ } Cl_2} = 1.4 \text{ L HCl}$$

EXERCISE How many liters of O_2 can be produced from 0.22 L O_3 in the following reaction? Assume temperature and pressure remain constant.

$$2O_3(g) \longrightarrow 3O_2(g)$$

[*Answer:* 0.33 L]

5.13   GAS DENSITY

KEY CONCEPT   The density of an ideal gas

By using the definition of density ($d = m/V$), the fact that the moles of a sample equal the mass of the sample divided by the molar mass ($n = m/$molar mass), and the ideal gas law, we can relate the density of an ideal gas to the temperature, pressure, and molar mass of the gas:

$$d = \frac{(\text{molar mass})P}{RT}$$

Substitution of the molar mass in grams per mole, pressure in atmospheres, temperature in kelvins, and $R = 0.08206 \; \text{L} \cdot \text{atm} \cdot \text{mol}^{-1} \cdot \text{K}^{-1}$ gives the density in grams per liter, the usual density unit for gases.

EXAMPLE 1   Calculating the density of an ideal gas

What is the density of carbon tetrafluoride (CF_4) at 1.0 atm and 298 K?

SOLUTION The molar mass of CF_4 is calculated first and then substituted into the preceding equation:

$$
\begin{array}{ll}
1 \; \text{C}: & 1 \times 12.01 = 12.01 \\
4 \; \text{F}: & 4 \times 19.00 = \underline{76.00} \\
& \qquad\qquad 88.01 \; \text{g} \cdot \text{mol}^{-1}
\end{array}
$$

$$d = \frac{(\text{molar mass})P}{RT}$$

$$= \frac{(88.01 \; \text{g} \cdot \text{mol}^{-1})(1.0 \; \text{atm})}{(0.082 \; 06 \; \text{L} \cdot \text{atm} \cdot \text{mol}^{-1} \cdot \text{K}^{-1})(298 \; \text{K})}$$

$$= 3.6 \; \text{g} \cdot \text{L}^{-1}$$

EXERCISE What is the density of acetylene (C_2H_2) at 50.0°C and 720 Torr?

[*Answer:* $0.930 \; \text{g} \cdot \text{L}^{-1}$]

EXAMPLE 2   Calculating the molar mass of a gas from its density

The density of an ideal gas is $2.56 \; \text{g} \cdot \text{L}^{-1}$ at 100.0°C and 744 Torr. What is the molar mass of the gas?

SOLUTION For this question, we have the following data: $d = 2.56 \; \text{g} \cdot \text{L}^{-1}$, $P = 744 \; \text{Torr} = 0.979 \; \text{atm}$, and $T = 100.0°\text{C} = 373.2 \; \text{K}$. Substituting into the relevant equation gives

$$d = \frac{(\text{molar mass})P}{RT}$$

$$2.56 \; \text{g} \cdot \text{L}^{-1} = \frac{(\text{molar mass})(0.979 \; \text{atm})}{(0.082 \; 06 \; \text{L} \cdot \text{atm} \cdot \text{mol}^{-1} \cdot \text{K}^{-1})(373.2 \; \text{K})}$$

Multiplying both sides of the equation by $(0.082 \; 06 \; \text{L} \cdot \text{atm} \cdot \text{mol}^{-1} \cdot \text{K}^{-1})(373.2 \; \text{K})$ and dividing both sides by 0.979 atm gives the desired answer:

$$\text{Molar mass} = \frac{(2.56 \; \text{g} \cdot \text{L}^{-1})(0.082 \; 06 \; \text{L} \cdot \text{atm} \cdot \text{mol}^{-1} \cdot \text{K}^{-1})(373.2 \; \text{K})}{(0.979 \; \text{atm})}$$

$$= 80.1 \; \text{g} \cdot \text{mol}^{-1}$$

EXERCISE The density of an ideal gas is $2.99 \; \text{g} \cdot \text{L}^{-1}$ at 120.0°C and 829 Torr. What is the molar mass of the gas?

[*Answer:* $88.4 \; \text{g} \cdot \text{mol}^{-1}$]

5.14   MIXTURES OF GASES

KEY CONCEPT   Dalton's law of partial pressures

Imagine we have a mixture of gases in a flask. The **partial pressure** of a gas in the mixture is the pressure the gas would have if it were alone in the flask. **Dalton's law of partial pressures** states that the total pressure of the mixture of gases is the sum of the partial pressures of the individual gases in the mixture. Dalton's law is a result of the fact that, for ideal gases, the presence of one gas has no effect on the other gases present.

EXAMPLE   Using Dalton's law of partial pressures

A sample of nitrogen that exerts a pressure of 562 Torr in a 5.00-L flask and a sample of oxygen that exerts a pressure of 444 Torr in a 10.0-L flask are both placed in a 10.0-L flask. What is the total pressure of the mixture of gases in the 10.0-L flask? No reaction occurs.

SOLUTION According to Dalton's law of partial pressures, the total pressure of the two gases is the sum of the pressures each would exert if it were alone in the flask:

$$P_{total} = P_{O_2} + P_{N_2}$$

Because the O_2 exerts a pressure of 444 Torr when in a 10.0-L flask by itself, its partial pressure is 444 Torr. The N_2 exerts a pressure of 562 Torr when alone in a 5.00-L flask. When the N_2 is placed in a 10.0-L flask, Boyle's law tells us its pressure is halved, so the partial pressure of N_2 in the 10.0-L flask is 281 Torr ($\frac{1}{2}$ of 562 Torr). The total pressure in the flask is the sum of the partial pressures:

$$P_{O_2} = 444 \text{ Torr} \qquad P_{N_2} = 281 \text{ Torr}$$

$$P_{total} = 444 \text{ Torr} + 281 \text{ Torr}$$

$$= 725 \text{ Torr}$$

EXERCISE When 0.50 g of neon gas (Ne) and 0.50 g of argon gas (Ar) are placed in a 3.0-L flask at 27°C, what are the partial pressures of each gas and the total pressure of the mixture of gases, in atmospheres?

[*Answer:* Ne, 0.21 atm; Ar, 0.10 atm; total, 0.31 atm]

PITFALL   Partial quantities in a mixture of gases

In a mixture of gases, each gas has its own partial pressure, but not its own partial temperature or partial volume. In the mixture, each gas occupies the same volume and has the same temperature. Of the three parameters P, V, and T, only the pressure of each gas has its own partial value, and only for the pressure is the total value the sum of the individual partial values.

KEY WORDS   Define or explain each term in a written sentence or two.

Avogadro's principle
Boyle's law
Charles's law
Dalton's law of partial pressures

extrapolation
gas constant
ideal gas
ideal gas law

molar volume
partial pressure
vapor pressure

MOLECULAR MOTION

5.15   DIFFUSION AND EFFUSION

KEY CONCEPT   Graham's law of diffusion and effusion

The rate of diffusion and rate of effusion of a gas are each proportional to the speed of the gas. The time it takes for effusion or diffusion to occur is proportional to 1/rate. These facts lead directly to **Graham's law of effusion,** which relates the relative times it takes for an equal number of moles of two gases at the same temperature and pressure to effuse:

$$\frac{t_A}{t_B} = \sqrt{\frac{\text{molar mass of gas A}}{\text{molar mass of gas B}}}$$

In this equation, t_A is the time it takes for gas A to effuse and t_B is the time for gas B. This equation also approximately describes the relative times of diffusion for two gases at the same temperature and pressure.

EXAMPLE Calculating the time needed for effusion

At a given temperature and pressure, a certain amount of propane (C_3H_8) requires 235 s to effuse through a porous plug. How long will it take an equivalent number of moles of carbon monoxide (CO) to diffuse at the same conditions?

SOLUTION The relationship that expresses the relative times needed for effusion of an equal number of moles of two gases at the same experimental conditions is

$$\frac{t_A}{t_B} = \sqrt{\frac{\text{molar mass of gas A}}{\text{molar mass of gas B}}}$$

The molar mass of propane is $44.09 \ \text{g} \cdot \text{mol}^{-1}$ and that of carbon monoxide is $28.01 \ \text{g} \cdot \text{mol}^{-1}$. Substituting into the equation gives

$$\frac{t_{CO}}{t_{C_3H_8}} = \sqrt{\frac{28.01 \ \cancel{\text{g} \cdot \text{mol}^{-1}}}{44.09 \ \cancel{\text{g} \cdot \text{mol}^{-1}}}} = 0.797$$

Because $t = 235$ s,

$$t_{CO} = (0.797)(235 \ \text{s}) = 187 \ \text{s}$$

As we expect, the lighter carbon monoxide diffuses in less time than the heavier propane.

EXERCISE At a given temperature and pressure, a certain amount of argon (Ar) requires 277s to diffuse through a porous plug. How long will it take an equivalent number of moles of nitrogen (N_2) to diffuse under the same conditions?

[*Answer:* 232 s]

PITFALL Square roots on a calculator

Be certain you know how to get a square root and a square on your calculator before trying the calculations in this section. Many calculators use the same key for these two functions, with one or the other activated by a second function or inv (inverse) key. Try a few simple examples such as the square and square root of 4 and 9 to be certain you understand how your calculator handles these functions.

5.16   THE KINETIC MODEL OF GASES

KEY CONCEPT   The kinetic theory

The kinetic theory has four main postulates:

1. A gas consists of a collection of molecules in continuous chaotic motion.
2. Molecules are infinitely small with much space between them.
3. Molecules move in a straight line until they collide.
4. Molecules do not affect one another except when they collide.

Postulate 2 indicates that a molecule of an ideal gas has approximately zero volume. If it were possible to take a snapshot of a flask containing an ideal gas, most of the volume of the flask would be "empty space." The fact that molecules travel in straight lines until the moment of collision means that the molecules do not attract or repel one another except at the instant of collision. In essence, the motion and behavior of each molecule are unaffected by the presence of the other molecules, except for occasional collisions. The kinetic theory, after much manipulation based on classical physics, predicts that the average speed of molecules (v) is directly proportional to the square root of the absolute temperature:

$$v \propto \sqrt{T}$$

If we wish to use this relationship to calculate the relative speed of gas molecules as a function of temperature, we must remember that it can only be used for one gas at a time. That is, we could not use it to relate the speed of N_2 at 300 K to that of O_2 at 325 K; molecular speed also depends on the molar mass of the gas!

EXAMPLE 1   Interpreting the postulates of the kinetic theory

Of the three gases listed below, which should most closely follow the ideal gas equation?

He: small size, weak intermolecular attractions
CH_4:   medium size, intermediate intermolecular attractions
HCl: large size, strong intermolecular attractions

SOLUTION Based on the kinetic molecular theory, an ideal gas is one that is infinitely small in size and that is not affected by other molecules. A gas with small molecules and weak intermolecular forces is most likely to act ideally. Conversely, when intermolecular attractions are large, molecules have a substantial effect on the motion of nearby molecules. By both criteria (size and effect on nearby molecules), helium should follow the ideal gas law most closely.

EXERCISE As the volume of a sample of gas is decreased, the gas molecules are forced nearer one another. Will a gas act more like an ideal gas when it is enclosed in a large volume or a small volume?

[*Answer:* Large volume]

EXAMPLE 2   Calculating average speeds

A sample of a gas has an average speed of 330 m·s^{-1} at 25°C. What is its average speed at 50°C?

SOLUTION The relationship $v \propto \sqrt{T}$ can be used to solve this problem, but we must remember that the absolute temperature should be used. If we denote the two speeds v_1 and v_2, then the relationship $v \propto \sqrt{T}$ means that

$$\frac{v_2}{v_1} = \sqrt{\frac{T_2}{T_1}}$$

Because $v_1 = 330$ m·s^{-1}, $T_1 = 273 + 50 = 323$ K, $T_2 = 273 + 25 = 298$ K,

$$\frac{v_2}{330 \text{ m·s}^{-1}} = \sqrt{\frac{323 \text{ K}}{298 \text{ K}}} = \sqrt{0.923} = 1.08 \quad \text{and}$$

$$v_2 = 330 \text{ m·s}^{-1} \times 1.08 = 358 \text{ m·s}^{-1}$$

EXERCISE Imagine now that the gas in Example 2 has an average speed of 400 m·s^{-1}. What is its temperature?

[*Answer:* 438 K]

5.17   THE MAXWELL DISTRIBUTION OF SPEEDS

KEY CONCEPT   Gas molecules move at different speeds

For a sample of gas molecules, we have already discussed the fact that the molecules in the sample do not all move at the same speed. Figure 5.30 of the text indicates the nature of the distribution of speeds in typical situations. Be certain that you can interpret the graphs in this figure. The *x*-axis gives the speed; the *y*-axis gives the number of molecules in a particular sample that has that speed. Often the number is a relative number. So, for instance, you might pick out 433 m·s^{-1} on the *x*-axis and find that 1.4% of the molecules in a sample have that speed. These graphs illustrate three important facts about the distribution of molecular speeds in a gas:

1. There is a fairly wide range of speeds.
2. The curve has a tail at high speeds; it is not bell-shaped. This means there are proportionally more molecules with very high speeds than with very low speeds. This effect is most pronounced for light molecules.
3. As temperature increases, a higher fraction of molecules move with above-average speeds.

EXAMPLE   Interpreting the Maxwell distribution of speeds

Assume for a sample of molecules that the peak of the Maxwell curve is at 376 m·s^{-1}. Which of the following might be a reasonable estimate of the average speed of the molecules: 288 m·s^{-1}, 376 m·$^{-1}$s, 479 m·s^{-1}?

SOLUTION From the shape of the Maxwell curve, with a tail at high speeds, we can conclude that the average speed must be above where the curve peaks. Of the three choices given, 479 m·s^{-1} is the only reasonable one.

EXERCISE Do light or heavy molecules have the widest range of speeds at a particular temperature? You may use Figure 5.30 of the text to answer this question.

[*Answer:* Light]

KEY WORDS   Define or explain each term in a written sentence or two.

convection   Graham's law
diffusion   Maxwell distribution of speeds
effusion   mean free path

REAL GASES

5.18   THE IDEAL GAS LAW AS A LIMITING LAW

KEY CONCEPT   Real gases follow ideal gas behavior at low pressures

A comparison between an ideal gas and a **real gas** is presented in the following table.

Ideal Gas	Real Gas
molecules are infinitely small	molecules have a small but finite volume
molecules do not affect one another except during collisions	molecules experience attractive forces at relatively long distances

Although it is true that all gas molecules attract one another to some extent, as the molecules move farther away from one another, the attractive forces become weaker and weaker. At very low pressures, there are very few gas molecules present and the molecules are, therefore, very far apart. Thus, they do not experience any significant intermolecular attractions; in addition, the actual volume of the gas molecules is insignificant relative to the enclosing volume. These are the conditions required (no intermolecular interactions, molecules have "zero" volume) for ideal gas behavior. Thus, as the pressure gets lower, real gases behave very much like an ideal gas; they follow the gas laws.

EXAMPLE   Understanding real gases and limiting behavior

For a (real) gas in which the intermolecular attractions are strong, will the pressure of the gas be larger or smaller than that predicted by the ideal gas law?

SOLUTION Because the molecules attract one another, those molecules about to collide with the walls of the enclosing vessel will be slightly attracted toward all of the molecules in the interior of the flask. This attraction means the molecules about to collide do not hit the wall as hard as they otherwise would. The slightly less forceful collision leads to a lower pressure.

EXERCISE Will a gas consisting of molecules with a large volume have a lower or higher pressure than that predicted by the ideal gas law?

[*Answer:* Higher]

5.19   THE JOULE-THOMSON EFFECT

KEY CONCEPT   At low temperatures, gases cool when they expand

Because of the interactions between molecules, when real gases expand, they cool. For this to occur, the gas must be at a relatively low temperature to begin with. For most gases, room temperature is generally a low enough temperature for the cooling to be observed.

EXAMPLE   Understanding the Joule-Thomson effect

The noble gases tend to act less and less ideally as they get larger. That is, He acts most like an ideal gas and Rn the least. Which of the noble gases is most likely to show a large Joule-Thomson effect?

SOLUTION The Joule-Thomson effects occurs because real gas molecules attract one another; this is non-ideal behavior. The stronger the attraction, the larger the Joule-Thomson effect is likely to be. Radon (Rn), the least ideal of the noble gases, should show the largest Joule-Thomson effect.

EXERCISE Qualitatively, what happens to the average speed of gas molecules when the Joule-Thomson effect is being observed?

[*Answer:* The average speed gets lower as the gas cools.]

KEY WORDS   Define or explain each term in a written sentence or two.

intermolecular forces
Joule-Thomson effect
limiting law

DESCRIPTIVE CHEMISTRY TO REMEMBER

- 1 Torr = 1 mmHg
- 1 atm = 760 Torr exactly
- The escape of a substance through a small hole is called **effusion.**
- The spreading of one substance through another is called **diffusion.**
- **Uranium** is enriched in U-235 by producing the volatile solid UF_6 and allowing the vapor to

diffuse through a series of porous barriers. The molecules containing the lighter U-235 diffuse slightly faster than those containing U-238, so an enrichment is accomplished.
- Solid carbon dioxide is called **dry ice.**
- The cooling of a gas as it expands is called the **Joule-Thomson effect.**
- The volume of 1 mol of an ideal gas at standard temperature and pressure (STP, 273.15 K, 1 atm) is **22.4 L.**
- The volume of 1 mol of an ideal gas at 25.00°C and 1.000 atm is 24.47 L.

CHEMICAL EQUATIONS TO KNOW

- **Sulfur dioxide** is a product when solid sulfur is burned in oxygen:

$$S_8(s) + 8O_2(g) \longrightarrow 8SO_2(g)$$

- **Potassium superoxide** reacts with carbon dioxide to form potassium carbonate and oxygen:

$$4KO_2(s) + 2CO_2(g) \longrightarrow 2K_2CO_3(s) + 3O_2(g)$$

- An explosion of **nitroglycerin** ($C_3H_5N_3O_9$) generates a large amount of gaseous products, 29 mol of gas for every 4 mol of nitroglycerin:

$$4C_3H_5N_3O_9(l) \longrightarrow 6N_2(g) + O_2(g) + 12CO_2(g) + 10H_2O(g)$$

- Lead azide [$Pb(N_3)_2$] explodes to release three moles of gas per mole of lead azide:

$$Pb(N_3)_2(s) \longrightarrow Pb(s) + 3N_2(g)$$

MATHEMATICAL EQUATIONS TO KNOW AND UNDERSTAND

$P = \dfrac{\text{force}}{\text{area}}$	definition of pressure
$P_1V_1 = P_2V_2$	Boyle's law for a sample of gas at constant temperature
$\dfrac{V_1}{T_1} = \dfrac{V_2}{T_2}$	Charles's law, for a sample of gas at constant pressure
$\dfrac{V_1}{n_1} = \dfrac{V_2}{n_2}$	Avogadro's principle
$\dfrac{P_1V_1}{T_1} = \dfrac{P_2V_2}{T_2}$	combined gas law
$PV = nRT$	ideal gas law
$d = \dfrac{P \times \text{molar mass}}{RT}$	density of an ideal gas
$\dfrac{t_A}{t_B} = \sqrt{\dfrac{\text{molar mass of gas A}}{\text{molar mass of gas B}}}$	Graham's law of effusion
$v \propto \sqrt{T}$	average speed of a gas

SELF-TEST EXERCISES

The nature of gases

1. In which state of matter are molecules very far apart and free to move from place to place?

(a) solid   (b) liquid   (c) gas   (d) none of these

2. Which state of matter is the least compressible?

(a) solid   (b) liquid   (c) gas   (d) none of these

3. Which of the following is a measure of the average speed of molecules in a gas?

(a) pressure   (b) volume   (c) moles

(d) temperature   (e) gas constant

4. At which temperature do the molecules in a gas move the slowest?

(a) 0°C   (b) 0 K   (c) 0°F   (d) 32°F   (e) 273 K

5. Pressure is defined as

(a) force   (b) mass $\times$ acceleration

(c) change in speed/time   (d) force/area

6. A pressure of 683 Torr (760 Torr = 1 atm) is equivalent to

(a) 0.899 atm   (b) 1.00 atm   (c) 5.19×10^5 atm

(d) 1.93×10^{-6} atm   (e) 1.11 atm

7. What is atmospheric pressure for the mercury barometer pictured?

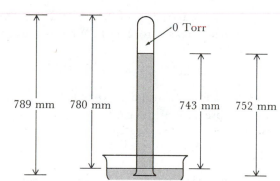

(a) 780 Torr   (b) 752 Torr   (c) 789 Torr

(d) 37 Torr   (e) 743 Torr

8. Which of the following is the SI unit of pressure?

(a) Torr   (b) atmosphere   (c) pascal   (d) mmHg   (e) $lb \cdot in^{-2}$

The gas laws

9. A sample of gas with a volume of 163 mL at 500 Torr undergoes a change in pressure to 600 Torr at constant temperature. What is the new volume of the gas (in mL)?

(a) 136 mL   (b) 196 mL   (c) 163 mL   (d) 201 mL   (e) 144 mL

10. What is the new volume of gas if 24 L, originally at 3.0 atm, undergo a change in pressure to 2.0 atm? Assume the temperature is constant.

(a) 24 L   (b) 12 L   (c) 36 L   (d) 16 L   (e) 8.0 L

11. 1.00 L of an ideal gas, originally at 710 Torr, is compressed to 0.830 L at constant temperature. What is the new pressure of the gas?

(a) 630 Torr   (b) 855 Torr   (c) 710 Torr

(d) 589 Torr   (e) 760 Torr

12. 2.22 mL of a gas at 23.0°C are heated to 100°C at constant pressure. What is the new volume of the gas?

(a) 2.78 mL   (b) 1.76 mL   (c) 9.65 mL

(d) 2.24 mL   (e) 1.89 mL

13. When 125 mL of an ideal gas at 300°C undergo a change in volume to 200 mL at constant pressure, what is the new temperature of the gas?

(a) 300°C   (b) 644°C   (c) 480°C   (d) −14.9°C   (e) 188°C

14. A gas with volume V_1 undergoes a doubling of both the pressure and absolute temperature. What is the new volume of the gas?

(a) V_1 (b) $2V_1$ (c) $V_1/4$ (d) $V_1/2$ (e) $4V_1$

15. A gas with volume V_1 undergoes a change in temperature from 20.0°C to 40.0°C. What is the new volume of the gas?

(a) $2.00V_1$ (b) V_1 (c) $V_1/2.00$ (d) $2.22V_1$ (e) $1.07V_1$

16. 6.00 L of an ideal gas at 1.00 atm pressure and 273 K undergo a change in temperature and pressure to 1.63 atm and 255 K. What is the new volume of the gas?

(a) 3.94 L   (b) 9.14 L   (c) 10.5 L   (d) 3.44 L   (e) 3.59 L

17. A sample of gas at 735 Torr occupies a volume of 2.55 L at 25°C. The pressure is suddenly increased to 800 Torr and the volume decreased to 2.45 L. What is the new temperature of the gas?

(a) −10°C   (b) 12°C   (c) 26°C   (d) 65°C   (e) 39°C

18. 12.0 L of an ideal gas at 60.0°C and 2.25 atm are expanded to 16.0 L and increased in temperature to 120°C. What is the new pressure of the gas?

(a) 1.99 atm   (b) 3.38 atm   (c) 1.43 atm

(d) 2.54 atm   (e) 3.54 atm

19. 2.00 mol of O_2 occupy 40.0 L at a certain temperature and pressure. What volume do 2.00 mol of CH_4 occupy at the same conditions?

(a) 80.0 L   (b) 20.0 L   (c) 40.00 L   (d) 10.0 L   (e) 160 L

20. What does the symbol n stand for in the ideal gas law, $PV = nRT$?

(a) temperature   (b) pressure   (c) moles of sample

(d) gas constant   (e) volume

21. What is the volume of 113 g of oxygen gas (O_2) at 25°C and 770 Torr?

(a) 0.00941 L   (b) 85.2 L   (c) 7.15 L   (d) 0.112 L   (e) 22.4 L

22. What is the pressure of 263 g of ammonia gas (NH_3) enclosed in a 12.0-L flask at 250°C?

(a) 1.0 atm   (b) 55.3 atm   (c) 318 atm

(d) 26.5 atm   (e) 665 atm

23. What volume is occupied by 7.72 g of methane (CH_4) at 35.2°C and 742 Torr?

(a) 0.0165 L   (b) 12.5 L   (c) 1.42 L   (d) 4.43 L   (e) 7.77 L

24. 0.308 g of acetone (C_3H_6O) occupies 175 mL at 752 Torr. What is the temperature of the sample?

(a) 671°C   (b) 125°C   (c) 303°C   (d) 223°C   (e) 37°C

25. What is the density of acetylene (C_2H_2) at 125°C and 0.850 atm?

(a) $1.48 \text{ g} \cdot \text{L}^{-1}$ (b) $2.16 \text{ g} \cdot \text{L}^{-1}$ (c) $1.88 \text{ g} \cdot \text{L}^{-1}$

(d) $0.225 \text{ g} \cdot \text{L}^{-1}$ (e) $0.678 \text{ g} \cdot \text{L}^{-1}$

26. What is the molar mass of a gas that has a density of $3.22 \text{ g} \cdot \text{L}^{-1}$ at 423 K and 1.11 atm?

(a) $223 \text{ g} \cdot \text{mol}^{-1}$ (b) $34.7 \text{ g} \cdot \text{mol}^{-1}$ (c) $154 \text{ g} \cdot \text{mol}^{-1}$

(d) $101 \text{ g} \cdot \text{mol}^{-1}$ (e) $274 \text{ g} \cdot \text{mol}^{-1}$

27. 0.447 g of a common chlorofluorocarbon in a 100-mL flask exerts a pressure of 758 Torr at 100°C. What is its molar mass?

(a) $1.00 \times 10^3 \text{ g} \cdot \text{mol}^{-1}$ (b) $186 \text{ g} \cdot \text{mol}^{-1}$ (c) $137 \text{ g} \cdot \text{mol}^{-1}$

(d) $91.3 \text{ g} \cdot \text{mol}^{-1}$ (e) $121 \text{ g} \cdot \text{mol}^{-1}$

28. Use the fact that the molar volume of an ideal gas is 22.4 L to calculate the volume occupied by 60.0 g of hydrogen sulfide at STP.

(a) 39.4 L   (b) 23.9 L   (c) 22.4 L   (d) 31.6 L   (e) 18.0 L

29. A 0.643-g sample of a gas occupies 100.0 mL at STP. What is the molar mass of the gas?

(a) $64.3 \text{ g} \cdot \text{mol}^{-1}$ (b) $144 \text{ g} \cdot \text{mol}^{-1}$ (c) $162 \text{ g} \cdot \text{mol}^{-1}$

(d) $2.87 \text{ g} \cdot \text{mol}^{-1}$ (e) $187 \text{ g} \cdot \text{mol}^{-1}$

30. A 0.777-g sample of a gas occupies 100.0 mL at 25.0°C and 1 atm. Use the fact that 1.000 mol of an ideal gas occupies 24.47 L at these conditions to calculate the molar mass of the gas, in $\text{g} \cdot \text{mol}^{-1}$.

(a) $318 \text{ g} \cdot \text{mol}^{-1}$ (b) $63 \text{ g} \cdot \text{mol}^{-1}$ (c) $190 \text{ g} \cdot \text{mol}^{-1}$

(d) $121 \text{ g} \cdot \text{mol}^{-1}$ (e) $224 \text{ g} \cdot \text{mol}^{-1}$

31. Use the fact that 1.000 mol of an ideal gas occupies 24.47 L at 25.00°C and 1 atm to calculate the volume (in mL) occupied by 0.567 g of N_2 at these conditions.

(a) 13.9 mL   (b) 43.1 mL   (c) 127 mL   (d) 495 mL   (e) 56.7 mL

32. 0.594 g of carbon is burned in oxygen to produce CO_2. What volume of CO_2, measured at 25.0°C and 1 atm, is formed?

$$C(s) + O_2(g) \longrightarrow CO_2(g)$$

(a) 14.6 L   (b) 1.21 L   (c) 8.88 L   (d) 22.4 L   (e) 15.3 L

33. How many liters of gaseous H_2O are formed by the reaction of 0.36 L O_2? All volumes are measured at 25°C and 1 atm.

$$2H_2(g) + O_2(g) \longrightarrow 2H_2O(g)$$

(a) 0.36 L   (b) 0.72 L   (c) 0.18 L   (d) 0.090 L   (e) 1.44 L

34. How many liters of N_2 (measured at STP) are produced by the reaction of 1.22 kg lead azide, $Pb(N_3)_2$?

$$Pb(N_3)_2(s) \longrightarrow Pb(s) + 3N_2(g)$$

(a) 93.8 L   (b) 281 L   (c) 8.23 L   (d) 31.3 L   (e) 162 L

35. A 25.0-L flask contains 20.0 g of carbon monoxide and 20.0 g of carbon dioxide at 298 K. What is the partial pressure of the carbon monoxide?

(a) 622 Torr   (b) 380 Torr   (c) 442 Torr

(d) 245 Torr   (e) 530 Torr

36. A flask containing two gases, A and B, is at a temperature such that the pressure of A, if it were in the flask alone, would be 655 Torr, and the pressure of B, if it were in the flask alone, would be 22 Torr. What is the pressure of the mixture of the two gases in the flask?

(a) 1.3×10^5 Torr   (b) 633 Torr   (c) 677 Torr

(d) 33 Torr   (e) 760 Torr

37. 115 g each of fluorine and chlorine are enclosed in a 250-mL flask at $-30°C$. What is the partial pressure of each of the gases? Answers are given as "partial pressure F_2, partial pressure Cl_2." Assume ideal gas behavior.

(a) 242 atm, 129 atm   (b) 302 atm, 161 atm   (c) 242 atm, 161 atm

(d) 302 atm, 129 atm   (e) 185 atm, 185 atm

38. For the mixture described in Exercise 37, what is the total pressure?

(a) 113 atm   (b) 141 atm   (c) 371 atm   (d) 463 atm   (e) 760 atm

39. A sample of oxygen gas is collected in a bottle, over water, at $25°C$ at a total pressure of 748.2 Torr. What is the actual pressure of the oxygen molecules in the bottle? The vapor pressure of water at $25°C$ is 23.8 Torr.

(a) 760.0 Torr   (b) 748.2 Torr   (c) 736.2 Torr

(d) 724.4 Torr   (e) 772.0 Torr

40. 285 mL of oxygen gas were collected in a gas bottle, over water, at $16.0°C$ at a total pressure of 761.1 Torr. How many millimoles of oxygen were collected? The vapor pressure of water at $16.0°C$ is 13.6 Torr.

(a) 12.2 mmol   (b) 11.8 mmol   (c) 221 mmol

(d) 12.0 mmol   (e) 213 mmol

Molecular motion

41. Assume it takes 122 s for 0.100 mol of Ne to diffuse from one point to another. How long will it take for 0.100 mol of Ar to diffuse the same distance?

(a) 242 s   (b) 263 s   (c) 12.2 s   (d) 478 s   (e) 172 s

42. A sample of an unknown gas takes 434 s to diffuse through a porous plug. An equal amount of N_2 takes 176 s to diffuse through the same plug at the same conditions. What is the molar mass of the unknown?

(a) $6.30 \text{ g} \cdot \text{mol}^{-1}$ (b) $115 \text{ g} \cdot \text{mol}^{-1}$ (c) $161 \text{ g} \cdot \text{mol}^{-1}$

(d) $170 \text{ g} \cdot \text{mol}^{-1}$ (e) $69.0 \text{ g} \cdot \text{mol}^{-1}$

43. By what factor does the average speed of any ideal gas change when the temperature is increased from $100°C$ to $200°C$?

(a) 1.00   (b) 2.00   (c) 1.13   (d) 1.28   (e) 4.00

44. Four of the following assumptions are used to model gases. Which one is not used?

(a) Gas molecules never change their direction of motion.

(b) Gas molecules are infinitely small.

(c) Gas molecules undergo random motion.

(d) Gas molecules move in straight lines until they collide.

(e) Gas molecules do not influence one another, except during collisions.

45. Imagine the molecules of a gas have average speed v. If the absolute temperature of the gas is tripled, what is the new average speed of the molecules?
(a) $9.00v$ (b) $3.00v$ (c) $1.73v$ (d) v (e) $v/3.00$

46. The average speed of the molecules in a gas is $651 \text{ m} \cdot \text{s}^{-1}$ at 25.0°C. What temperature must the gas be changed to in order to change the average speed to $733 \text{ m} \cdot \text{s}^{-1}$?
(a) 28.1°C   (b) 62.5°C   (c) 31.7°C   (d) 19.7°C   (e) 105°C

47. Only one of the following statements regarding the speed of the molecules in a gas is false. Which one?
(a) The molecules have a range of speeds.
(b) As the temperature is increased, the range of speeds increases.
(c) At low temperatures, most molecules have speeds close to the average speed.
(d) At very high partial pressures, all the molecules travel at the same speed.
(e) Hot gases have a high proportion of very fast molecules.

Real gases

48. Which type of molecules is most likely to behave like an ideal gas?
(a) Large molecules with large intermolecular interactions.
(b) Small molecules with large intermolecular interactions.
(c) Large molecules with small intermolecular interactions.
(d) Small molecules with small intermolecular interactions.

49. Under normal circumstances, when a gas expands, it
(a) cools   (b) warms   (c) stays at the same temperature

Descriptive chemistry

50. The volume of 1 mol of an ideal gas at 25.00°C and 1.000 atm is
(a) 1.000 L   (b) 22.41 L   (c) 24.47 L
(d) 10.00 L   (e) depends on the gas

51. What product is formed when sulfur is oxidized in air?
$$S_8(s) + O_2(g) \longrightarrow ?$$
(a) SO_2 (b) SO_3 (c) SO (d) S_2O_8 (e) S_2O_2

52. Which compound, when reacted with carbon dioxide, produces oxygen (O_2) as a product?
(a) K_2O_2 (b) KCl (c) KO_2 (d) K_2O (e) KOH

53. What is the formula of lead(II) azide?
(a) PbN (b) PbN_3 (c) $Pb_2(N_3)_2$ (d) Pb_2N_3 (e) $Pb(N_3)_2$

54. What low temperature can be reached by adding solid chips of carbon dioxide to a low freezing liquid?
(a) -77°C   (b) -35°C   (c) 0°C   (d) -196°C   (e) -105°C

CHAPTER 6

THERMOCHEMISTRY: THE FIRE WITHIN

Human technology makes use of fuels for transportation, communication, and for life itself. To make wise use of the fuels that nature provides and to manage the energy supply we have available requires an understanding of how energy is released during a chemical reaction. The energetics of chemical reactions are investigated in this chapter.

ENERGY, HEAT, AND ENTHALPY

6.1 TRANSFER OF ENERGY AS HEAT

KEY CONCEPT Heat flows from hot objects to cool objects

Imagine that you take a hot copper rod and drop it into a bucket of cold water. The result is familiar; **heat** flows from the hot rod to the cold water. The temperature of the rod decreases and the temperature of the water increases until both reach the same temperature. Because we know that the temperature of an object is related to the motion of molecules, we can conclude that the copper atoms start moving (vibrating) more slowly as the water molecules start moving more rapidly. Heat, then, is the flow of energy (as molecular motion) from hot objects to cold objects. When we use the term *heat*, we are speaking not of energy itself, but of the flow of energy from one place to another.

EXAMPLE Understanding heat flow

In the process of putting the hot copper rod into the cold water, let's assume that the rod loses 25 units of energy. What can we say about the amount of energy gained by the water? Assume no energy is lost to the atmosphere or bucket containing the water.

SOLUTION The law of conservation of energy, one of the most important laws in all of science, demands that the 25 units of energy must go somewhere; it cannot mysteriously disappear. The most reasonable (and correct) assumption is that it goes into the water. Thus, we can say that the water must gain 25 units of energy.

EXERCISE What is an important difference between internal energy and temperature?

[*Answer:* Internal energy is an extensive property; temperature is intensive. They are related, but they are not the same.]

6.2 EXOTHERMIC AND ENDOTHERMIC PROCESSES

KEY CONCEPT Chemical reactions may release or consume heat

One way to classify a chemical reaction (or physical process) is based on whether the matter surrounding the chemical reaction (for instance, the atmosphere or the solution in which a reaction occurs) warms up or cools down as the reaction occurs. When the surroundings warm up, the chemical reaction, as it occurs, releases energy to the surroundings as heat; the molecular motion of the molecules in the surroundings increases. Such a reaction is called exothermic. When the surroundings cool down, the chemical reaction, as it occurs, consumes energy from the surroundings as heat; the molecular motion of the molecules in the surroundings decreases. Such a reaction is called endothermic. It should be understood that these heat transfers are required if the reaction is to occur; without the release (exothermic) or consumption (endothermic) of energy, the reaction would not occur.

EXAMPLE   Determining the exo- or endothermicity of a reaction

Is the burning of a fuel, such as gasoline or natural gas, exothermic or endothermic?

SOLUTION When a fuel is burned, the surroundings warm up significantly. This outcome means that the chemical reaction releases heat to the surroundings as it occurs. The burning of a fuel is, therefore, exothermic.

EXERCISE Is the freezing of water to ice an exothermic or endothermic process?

[*Answer:* Exothermic]

EXAMPLE 2   Recognizing the direction of energy transfer

A refrigerator does its job by taking advantage of how energy is transferred when a refrigerant changes its physical state. Use the term *exothermic* or *endothermic* to describe the change in the physical state of the refrigerant that occurs in the coils inside a refrigerator.

SOLUTION The purpose of the refrigerant is to remove energy from the inside of the refrigerator and release it (the energy) to the outside. An *endothermic* process (usually a vaporization) occurs in the coils inside the refrigerator; energy (as heat) is absorbed from the contents of the refrigerator as the vaporization occurs, so the refrigerator cools. This energy is later released to the outside of the refrigerator by an exothermic process (usually a condensation); you can sense this release by touching the coils on the back of your refrigerator.

EXERCISE When coming out of the water after a swim, you normally feel cool. This feeling is due to the evaporation of the water on your skin. Is evaporation an endothermic or exothermic process?

[*Answer:* Endothermic]

6.3   MEASURING HEAT TRANSFER

KEY CONCEPT A   The units of energy

Two common units of energy are the **joule** and the **calorie.** The relation between these units is

$$1 \text{ cal} = 4.184 \text{ J}   (\text{exactly})$$

The kilojoule (1 kJ = 1000 J) and kilocalorie (1 kcal = 1000 cal) are also commonly used. Because 4.184 J is the heat required to raise the temperature of 1 g of water by 1°C, 1 cal is the heat required to raise the temperature of 1 g of water by 1°C.

EXAMPLE   Calculating an energy transfer

When 50 mL of an aqueous solution of sodium hydroxide are mixed with 50 mL of an aqueous solution of hydrochloric acid, a reaction occurs and the temperature of the resulting 100 mL of solution increases by 3.8°C. How much heat (in joules) was released by the reaction. Is the reaction exothermic or endothermic? Assume the solution is very dilute, so it behaves like water.

SOLUTION We know that when 1 g of water absorbs 4.184 J of heat, the temperature is increased by 1°C. This fact translates into the conversion factor:

$$\frac{4.184\ \text{J}}{1\ \text{g °C}}$$

Thus, if we know the mass of water and the temperature change, we can calculate the heat absorbed. The volume of water is 100 mL, and we assume it has a density of $1.00\ \text{g}\cdot\text{mL}^{-1}$; so the mass of water is 100 g. Thus, the heat absorbed is

$$100\ \cancel{\text{g}} \times 3.8\,\cancel{°C} \times \frac{4.184\ \text{J}}{1\ \cancel{\text{g}}\,\cancel{°C}} = 1.6 \times 10^3\ \text{J}$$

$$= 1.6\ \text{kJ}$$

Because the temperature of the water increases, energy (as heat) is released, and the reaction is exothermic.

EXERCISE When a beaker containing 750.0 g of water is placed inside a refrigerator, the temperature of the water changes from 23.2°C to 14.8°C. How much energy did the water lose?

[*Answer:* 26 kJ]

KEY CONCEPT B Heat capacity and specific heat capacity

The **heat capacity** is an extensive property (that is, it depends on the size of the sample) that relates the heat transfer of a specific sample to the temperature change the sample undergoes:

$$\text{Heat transfer} = \text{heat capacity} \times \text{change in temperature}$$
$$\text{Joule} \qquad\qquad \text{Joule}\cdot\text{K}^{-1} \qquad\qquad \text{K}$$

The **specific heat capacity** is an intensive property (that is, it does not depend on the size of the sample) that relates the heat transfer of a sample of a substance of given mass to the temperature change the sample undergoes:

$$\text{Heat transfer} = \text{change in temperature} \times \text{mass of sample} \times \text{specific heat capacity}$$
$$\text{Joule} \qquad\qquad \text{K} \qquad\qquad\qquad \text{g} \qquad\qquad \text{Joule}\cdot\text{g}^{-1}\cdot\text{K}^{-1}$$

Because the Celsius degree and the kelvin are the same size, the change in °C is the same as the change in absolute temperature, and either scale may be used to express the change in temperature.

EXAMPLE 1 Using the specific heat capacity

How much heat must be supplied to a 870-g piece of copper to warm it from 25°C to 432°C? The specific heat capacity of copper is $0.38\ \text{J}\cdot\text{g}^{-1}\cdot\text{K}^{-1}$.

SOLUTION The relationship that relates the heat gain to the temperature change of a sample is

$$\text{Heat} = \text{change in temperature} \times \text{mass of sample} \times \text{specific heat capacity}$$

In this problem, mass = 870 g, specific heat capacity = $0.38\ \text{J}\cdot\text{g}^{-1}\cdot\text{K}^{-1}$, and change in temperature (ΔT) = 432°C − 25°C = 407°C. A ΔT of 407°C corresponds to a ΔT of 407 K, so

$$\text{Heat} = 407\ \cancel{\text{K}} \times 870\ \cancel{\text{g}} \times 0.38\ \text{J}\cdot\cancel{\text{g}^{-1}}\cdot\cancel{\text{K}^{-1}}$$

$$= 1.3 \times 10^5\ \text{J}$$

$$= 1.3 \times 10^2\ \text{kJ}$$

EXERCISE How much heat is required to raise the temperature of 25.0 g of ethanol from 0.0°C to 100.0°C? The specific heat capacity of ethanol is $2.42\ \text{J}\cdot\text{g}^{-1}\cdot\text{K}^{-1}$.

[*Answer:* 6.05 kJ]

EXAMPLE 2   Using the specific heat capacity

A 225-g sample of benzene at 55.3°C is warmed by the addition of 432 J of heat. What is the new temperature of the sample? The specific heat of benzene is $1.05 \text{ J} \cdot \text{g}^{-1} \cdot \text{K}^{-1}$.

SOLUTION The relationship between the heat withdrawn from or added to a sample and the change in temperature of the sample is

$$\text{Heat transfer} = \text{change in temperature} \times \text{mass of sample} \times \text{specific heat capacity}$$

In this problem, mass = 225 g, specific heat capacity = $1.05 \text{ J} \cdot \text{g}^{-1} \cdot \text{K}^{-1}$, and heat transfer = 432 J; the change in temperature is the unknown in the equation:

$$432 \text{ J} = \text{change in temperature} \times 225 \text{ g} \times 1.05 \text{ J} \cdot \text{g}^{-1} \cdot \text{K}^{-1}$$

$$\text{Change in temperature} = \frac{432 \cancel{\text{J}}}{1.05 \cancel{\text{J}} \cdot \cancel{\text{g}^{-1}} \cdot \cancel{\text{K}^{-1}} \times 225 \cancel{\text{g}}}$$

$$= 1.83 \text{ K}$$

EXERCISE A 166-g sample of water at 88.8°C is warmed by the addition of 752 J of heat. What is the change in temperature and the new temperature of the sample?

[*Answer:* $\Delta T = 1.08°C$; new temperature = 89.9°C]

PITFALL   The units used for the change in temperature ΔT

The kelvin and Celsius scales use different numbers for the temperature of a sample, but because the size of the Celsius degree is the same as the size of the kelvin, the change in temperature in Celsius degrees is the same as the change in kelvins:

$$T(\text{Celsius}) \neq T(\text{kelvins}) \quad \text{but} \quad \Delta T \text{ (celsius)} = \Delta T \text{ (kelvins)}$$

6.4   THERMAL ACCOUNTING: ENTHALPY

KEY CONCEPT A   Heat transfer: Exothermic versus endothermic reactions

It is common practice to run reactions in a vessel open to the atmosphere so that the pressure on the reaction system stays constant at atmospheric pressure. Under conditions of constant pressure, the heat transfer that occurs is attributed to a change in **enthalpy** (ΔH) of the chemical system.

$$\underset{\Delta H}{\text{Change in enthalpy of system}} = \underset{q_\text{p}}{\text{heat transfer that accompanies reaction}}$$

It is important to realize that the enthalpy is a type of energy and, therefore, has units of energy. The change in enthalpy is defined as

$$\Delta H = \text{final value of enthalpy} - \text{initial value of enthalpy}$$

When the initial enthalpy is greater than the final enthalpy, the preceding equation tells us that ΔH is negative; in this case, heat is lost to the surroundings. When the initial enthalpy is smaller than the final enthalpy, ΔH is positive and the surroundings transfer heat to the chemical system. Thus, a negative change in enthalpy corresponds to an exothermic process and a positive change in enthalpy to an endothermic process.

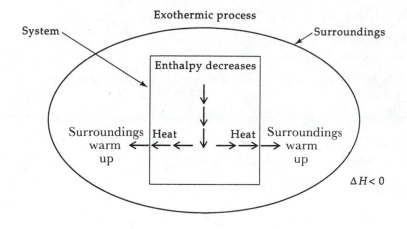

Exothermic process

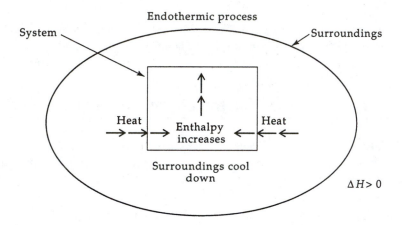

Endothermic process

EXAMPLE 1 Using the definition of enthalpy

A chemical reaction run in an open beaker releases 1.23 kJ of heat. The same reaction run in a closed stainless steel cylinder releases 1.25 kJ. What is ΔH for the reaction?

SOLUTION The change in enthalpy for a chemical reaction is the heat released or absorbed when the reaction is run at constant pressure. When a reaction is run in an open container, the pressure is constant at atmospheric pressure, and the heat released or absorbed is ΔH. Because in this example heat is released, a negative sign is associated with the change. Thus, $\Delta H = -1.23$ kJ.

EXERCISE A reaction run at constant volume absorbs 3.55 kJ of heat from the surroundings. When run in an open container at atmospheric pressure, the same reaction absorbs 3.47 kJ of heat. What is ΔH for the reaction?

[*Answer:* +3.47 kJ]

EXAMPLE 2 Classifying a reaction as exothermic or endothermic

A chemical reaction occurs at constant pressure in which 11 kJ of energy (as heat) leaves the system and enters the surroundings. What is ΔH for the reaction? Classify the reaction as endothermic or exothermic and discuss what happens to the temperature of the surroundings.

SOLUTION Because the reaction is run at constant pressure, the heat gained or lost is equal to the change in the enthalpy of the system, ΔH. A loss of energy from the system as heat corresponds to a decrease in the enthalpy and requires a negative ΔH, so $\Delta H = -11$ kJ. Because energy is transferred to the surroundings, the reaction is exothermic; the surroundings warm up.

EXERCISE A reaction run a constant pressure causes 7.8 kJ of energy as heat to leave the surroundings and enter the system. Is the reaction exothermic or endothermic? What happens to the temperature of the surroundings? What is ΔH?

[*Answer:* Endothermic; temperature of surroundings decreases; $\Delta H = +7.8$ kJ]

KEY CONCEPT B   State properties

A **state property** of a system is a property that depends only on the present condition of the system, not on how it got there. The enthalpy, volume, temperature, and pressure of a system are state properties. As an example, let's consider starting with a gas at a pressure of 700 Torr, and changing the pressure to 750 Torr in two different ways, by path (A + B) and by path C:

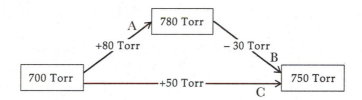

By path (A + B), we change the pressure in two steps. First, we increase the pressure by 80 Torr to get 780 Torr; second, we decrease it by 30 Torr to arrive at the final pressure of 750 Torr. By path C, we simply increase the pressure by 50 Torr to get to the final pressure of 750 Torr. The final pressure is 750 Torr in both cases and does not depend on the path taken. We express the change in a state property by subtracting the value of the property before the change (the initial value) from the value of the property after the change (the final value). A Greek upper case delta (Δ) is used to symbolize a change, so ΔP means change in pressure:

$$\Delta P = P_{final} - P_{initial}$$

Note that, with this definition, a change from a higher to a lower value for a property results in a negative change, and that a change from a lower to a higher value gives a positive change.

Change	Initial Value	Final Value	Δ Value
change T from 350 K to 325 K	350 K	325 K	-25 K
change P from 700 Torr to 750 Torr	700 Torr	750 Torr	$+50$ Torr

EXAMPLE   Calculating changes in state properties

A sample of copper at a temperature of 250 K undergoes the following changes in temperature, in the order given: (1) an increase of 23 K, to 273 K; (2) a decrease of 7 K, to 266 K; (3) a decrease of 12 K, to 254 K; (4) an increase of 9 K, to 263 K; (5) and an increase of 16 K, to 279 K. What is ΔT for the copper?

SOLUTION Because temperature is a state property, if we know the initial and final temperatures, we need not be concerned with the individual changes that got the copper to its final temperature. We may use the formula

$$\Delta T = T_{final} - T_{initial}$$

From the data given, $T_{final} = 279$ K and $T_{initial} = 250$ K; thus,

$$\Delta T = 279 \text{ K} - 250 \text{ K}$$

$$= +29 \text{ K}$$

6.5 VAPORIZATION

KEY CONCEPT   A substance's enthalpy depends on its physical state

Although it is impossible to measure the enthalpy of a substance (only enthalpy *changes* can be measured), it is safe to say that for any substance at a given temperature

$$H_{vapor} > H_{liquid} > H_{solid}$$

Thus, a phase change is accompanied by a change in enthalpy, which is manifested by the consumption or release of heat. The change in enthalpy for a phase change is given by

$$\Delta H = H_{final} - H_{initial}$$

The following table summarizes the enthalpies of phase change.

Phase Change	Enthalpy Change	Exo- or Endothermic	Value for Water (kJ·mol⁻¹)
vaporization	$\Delta H_{vap} = H_{vapor} - H_{liquid}$	endothermic	+40.7
condensation*	$\Delta H = H_{liquid} - H_{vapor}$	exothermic	−40.7
melting	$\Delta H_{fus} = H_{liquid} - H_{solid}$	endothermic	+6.01
freezing	$\Delta H = H_{solid} - H_{liquid}$	exothermic	−6.01
sublimation	$\Delta H_{sub} = H_{vapor} - H_{solid}$	endothermic	+46.7
condensation*	$\Delta H = H_{solid} - H_{vapor}$	exothermic	−46.7

*The term *condensation* refers to the change from vapor to liquid and to the change from vapor to solid.

EXAMPLE   Measuring the molar enthalpy of melting

Suppose it takes 15.0 kJ of heat to melt 45.0 g of ice at its melting temperature of 0°C. What is the enthalpy of fusion of water at 0°C in kilojoules per mole?

SOLUTION The enthalpy of fusion is the heat needed to melt 1 mol of H_2O at its melting point. The problem does not give us the heat for 1 mol, but for 45.0 g. To get the enthalpy for 1 mol, we must first find the number of moles of H_2O in 45.0 g H_2O. For this, we use the molar mass of water, 18.02 g·mol⁻¹:

$$45.0 \text{ g} \times \frac{1 \text{ mol } H_2O}{18.02 \text{ g}} = 2.50 \text{ mol } H_2O$$

To calculate the heat per mole, we divide the heat required for the 2.50 mol (45.0 g) by 2.50 mol:

$$\Delta H_{fus} = \frac{15.0 \text{ kJ}}{2.50 \text{ mol}}$$

$$= 6.00 \text{ kJ·mol}^{-1}$$

EXERCISE It takes 3.4 kJ of heat to melt 56.3 g of iodine (I_2) at its melting point of 387°C. Calculate the enthalpy of melting of iodine in kilojoules per mole.

[*Answer:* 15 kJ·mol⁻¹]

KEY CONCEPT   The enthalpy for a reverse process

Because enthalpy is a state property, the enthalpy change of the reverse of a process is the negative of the enthalpy change for the forward process. For instance, the enthalpy of the melting of water is $+6.01 \text{ kJ} \cdot \text{mol}^{-1}$; therefore, the enthalpy of freezing is $-6.01 \text{ kJ} \cdot \text{mol}^{-1}$.

$$H_2O(s) \longrightarrow H_2O(l) \qquad \Delta H = +6.01 \text{ kJ}$$

$$H_2O(l) \longrightarrow H_2O(s) \qquad \Delta H = -6.01 \text{ kJ}$$

This change of sign for the enthalpy change for a reverse process works for chemical reactions as well as for phase changes. In fact, it works for any process.

EXAMPLE   Calculating the enthalpy change for a reverse process

Table 6.2 of the text states that it takes 8.2 kJ to vaporize 1 mol of methane (CH_4) at its boiling temperature. How much heat is released when 1 mol of methane condenses from the vapor to the liquid at the same temperature?

SOLUTION   Condensation from the vapor to the liquid is the reverse of vaporization. Thus, the change in enthalpy for one process will be the negative of the change in enthalpy for the other, and 8.2 kJ of heat is released when 1 mol of methane condenses. This is shown by the following equations:

$$CH_4(l) \longrightarrow CH_4(g) \qquad \Delta H = +8.2 \text{ kJ}$$

$$CH_4(g) \longrightarrow CH_4(l) \qquad \Delta H = -8.2 \text{ kJ}$$

EXERCISE   The decomposition of HF(g) requires $542.2 \text{ kJ} \cdot \text{mol}^{-1}$ of heat:

$$2HF(g) \longrightarrow H_2(g) + F_2(g) \qquad \Delta H = +542.2 \text{ kJ}$$

What is ΔH for the reaction

$$H_2(g) + F_2(g) \longrightarrow 2HF(g) \qquad \Delta H = ?$$

[*Answer:* −542.2 kJ]

PITFALL   Enthalpies of physical change and temperatures of physical change

Values for enthalpies of physical change, such as those given in Table 6.2 of the text, are correct only when the phase change occurs at the normal temperature for the change, such as 0°C for the melting or freezing of water. If the phase change occurs at a different temperature, the value for the enthalpy change is different. Many problems involving the change in enthalpy for a phase change will state the temperature to let us know when it is permissible to use the type of data given in Table 6.2. However, the temperature itself is not used in the calculation.

6.7   HEATING CURVES

KEY CONCEPT   The temperature is constant during a phase change

Is it possible to get the temperature of boiling water above 100°C by applying more heat to it? Most of us know that the answer to this question is "no." Applying more heat by turning up your stove or Bunsen burner, for instance, makes the water boil faster, but it will not increase the temperature above the boiling point. Similarly, one could not lower the temperature of a water-ice mixture below

0°C by adding more ice. The energy that is added (or removed) to a sample during a phase change is used to accomplish the phase change, not to change the temperature.

EXAMPLE   Understanding energy effects during a phase change

Explain, on a molecular level, how it is possible to add heat to boiling water and not increase the temperature of the water.

SOLUTION Remember that conservation of energy must occur. The added heat cannot mysteriously disappear; it must do something or go somewhere. In this case, we must recall that the molecules of a liquid are very close together and those of a gas very far apart. Because the water molecules attract one another, it takes energy to move them away from one another. The heat added to the boiling water accomplishes this process. It is used to separate the molecules from one another rather than to make them move faster (and, thereby, increase the temperature).

EXERCISE An air conditioner uses phase changes to cool. What type of phase change must occur inside the room being cooled for the air conditioner to work?

[*Answer:* An evaporation (which is endothermic) must occur. (It cannot be melting, the other endothermic phase change, because a solid could not circulate through the air conditioner's coils. A fluid must be used).]

KEY WORDS   Define or explain each term in a written sentence or two.

calorimeter	joule	sublimation
endothermic reaction	heat capacity	surroundings
energy	law of conservation of energy	system
enthalpy	specific heat capacity	thermochemistry
exothermic reaction	state property	

THE ENTHALPY OF CHEMICAL CHANGE

6.8   REACTION ENTHALPIES

KEY CONCEPT   Enthalpy change for a chemical equation

In general, a chemical reaction is accompanied by a change in enthalpy. The enthalpy change is called the **reaction enthalpy**. The reaction enthalpy is usually written next to the equation for the reaction:

$$C_2H_4(g) + 3O_2(g) \longrightarrow 2CO_2(g) + 2H_2O(g) \qquad \Delta H = -1411 \text{ kJ}$$

This way of indicating the enthalpy has a very specific meaning: It reflects the enthalpy change for the exact reaction that is written. For the preceding reaction, the interpretation is as follows: when 1 mol of C_2H_4 (ethene) in the gaseous state reacts with 3 mol of O_2 in the gaseous state, resulting in the production of 2 mol of CO_2 in the gaseous state and 2 mol of H_2O in the gaseous state, the enthalpy of the reaction system decreases by 1411 kJ, and, therefore, 1411 kJ of heat are released.

EXAMPLE   Using reaction enthalpies

How much heat is released when 25.0 g of propane (C_3H_8) are completely burned?

$$C_3H_8(g) + 5O_2(g) \longrightarrow 3CO_2(g) + 4H_2O(l) \qquad \Delta H = -2220 \text{ kJ}$$

SOLUTION The chemical equation tells us that when 1 mol of propane is burned, 2220 kJ of heat are released:

$$1 \text{ mol } C_3H_8 = 2220 \text{ kJ}$$

This relation can be used as a conversion factor to calculate the heat produced by any amount of propane. The moles of propane are calculated by using its molar mass, $44.09 \text{ g} \cdot \text{mol}^{-1}$:

$$25.0 \text{ g } \cancel{C_3H_8} \times \frac{1 \text{ mol } C_3H_8}{44.09 \cancel{g}} = 0.567 \text{ mol } C_3H_8$$

$$0.567 \cancel{\text{mol } C_3H_8} \times \frac{2220 \text{ kJ}}{1 \cancel{\text{mol } C_3H_8}} = 1260 \text{ kJ}$$

$$= 1.26 \times 10^3 \text{ kJ}$$

EXERCISE How much heat is released when 75.0 g of methanol (CH_3OH) are burned completely in air?

$$2CH_3OH(l) + 3O_2(g) \longrightarrow 2CO_2(g) + 4H_2O(l) \qquad \Delta H = -726 \text{ kJ}$$

[*Answer:* 850 kJ]

6.9   STANDARD REACTION ENTHALPIES

KEY CONCEPT   Standard states and standard reaction enthalpies

The reaction enthalpy depends on the temperature and pressure of the reactants and products, as well as on their physical states. It is easy to imagine that chemists might have a great deal of difficulty comparing results from different laboratories if each laboratory chose a different set of conditions for reporting the reaction enthalpy. In response to this concern, **standard states** have been defined for all substances, and reaction enthalpies are frequently reported with all of the reactants and products in their standard states and at 298.15 K. The standard state of a substance is its most stable physical state at 1 atm pressure. The reaction enthalpy for a reaction with all of the reactants and products in their respective standard states is called the **standard reaction enthalpy** and is symbolized by $\Delta H°$ rather than ΔH. Thus,

Reactants $\longrightarrow$ products   reaction enthalpy $= \Delta H$

Reactants in standard states $\longrightarrow$ products in standard states   reaction enthalpy $= \Delta H°$

EXAMPLE 1   Understanding the standard reaction enthalpy

When methane is burned in air in a Bunsen burner, is the observed enthalpy change the standard reaction enthalpy?

$$CH_4(g) + 2O_2(g) \longrightarrow CO_2(g) + 2H_2O(g) \qquad \Delta H \stackrel{?}{=} \Delta H°$$

SOLUTION For the reaction enthalpy to be the standard reaction enthalpy, all of the reactants and products must be in their standard states. This means that gases must be present at a partial pressure of 1 atm. The oxygen in the air has a partial pressure of 0.2 atm and is not in its standard state; thus, the reaction enthalpy cannot be the standard reaction enthalpy and $\Delta H \neq \Delta H°$. In addition, some reflection should indicate that it is unlikely that any of the other gaseous products exists at exactly 1 atm partial pressure, bolstering the fact that the observed reaction enthalpy is not the standard reaction enthalpy. A set of well-established, albeit complex, calculations is required to convert the observed reaction enthalpy to the standard reaction enthalpy.

EXERCISE A sample of water is boiled at an atmospheric pressure of 1 atm, and the water vapor at 1 atm pressure is allowed to come to equilibrium with the liquid water. Is the observed enthalpy the standard enthalpy of boiling?

[*Answer:* Yes, because the water is pure, in its most stable state and at 1 atm pressure, and the gaseous water has a pressure of 1 atm.]

EXAMPLE 2 Understanding standard reaction enthalpies

For which of the following reactions is the reaction enthalpy the standard reaction enthalpy? All the reactions are at 25°C, and all gases are at 1 atm pressure.

(a) $\qquad$ $C(g) + O_2(g) \longrightarrow CO_2(g)$ $\qquad$ $\Delta H = -1110$ kJ

(b) $\qquad$ $CH_4(g) + 2O_2(g) \longrightarrow CO_2(g) + H_2O(l)$ $\qquad$ $\Delta H = -890$ kJ

SOLUTION For a reaction enthalpy to be a standard reaction enthalpy, all products and reactants must be present in their standard states. In (a), carbon is not in its standard state, so the reaction enthalpy of −1110 kJ is not the standard reaction enthalpy. In (b), all the reactants and products are in their standard state, so the reaction enthalpy is the standard reaction enthalpy.

EXERCISE For which of the following reactions is the reaction enthalpy the standard reaction enthalpy? All the reactions are at 25°C, and all gases are at 1 atm pressure.

(a) $\qquad$ $C_2H_4(g) + 3O_2(g) \longrightarrow 2CO_2(g) + 2H_2O(l)$ $\qquad$ $\Delta H = -1411$ kJ

(b) $\qquad$ $C(diamond) + O_2(g) \longrightarrow CO_2(g)$ $\qquad$ $\Delta H = -395$ kJ

[*Answer:* (a) Yes; (b) no]

6.10 COMBINING REACTION ENTHALPIES: HESS'S LAW

KEY CONCEPT Hess's law and combining reaction enthalpies

Enthalpy is a state property; therefore, the change in enthalpy for any process is independent of the steps taken to accomplish the process. Hess's law restates this proposition in slightly different terms. It states that the overall reaction enthalpy is the sum of the reaction enthalpies of any sequence of reactions that accomplishes the same overall reaction. Hess's law makes it possible to determine the reaction enthalpy for any reaction, using the following three steps:

Step 1. Write any number of intermediate reactions that go from the original reactants to the desired products; even imaginary reactions are all right to use.
Step 2. Write the reaction enthalpies for each of the intermediate reactions.
Step 3. Sum all of the intermediate reaction enthalpies to get the reaction enthalpy for the overall reaction.

EXAMPLE Using Hess's law

The standard enthalpy of combustion of $S_8(s)$ to $SO_3(g)$ is -3166 kJ·mol^{-1}; for combustion of $S_8(s)$ to $SO_2(g)$, the standard enthalpy of combustion is -2374 kJ·mol^{-1}. Use this information to calculate the standard reaction enthalpy for

$$2SO_2(g) + O_2(g) \longrightarrow 2SO_3(g) \qquad \Delta H° = ?$$

SOLUTION Because there are so many details to keep track of in this kind of calculation, it is advisable to write down all the information we have to see how to proceed. From the problem, we have these two reactions:

$$S_8(s) + 12O_2(g) \longrightarrow 8SO_3(g) \qquad \Delta H° = -3166 \text{ kJ} \qquad (1)$$

$$S_8(s) + 8O_2(g) \longrightarrow 8SO_2(g) \qquad \Delta H° = -2374 \text{ kJ} \qquad (2)$$

We now envision a two-step process that will accomplish the same overall reaction as that given in the problem. First, we use reaction (2) to write a reaction with SO_2 as a reactant, because SO_2 is one of the reactants in the original problem. The result, reaction (3) (which follows), is the reverse of reaction (2); so $\Delta H = +2374$ kJ. Second, we note that the products of reaction (3) are S_8 and O_2; so we write a reaction that has S_8 and O_2 as reactants and SO_3 (the desired product in the original problem) as a product. This is reaction (4), which is the same as reaction (1). We arrive at the desired reaction, reaction (5), by summing reactions (3) and (4) and canceling the substances that appear on both sides (S_8 and $8O_2$):

$$8SO_2(g) \longrightarrow S_8(s) + 8O_2(g) \qquad \Delta H° = +2374 \text{ kJ} \qquad (3)$$
$$S_8(s) + 12O_2(g) \longrightarrow 8SO_3(g) \qquad \Delta H° = -3166 \text{ kJ} \qquad (4)$$
$$8SO_2(g) + 4O_2(g) \longrightarrow 8SO_3(g) \qquad \Delta H° = -792 \text{ kJ} \qquad (5)$$

However, this is not quite the result we seek. If we divide all the coefficients in reaction (5) by 4, and also divide the reaction enthalpy by 4, we arrive at the desired answer:

$$2SO_2(g) + O_2(g) \longrightarrow 2SO_3(g) \qquad \Delta H° = -198 \text{ kJ}$$

EXERCISE Use the given reaction enthalpies to calculate the unknown reaction enthalpy. (The unknown reaction is a hypothetical one.)

$$2N_2 + O_2 \longrightarrow 2N_2O \qquad \Delta H° = 164.1 \text{ kJ}$$

$$N_2 + O_2 \longrightarrow 2NO \qquad \Delta H° = 180.5 \text{ kJ}$$

$$N_2 + 2O_2 \longrightarrow N_2O_4 \qquad \Delta H° = 9.2 \text{ kJ}$$

$$2NO + 2O_2 + 2N_2O \longrightarrow 2N_2O_4 + N_2 \qquad \Delta H° = ?$$

[*Answer*: -326.2 kJ]

KEY WORDS
Define or explain each term in a written sentence or two.

bar	standard reaction enthalpy
Hess's law	standard state
reaction enthalpy	thermochemical equation

THE HEAT OUTPUT OF REACTIONS

6.11 ENTHALPIES OF COMBUSTION

KEY CONCEPT A   Standard enthalpies of combustion

The **standard enthalpy of combustion** of a substance is the reaction enthalpy when 1 mol of the substance undergoes complete combustion under standard conditions. When you write the combustion reaction, CO_2, H_2O, N_2, and SO_2 should be used as the products for carbon, hydrogen, nitrogen, and sulfur, respectively. Two combustion reactions that follow this stipulation are

$$2C_2H_6S(g) + 9O_2(g) \longrightarrow 4CO_2(g) + 6H_2O(l) + 2SO_2(g)$$
$$2C_6H_8N(l) + 16O_2(g) \longrightarrow 12CO_2(g) + 8H_2O(l) + N_2(g)$$

EXAMPLE   Using standard enthalpies of combustion

Use the data in Table 6.3 of the text to calculate the reaction enthalpy for the hypothetical reaction $C_6H_6(l) + CH_4(g) \rightarrow C_7H_8(l) + H_2(g)$. The standard enthalpy of combustion of toluene, $C_7H_8(l)$, is -3910 kJ·mol^{-1}.

SOLUTION We anticipate that we shall need the combustion reactions for each of the reactants and products in the given equation, so we first write the correct equations with their enthalpies:

$$2C_6H_6(l) + 15O_2(g) \longrightarrow 12CO_2(g) + 6H_2O(l) \qquad \Delta H° = -6536 \text{ kJ} \qquad (1)$$

$$CH_4(g) + 2O_2(g) \longrightarrow CO_2(g) + 2H_2O(l) \qquad \Delta H° = -890 \text{ kJ} \qquad (2)$$

$$2C_7H_8(l) + 18O_2(g) \longrightarrow 14CO_2(g) + 8H_2O(l) \qquad \Delta H° = -7820 \text{ kJ} \qquad (3)$$

$$2H_2(g) + O_2(g) \longrightarrow 2H_2O(l) \qquad \Delta H° = -572 \text{ kJ} \qquad (4)$$

Notice that for all but equation (2) the enthalpy is twice that given in Table 6.3 because 2 mol of each reactant are required for a balanced equation. We now use Hess's law to add these equations in order to generate the desired equation. Because we want the CO_2 and H_2O to completely cancel, we must reverse equations (3) and (4) and multiply equation (2) by 2. This also results in the proper ratios of $C_6H_6(l)$, $CH_4(g)$, $C_7H_8(l)$, and $H_2(g)$ in the final equation:

$$
\begin{aligned}
2C_6H_6(l) + 15O_2(g) &\longrightarrow 12CO_2(g) + 6H_2O(l) & \Delta H° = 1 \times (-6536 \text{ kJ}) = -6536 \text{ kJ} \\
2CH_4(g) + 4O_2(g) &\longrightarrow 2CO_2(g) + 4H_2O(l) & \Delta H° = 2 \times (-890 \text{ kJ}) = -1.78 \times 10^3 \text{ kJ} \\
14CO_2(g) + 8H_2O(l) &\longrightarrow 2C_7H_8(l) + 18O_2(g) & \Delta H° = 2 \times (+7820 \text{ kJ}) = +7820 \text{ kJ} \\
2H_2O(l) &\longrightarrow 2H_2(g) + O_2(g) & \Delta H° = 1 \times (+572 \text{ kJ}) = +572 \text{ kJ} \\
\hline
2C_6H_6(l) + 2CH_4(g) + 19O_2(g) + 10H_2O(l) + 14CO_2(g) &\longrightarrow & +76 \text{ kJ}
\end{aligned}
$$

$$C_7H_8(l) + 2H_2(g) + 19O_2(g) + 10H_2O(l) + 14CO_2(g)$$

$$2C_6H_6(l) + 2CH_4(g) \longrightarrow 2C_7H_8(l) + 2H_2(g) \qquad \Delta H° = +76 \text{ kJ}$$

This equation is the one asked for in the problem, multiplied by 2. We arrive at the desired result by dividing the equation (and its enthalpy) by 2.

$$C_6H_6(l) + CH_4(g) \longrightarrow C_7H_8(l) + H_2(g) \qquad \Delta H° = +38 \text{ kJ}$$

EXERCISE Use the data in Table 6.3 of the text to calculate the standard reaction enthalpy for the reaction of 3 mol of acetylene (C_2H_2) to form 1 mol of benzene (C_6H_6).

[**Answer:** −632 kJ]

KEY CONCEPT B   Enthalpy as a resource, specific enthalpy, and enthalpy density

We use the word enthalpy here instead of the more common term *energy* because most of the chemical reactions used to extract heat from fuels are run at constant pressure.

Whenever humans use fuel (either as food or as fuel), a chemical oxidation reaction occurs. In most cases, fuels are oxidized through combustion reactions. A fuel with a high specific enthalpy is used when a lot of heat is needed from a small mass of fuel: a high **specific enthalpy** for a fuel means that a relatively small mass of the fuel releases a relatively large amount of heat when it is combusted. A fuel with a high energy density is used when a lot of heat is needed from a small volume of fuel; a high **energy density** for a fuel means that a relatively small volume of the fuel releases a relatively large amount of heat when it is combusted.

EXAMPLE   Calculating the energy stored by plants

The photosynthesis of 1 g of glucose needs 16 kJ of energy. Show that the production of 6×10^{14} kg of glucose results in the storage of about 10^{19} kJ of energy.

SOLUTION We first use the fact that it takes 16 kJ of energy to produce 1 g of glucose to calculate the energy required to produce 1 kg of glucose:

$$\frac{16 \text{ kJ}}{1 \text{ g glucose}} \times \frac{1000 \text{ g}}{1 \text{ kg}} = 16 \times 10^3 \frac{\text{kJ}}{\text{kg glucose}}$$

We can then proceed, using unit analysis, to calculate the energy required to produce 6×10^{14} kg of glucose:

$$6 \times 10^{14} \text{ kg glucose} \times \frac{16 \times 10^3 \text{ kJ}}{1 \text{ kg glucose}} = 1 \times 10^{19} \text{ kJ}$$

EXERCISE Give an example, other than those in the text, of a situation in which a fuel with both a high specific enthalpy and a high energy density would be desirable.

[*Answer:* A hiker, who must carry as much equipment as possible with the lowest possible weight and volume, would appreciate such a fuel.]

6.12   STANDARD ENTHALPIES OF FORMATION

KEY CONCEPT   Standard enthalpies of formation

The **standard enthalpy of formation** ΔH_f° of a substance is the standard reaction enthalpy, per mole of the substance, for formation of the substance from its elements in their most stable form. This definition assumes that the standard enthalpy of formation of an element is zero (if the element is in its most stable form). We can calculate the standard reaction enthalpy for any reaction for which we know the standard enthalpies of formation of all the reactants and products. To do this, we substitute into the formula

ΔH° (for a reaction) = (sum of ΔH_f° of all products) − (sum of all ΔH_f° of all reactants)

Values of ΔH_f° are tabulated in Table 6.5 and Appendix 2A of the text.

EXAMPLE   Using enthalpies of formation

Use enthalpies of formation to calculate the standard enthalpy of combustion of pentane $C_5H_{12}(g)$. Assume the water formed is in the liquid state.

SOLUTION Before doing any calculations, we write the chemical equation involved:

$$C_5H_{12}(g) + 8O_2(g) \longrightarrow 5CO_2(g) + 6H_2O(l) \qquad \Delta H^\circ = ?$$

Next, we look up the standard enthalpy of formation for each of the reactants and products and write it above each of the substances in the equation:

$$
\begin{array}{ccccccc}
-146.44 \text{ kJ·mol}^{-1} & & 0 \text{ kJ·mol}^{-1} & & -393.51 \text{ kJ·mol}^{-1} & & -285.83 \text{ kJ·mol}^{-1} \\
C_5H_{12}(g) & + & 8O_2(g) & \longrightarrow & 5CO_2(g) & + & 6H_2O(l)
\end{array}
$$

Next, we insert the standard enthalpies of formation into the equation for calculating ΔH°, being certain to multiply each standard enthalpy of formation by the stoichiometric coefficient in front of each respective substance in the equation:

ΔH° (for a reaction) = [sum of ΔH_f° of all products] − [sum of ΔH_f° of all reactants]

$$= [5 \text{ mol} \times \Delta H_f^\circ \{CO_2(g)\} + 6 \text{ mol} \times \Delta H_f^\circ \{H_2O(l)\}]$$

$$- [1 \text{ mol} \times \Delta H_f^\circ \{C_5H_{12}(g)\} + 8 \text{ mol} \times \Delta H_f^\circ \{O_2(g)\}]$$

$$= [(5 \text{ mol} \times -393.51 \text{ kJ·mol}^{-1}) + (6 \text{ mol} \times -285.83 \text{ kJ·mol}^{-1})]$$

$$- [(1 \text{ mol} \times -146.44 \text{ kJ·mol}^{-1}) + (8 \text{ mol} \times 0 \text{ kJ·mol}^{-1})]$$

Next, perform all the calculations inside the parentheses:

$$\Delta H^\circ = [-3682.53 \text{ kJ}] - [-146.44 \text{ kJ}]$$

Finally, perform the subtraction indicated. Do not forget that minus × minus = plus.

$$\Delta H^\circ = -3536.09 \text{ kJ}$$

This reaction corresponds to the combustion of 1 mol of pentane, as can be seen from the equation

$$C_5H_{12}(g) + 8O_2(g) \longrightarrow 5CO_2(g) + 6H_2O(l) \qquad \Delta H° = -3536.09 \text{ kJ}$$

Thus, the enthalpy of combustion of pentane is $-3536.09 \text{ kJ} \cdot \text{mol}^{-1}$ C_5H_{12}.

EXERCISE Repeat the calculation of the enthalpy of combustion of pentane, but assume that the water formed is in the gaseous state.

[*Answer:* $\Delta H = -3272.03 \text{ kJ} \cdot \text{mol}^{-1}$]

PITFALL   Reaction enthalpies and the physical state of substances

We have seen many examples in the text and study guide that indicate that the reaction enthalpy depends on the physical state (solid, liquid, or gas) of the reactants and products. You should take great care when using a table like Table 6.5, not only to choose the correct compound but also to make sure the compound is in the correct physical state.

PITFALL   Negative signs in calculating ΔH with $\Delta H_f°$

When using $\Delta H°$ (for a reaction) = [sum of $\Delta H_f°$ of all products] − [sum of $\Delta H_f°$ of all reactants), be careful to (1) use the correct sign for the enthalpies of formation and (2) properly take into account the negative sign in the equation itself. By using the calculational technique illustrated in the preceding example—substituting in each individual $\Delta H_f°$ with its own sign and then performing the calculation *inside* the brackets first—you can avoid errors.

KEY WORDS   Define or explain each term in a written sentence or two.

enthalpy density	standard enthalpy of combustion
specific enthalpy	standard enthalpy of formation

DESCRIPTIVE CHEMISTRY TO REMEMBER

- The salt ammonium nitrate (NH_4NO_3) has a large **endothermic heat of solution** and produces a pronounced cooling effect when it is dissolved in water.
- **Graphite** is the most stable form of carbon at 25°C.
- One of the simplest carbohydrates is **glucose** ($C_6H_{12}O_6$), the main energy source of the human body.
- **Hydrogen** condenses to a liquid only at a very low temperature, and the liquid form has a very low density ($0.089 \text{ g} \cdot \text{mL}^{-1}$).
- Compounds with approximate formulas such as **$FeTiH_2$** release hydrogen gas when heated or treated with acid. They are being studied as possible fuels that would release hydrogen when needed, but they suffer from the problem of having a low specific enthalpy.
- The low melting point (29°C) of **calcium chloride hexahydrate** ($CaCl_2 \cdot 6H_2O$) makes it useful in solar energy collectors.
- The dissolution of **H_2SO_4** in water is an extremely exothermic process.
- Solid carbon dioxide is called **dry ice.**
- **Sublimation** is the direct conversion of a solid to a vapor, with no intermediate formation of a liquid.

CHEMICAL EQUATIONS TO KNOW

- The reaction of the white crystalline solid **barium hydroxide octahydrate** [$Ba(OH)_2 \cdot 8H_2O$] with the white crystalline solid ammonium thiocyanate (NH_4SCN) is very **endothermic**:

$$Ba(OH)_2 \cdot 8H_2O(s) + 2NH_4SCN(s) \longrightarrow Ba(SCN)_2(aq) + 2NH_3(g) + 10H_2O(l)$$

- **Hydrocarbon fuels** release a great deal of heat when they are combusted. Complete combustion results in the formation of carbon dioxide and water. The water formed may be in the gaseous or liquid state, so its state is not specified in the following equations:

methane:	$CH_4(g) + 2O_2(g) \longrightarrow CO_2(g) + 2H_2O$
propane:	$C_3H_8(g) + 5O_2(g) \longrightarrow 3CO_2(g) + 4H_2O$
butane:	$2C_4H_{10}(g) + 13O_2(g) \longrightarrow 8CO_2(g) + 10H_2O$
octane:	$2C_8H_{18}(l) + 25O_2(g) \longrightarrow 16CO_2(g) + 18H_2O$
benzene:	$2C_6H_6(l) + 15O_2(g) \longrightarrow 12CO_2(g) + 6H_2O$

- In a **thermite reaction**, aluminum metal reacts with a metal oxide such as iron(III) oxide in a fiery exothermic reaction:

$$2Al(s) + Fe_2O_3(s) \longrightarrow Al_2O_3(s) + 2Fe(s)$$

- **Phosphorus trichloride** can be synthesized by the direct reaction of phosphorus with chlorine:

$$P_4(s) + 6Cl_2(g) \longrightarrow 4PCl_3(l)$$

- In an **exothermic** reaction, solid zinc reacts with iodine to produce zinc iodide:

$$Zn(s) + I_2(s) \longrightarrow ZnI_2(s)$$

- The **oxidation of glucose** is an important biochemical energy source for humans; because it is a source of energy, it must be exothermic:

$$C_6H_{12}O_6(aq) + 6O_2(g) \longrightarrow 6CO_2(g) + 6H_2O(l)$$

- The **photosynthetic** formation of glucose in plants is endothermic; the energy needed for the reaction is provided by the Sun:

$$6CO_2(g) + 6H_2O(l) \longrightarrow C_6H_{12}O_6(aq) + 6O_2(g)$$

- Ethanol trapped in a gel (known as **Sterno**) undergoes an exothermic oxidation in which it is used as a source of heat:

$$2C_2H_5OH(l) + 7O_2(g) \longrightarrow 4CO_2(g) + H_2O(l)$$

MATHEMATICAL EQUATIONS TO KNOW AND UNDERSTAND

$\Delta X = X_{final} - X_{initial}$	change or difference in any state property X
heat = heat capacity $\times \Delta T$	heat absorbed or release
heat = $\Delta T \times$ mass $\times$ specific heat capacity	heat absorbed or released
$\Delta H = q_p$	definition of enthalpy change
$\Delta H° =$ (sum of all $\Delta H_f°$ of all products) $-$ (sum of all $\Delta H_f°$ of all reactants)	use of $\Delta H_f°$ to calculate $\Delta H°$

SELF-TEST EXERCISES

Energy, heat, and enthalpy

1. The energy of a system is a measure of its
(a) capacity to do work or supply heat
(b) temperature
(c) speed of movement
(d) height above the ground
(e) tendency to emit electromagnetic radiation

2. Imagine a chemical reaction loses 203 kJ of energy as the reaction proceeds. Which of the following is true?
(a) The universe contains 203 kJ more energy.
(b) The surroundings must lose 203 kJ.
(c) The universe contains 203 kJ less energy.
(d) The surroundings must gain 203 kJ.
(e) none of these

3. Energy as heat always flows from an area of _____ to an area of _____.
(a) high temperature, low temperature
(b) low temperature, high temperature
(c) high pressure, low pressure
(d) low pressure, high pressure
(e) low volume, high volume

4. Which of the following is an exothermic process?
(a) freezing of water
(b) sublimation of dry ice
(c) melting of ice
(d) boiling of water
(e) none of these

5. How many joules of heat are required to increase the temperature of 25.0 g of water by 3.40°C?
(a) 105 J
(b) 85.0 J
(c) 356 J
(d) 30.8 J
(e) 20.3 J

6. How many joules are there in 123 cal? 4.184 J = 1 cal.
(a) 127 J
(b) 515 J
(c) 119 J
(d) 29.4 J
(e) 632 J

7. How much heat is lost from a 25.0-g block of copper when it cools by 7.1°C? The specific heat capacity of copper is $0.38 \ J \cdot g^{-1} \cdot K^{-1}$.
(a) 4.7 kJ
(b) 9.5 J
(c) 1.3 J
(d) 11 J
(e) 67 J

8. How much heat is required to raise the temperature of 2.00 kg of water from 22.0°C to 26.3°C?
(a) 8.6 kJ
(b) 8.2 kJ
(c) 1.2×10^3 kJ
(d) 2.2×10^2 kJ
(e) 36 kJ

9. What is the final temperature of a 40-g sample of ethanol (C_2H_6O) at 25.0°C that absorbs 355 J of heat? The specific heat capacity of ethanol is $2.42 \ J \cdot g^{-1} \cdot K^{-1}$.
(a) 46.5°C
(b) 59.4°C
(c) 28.7°C
(d) 25.5°C
(e) 21.3°C

10. How much heat is required to increase the temperature of 2.0 mol of benzene (C_6H_6) from 10.0°C to 25.0°C? The specific heat capacity of benzene is $1.05 \ J \cdot g^{-1} \cdot K^{-1}$.
(a) 41 J
(b) 2.5 kJ
(c) 32 J
(d) 4.1 kJ
(e) 9.8 kJ

11. A 34.2-g block of aluminum increases in temperature from 22.3°C to 27.7°C when 166 J of heat is added to it. What is the specific heat capacity of aluminum, in $J \cdot g^{-1} \cdot K^{-1}$?
(a) 26
(b) 1.0
(c) 0.90
(d) 31
(e) 0.18

12. What is the molar heat capacity of ethanol (C_2H_6O), which has a specific heat capacity of $2.42 \text{ J} \cdot \text{g}^{-1} \cdot \text{K}^{-1}$? All answers are given in $\text{J} \cdot \text{mol}^{-1} \cdot \text{K}^{-1}$.

(a) 111   (b) 5.25×10^{-2} (c) 19.0   (d) 9.01×10^{-3} (e) 72.6

13. What is the heat capacity of 500 g of copper, in $\text{J} \cdot \text{K}^{-1}$? The specific heat capacity of copper is $0.38 \text{ J} \cdot \text{g}^{-1} \cdot \text{K}^{-1}$.

(a) 5.3×10^{-3} (b) 1.3×10^3 (c) 1.9×10^2
(d) 1.2×10^4 (e) 3.0

14. In an experiment, 42.6 g of zinc at 112.0°C is placed in 50.0 mL of water, initially at 25.0°C, in a styrofoam calorimeter. The final temperature of the water and zinc is 31.4°C. What is the specific heat capacity of zinc, in $\text{J} \cdot \text{g}^{-1} \cdot \text{°C}^{-1}$?

(a) 0.36   (b) 0.39   (c) 1.4   (d) 0.89   (e) 0.77

15. In an experiment, 6.77 g of silver at 98.1°C is placed in 25.0 mL of water, initially at 25.0°C, in a Styrofoam calorimeter. The final temperature of the water and silver is 26.1°C. What is the specific heat capacity of silver, in $\text{J} \cdot \text{g}^{-1} \cdot \text{°C}^{-1}$?

(a) 0.19   (b) 0.11   (c) 0.078   (d) 0.24   (e) 0.35

16. When a reaction that is known to release 19.1 kJ of heat is run in a calorimeter containing 100.0 mL of water, the temperature increases by 3.95°C. What is the heat capacity of the calorimeter (including the water)?

(a) $4.85 \text{ kJ} \cdot \text{°C}^{-1}$ (b) $75.4 \text{ kJ} \cdot \text{°C}^{-1}$ (c) $0.0754 \text{ kJ} \cdot \text{°C}^{-1}$
(d) $0.206 \text{ kJ} \cdot \text{°C}^{-1}$ (e) $0.191 \text{ kJ} \cdot \text{°C}^{-1}$

17. A chemical reaction performed at constant pressure in a simple foam cup calorimeter containing 200.0 mL of solution increases the temperature of the solution by 6.82°C. The heat capacity of the calorimeter (including cup and contents) is $878.8 \text{ J} \cdot \text{K}^{-1}$. What is the enthalpy change for the reaction system?

(a) -5.17 kJ (b) $+5.99 \text{ kJ}$ (c) $+1.20 \times 10^3 \text{ kJ}$
(d) -5.99 kJ (e) $+5.71 \text{ kJ}$

18. A simple calorimeter is calibrated by performing a reaction in the calorimeter that is known to release 6.225 kJ of heat. The temperature of the calorimeter increases from 22.22°C to 28.48°C during the course of the reaction. What is the heat capacity of the calorimeter?

(a) $994 \text{ J} \cdot \text{K}^{-1}$ (b) $4.18 \text{ J} \cdot \text{K}^{-1}$ (c) $6.26 \text{ J} \cdot \text{K}^{-1}$
(d) $280 \text{ J} \cdot \text{K}^{-1}$ (e) $218 \text{ J} \cdot \text{K}^{-1}$

19. 65.5 kJ of heat are required to vaporize a sample of carbon tetrachloride inside a closed vessel. Because the vessel is closed, the pressure increases from 1.0 atm to 6.4 atm. What is ΔH for this process?

(a) $+65.5 \text{ kJ}$ (b) -65.5 kJ (c) $+354 \text{ kJ}$
(d) -354 kJ (e) need more information to tell

20. A reaction run at constant pressure releases 512 kJ of heat. At constant volume, the same reaction releases 518 kJ of heat. What is ΔH for the reaction?

(a) $+518 \text{ kJ}$ (b) -6 kJ (c) -512 kJ (d) -506 kJ (e) $+6 \text{ kJ}$

21. A chemical system releases heat as it undergoes a reaction at constant pressure. What happens to the enthalpy of the system as the reaction occurs?

(a) decreases   (b) increases   (c) stays the same

22. Which of the following values for ΔH corresponds to an endothermic reaction?
(a) +612 kJ   (b) −667 kJ   (c) −12 kJ   (d) 0 kJ   (e) −344 kJ

23. You drive from your home to Washington, DC. Which of the following is a "state property" for your trip?
(a) time it takes to make the trip
(b) actual distance, in miles, between your home and Washington, DC
(c) number of miles you actually drive to get to Washington
(d) gallons of gasoline required to make the trip
(e) average speed you traveled, including time off the road

24. The pressure of a system is changed from 750 Torr to 700 Torr. What is ΔP?
(a) 725 Torr   (b) 700 Torr   (c) 750 Torr
(d) −50 Torr   (e) 50 Torr

25. When 15.0 g of water freezes, 5.00 kJ of heat is released. How much heat is required to melt 30.0 g of water?
(a) 2.50 kJ   (b) 6.00 kJ   (c) 75.0 kJ   (d) 5.00 kJ   (e) 10.0 kJ

26. What is the change in enthalpy of the system when 125 g of water vapor at 100°C condenses to the liquid at the same temperature? (See Table 6.2 of the text.)
(a) −283 kJ   (b) −91.6 kJ   (c) −5.09 kJ
(d) +283 kJ   (e) +5.09 kJ

27. What is ΔH when 100 g of liquid ammonia freezes to the solid at its freezing point? (See Table 6.2 of the text.)
(a) $+9.62 \times 10^3$ kJ   (b) −565 kJ   (c) +0.960 kJ
(d) −301 kJ   (e) −33.2 kJ

28. As heat is added to a sample of a liquid, the added energy causes molecules to change from the liquid state to the vapor state. What happens to the temperature of the liquid at the same time?
(a) it increases   (b) it decreases   (c) it stays the same

29. What determines, on a heating curve, the slope of the line that corresponds to an increase in the temperature of the liquid?
(a) enthalpy of melting   (b) enthalpy of vaporization
(c) heat capacity of the gas   (d) heat capacity of the liquid
(e) heat capacity of the solid

30. The dissolution of sodium chloride is endothermic:

$$NaCl(s) \longrightarrow Na^+(aq) + Cl^-(aq) \qquad \Delta H = +3.9 \text{ kJ}$$

What is the reaction enthalpy for the crystallization of sodium chloride?

$$Na^+(aq) + Cl^-(aq) \longrightarrow NaCl(s) \qquad \Delta H = ?$$

(a) 0 kJ   (b) +3.9 kJ   (c) −3.9 kJ   (d) +2.6 kJ   (e) −2.6 kJ

The enthalpy of chemical change

31. The thermochemical equation for the combustion of acetylene (C_2H_2) is shown below. What heat effect occurs when exactly 1 mol of acetylene is burned?

$$2C_2H_2(g) + 5O_2(g) \longrightarrow 4CO_2(g) + 2H_2O(l) \qquad \Delta H = -2600 \text{ kJ}$$

(a) 1300 kJ from the surroundings are consumed

(b) 2600 kJ from the surroundings are consumed

(c) 1300 kJ are lost to the surroundings

(d) 2600 kJ are lost to the surroundings

32. When 0.1016 g of octane (C_8H_{18}) is burned in a calorimeter with heat capacity 994 $J \cdot {}^{\circ}C^{-1}$, the temperature of the calorimeter increases by 4.88°C. What is the reaction enthalpy for the combustion of octane (in kJ)?

$$2C_8H_{18}(l) + 25O_2(g) \longrightarrow 16CO_2(g) + 18H_2O(l) \qquad \Delta H = ?$$

(a) -5.45×10^3 (b) -9.21×10^3 (c) -5.68×10^3

(d) -4.61×10^3 (e) -1.09×10^4

33. For reaction (1), $\Delta H = -286$ kJ. What is ΔH for reaction (2)?

$$H_2(g) + \tfrac{1}{2}O_2(g) \longrightarrow H_2O(g) \qquad\qquad (1)$$

$$2H_2(g) + O_2(g) \longrightarrow 2H_2O(g) \qquad\qquad (2)$$

(a) -286 kJ   (b) $+286$ kJ   (c) -572 kJ

(d) $+572$ kJ   (e) -322 kJ

34. ΔH for the reaction that follows is -940 kJ:

$$2Na(s) + 2H_2O(l) \longrightarrow 2NaOH(aq) + H_2(g)$$

What is ΔH when 0.500 mol of sodium reacts with water according to this equation?

(a) -940 kJ   (b) -235 kJ   (c) -470 kJ

(d) -1.88×10^3 kJ   (e) -3.76×10^3 kJ

35. ΔH for the reaction that follows is -890 kJ:

$$CH_4(g) + 2O_2(g) \longrightarrow CO_2(g) + H_2O(l)$$

How many grams of methane must be burned in order to generate 2.0 MJ of heat?

(a) 3.2×10^4 g   (b) 0.14 g   (c) 36 g   (d) 2.2 g   (e) 2.9×10^7 g

36. For which of the following reactions is ΔH the standard reaction enthalpy? Assume all reactions are run at 25°C and all gases are at 1 atm pressure.

(a) $CH_4(g) + 2O_2(g) \longrightarrow CO_2(g) + 2H_2O(g)$ (b) $C(graphite) + 2Cl_2(g) \longrightarrow CCl_4(l)$

(c) $C(diamond) + O_2(g) \longrightarrow CO_2(g)$ (d) $C(g) + 2H_2(g) \longrightarrow CH_4(g)$

37. How much heat is produced by burning 25 kg of liquid methanol at constant pressure? Assume all reactants and products are at 25°C. (See Table 6.4 of the text.)

$$2CH_3OH(l) + 3O_2(g) \longrightarrow 2CO_2(g) + 4H_2O(l)$$

(a) 0.93 MJ   (b) 5.8×10^2 MJ   (c) 0.57 MJ

(d) 9.1 kJ   (e) 1.8 MJ

38. Which of the following indicates that a substance is in its standard state at 25°C?

(a) NaCl(aq)   (b) H_2O (g, 20 Torr)   (c) $H_2O(l, 1$ atm)

(d) NaCl(g, 0.05 Torr)   (e) $O_2(g, 0.95$ atm)

39. Given the reaction enthalpies

$$2P(s) + 3Cl_2(g) \longrightarrow 2PCl_3(g) \qquad \Delta H = -574 \text{ kJ}$$

$$2P(s) + 5Cl_2(g) \longrightarrow 2PCl_5(l) \qquad \Delta H = -887 \text{ kJ}$$

what is the reaction enthalpy for the reaction

$$PCl_3(g) + Cl_2(g) \longrightarrow PCl_5(l) \qquad \Delta H = ?$$

(a) -157 kJ   (b) -313 kJ   (c) -1461 kJ

(d) $+1461$ kJ   (e) $+222$ kJ

40. The density of methanol is 0.791 g·mL^{-1}. Calculate the heat released when 10.0 mL of methanol burns under standard conditions at 25°C. (See Table 6.4 of the text.)

(a) 179 kJ   (b) 196 kJ   (c) 227 kJ   (d) 184 kJ   (e) 5.56×10^4 kJ

The heat output of reactions

41. Given the standard reaction enthalpy at 25°C for the reaction

$$N_2(g) + 3H_2(g) \longrightarrow 2NH_3(g) \qquad \Delta H = -92.22 \text{ kJ}$$

what is the standard enthalpy of formation at 25°C for $NH_3(g)$, in kJ·mol^{-1}?

(a) -92.22 (b) -46.11 (c) $+92.22$ (d) $+30.74$ (e) $+46.11$

42. Use standard enthalpies of formation to calculate ΔH for the reaction

$$2H_2O(l) + 2F_2(g) \longrightarrow 4HF(aq) + O_2(g)$$

(a) -1902 kJ   (b) -758.9 kJ   (c) -1814 kJ

(d) -846.9 kJ   (e) -46.8 kJ

43. What is the standard reaction enthalpy at 25°C for the decomposition of calcium carbonate (calcite) to calcium oxide and carbon dioxide?

$$CaCO_3(s) \longrightarrow CaO(s) + CO_2(g) \qquad \Delta H = ?$$

(a) -2235.5 kJ   (b) -965.3 kJ   (c) $+1206.9$ kJ

(d) $+178.3$ kJ   (e) $+654.8$ kJ

44. What is the standard enthalpy of combustion at 25°C for cyclopropane $[C_3H_6(g)]$? The standard enthalpy of formation of cyclopropane is $+53.30$ kJ·mol^{-1} at 25°C.

(a) -161 kJ·mol^{-1} (b) -108 kJ·mol^{-1} (c) -2091 kJ·mol^{-1}

(d) -3325 kJ·mol^{-1} (e) -733 kJ·mol^{-1}

45. The energy of the solar radiation absorbed by vegetation on Earth is enough to produce about 6×10^{14} kg of glucose a year. Given that the photosynthesis of 1 g of glucose requires 16 kJ of energy, calculate the energy absorbed by the vegetation per year.

(a) 1×10^{19} kJ   (b) 1×10^{16} kJ   (c) 4×10^{16} kJ   (d) 4×10^{19} kJ

46. The standard enthalpy of combustion of ethene $[C_2H_4(g)]$ is -1411 kJ·mol^{-1}. What is the specific enthalpy of ethene?

(a) 8.88 kJ·g^{-1} (b) 19.1 kJ·g^{-1} (c) 50.3 kJ·g^{-1}

(d) 39.6 kJ·g^{-1} (e) 26.2 kJ·g^{-1}

47. The density of heptane $[C_7H_{16}(l)]$ is 0.684 g·mL^{-1} and its enthalpy of combustion is 4.85×10^3 kJ·mol^{-1}. What is the enthalpy density of heptane, in MJ·L^{-1}?

(a) 48.4   (b) 26.5   (c) 19.2   (d) 81.1   (e) 33.1

Descriptive chemistry

48. Which of the following is not a fossil fuel?

(a) coal   (b) hydrogen gas   (c) oil   (d) natural gas

49. Which two products are always obtained when a hydrocarbon is burned in an excess of oxygen?

(a) CO_2 + H_2O (b) CO + H_2O (c) C + H_2

(d) CO_2 + H_2 (e) CO + H_2

50. The major component of natural gas is

(a) H_2 (b) CO_2 (c) CH_4 (d) C_2H_4 (e) CO

51. The main biochemical energy source for humans is

(a) $C_6H_{12}O_6$ (b) C_6H_6 (c) CO_2 (d) C_8H_{18} (e) $C_{12}H_{22}O_{11}$

52. Dry ice is composed of solid

(a) CH_4 (b) CO_2 (c) C_8H_{18} (d) H_2O (e) N_2

53. The two reactants in a common thermite reaction are

(a) H_2, O_2 (b) Al, MgO (c) Al, O_2 (d) Al, Fe_2O_3 (e) CH_4, Fe

CHAPTER 7

INSIDE THE ATOM

Chemists are deeply interested in the organization of the electrons in the atom. The electrons are responsible for the colors emitted when atoms are heated or subjected to an electric field; they are also, as we shall see in later chapters, intimately involved in chemical bonding.

OBSERVING ATOMS

7.1 THE CHARACTERISTICS OF LIGHT

KEY CONCEPT Electromagnetic radiation as waves

Electromagnetic radiation can be considered to be a wavelike disturbance in space (not in air), much like a water wave is a disturbance in water. Like any wave, it is described by its **wavelength** (λ), **frequency** (ν), and **speed of propagation** (c). The wavelength has units of length, such as meters; the frequency has units of 1/s (1/second), or s^{-1}. The unit s^{-1} is also called the **hertz** ($Hz = s^{-1}$). The wavelength of a wave times its frequency is the speed of propagation of the wave. The symbol c is used for the speed of electromagnetic radiation (or speed of light), 3.00×10^8 m·s^{-1}.

$$\lambda \times \nu = c$$

Because the speed of light is a constant, the wavelength and frequency are not independent of each other. The value of one determines the other. Finally, we note that for visible light, the frequency (or wavelength) determines the color of the light.

EXAMPLE Calculating the frequency of electromagnetic radiation

The wavelength of the light to which the eye is most sensitive is 556 nanometers (nm). What is the frequency of this light? What color is it?

SOLUTION The relationship between the wavelength and frequency of light is $\lambda \times \nu = c$. $\lambda = 556$ nm and $c = 3.00 \times 10^8$ m·s^{-1}. Substitution of these values into the equation will allow us to solve for the frequency ν. Because c is given in m·s^{-1}, we must first convert the wavelength from nanometers to meters:

$$556 \text{ nm} \times \frac{1 \text{ m}}{10^9 \text{ nm}} = 5.56 \times 10^{-7} \text{ m}$$

Now, we substitute into the equation:

$$\lambda \times \nu = c$$

$$(5.56 \times 10^{-7} \text{ m}) \times \nu = 3.00 \times 10^8 \text{ m·s}^{-1}$$

$$\nu = 5.40 \times 10^{14} \text{ s}^{-1} = 5.40 \times 10^{14} \text{ Hz}$$

It is possible to determine the color by noting that 556 nm is between the 580 nm of yellow and 530 nm of green. We conclude that 556 nm is a yellow-green (Table 7.1 of the text); this is the color of many modern emergency vehicles.

EXERCISE Electromagnetic radiation with a frequency of 4.0×10^{14} Hz is in the infrared region of the electromagnetic spectrum. What is the wavelength of this electromagnetic radiation?

[*Answer:* 7.5×10^{-7} m]

7.2 QUANTA AND PHOTONS

KEY CONCEPT   Photons are massless packets of energy

Electromagnetic radiation often behaves like a wave. Rainbows are a spectacular reminder of the wave nature of light because only waves can be refracted and dispersed. However, in many experiments, light acts like a stream of massless particles, called **photons**. Each particle has a frequency and a wavelength. The brightness of the light is determined by the number of photons; a larger number of photons results in a more intense beam of light. By feeling the warmth of sunlight on a cool spring day, we can observe that electromagnetic radiation carries energy. The energy of one photon is given by

$$E = h \times \nu$$

where ν is the frequency in Hz (s^{-1}) and h is the **Planck constant**, 6.63×10^{-34} J·Hz^{-1}.

EXAMPLE   Calculating the energy of a photon

What is the energy of a photon of blue light of wavelength 455 nm?

SOLUTION The energy of a photon is given by the equation $E = h \times \nu$, with $h = 6.63 \times 10^{-34}$ J·Hz^{-1}, so we need the frequency of the light to calculate the energy of the photon. To get the frequency from the wavelength, we substitute into the equation

$$\nu \times \lambda = c$$

$$\nu = \frac{c}{\lambda} = \frac{3.00 \times 10^8 \text{ m·s}^{-1}}{4.55 \times 10^{-7} \text{ m}}$$

$$= 6.59 \times 10^{14} \text{ s}^{-1}$$

$$= 6.59 \times 10^{14} \text{ Hz} \qquad \text{(recall that } s^{-1} = \text{Hz)}$$

Finally, we use the value of the frequency in the equation for the energy of a photon:

$$E = h \times \nu$$

$$= 6.63 \times 10^{-34} \text{ J·Hz}^{-1} \times 6.59 \times 10^{14} \text{ Hz}$$

$$= 4.37 \times 10^{-19} \text{ J}$$

EXERCISE What is the wavelength of a photon that has an energy equal to 5.33×10^{-19} J?

[*Answer* 373 nm (3.73×10^{-7} m)]

7.3   ATOM SPECTRA AND ENERGY LEVELS

KEY CONCEPT A   The hydrogen atom spectrum

When hydrogen gas is heated or placed in an intense electric field, it glows with a purplish color as a result of light emitted from hydrogen atoms. If all the electromagnetic radiation emitted in such an experiment (not only the visible light) is analyzed, it is found that only certain specific frequencies and, therefore, only certain colors (in the visible region) are present. Because it is the electron that is responsible for the emission of light, these observations suggest that the emission occurs when the electron drops from a specific high-energy state to a specific low-energy one. In doing so, the atom loses energy, which equals the difference of energy between the two specific energy states; this energy leaves the atom as a photon. As a result of conservation of energy, the energy of the photon must equal the energy loss of the atom.

$$\Delta E = h\nu$$

ΔE should be expressed as a positive energy, because it represents the actual decrease in energy.

EXAMPLE Using the equation $\Delta E = h\nu$

A hydrogen atom undergoes a change in energy from -8.72×10^{-20} J to -5.45×10^{-19} J. What is the energy of the emitted photon? What is the frequency of the photon?

SOLUTION First we calculate ΔE for the atom. We must be careful to keep track of the negative signs associated with the energies:

$$\Delta E = E_f - E_i$$
$$= (-5.45 \times 10^{-19} \text{ J}) - (-8.72 \times 10^{-20} \text{ J})$$
$$= -4.58 \times 10^{-19} \text{ J}$$

This energy (expressed as a positive number because photons always have positive energy) is equal to the energy of the emitted photon,

$$E_{photon} = 4.58 \times 10^{-19} \text{ J}$$

and the energy of the photon is equal to $h\nu$, which allows us to solve for the frequency ν:

$$h\nu = 4.58 \times 10^{-19} \text{ J}$$
$$\nu = \frac{4.58 \times 10^{-19} \text{ J}}{6.63 \times 10^{-34} \text{ J} \cdot \text{Hz}^{-1}}$$
$$= 6.91 \times 10^{14} \text{ Hz}$$

EXERCISE An atom with energy equal to -1.35×10^{-19} J loses energy by emission of a photon with a frequency of 3.08×10^{15} Hz. What is the energy of the atom after emission of the photons?

[*Answer:* -2.18×10^{-18} J]

KEY CONCEPT B The energy states of an atom

One of the remarkable features of the loss of energy by atoms is that only a few of the seemingly infinite possible photon energies occur, so each atom has a unique observed line spectra. The reason that only a few specific photon energies are observed is that the atom can exist in only a few specific energy states; and when it loses energy, it changes from one specific energy to another specific energy, releasing a photon with an energy that is the difference between the two specific energy states.

EXAMPLE Understanding energy states

A hypothetical atom can exist in only the energy states shown below. List all the energies of all the photons that can be emitted from this atom when it is heated.

$\underline{\hspace{2cm}}$ -0.97×10^{-18} J

$\underline{\hspace{2cm}}$ -2.18×10^{-18} J

$\underline{\hspace{2cm}}$ -8.72×10^{-18} J

SOLUTION An atom releases energy as a photon when it goes from a higher energy state to a lower energy state. The energy of the photon equals the difference in energies between the higher energy state and the lower one. The hypothetical atom pictured in the problem can go from a higher to a lower state in three possible ways, indicated by A, B, and C in the following diagram:

$$
\begin{array}{cl}
\underline{} & -0.97 \times 10^{-18} \text{ J} \\
A\downarrow & \\
\underline{} & -2.18 \times 10^{-18} \text{ J} \\
B\ \big\downarrow\ \big\downarrow C & \\
\underline{} & -8.72 \times 10^{-18} \text{ J}
\end{array}
$$

The energy of the photon for each transition is the difference in the energies of the two states involved.

Transition	Difference in Energies of Two States	= Energy of Emitted Photon
A	$(-0.97 \times 10^{-18}$ J$) - (-2.18 \times 10^{-18}$ J$)$	1.21×10^{-18} J
B	$(-0.97 \times 10^{-18}$ J$) - (-8.72 \times 10^{-18}$ J$)$	7.75×10^{-18} J
C	$(-2.18 \times 10^{-18}$ J$) - (-8.72 \times 10^{-18}$ J$)$	6.54×10^{-18} J

The last column lists the energies asked for in the problem.

EXERCISE A hypothetical atom can exist in only the energy states shown. List all of the energies of all of the photons that will be released from this atom when it is heated.

$$
\begin{array}{cl}
\underline{} & -0.36 \times 10^{-18} \text{ J} \\
\\
\underline{} & -0.81 \times 10^{-18} \text{ J} \\
\\
\underline{} & -3.22 \times 10^{-18} \text{ J}
\end{array}
$$

[*Answer:* 0.45×10^{-18} J; 2.86×10^{-18} J; 2.41×10^{-18} J]

7.4   THE WAVELIKE PROPERTIES OF ELECTRONS

KEY CONCEPT   The de Broglie relation

The de Broglie relation quantitatively expresses the observation that all matter exhibits both a particle character and a wave character. Whether the matter exhibits its wave character or particle character depends on the experiment we use to observe the matter. Certain experiments "force" the particle character to appear, whereas others "force" the wave character to appear. The quantitative expression of this wave-particle duality is

$$
\text{Wavelength of particle} = \frac{h}{\text{mass of particle} \times \text{velocity of particle}}
$$

$$
\lambda = \frac{h}{\text{mass} \times \text{velocity}}
$$

EXAMPLE   Using the de Broglie relation

What is the de Broglie wavelength of a 4.0-oz (114-g) rock moving at 73 m·s^{-1}?

SOLUTION If we know the mass and velocity of an object, we can calculate the de Broglie wavelength, using the equation just given. In this problem, mass = 114 g, velocity = 73 m·s^{-1}, $h = 6.63 \times 10^{-34}$ J·s. In the solution we use the fact that 1 J = 1 kg·m^2·s^{-2}

$$
\lambda = \frac{h}{\text{mass} \times \text{velocity}}
$$

$$
= \frac{6.63 \times 10^{-34} \text{ J·s}}{114 \text{ g} \times 73 \text{ m·s}^{-1}}
$$

$$
= \frac{6.63 \times 10^{-34} \ \cancel{\text{kg}} \cdot \cancel{\text{m}^2} \cdot \cancel{\text{s}^{-2}} \cdot \cancel{\text{s}} \times \dfrac{1000 \ \cancel{\text{g}}}{\cancel{\text{kg}}}}{8.32 \times 10^3 \ \cancel{\text{g}} \cdot \cancel{\text{m}} \cdot \cancel{\text{s}^{-1}}}
$$

Because the units in the numerator contain kilograms and those in the denominator contain grams, we have included a conversion from kilograms to grams. Thus,

$$\lambda = 7.96 \times 10^{-35} \text{ m}$$

EXERCISE What is the mass, in grams, of a particle with a de Broglie wavelength of 450 picometers (pm) and a velocity of 2500 km·s^{-1}?

[*Answer:* 5.89×10^{-31} g]

KEY WORDS   Define or explain each term in a written sentence or two.

Bohr frequency condition	frequency	spectrum
de Broglie relation	photon	transition
electromagnetic radiation	quanta	wavelength
energy level	quantized	wave-particle duality

MODELS OF ATOMS

7.5   ATOMIC ORBITALS

KEY CONCEPT   Orbitals define the electron's energy and probability

An atomic orbital is a volume of space. Two important meanings are attached to the orbital concept. First, an orbital defines the region of space where there is a high **probability** of finding an electron (usually a 95% probability). It must be remembered that an electron in an orbital is found not only on the surface of the orbital but also in (almost) any part of the whole volume of space defined by the orbital. In naive terms, the electron can be found anywhere "inside" the orbital as well as on its surface. The second important aspect of an orbital is that it defines the **energy** of the electron. Orbitals have specific labels and shapes; each main type of orbital—*s*, *p*, *d*, or *f*—has its own particular shape, as indicated in Figures 7.14–7.17 of the text for the *s*, *p* and *d*.

EXAMPLE   Understanding atomic orbitals

Assume an orbital is defined as the volume of space in which there is a 95% chance of finding an electron. What is the probability of finding an electron in one of the lobes of the p_z-orbital (Figure 7.16).

SOLUTION Each *p*-orbital has two lobes; in the picture in the text, we see that the lobes are completely equivalent. Thus, we can conclude that an electron in a *p*-orbital spends half its existence in each lobe. The probability of finding the electron in one lobe is one-half of 95%, or 47.5%.

EXERCISE What is the probability of finding an electron in a d_{xy}-orbital at the nucleus? (Use Figure 7.17 of the text.)

[*Answer:* Zero]

7.6   QUANTUM NUMBERS AND ATOMIC ORBITALS

KEY CONCEPT A   The importance of the principal quantum number

The model of the atom we are developing was first proposed by Schrödinger. In this model, the electron is allowed to exist only in specific orbitals with specific energies. Each orbital is labeled by three quantum numbers. In the hydrogen atom, the principal quantum number, n, gives the energy of the electron through the equation,

$$E_n = -h \times \frac{\mathscr{R}}{n^2}$$

To find the energies of all the possible energy states (and orbitals) in hydrogen, we set $n = 1$ for the lowest energy, $n = 2$ for the next highest, $n = 3$ for the next highest, and so on. The energy loss that occurs when an electron drops from a higher energy to a lower energy is given by the following equation. In the equation, n_2 is the principal quantum number of the orbital to which the electron drops (not necessarily $n = 2$) and n_1 the principal quantum number of the orbital from which the electron falls (not ever $n = 1$).

$$\text{Energy lost} = \mathscr{R} \times h \times \left(\frac{1}{n_2^2} - \frac{1}{n_1^2} \right)$$

The energy lost is the same as the energy of the emitted photon.

EXAMPLE 1 Calculating the energy of a transition

What is the energy of the photon emitted when an electron is hydrogen drops from the $n = 5$ orbit to the $n = 2$ orbital?

SOLUTION The energy of the emitted photon is the same as the energy lost by the electron. We substitute $n_1 = 5$ and $n_2 = 2$ and the universal constants $\mathscr{R} = 3.29 \times 10^{15}$ Hz and $h = 6.63 \times 10^{-34}$ J·Hz^{-1} into the equation just given:

$$\text{Energy of emitted photon} = (6.63 \times 10^{-34} \text{ J·Hz}^{-1}) \times (3.29 \times 10^{15} \text{ Hz}) \times \left(\frac{1}{2^2} - \frac{1}{5^2} \right)$$

$$= (2.18 \times 10^{-18} \text{ J})\left(\frac{1}{4} - \frac{1}{25} \right)$$

$$= (2.18 \times 10^{-18} \text{ J})\frac{21}{100}$$

$$= 4.58 \times 10^{-19} \text{ J}$$

EXERCISE What is the energy of the photon emitted when an electron in hydrogen drops from the $n = 6$ state to the $n = 3$ state?

[*Answer*: 1.82×10^{-19} J]

EXAMPLE 2 Calculating the frequency of light emitted from hydrogen

What frequency of light emitted from a heated hydrogen gas is associated with $n_2 = 1$ and $n_1 = 3$?

SOLUTION To solve this problem, we note that $\Delta E = h\nu$ and $\Delta E = h\mathscr{R} \times \left(\frac{1}{n_2^2} - \frac{1}{n_1^2} \right)$. Thus,

$$h\nu = h\mathscr{R} \times \left(\frac{1}{n_2^2} - \frac{1}{n_1^2} \right)$$

By canceling the Planck constant from both sides of the equation, we obtain the following equation for the frequency of a transition. $n_1 = 3$ and $n_2 = 1$ must be substituted into this equation.

$$\nu = \mathscr{R} \times \left(\frac{1}{n_2^2} - \frac{1}{n_1^2} \right)$$

The value of the Rydberg constant, $\mathcal{R} = 3.29 \times 10^{15}$ Hz, must also be used.

$$\nu = \mathcal{R} \times \left(\frac{1}{1^2} - \frac{1}{3^2}\right)$$

$$= \mathcal{R} \times \left(\frac{1}{1} - \frac{1}{9}\right)$$

$$= (3.29 \times 10^{15} \text{ Hz})\left(\frac{8}{9}\right)$$

$$= 2.92 \times 10^{15} \text{ Hz}$$

It should be noted that 1 and 3 are exact numbers, so the number of significant figures in the answer is determined by the three significant figures used for $\mathcal{R}$.

EXERCISE What frequency and color of light from heated hydrogen is associated with $n_2 = 2$ and $n_1 = 4$?

[*Answer:* 6.17×10^{14}; blue-green]

PITFALL   Evaluating terms such as $\left(\dfrac{1}{n_2^2} - \dfrac{1}{n_1^2}\right)$

When evaluating such an expression, the lowest common denominator must be found and the subtraction then completed. For example, if $n_2 = 2$ and $n_1 = 3$, the lowest common denominator is 36 (4×9). The preceding expression is

$$\frac{1}{2^2} - \frac{1}{3^2} = \frac{1}{4} - \frac{1}{9} = \frac{9}{36} - \frac{4}{36} = \frac{5}{36}$$

KEY CONCEPT B   Shells, subshells, and orbitals

The quantum mechanical picture of the atom results in a probabilistic picture of an electron in the atom, in which it is impossible to know, with absolute certainty, where the electron is. The best that we can do in locating an electron is to state the probability of finding it at some position. An **atomic orbital** is a volume in space in which there is a high probability of finding an electron. Every orbital is described by the three **quantum numbers**—n, l, and m_l—which (with the spin quantum number described below) label the state of an electron in the orbital and specify the value of every property associated with the electron. Only certain quantum numbers are allowed with one another, so a specific pattern of orbitals arises. The orbitals for the three lowest values of n are shown in the following figure. Each orbital is represented by a square.

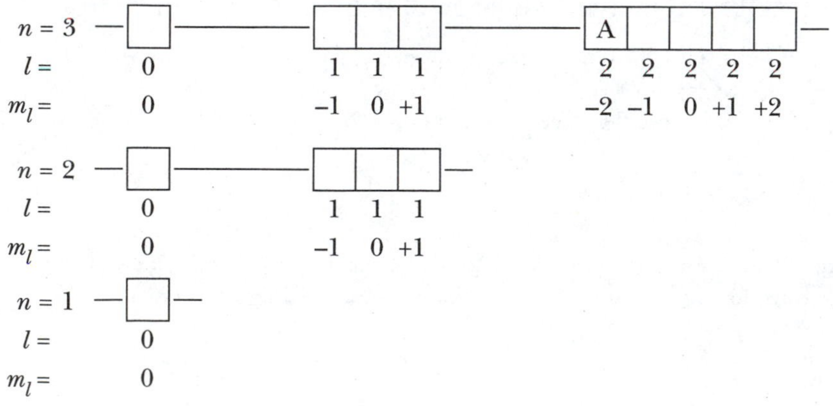

$$
\begin{array}{llll}
n = 3 & \square & \square\,\square\,\square & \boxed{A}\,\square\,\square\,\square\,\square \\
l = & 0 & 1\ \ 1\ \ 1 & 2\ \ 2\ \ 2\ \ 2\ \ 2 \\
m_l = & 0 & -1\ \ 0\ +1 & -2\ -1\ \ 0\ +1\ +2 \\
\\
n = 2 & \square & \square\,\square\,\square \\
l = & 0 & 1\ \ 1\ \ 1 \\
m_l = & 0 & -1\ \ 0\ +1 \\
\\
n = 1 & \square \\
l = & 0 \\
m_l = & 0
\end{array}
$$

The values of the quantum numbers for the orbital labeled A are $n = 3$, $l = 2$, and $m_l = -2$. All the orbitals with the same value of n are said to constitute a **shell**. All the orbitals with the same value of n and the same value of l constitute a **subshell**. For example, in the $n = 3$ shell, there are three subshells, one with $l = 0$ (1 orbital), one with $l = 1$ (3 orbitals), and one with $l = 2$ (5 orbitals). Subshells are often labeled according to the value of l: an s subshell has $l = 0$, a p subshell has $l = 1$, a d subshell has $l = 2$, and an f subshell has $l = 3$. Orbitals are designated with the same symbols; the orbital labeled A in the preceding figure is called a $3d$-orbital.

EXAMPLE Understanding the quantum numbers or orbitals

What quantum numbers are possible for a $3p$-orbital?

SOLUTION By definition, a $3p$-orbital has $n = 3$. The p in the label tells us that $l = 1$ for the orbital. Finally, the preceding illustration indicates that for $l = 1$, m_l may be -1 or 0 or $+1$. Thus, there are three possible sets of quantum numbers for the orbital. These are

$$n = 3, \quad l = 1, \quad m_l = -1$$

$$n = 3, \quad l = 1, \quad m_l = 0$$

$$n = 3, \quad l = 1, \quad m_l = +1$$

EXERCISE Which two orbitals in the first three shells have the same three quantum numbers?

[*Answer:* No two orbitals have the same three quantum numbers.]

7.7 ELECTRON SPIN

KEY CONCEPT Electrons act like a spinning top

Experimental evidence indicates that electrons have what might be considered an intrinsic spin. The spin is labeled with a quantum number called the **spin magnetic quantum number** (m_s). Because experiments indicate that there are only two possible spin states for any electron, there are only two possible spin magnetic quantum numbers: $m_s = +\frac{1}{2}$ and $m_s = -\frac{1}{2}$. An electron with $m_s = +\frac{1}{2}$ is represented by an up arrow ($\uparrow$) and one with $m_s = -\frac{1}{2}$ by a down arrow ($\downarrow$). Every electron must have one of these two spins. For a very large number of electrons in any sample of matter, about one-half have $m_s = +\frac{1}{2}$ and about one-half have $m_s = -\frac{1}{2}$.

EXAMPLE Understanding quantum numbers including spin

What four quantum numbers are consistent with an electron represented by an up arrow in a $2p$-orbital?

$2p$

SOLUTION The quantum numbers for the orbital are associated with an electron in the orbital, so we must first deduce which quantum numbers are allowed for the orbital. By definition, $n = 2$ for a $2p$-orbital and for any p-orbital, $l = 1$. The three allowed values for m_l for $l = 1$ are $m_l = -1$, $m_l = 0$, and $m_l = +1$. Finally, the up arrow indicates that $m_s = +\frac{1}{2}$ for this electron. Any one of the following sets of four quantum numbers is allowed for this $2p$-electron:

$$n = 2, \quad l = 1, \quad m_l = -1, \quad m_s = +\frac{1}{2}$$

$$n = 2, \quad l = 1, \quad m_l = 0, \quad m_s = +\frac{1}{2}$$

$$n = 2, \quad l = 1, \quad m_l = +1, \quad m_s = +\frac{1}{2}$$

Which four quantum numbers are consistent with an electron represented by a down arrow in a 3d-orbital?

$$\boxed{\downarrow}$$
3d

Answer: Five sets of quantum numbers are consistent. These are

$$n = 3, \quad l = 2, \quad m_l = -2, \quad m_s = -\tfrac{1}{2}$$
$$n = 3, \quad l = 2, \quad m_l = -1, \quad m_s = -\tfrac{1}{2}$$
$$n = 3, \quad l = 2, \quad m_l = 0, \quad m_s = -\tfrac{1}{2}$$
$$n = 3, \quad l = 2, \quad m_l = +1, \quad m_s = -\tfrac{1}{2}$$
$$n = 3, \quad l = 2, \quad m_l = +2, \quad m_s = -\tfrac{1}{2}$$

7.8 THE ELECTRONIC STRUCTURE OF HYDROGEN

KEY CONCEPT   Quantized energy in the hydrogen atom

A hydrogen atom in its lowest electronic energy state (the ground state) has its one electron in the lowest energy orbital in hydrogen, the 1s-orbital. There are two possible sets of quantum numbers for this electron, listed as (n, l, m_l, m_s): $(1, 0, 0, +\tfrac{1}{2})$ or $(1, 0, 0, -\tfrac{1}{2})$. If the electron gains more energy (to form an excited state), it can move only into one of the higher energy orbitals; it cannot have a stable existence except in one of the quantized energy states. So, for instance, if the electron gains enough energy to get into one of the next set of orbitals (those with $n = 2$), it can have any one of the eight sets of quantum numbers in the following table (work these through to be certain you understand them). In hydrogen, the energy of all these orbitals is the same:

	1	2	3	4
A	$(2, 0, 0, +\tfrac{1}{2})$	$(2, 1, -1, +\tfrac{1}{2})$	$(2, 1, 0, +\tfrac{1}{2})$	$(2, 1, +1, +\tfrac{1}{2})$
B	$(2, 0, 0, -\tfrac{1}{2})$	$(2, 1, -1, -\tfrac{1}{2})$	$(2, 1, 0, -\tfrac{1}{2})$	$(2, 1, +1, -\tfrac{1}{2})$

As an example, the electron with the quantum numbers corresponding to location B3 in the table would be a 2p-electron with spin $-\tfrac{1}{2}$. It is possible to impart enough energy to the electron to force it to leave the atom entirely; for hydrogen, this requires 1.31×10^3 kJ. When this occurs, the atom is said to be **ionized,** and the process is called **ionization.** Finally, we should note that if some energy is imparted to the atom, but not enough to get the electron to jump to the next highest energy shell, the electron will stay in the ground state; the energy will just pass through the atom.

EXAMPLE   Understanding quantized energy states

Let's assume that when a particular ground-state electron (with spin $+\tfrac{1}{2}$) is excited to the energy level two levels up from the ground state, it changes only its principal quantum number. What quantum numbers will the electron have in the excited state and in what type of orbital is the electron located?

SOLUTION In the ground state, the electron will have the quantum numbers $(1, 0, 0, +\tfrac{1}{2})$. If only the principal quantum number changes when it is excited two levels up from the ground state, the new principal quantum number must be $n = 3$. Thus, the new quantum numbers are $(3, 0, 0, +\tfrac{1}{2})$ and the electron is in a 3s-orbital.

EXERCISE Let's assume that when the electron in the example just given loses energy, it falls to the next lower energy state and its azimuthal quantum number also increases by 1. What are the quantum numbers for the electron, and what type of orbital is it in?

[*Answer:* $(2, 1, 0, +\tfrac{1}{2})$, 2p]

KEY WORDS   Define or explain each term in a written sentence or two.

atomic orbital	magnetic quantum number	s-, p-, d-, f-orbital
azimuthal quantum number	nodal plane	shell
electron energy	principal quantum number	spin magnetic quantum number
ionized	Rydberg constant	subshell

THE STRUCTURES OF MANY-ELECTRON ATOMS

7.9 ORBITAL ENERGIES

KEY CONCEPT   Effective nuclear charge in a many-electron atom

When there is more than one electron in an atom, electron-electron repulsions are present. Any electron in an atom is attracted toward the positive nucleus, but electrons closer to the nucleus repel outer electrons away from the center of the atom, in effect canceling some of the effect of the positive nuclear charge. This effect is called **shielding,** because the closer-in electron essentially shields the farther-out electron from some of the nuclear charge. Orbital shapes determine, to some extent, the shielding ability of an electron. For instance, a 4s-electron, on average, is closer to the nucleus than a 4p-electron. It is said to **penetrate** closer to the nucleus and, therefore, shields other electrons from the nucleus more effectively than a 4p-electron. In addition, because the 4s-electron is closer to the nucleus and less shielded than a 4p-electron, it feels a higher **effective nuclear charge** than the 4p-electron. When an electron feels a higher nuclear charge, it is lower in energy; so a 4s-electron has a lower energy than a 4p-electron. For any shell, penetration and shielding effects result in the following order of subshell energies:

$$\text{Lowest energy} \quad ns < np < nd < nf \quad \text{Highest energy}$$

EXAMPLE   Predicting subshell energies

Without looking at any references, predict which of the two subshells, 2s or 2p, has the lower energy.

SOLUTION For any shell, s-orbitals penetrate closer to the nucleus and shield other electrons more than p-orbitals. Therefore, an electron in a 2s-orbital feels a higher effective nuclear charge and is lower in energy than an electron in a 2p-orbital. This is equivalent to saying that the 2s-subshell is lower in energy than the 2p-subshell.

EXERCISE Which subshell is higher in energy, the 3d or the 3p?

[*Answer:* 3d]

PITFALL   Penetration and shielding for electrons in different shells

The discussions in the text and study guide concern the relative energies of orbitals and subshells in the same shell. The same concepts can be used to understand the relative energies of subshells in different shells, but the details are quite different. At this point, we have not discussed the relative penetration and shielding of subshells in different shells, so we could not easily decide, for example, whether the 3d-subshell is higher or lower in energy than the 4s-subshell.

7.10   THE BUILDING-UP PRINCIPLE

KEY CONCEPT A   The exclusion principle

The exclusion principle consists of two related observations that have been verified experimentally countless times:

1. An orbital can hold a maximum of two electrons.
2. When there are two electrons in an orbital, they must have opposite spins; that is, the spins must be paired.

There are no exceptions to these rules. When two electrons in an orbital have opposite spins, the spins are said to be **paired**. Paired spins are represented by an up arrow next to a down arrow, ↑↓. Another, more common, way of stating these two rules is

• No two electrons in an atom can have the same four quantum numbers.

EXAMPLE   Using the exclusion principle

Assume two electrons must be placed in a $2p$-orbital. Which of the following represents a permissible way of placing the electrons in the orbital?

$$-\boxed{\uparrow\downarrow}- \quad -\boxed{\uparrow\uparrow}- \quad -\boxed{\downarrow\downarrow}-$$

(a)   (b)   (c)

SOLUTION According to the exclusion principle, two electrons in any orbital must have paired spins, that is, they must have opposite spins. In (b) and (c), the two electrons have the same spin; thus, the spins are not paired and the configuration is not allowed. In (a), the spins are paired, so (a) represents the only way to place two electrons in an orbital.

EXERCISE Is it possible for all the electrons in a lithium atom to be in the $1s$-orbital?

[*Answer:* No, because lithium has three electrons, and an orbital can only hold a maximum of two.]

KEY CONCEPT B   The electron configurations of hydrogen, helium, and lithium

We now turn our attention to the **electron configurations** of atoms, in which the number of electrons in each orbital or subshell of an atom is listed. We are interested in atoms that are in their lowest energy state, which is called the **ground state.** In the ground state, electrons are always in the lowest energy orbitals possible, subject to the exclusion principle. As an example, for hydrogen in its ground state, the single electron must be placed in the lowest energy orbital available, which is the $1s$-orbital. We symbolize this by $1s^1$, indicating that one electron is in the $1s$-orbital. For lithium, with atomic number 3, the first two electrons go into the $1s$-orbital. The third electron cannot fit into the $1s$-orbital because this would violate the exclusion principle; it goes in the next lowest energy orbital available, the $2s$-orbital, thus producing a $1s^2 2s^1$ configuration. When a shell in an atom contains the maximum number of electrons it can hold, it is called a **closed shell.** Similarly, a filled subshell is called a **closed subshell.** Helium and lithium each have a closed $1s$-shell, because the $1s$-shell is filled when it contains two electrons.

EXAMPLE   Understanding the electron configuration of lithium

Is it possible for lithium in the ground state to have the electron configuration $1s^1 2s^2$?

SOLUTION No. For an atom to have its ground-state (lowest energy) configuration, the lower energy orbitals must fill with electrons (two electrons per orbital) before electrons go into higher energy orbitals. The $1s^1 2s^2$ configuration for lithium is an excited-state (high-energy) configuration.

EXERCISE How many closed shells are there in ground-state lithium?

[*Answer:* One]

KEY CONCEPT C   Hund's Rule

As we build up the electron configurations of larger atoms, we are occasionally faced with the problem of where to place electrons when orbitals of equal energy are available. For instance, how do we place two electrons into a p-subshell, in which there are three equal-energy orbitals available?

Hund's rule supplies the answer to this question by stating that electrons in a subshell will occupy different orbitals in the subshell before filling any orbital, and that the electrons in the singly occupied orbitals all have the same spin. This rule is illustrated in the following figure for the problem of placing two electrons in three *p*-orbitals.

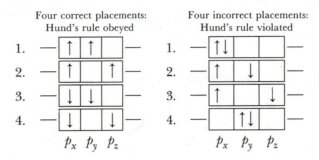

When electrons have their spins in the same direction, as in all the correct configurations just shown, the spins are referred to as **parallel spins.**

EXAMPLE   Using Hund's rule

Draw a correct electron configuration, different from those just shown for two electrons in a *p*-subshell.

SOLUTION To obey Hund's rule, we must place the two electrons in separate orbitals with their spins parallel (in the same direction). We can choose any two orbitals and can use either up spins (↑) or down spins (↓). The following configuration follows these restrictions and is different:

$$-\boxed{\,|\,\downarrow\,|\,\downarrow}-$$
$$p_x \quad p_y \quad p_z$$

EXERCISE When four electrons are placed into a *p*-subshell, how many of the electrons will have their spins unpaired?

[*Answer:* Two]

7.11   THE ELECTRON CONFIGURATIONS OF ATOMS

KEY CONCEPT A   The building-up (Aufbau) principle

The **building-up principle** gives us a straightforward way to predict the electron configuration of many of the atoms in the periodic table. To get an electron configuration, we first determine the number of electrons in the atom from its atomic number. Then, with care to follow the exclusion principle and Hund's rule, we place the electrons into orbitals, starting with the lowest energy orbital and moving to higher energy orbitals until all the electrons have been placed. The first 10 subshells in a large number of neutral many-electron atoms are ordered as follows. (It should be noted that there are some slight changes in this ordering in many ions.)

Lowest energy $1s < 2s < 2p < 3s < 3p < 4s < 3d < 4p < 5s < 4d$ Highest energy

Thus, the order of filling of orbitals, using the building-up principle for atoms of up to 38 electrons, is usually

Fill first $1s, 2s, 2p, 3s, 3p, 4s, 3d, 4p, 5s, 4d$ Fill last

Outer shell electrons are known as **valence electrons.** To save space, inner electrons that mimic a noble-gas configuration may be symbolized by the noble-gas elemental symbol enclosed in brackets [ ]. The electron configurations of five elements are shown in the following table, with valence electrons indicated in boldface.

Element	$1s$	$2s$	$2p_x$	$2p_y$	$2p_z$	$3s$	Configuration
lithium, $Z = 3$	2	1					$1s^2 2s^1$ or $[\text{He}]2s^1$
carbon, $Z = 6$	2	2	1	1			$1s^2 2s^2 2p^2$ or $[\text{He}]2s^2 2p^2$
oxygen, $Z = 8$	2	2	2	1	1		$1s^2 2s^2 2p^4$ or $[\text{He}]2s^2 2p^4$
fluorine, $Z = 9$	2	2	2	2	1		$1s^2 2s^2 2p^5$ or $[\text{He}]2s^2 2p^5$
sodium, $Z = 11$	2	2	2	2	2	1	$1s^2 2s^2 2p^6 3s^1$ or $[\text{Ne}]3s^1$

EXAMPLE Predicting electron configurations

Predict the electron configuration of silicon.

SOLUTION Silicon has atomic number 14, so 14 electrons must be placed into the many-electron atomic orbitals. The order of filling is

$$\text{Fill first} \qquad 1s, \ 2s, \ 2p, \ 3s, \ 3p, \ 4s, \ 3d, \ 4p, \ 5s, \ 4d \qquad \text{Fill last}$$

Care must be taken to obey the exclusion principle and Hund's rule as the 14 electrons are placed into the orbitals. The filling proceeds as follows:

1. The first and second electrons go into the $1s$ orbital with opposite spins, filling the $1s$ orbital.
2. The third and fourth electrons go into the $2s$ orbital with opposite spins, filling the $2s$ orbital.
3. The fifth–seventh electrons go, one each, into the three $2p$ orbitals with parallel spins.
4. The eighth–tenth electrons complete the $2p$ subshell.
5. The eleventh and twelfth electrons go into the $3s$ orbital with their spins paired; this fills the $3s$ orbital.
6. The last two electrons, the thirteenth and fourteenth, go into two different $3p$ orbitals with parallel spins.

The final electron configuration is $1s^2 2s^2 2p^6 3s^2 3p^2$ or $[\text{Ne}]3s^2 3p^2$.

EXERCISE Predict the electron configuration of Cl.

[*Answer:* $1s^2 2s^2 2p^6 3s^2 3p^5$ or, equivalently, $[\text{Ne}]3s^2 3p^5$]

KEY CONCEPT B The filling of *d*-orbitals

The third period of the periodic table ends with argon, which has the configuration $1s^2 2s^2 2p^6 3s^2 3p^6$. Following the order of filling $1s$, $2s$, $2p$, $3s$, $3p$, $4s$, $3d$, $4p$, $5s$, $4d$, the $4s$-orbital fills next, thus producing potassium (K) and calcium (Ca). The $3d$-subshell then fills; 10 electrons are required to fill this subshell, corresponding to the 10 transition elements scandium (Sc) to zinc (Zn). Scandium, titanium, and vanadium have the configurations predicted by the building-up principle. Vanadium, for example, has the configuration $1s^2 2s^2 2p^6 3s^2 3p^6 3d^3 4s^2$ or $[\text{Ar}]3d^3 4s^2$. Chromium, however, has an unexpected configuration as a result of the relatively low energy of half-filled subshells; its configuration is $1s^2 2s^2 2p^6 3s^2 3p^6 3d^5 4s^1$ or $[\text{Ar}]3d^5 4s^1$. As we continue across the period, cobalt, for example, has the configuration $1s^2 2s^2 2p^6 3s^2 3p^6 3d^7 4s^2$ or $[\text{Ar}]3d^7 4s^2$. Copper, like chromium, has an unexpected configuration. In this case, the relatively low energy of a completed subshell is responsible for the configuration $1s^2 2s^2 2p^6 3s^2 3p^6 3d^{10} 4s^1$ or $[\text{Ar}]3d^{10} 4s^1$. One more electron completes the fourth-period transition elements with zinc, which has the configuration $1s^2 2s^2 2p^6 3s^2 3p^6 3d^{10} 4s^2$ or $[\text{Ar}]3d^{10} 4s^2$. Even though the $3d$-subshell fills after the $4s$-subshell, it is quite common to write electron configurations with the $4s$-electrons written after the $3d$, because the $4s$-electrons are lost first when the transition elements ionize.

EXAMPLE Predicting the electron configuration of a transition element

Predict the electron configuration of the fifth-period transition element niobium (Nb) from the building-up principle. Is this the correct configuration for Nb? How many unpaired electrons does Nb have?

SOLUTION The fifth period starts with filling the $5s$-orbital in rubidium (Rb) and strontium (Sr). The transition elements that follow involve filling the $4d$-subshell. Nb is the third transition element in this period and is therefore, predicted to have three $4d$-electrons and the configuration $[Kr]4d^35s^2$. Reference to Appendix 2C of the text indicates that the predicted configuration is not the actual one. Nb has the actual configuration $[Kr]4d^45s^1$. It turns out that there are many such irregular configurations in the fifth-period transition series. Also, Nb has five unpaired electrons because Hund's rule must be obeyed in placing electrons in the $4d$ subshell. This configuration is shown in the following figure.

$$-\boxed{\uparrow}\boxed{\uparrow}\boxed{\uparrow}\boxed{\uparrow}\boxed{\ }-\boxed{\uparrow}-$$
$$\quad\quad 4d \quad\quad\quad\quad 5s$$

EXERCISE What is the predicted and the actual electron configuration of technetium (Tc)? How many unpaired electrons does it have?

[*Answer:* Predicted and actual (from text), $[Kr]4d^55s^2$; five unpaired electrons.]

7.12   THE ELECTRON CONFIGURATIONS OF IONS

KEY CONCEPT   The configurations of positive monatomic ions

Metals lose their most loosely bound electrons to form ions. The electron configuration of an atom determines how tightly electrons are bound and, therefore, affects the electron configuration of the ions formed by the atom. For the transition elements, many ions are formed by loss of one or two loosely held outer s-electrons. As an example, four ions formed in the first transition series by loss of the outer $4s$-electrons are shown in the following table.

Element	Atomic Configuration	Electron Change	Ion	Ion Configuration
Ti	$[Ar]3d^24s^2$	loses two $4s$-electrons	Ti^{2+}	$[Ar]3d^2$
Fe	$[Ar]3d^64s^2$	loses two $4s$-electrons	Fe^{2+}	$[Ar]3d^6$
Ni	$[Ar]3d^84s^2$	loses two $4s$-electrons	Ni^{2+}	$[Ar]3d^8$
Cu	$[Ar]3d^{10}4s^1$	loses one $4s$-electron	Cu^+	$[Ar]3d^{10}$

EXAMPLE   Predicting the electron configuration of an ion

State the charge and configuration of a common ion of silver (Ag).

SOLUTION Ag has the electron configuration $[Kr]4d^{10}5s^1$. Because all the elements in the first transition series (except scandium) can form an ion by loss of their outer s-electrons, we expect Ag to behave similarly and predict it will form an Ag^+ ion by loss of its $5s$-electron. This results in the configuration $[Kr]4d^{10}$.

EXERCISE State the charge and configuration of a common ion formed by Pb.

[*Answer:* Pb^{4+}; $[Xe]4f^{14}5d^{10}$]

PITFALL   The oxidation states of the transition elements

More than 50 different simple ions have been observed for the 10 elements Sc to Zn. We have mentioned only a few of these in the text and study guide up to now. More will be described later in the course.

7.13 ELECTRONIC STRUCTURE AND THE PERIODIC TABLE

KEY CONCEPT Categorizing the elements in the periodic table

Chemists have adopted a variety of ways of categorizing the periodic table. The different ways complement each other and allow us to better understand the various relationships in the periodic table. It would be wise to refer to a periodic table as you read the following list of categorizations.

Blocks: The elements are divided according to the subshell that received the last electron in the building-up process. The periodic table has an *s-block*, a *p-block*, a *d-block*, and an *f-block*. Refer to Figure 7.25 of the text to see where each is.

Main-group elements and others: The elements are also divided into three broad classes called the **main-group elements,** the **transition elements,** and the **inner transition elements.** The *s*-block elements and *p*-block elements grouped together are collectively called the main-group elements. In this classification scheme, the *d*-block elements are called the transition elements and the *f*-block elements are the inner transition elements.

Periods: Each row of elements in the periodic table is called a **period.** The period number is the same as the shell number of the valence shell. For example, selenium is a fourth-period element with electron configuration $[Ar]3d^{10}4s^24p^4$ and its $n = 4$ shell is its valence shell.

Groups: Columns of elements in the periodic table are called **groups.** There is currently no universally accepted way to number the groups. The periodic table we use labels the *s*-block elements Groups 1 and 2 and the *p*-block Groups 13 to 18. Some of the groups also have traditional, but still commonly used, names. These are Group 1, alkali metals; Group 2, alkaline earth metals; Group 17, halogens; and Group 18, noble gases.

EXAMPLE Categorizing the periodic table

What *s*-block element has two electrons in its valence shell and is in the fifth period?

SOLUTION The *s*-block elements are the elements of Group 1 and Group 2. Because the element we are seeking has two electrons in its valence shell, it must be a Group 2 element. The fifth period starts with Rb and ends with Xe. The only element that is in Group 2 and in the fifth period is Sr.

EXERCISE Name the main-group element that has only two 6*p*-electrons.

[*Answer:* Pb (lead)]

KEY WORDS Define or explain each term in a written sentence or two.

building-up principle	exclusion principle	many-electron atom	valence electron
closed shell	ground state	paired spins	
effective nuclear charge	Hund's rule	shielding	
electron configuration	main-group element	transition element	

THE PERIODICITY OF ATOMIC PROPERTIES

7.14 ATOMIC RADIUS

KEY CONCEPT The atomic radius is a periodic property

If we think of atoms as small spheres, then the **radius** of an atom is a measure of its **size.** The atomic radius is determined by experiment. We should note the following facts and trends regarding atomic radii.

1. As we go from left to right in a period in the periodic table, atomic radii tend to decrease.

2. As we go from top to bottom in a group, atomic radii tend to increase. This trend is illustrated by the following examples (and Figures 7.26 and 7.27 of the text).

Period	Group 1	Group 14	Group 17
third	Na ($r = 191$ pm)	Si ($r = 118$ pm)	Cl ($r = 99$ pm)
fifth	Rb ($r = 250$ pm)	Sn ($r = 158$ pm)	I ($r = 133$ pm)

EXAMPLE Understanding periodic trends in atomic radii

Use the periodic table to decide whether Se or Cl has the larger atomic radius.

SOLUTION The main tools we have to solve this problem are the periodic trends of atomic radii. As we go down a group, atomic radii tend to increase. Even though Se and Cl are not in the same group, Se is below Cl, so we suspect that it may have a larger radius than Cl. The second periodic trend is seen as we go from left to right in a period: Atomic radii tend to decrease. Again, Se and Cl are not in the same period, but Se is to the left of Cl, so it probably has a larger radius than Cl. Both trends indicate that Se should be larger than Cl, and, indeed, Se has the larger radius.

EXERCISE Which has the larger radius, Rb or Ca?

[*Answer:* Rb]

7.15   IONIC RADIUS

KEY CONCEPT   The ionic radius is a periodic property

If we think of ions as small spheres, then the radius of an ion is a measure of its size. The **ionic radius** is determined by experiment. We should note the following facts and trends regarding ionic radii.

1. Cations are smaller than the parent atom from which they are formed. For example,

    K $r = 235$ pm   Ca $r = 197$ pm

    K^+ $r = 138$ pm Ca^{2+} $r = 100$ pm

2. Anions are larger than the parent atom from which they are formed. For example,

    Cl $r = 99$ pm   S $r = 104$ pm

    Cl^- $r = 181$ pm S^{2-} $r = 184$ pm

For a series of isoelectronic species (species with the same electronic configuration), the radius decreases as the charge increases (gets more positive) and, conversely, the radius increases as the charge gets more negative. For example, in the following table, all of the species shown are **isoelectronic**, with electron configuration $1s^2 2s^2 2p^6$.

Species	Al^{3+}	Mg^{2+}	Na^+	F^-	O^{2-}	N^{3-}
Radius (pm)	53	72	102	133	140	171

EXAMPLE Predicting the radii of ions

Without looking at any references, predict which is larger, I or I^-.

SOLUTION An anion is always larger than the atom from which it is formed. Thus I^- must have a larger radius than I. (Check the text to find the respective radii and confirm this answer.)

[*Answer:* smallest $K^+ < S^{2-} < P^{3-}$ largest (these species are isoelectronic)]

7.16 IONIZATION ENERGY

KEY CONCEPT   Ionization energies

It is possible to knock the outermost electron off an atom. This is called **ionization**; the minimum energy needed to do this is called the **ionization energy** of the atom. Electrons can also be knocked off molecules and ions, so ionization energies can also be measured for ions and molecules. We can express the first two ionizations of nitrogen, as

$$N(g) \longrightarrow N^+(g) + e^- \qquad I_1 = 1400 \text{ kJ}$$
$$N^+(g) \longrightarrow N^{2+}(g) + e^- \qquad I_2 = 2860 \text{ kJ}$$

I_1 is called the first ionization energy and is the minimum energy required to remove an electron from a gaseous nitrogen atom. I_2 is the second ionization energy and is the minimum energy required to remove an electron from a gaseous N^+ ion. For atoms, the ionization energy is largely (but not exclusively) determined by the size of the atom because it is the outer electron that is removed. Ionization energies follow the periodic trend expected from atomic size. As we go across the periodic table from left to right in a period, ionization energy tends to increase. As we go from top to bottom down a group, ionization energy tends to decrease. Thus, it is quite difficult to remove an electron from O, F, or N. Conversely, atoms on the bottom of the periodic table and to the left have low ionization energies; for example, it is *relatively* easy to remove an electron from Cs, Fr, or Ra. Metals, which form positive ions in their ionic compounds, have low ionization energies. Nonmetals, which form negative ions in their ionic compounds, have high ionization energies; they do not readily form positive ions.

EXAMPLE   Understanding periodic trends in ionization energy

Three elements and their atomic radii are Ca (180 pm), Fe (140 pm), Br (115 pm). Without consulting any references, list these in order of increasing ionization energy. Does the order properly reflect expected periodic trends?

SOLUTION When an atom undergoes ionization, the outer electron is removed. When this electron is far away from the positively charged nucleus, it is generally easier to remove than when it is closer to the nucleus. Thus, all other things being equal, a smaller atom should have a higher ionization energy than a larger atom. The order of ionization energies should be

Highest ionization energy   Br > Fe > Ca   Lowest ionization energy

This order does follow the expected trend in the periodic table. All three elements are in the fourth period, with calcium furthest to the left and bromine furthest to the right. Ionization energies tend to increase in going from left to right in the periodic table.

EXERCISE Without consulting any references except a periodic table, list the three atoms Li, K, and Cs in order of decreasing ionization energy.

[*Answer:* Largest Li > K > Cs Smallest]

7.17 IONIZATION ENERGY AND METALLIC CHARACTER

KEY CONCEPT   Ionization energy is related to metallic character

Metals tend to lose electrons relatively easily. This makes them good electrical conductors and allows them to form the cations that we normally associate with metallic character. Because an element with a low ionization energy loses electrons relatively easily, we can associate a low ionization energy with a more metallic character. That is, metals have low ionization energies and nonmetals have high ionization energies.

EXAMPLE   Using ionization energies and the periodic table

Using a periodic table only, predict whether Na or Cs is more metallic in character.

SOLUTION Both of these elements are metals, but Cs is below Na in Group 1. Because Cs is lower in the group, it will have a smaller ionization energy than Na. It will lose an electron more easily (to form a +1 ion) and will, therefore, be more metallic.

EXERCISE Based on position in the periodic table, would you expect Li or B to be more metallic?

[*Answer:* Li]

7.18   THE INERT-PAIR EFFECT

KEY CONCEPT   A pair of *s*-electrons may not easily form bonds

Elements in Groups 13, 14, and 15 are expected to form the +3, +4, and +5 oxidation states, respectively, corresponding to loss of all outer electrons. However, a pair of *s*-electrons in these elements has a strong tendency to remain attached to the atom, especially for those elements that are low in the group. For instance, in Group 13, elements lower down tend to lose 1 electron as well as the expected 3 electrons. In Group 14, the lower elements lose 2 electrons as well as the expected 4. Thus, for elements in Groups 13, 14, and 15, the +1, +2, and +3 oxidation states, respectively, tend to become more stable as we go down the group.

EXAMPLE   Applying the inert-pair effect

The only stable oxide of aluminum is Al_2O_3. Would you expect the same behavior for thallium, that is, that its only oxide would be Tl_2O_3?

SOLUTION No. We expect to encounter the inert-pair effect at the bottom of Groups 13, 14, and 15. Thus, even though the +3 oxidation state is the only stable state for aluminum, we expect to find both the +1 and +3 state stable for thallium. In fact, it would not be surprising to find the +1 state more stable in some circumstances. Thallium should (and does) form the oxide Tl_2O.

EXERCISE Write the formulas of all of the expected iodides of carbon and lead.

[*Answer:* CI_4, PbI_4, PbI_2 (in actuality, PbI_4 is very unstable)]

7.19   DIAGONAL RELATIONSHIPS

KEY CONCEPT   Diagonal neighbors are sometimes similar

For the paired elements Li and Mg, Be and Al, B and Si, and the metalloids, similarities in properties occur, even though the elements involved are not in the same group. The similarities are (usually) not as strong as for nearby members of the same group, but they are present. An example of this effect can be found, for instance, in the fact that both lithium and magnesium have almost identical atomic radii: 157 pm for Li and 160 pm for Mg. Similarities such as this are recognized as a **diagonal relationship** because the elements involved are diagonal to each other in the periodic table.

EXAMPLE   Using diagonal relationships

Lithium is the only element in Group 1 that reacts directly with nitrogen gas to form a nitride (Li_3N). Name another element that reacts in the same manner—there is at least one that will do so.

SOLUTION With the knowledge that there is at least one other element that forms a nitride by direct reaction with nitrogen and that no other Group 1 element will do so, we fall back on the existence of diagonal relationships to predict that magnesium will undergo the same reaction. The prediction is correct.

PITFALL   Diagonal relationships are of limited predictive value

It is very difficult to know when a diagonal relationship will occur. In the example problem and exercise just given, very specific hints were given to allow the inference that a diagonal relationship is present. In most cases, diagonal relationships are useful in codifying something we already know about the properties of the elements rather than predicting something we don't know.

7.20   ELECTRON AFFINITY

KEY CONCEPT   Attachment of an electron to a gas-phase atom

It is possible to conceive of a process in which an electron becomes attached to a neutral atom. In some cases it is necessary to force the atom to accept the electron; that is, it takes energy to place the electron on the atom. In other cases, energy is released when the electron is attached to the neutral atom, and the electron "wants" to attach to the atom. The **electron affinity** is the energy *released* when an electron becomes attached to a gaseous atom or other chemical species. Thus, the electron affinity is positive when energy is released and negative when it is consumed (this is opposite to the signs associated with enthalpy changes). So, for instance, when 1 mol of electrons are attached to 1 mol of gas-phase fluorine atoms to form 1 mol of gas-phase fluoride ions, 328 kJ of energy is released; the electron affinity of fluorine is $+328 \text{ kJ} \cdot \text{mol}^{-1}$ (see Figure 7.40 of the text). Similarly, 116 kJ is required to attach 1 mol of electrons to 1 mol of neon atoms to form 1 mol of neon anions; the electron affinity of neon is $-116 \text{ kJ} \cdot \text{mol}^{-1}$.

$$F(g) + e^- \longrightarrow F^-(g) \qquad EA = +328 \text{ kJ}$$
$$Ne(g) + e^- \longrightarrow Ne^-(g) \qquad EA = -116 \text{ kJ}$$

Electron affinities do not follow strong periodic trends, but elements in the upper right of the periodic table (nonmetals) have the most negative electron affinities; these elements, therefore, form anions more readily than the other elements in the periodic table.

EXAMPLE   Using electron affinities

What energy change occurs when 1 mol of gas-phase sulfur atoms accepts 1 mol of electrons to form 1 mol of gas-phase S^- ions? Refer to Figure 7.40 of the text.

SOLUTION The electron affinity of sulfur is $+200 \text{ kJ} \cdot \text{mol}^{-1}$. This means that when 1 mol of gas-phase sulfur atoms accepts 1 mol of electrons to form 1 mol of gas-phase S^- ions, 200 kJ of energy is released.

EXERCISE What energy change occurs when 1 mol of gas-phase S^- ions accepts 1 mol of electrons to form 1 mol of gas-phase S^{2-} ions? Refer to Figure 7.40 of the text.

[*Answer:* 532 kJ of energy is consumed from the surroundings.]

PITFALL   The physical state in electron gain

Ionization energies and electron affinities refer to processes that occur in the gas phase. They do not apply directly to similar processes that occur in aqueous solutions.

KEY WORDS   Define or explain each term in a written sentence or two.

atomic radius   electron affinity   ionization energy
covalent radius   ionic radius   isoelectric species
diagonal relationship inert-pair effect

CHEMISTRY AND THE PERIODIC TABLE

7.21–7.23   THE *s-*, *p-*, AND *d-*BLOCK ELEMENTS

KEY CONCEPT A   The relationship among properties, electronic structure, and position in the periodic table

The position of an element in the periodic table determines its properties because the properties are intimately related to the underlying electronic structure. There is a three-way connection among periodic properties, position in the periodic table, and electronic structure, as shown in the figure.

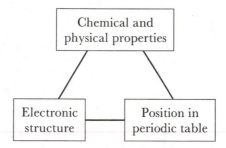

This relationship makes it possible to discuss the properties of the elements on the basis of whether the element is an *s*-block, a *p*-block, or a *d*-block element.

EXAMPLE   Predicting properties of the elements

Which of the elements Se, P, As, or Br is most similar to S in its properties?

SOLUTION   All of the elements in the question are *p*-block elements, so we expect to find some similarities among them. However, both S and Se have similar ns^2np^4 electronic configurations, and we would correctly suspect that Se is the closest to S in many of its properties. For instance, both form ions with a charge of -2 by gaining two electrons to form a noble-gas configuration (Ar configuration for S; Kr configuration for Se) and both are nonconducting solids. Bromine, on the other hand, is a liquid and arsenic is a semiconductor (it conducts electricity, but not well). Phosphorus forms an ion with a -3 charge and bromine an ion with a -1 charge.

EXERCISE   Of the elements Kr, I, Te and C, which is most similar to Cl? Why?

[*Answer:* I; both I and Cl have an ns^2np^5 configuration.]

KEY CONCEPT B   Gradual changes in properties in the periodic table

The position of an element in the periodic table is a clue to the properties of the element; in addition, as we move from one part of the table to another, we see gradual changes in properties, rather than abrupt changes. This has two important ramifications:

1. There are elements with properties intermediate between metals and nonmetals. As we move from right to left in the periodic table, properties gradually change from nonmetallic to metallic; the dividing line between metals and nonmetals is the diagonal

staircase starting to the left of boron and ending between polonium and astatine. The elements in this "gray" area possess some of the properties of metals and some of the properties of nonmetals.

2. *d*-block elements near the *p*-block tend to be similar to nearby *p*-block elements in their properties, whereas *d*-block elements near the *s*-block tend to be similar to nearby *s*-block elements in their properties.

EXAMPLE Understanding gradual changes in the periodic table

Of the elements Ni, Ge, and Br, all fourth-period elements, which possesses properties intermediate between those of a metal and nonmetal?

SOLUTION This is not a memorization question. If we look at a periodic table, we note that Ni is a *d*-block element, well to the left of the dividing line between metals and nonmetals, and is, therefore, a metal; all of the *d*-block elements are metals. Br is a *p*-block element, well to the right of the dividing line between metals and nonmetals; it is a nonmetal. Ge (germanium) is on the dividing line between metals and nonmetals and is, therefore, expected to have properties intermediate between a metal and nonmetal.

EXERCISE Of the elements Sc (scandium), Ru (ruthenium), and Ag (silver), which is expected to be most like Pb (lead) in its chemical reactivity?

[*Answer:* Ag]

PITFALL   How reliable are trends in the periodic table?

We have discussed a variety of similarities and trends that are based on position in the periodic table, but we must be careful not to overgeneralize these relationships. Depending on the elements involved, similarities and trends are observed for some properties but not for others. As an example, even though they are in the same group, Li and Na do not react similarly when burned in air. Sodium forms an oxide and peroxide, whereas lithium forms some nitride and some oxide.

DESCRIPTIVE CHEMISTRY TO REMEMBER

- **Visible light** has wavelengths from 700 nm (red light) to 420 nm (violet light).
- The members of the *s*- and *p*-blocks are called **main-group elements.**
- In, Tl, Sn, Sb, and Bi have an **inert pair of electrons** and, therefore, lose electrons in stages; this results in ions of two different charges for each element.
- **Group I** elements form cations with +1 charge, and **Group II** elements form cations with +2 charge.
- The following pairs of elements share a **diagonal relationship:** Li and Mg, Be and Al, B and Si.
- The *s*-block metals are all very reactive.
- Steel cans are **tin-plated** to protect them against corrosion.
- All of the *d*-block elements (transition elements) are metals.
- Elements on the left of the *p*-block are metals, those on the right are nonmetals.
- The **transition elements** are all metals; many form a variety of oxidation states.

MATHEMATICAL EQUATIONS TO KNOW AND UNDERSTAND

$\lambda \times \nu = c$ electromagnetic radiation

$E = h \times \nu$ energy of a photon

$\Delta E = h \times \nu$ frequency of light generated by an excited atom

$$E = \frac{-h \times \mathcal{R}}{n^2}$$ 　　　　energy of electron in the hydrogen atom

$$\text{Energy lost} = \mathcal{R} \times h \times \left(\frac{1}{n_2^2} - \frac{1}{n_1^2} \right)$$ 　　　　energy lost by electron in hydrogen atom

$$\lambda = \frac{h}{\text{mass} \times \text{velocity}}$$ 　　　　de Broglie wavelength

SELF-TEST EXERCISES

Observing atoms

1. What is the wavelength of electromagnetic radiation with a frequency of 4.4×10^{12} Hz?
(a) 1.5×10^4 m 　　　(b) 3.0×10^5 m 　　　(c) 6.8×10^{-5} m
(d) 1.3×10^{21} m 　　　(e) 4.4×10^{10} m

2. What is the frequency of light with $\lambda = 561$ nm?
(a) 1.87×10^{-15} Hz 　　　(b) 5.35×10^5 Hz 　　　(c) 5.35×10^{14} Hz
(d) 1.87×10^{-6} Hz 　　　(e) 168 Hz

3. What is the energy of a photon with a frequency of 6.5×10^9 Hz?
(a) 1.0×10^{-43} J 　　　(b) 9.8×10^{42} J 　　　(c) 5.2×10^{-19} J
(d) 4.3×10^{-24} J 　　　(e) 2.3×10^{23} J

4. What is the energy of a mole of photons with wavelength 3.3 cm?
(a) 6.6×10^{-25} J 　　　(b) 3.6 J 　　　(c) 1.3×10^{-10} J
(d) 5.4×10^{31} J 　　　(e) 21 J

5. Approximately 1.0% of the energy emitted from a $150\text{-J} \cdot \text{s}^{-1}$ 3200-K photoflood lamp is from photons with wavelength 450 nm. How many photons per second are emitted at 450 nm?
(a) 3.4×10^{18} 　　　(b) 8.2×10^{24} 　　　(c) 7.4×10^{21}
(d) 4.4×10^{20} 　　　(e) 3.40×10^9

6. Which of the following wavelengths of light falls within the visible region?
(a) 500 nm 　　　(b) 20 nm 　　　(c) 800 nm 　　　(d) 3.00 cm 　　　(e) 180 nm

7. Which type of photon has the highest energy?
(a) infrared 　　　(b) red 　　　(c) blue 　　　(d) x-ray 　　　(e) ultraviolet

8. What is the frequency of the light emitted when a hydrogen atom changes energy states from the $n = 7$ state to the $n = 5$ state?
(a) 1.37×10^{14} Hz 　　　(b) 2.65×10^{14} Hz 　　　(c) 1.88×10^{14} Hz
(d) 1.65×10^{15} Hz 　　　(e) 6.45×10^{13} Hz

9. When the electron in a hydrogen atom drops to the $n = 1$ level, light with frequency 3.223×10^{15} Hz is emitted. What is the value of n for the energy level the electron originated in?
(a) 2 　　　(b) 3 　　　(c) 7 　　　(d) 4 　　　(e) 5

10. An atom with energy -3.23×10^{-19} J changes its energy to -7.27×10^{-19} J by emitting a photon. What is the energy of the photon?
(a) 10.5×10^{-19} J 　　　(b) 0 J 　　　(c) 4.04×10^{-19} J
(d) 3.23×10^{-19} J 　　　(e) 7.27×10^{-19} J

11. A potassium atom emits a photon with energy 2.59×10^{-21} J in changing energy states. What is the frequency of the photon?
(a) 8.63×10^{-30} Hz   (b) 3.91×10^{12} Hz   (c) 1.71×10^{-54} Hz
(d) 2.56×10^{-13} Hz   (e) 5.44×10^{15} Hz

12. Calculate the energy of the $n = 3$ level of the hydrogen atom.
(a) -2.18×10^{-18} J   (b) -7.27×10^{-19} J   (c) -3.54×10^{-18} J
(d) -5.89×10^{-17} J   (e) -2.42×10^{-19} J

13. What is the change in energy of a hydrogen atom when an electron falls from the $n = 6$ state to the $n = 1$ state?
(a) -2.74×10^{15} J   (b) -1.82×10^{-18} J   (c) -3.20×10^{15} J
(d) -2.18×10^{-18} J   (e) -2.12×10^{-18} J

14. A mole of hydrogen atoms changes energy from the $n = 3$ state to the $n = 2$ state. How much energy is released?
(a) 4.36×10^{-19} J   (b) 3.03×10^{-19} J   (c) 262 kJ
(d) 182 kJ   (e) 109 kJ

15. Calculate the de Broglie wavelength of a proton moving at 350 km·s^{-1}.
(a) 1.13×10^{-12} m   (b) 1.39×10^5 nm   (c) 0.139 nm
(d) 1.39×10^{-4} nm   (e) 1.13 nm

16. The de Broglie wavelength of an electron is 410 nm. How fast is the electron moving? The mass of an electron is 9.109×10^{-31} kg; $h = 6.63 \times 10^{-34}$ J·Hz^{-1}.
(a) 2.54×10^3 m·s^{-1} (b) 1.78×10^3 m·s^{-1} (c) 550 m·s^{-1}
(d) 6.55×10^3 m·s^{-1} (e) 2.11×10^4 m·s^{-1}

Models of atoms

17. Which of the following best represents a p-orbital?

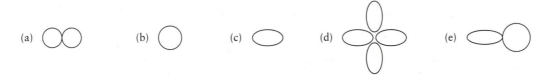

18. How many subshells are present in the $n = 3$ shell?
(a) 3   (b) 4   (c) 9   (d) 18   (e) 2

19. How many orbitals are present in the $n = 2$ shell?
(a) 4   (b) 2   (c) 8   (d) 16   (e) 1

20. What three quantum numbers are permissible for a $3p$-orbital? The answers are expressed as (n, l, m_l).
(a) (3, 0, 0)   (b) (3, 2, −2)   (c) (3, 2, −1)
(d) (2, 1, 0)   (e) (3, 1, −1)

21. How many orbitals are present in the $7d$-subshell?
(a) 10   (b) 5   (c) 4   (d) 7   (e) 14

22. What four quantum numbers are permissible for a $3d$-electron? The answers are expressed as (n, l, m_l, m_s).

(a) $(3, 0, 0, -\frac{1}{2})$ (b) $(3, 2, 2, -\frac{1}{2})$ (c) $(3, 2, 2, 1)$

(d) $(2, 1, 0, +\frac{1}{2})$ (e) $(3, 1, -1, \frac{1}{2})$

23. An electron with the quantum numbers $n = 4$, $l = 2$, $m_l = 0$, and $m_s = -\frac{1}{2}$ is designated as what kind of electron?

(a) $6s$ (b) $2s$ (c) $2p$ (d) $4d$ (e) $4s$

The structures of many-electron atoms

24. Which of the following orbitals penetrates most?

(a) $4s$ (b) $4p$ (c) $4d$ (d) $4f$

25. What is the maximum number of electrons permitted in a $4f$-orbital?

(a) 7   (b) 14   (c) 1   (d) 3   (e) 2

26. What is the electron configuration of N?

(a) $1s^2 2s^2 3p^6 2 3s^1$ (b) $1s^2 2s^2 3p^6 6 3s^2 3p^2$ (c) $1s^2 2s^2 3p6^3$

(d) $1s^6 2s^1$ (e) $1s^2 2s^2 3s^2 4f^1$

27. What is the electron configuration of Mn?

(a) $[Ar]3d^7$ (b) $[Ar]3d^6 4s^1$ (c) $[Ar]3d^5 4s^2$

(d) $[Ar]4s^2 4p^5$ (e) $[Ar]3d^3 4s^2 4p^2$

28. How many unpaired electrons are present in an oxygen atom?

(a) 4   (b) 8   (c) 1   (d) 6   (e) 2

29. What is the electron configuration of lithium in the ground state?

(a) $1s^3$ (b) $1s^1 2s^2$ (c) $2s^3$

(d) $1s^1 2s^1 2p^1$ (e) $1s^2 2s^1$

30. Which of the following configurations is permitted for four electrons in a p-subshell?

(a) $-\boxed{\uparrow\;|\;\uparrow\;|\;\uparrow\downarrow}-$ (b) $-\boxed{\uparrow\uparrow\;|\;\uparrow\;|\;\uparrow}-$ (c) $-\boxed{\uparrow\downarrow\;|\;\uparrow\downarrow\;|\;}-$

(d) $-\boxed{\uparrow\uparrow\uparrow\downarrow\;|\;}-$ (e) $-\boxed{\uparrow\downarrow\;|\;\uparrow\;|\;\uparrow}-$

31. How many valence electrons are present in Cl?

(a) 7   (b) 5   (c) 10   (d) 17   (e) 2

32. Which of the following atoms possesses only closed shells in the ground state?

(a) K   (b) F   (c) Ar   (d) Se   (e) Zn

33. What is the electron configuration of Pb?

(a) $[Xe]6s^2 6p^2$ (b) $[Xe]4f^{14} 5d^{10} 6s^2 6p^2$ (c) $[Xe]5d^{10} 6s^2 6p^2$

(d) $[Xe]6s^2 6p^6 6d^{10} 6f^{10}$ (e) $[Xe]6s^2 6p^2 6d^{10} 6f^{14}$

34. What is the electron configuration of Se^{2-}?

(a) $[Ar]3d^{10} 4s^2 4p^4$ (b) $[Ar]$ (c) $[Ar]3d^{10} 4s^2 4p^5$

(d) $[Kr]$ (e) $[Ar]4s^2 4p^4$

35. What is the electron configuration of Fe^{3+}?
(a) $[Ar]3d^34s^2$ (b) $[Ar]3d^5$ (c) $[Ar]3d^44s^1$
(d) $[Ar]4s^24p^3$ (e) $[Ar]4p^5$

36. Which of the following is a main-group element?
(a) Si   (b) Zn   (c) Hg   (d) Ce   (e) U

37. Use the periodic table to find which element has four p-electrons in its outer shell. It is not necessary to write out complete electron configurations.
(a) Mn   (b) Ti   (c) S   (d) C   (e) Cl

38. Which of the following is a p-block element?
(a) Ca   (b) Pd   (c) U   (d) Br   (e) V

The periodicity of atomic properties

39. Which of the following has the largest atomic radius?
(a) Rb   (b) K   (c) Ca   (d) I   (e) S

40. Which Group 1 element has the highest ionization energy?
(a) Li   (b) Na   (c) K   (d) Rb   (e) Cs

41. Which of the following has the highest ionization energy?
(a) Al^{2+} (b) Mg^+ (c) Mg^{2+} (d) Na   (e) Mg

42. All of the following ions are isoelectronic. Which is the smallest?
(a) N^{3-} (b) O^{2-} (c) F^- (d) Na^+ (e) Mg^{2+}

43. High metallic character is associated with _____ ionization energy.
(a) high   (b) low   (c) neither high nor low

44. The electron affinity is associated with which of the following processes? (E = element)
(a) $E(aq) + e^- \longrightarrow E^-(aq)$ (b) $E(g) \longrightarrow E^+(g) + e^-$ (c) $2E(g) \longrightarrow E_2(g)$
(d) $E(s) \longrightarrow E(g)$ (e) $E(g) + e^- \longrightarrow E^-(g)$

45. A hypothetical element Z that possesses an inert pair forms an ion with a +1 charge. What charge would be expected for a second ion of Z, based on the fact that it possesses an inert pair?
(a) +2   (b) +7   (c) +4   (d) +3   (e) no other ions

46. What element shares a diagonal relationship with Mg?
(a) Sc   (b) Li   (c) B   (d) Ca   (e) K

Chemistry and the periodic table

47. Which of the following is most likely to be a reactive metal?
(a) Cs   (b) Cu   (c) Pb   (d) Au   (e) Ni

48. Which of the following groups of the periodic table has a nonmetal at the top and a metal at the bottom?
(a) 1   (b) 14   (c) 17   (d) 10   (e) 18

49. Which of the following elements is most likely to form more than one oxidation state in its compounds?
(a) Rb   (b) Sr   (c) Cr   (d) Al   (e) F

Descriptive chemistry

50. What is the charge of the ion formed by potassium (Group 1)?
(a) +3   (b) −1   (c) −2   (d) +1   (e) +2

51. Which of the following is a colorless, odorless, unreactive gas?
(a) O_2 (b) Cl_2 (c) H_2 (d) Ar   (e) F_2

52. Which of the following is a relatively unreactive metal?
(a) K   (b) Pb   (c) Ba   (d) Cr   (e) Be

53. Which element reacts with both acids and bases?
(a) Na   (b) Be   (c) S   (d) He   (e) Au

54. Which of the following Group 1 elements is the most reactive?
(a) Li   (b) Na   (c) K   (d) Rb   (e) Cs

55. Which element is expected to have properties intermediate between a metal and nonmetal?
(a) Mn   (b) Si   (c) Xe   (d) O   (e) Ag

CHAPTER 8

INSIDE MATERIALS: CHEMICAL BONDS

If it is possible to do so, chemical systems will generally move from a higher energy state to a lower energy state. For atoms, one way to reach lower energy is to form chemical bonds; thus, chemical bond formation is a reflection of the desirability of low-energy states. The outer shell electrons (**valence** electrons) in an atom are intimately involved in the formation of chemical bonds. We will now put to use the information regarding electron configurations that we learned in the previous chapter.

IONIC BONDS

8.1   LEWIS SYMBOLS FOR ATOMS AND IONS

KEY CONCEPT   Lewis symbols show all valence electrons

Ionic compounds form through the loss of electrons by a metal and the gain of those electrons by a nonmetal. A conventional method of keeping track of the electrons in a compound starts with the **Lewis symbols** of atoms, in which each of the valence electrons of an atom is indicated by a dot next to the symbol of the element. In drawing Lewis symbols, it is helpful to remember that the number of valence electrons for the s-block elements is the same as the group number and, for the p-block elements, the same as the group number minus ten.

1	2	13	14	15	16	17	18
Na·	Mg:	Ȧl:	·Ṡi:	·P̈:	·S̈:	:C̈l:	: Är:

If two electrons are in the same orbital, the dots are placed next to each other, as illustrated for the two 3s-electrons in magnesium. Electrons in separate orbitals are placed away from each other; thus, the three p-electrons in phosphorus, which are each in separate orbitals, are represented by three single dots. Because all the atoms in a group have similar electron configurations, all the elements in a group have similar electron dot diagrams. For instance, in Group 2, we have

Element	Be	Mg	Ca	Sr	Ba	Ra
Configuration	[He]$2s^2$	[Ne]$3s^2$	[Ar]$4s^2$	[Kr]$5s^2$	[Xe]$6s^2$	[Rn]$7s^2$
Electron dot diagram	Be:	Mg:	Ca:	Sr:	Ba:	Ra:

The orientation of dots around the symbol is immaterial, as long as the number of dots and their presence in the same or different orbitals are correctly shown. For example, all of the following Lewis symbols are correct for oxygen (Group 16):

$$:\ddot{O}· \quad :\ddot{O}: \quad :\ddot{O}· \quad ·\ddot{O}· \quad ·\ddot{O}: \quad ·\ddot{O}:$$

EXAMPLE 1   Drawing Lewis Symbols

Draw the Lewis symbol of Te.

SOLUTION It is not necessary to write the entire electron configuration of an element to draw its Lewis symbol. Tellurium is a Group 16 element, so it has six valence electrons. In addition, because all Group 16 elements have an ns^2np^4 configuration, four of the electrons will be paired in two orbitals (an s-orbital and one of the three p-orbitals) and the fifth and sixth electrons are each alone in a p-orbital. Thus, any one of the Lewis symbols shown here is correct:

$$:\overset{\displaystyle ..}{\text{Te}}\cdot \quad :\text{Te}: \quad :\overset{\displaystyle .}{\underset{\displaystyle .}{\text{Te}}}\cdot \quad \cdot\overset{\displaystyle ..}{\text{Te}}\cdot \quad \cdot\overset{\displaystyle .}{\text{Te}}: \quad \cdot\underset{\displaystyle ..}{\text{Te}}:$$

EXERCISE Draw the Lewis symbol of Ga.

[*Answer:* $\overset{\displaystyle ..}{\text{Ga}}\cdot$]

EXAMPLE 2   Drawing Lewis formulas of ionic compounds

Use Lewis symbols to describe the formula of magnesium nitride.

SOLUTION We first note that magnesium is a Group 2 element and will, therefore, form an ion with charge +2; nitrogen is a Group 15 element and will form an ion with charge −3. Each magnesium attains the configuration of the previous noble gas (Ne) by losing two electrons and, in doing so, attains a +2 charge. Each nitrogen attains the configuration of the next noble gas (Ne) by gaining three electrons and, in doing so, gets a −3 charge. Electron balance is maintained by having three magnesiums each lose two electrons for a total of six electrons; these electrons are then gained by two nitrogens, which each gain three:

$$3\,\text{Mg}: + 2:\overset{\displaystyle .}{\underset{\displaystyle .}{\text{N}}}\cdot \longrightarrow 3\,\text{Mg}^{2+} + 2:\overset{\displaystyle ..}{\underset{\displaystyle ..}{\text{N}}}:^{3-} \longrightarrow \text{Mg}_3\text{N}_2$$

EXERCISE Use Lewis symbols to describe the formula of sodium sulfide.

[*Answer:* $2\,\text{Na}\cdot + :\overset{\displaystyle ..}{\underset{\displaystyle .}{\text{S}}}\cdot \longrightarrow 2\,\text{Na}^+ + :\overset{\displaystyle ..}{\underset{\displaystyle ..}{\text{S}}}:^{2-} \longrightarrow \text{Na}_2\text{S}$]

8.2   LATTICE ENTHALPIES

KEY CONCEPT A   Lattice enthalpies

We now consider a process in which gas-phase positive ions and negative ions come together to form a solid rather than a pair of gas-phase ions. For NaF, this would be

$$\text{Na}^+(g) + \text{F}^-(g) \longrightarrow \text{NaF}(s) \qquad \Delta H = -929\,\text{kJ}$$

We assume here that the ions on the left side of the equation are very far apart, so their energy of interaction is zero. Also, it should be recognized that every positive ion in the solid interacts with all of the other ions in the solid (both positive and negative), and every negative ion also interacts with every other ion in the solid; the enthalpy of reaction results from this myriad of interactions, not from the interaction of just one positive ion with one negative ion. Because the dominant interaction is the favorable one between cations and anions, forming the solid from the gas-phase ions is always highly exothermic. The enthalpy of the reverse reaction is called the **lattice enthalpy** of the ionic compound involved. For NaF(s), for instance, the lattice enthalpy is +929 kJ.

EXAMPLE   Understanding the magnitude of lattice enthalpies

According to Table 8.1 of the text, the lattice enthalpy of LiF is 1046 kJ·mol^{-1}, whereas that of KI is 645 kJ/mol. Suggest a reason for this difference.

SOLUTION The size of the lattice enthalpy depends predominantly on two factors: the size of the ions and the charge of the ions. Smaller ions can get closer together than larger ions. Thus, all other things the same, smaller ions experience stronger interactions and compounds with small ions have a larger lattice enthalpy than compounds with large ions. With all other things the same, ions of higher charge experience stronger interactions than ions of lower charge; this results in a higher lattice enthalpy for compounds with ions of higher charge. For our problem, the charge on all of the ions (Li$^+$, F$^-$, K$^+$, I$^-$) is +1 or −1, so differences in charge cannot account for the different lattice enthalpies. However, K$^+$ is significantly larger than Li^+, and I^- is significantly larger than F^-. The larger ions in KI, relative to those in LiF, are responsible for the higher lattice enthalpy of LiF.

EXERCISE Which has the larger lattice enthalpy, BaS or KCl?

[*Answer:* BaS (because of +2 and −2 charges on ions)]

KEY CONCEPT B   The Born-Haber cycle

When dealing with ionic solids, we are frequently interested in the value of the standard enthalpy of formation of the solid from the elements. For NaF, this is the enthalpy of reaction for the reaction

$$Na(s) + \tfrac{1}{2}F_2(g) \longrightarrow NaF(s) \qquad \Delta H = \text{standard enthalpy of formation}$$

The **Born-Haber cycle** breaks down the overall reaction into five steps, as illustrated below for NaF. We start with $Na(s)$ and $F_2(g)$ and use the various enthalpies from the text.

Step 1. Sublime 1 mol Na	$Na(s) \longrightarrow Na(g)$	$\Delta H° = +108.4$ kJ
Step 2. Ionize 1 mol Na	$Na(g) \longrightarrow Na^+(g) + e^-$	$\Delta H° = +494$ kJ
Step 3. Dissociate $\tfrac{1}{2}$ mol $F_2(g)$	$\tfrac{1}{2}F_2(g) \longrightarrow F(g)$	$\Delta H° = +78.99$ kJ
Step 4. Attach electron to 1 mol F(g)	$F(g) + e^- \longrightarrow F^-(g)$	$\Delta H° = -328$ kJ
Step 5. Form solid lattice from 1 mol $Na^+(g)$ and 1 mol $F^-(g)$	$Na^+(g) + F^-(g) \longrightarrow NaF(s)$	$\Delta H° = -929$ kJ

These reactions all sum to the reaction for formation of solid NaF from the elements, and the sum of the enthalpies is the standard enthalpy of formation of NaF:

$$Na(s) + \tfrac{1}{2}F_2(g) \longrightarrow NaF(s) \qquad \Delta H° = -576 \text{ kJ}$$

The large exothermicity of this reaction tells us that the formation of NaF from its elements is energetically a very favorable process.

EXAMPLE   Using the Born-Haber cycle

Would we expect $NaF_2(s)$ to form from its elements? Assume that the lattice enthalpy of $NaF_2(s)$ is about the same as that of $MgF_2(s)$, +2913 kJ. Comment on the result. (Experimentally, $NaF_2(s)$ does not exist.)

SOLUTION If the enthalpy of formation of $NaF_2(s)$ is negative, we would predict that it would form from its elements. On the other hand, a positive value for the enthalpy of formation indicates it would not form. The Born-Haber cycle can be used to calculate the standard energy of formation of this (hypothetical) ionic solid:

$$Na(s) + F_2(g) \longrightarrow NaF_2(s) \qquad \Delta H° = ?$$

From the Born-Haber cycle, we get these steps:

Step 1. Sublime metal	$Na(s) \longrightarrow Na(g)$	$\Delta H° = +108.4$ kJ
Step 2. Ionize metal to +1 ion	$Na(g) \longrightarrow Na^+(g) + e^-$	$\Delta H° = +494$ kJ
Step 3. Ionize +1 metal ion to +2 ion	$Na^+(g) \longrightarrow Na^{2+}(g) + e^-$	$\Delta H° = +4562$ kJ
Step 4. Dissociate 1 mol $F_2(g)$	$F_2(g) \longrightarrow 2F(g)$	$\Delta H° = +157.98$ kJ
Step 5. Attach electrons to 2 mol F(g)	$2F(g) + 2e^- \longrightarrow 2F^-(g)$	$\Delta H° = -656$ kJ
Step 6. Form solid lattice from $Na^{2+}(g)$ and $F^-(g)$	$Na^+(g) + 2F^-(g) \longrightarrow NaF_2(s)$	$\Delta H° = -2913$ kJ

The sum of the enthalpies of these reactions is +1753 kJ:

$$Na(s) + F_2(g) \longrightarrow NaF_2(s) \qquad \Delta H° = +1753$$

The positive enthalpy of formation indicates that $NaF_2(s)$ will not form when $Na(s)$ and $F_2(g)$ react. The main reason for this result is the enormous amount of energy required to form the Na^{2+} ion in Step 3.

EXERCISE Use the Born-Haber cycle to calculate the enthalpy of formation of $AgCl(s)$.

[*Answer:* $-127 \text{ kJ} \cdot \text{mol}^{-1}$]

8.3   THE PROPERTIES OF IONIC COMPOUNDS

KEY CONCEPT   Ionic solids

The properties of ionic solids are a reflection of the underlying structure. Lattice enthalpies are very large, and the oppositely charged ions in an ionic compound are held together with great tenacity. In addition, the ions cannot slide by each other very easily because, to do so, ions of the same charge would encounter each other. Thus, ionic solids typically have high melting and high boiling points; in addition, they tend to be brittle. Many ionic solids can be broken up into separate ions by dissolution in water; however, for many other ionic solids, the ions are held together so strongly in the crystal lattice that they will not break up into separate ions and will not dissolve.

EXAMPLE   Understanding ionic solids

Which familiar substance is most likely ionic, candle wax or chalk?

SOLUTION Candle wax melts at a low temperature and is not very brittle (it bends easily at slightly elevated temperatures). Chalk is very brittle and has a very high melting point, as you would discover if you tried to melt it. Chalk has the properties of a typical ionic solid; it is, in fact, the ionic compound calcium carbonate.

EXERCISE Would it be possible to have an ionic compound composed of a lattice of Ba^{2+} ions combined with Li^+ ions only?

[*Answer:* No]

KEY WORDS   Define or explain each term in a written sentence or two.

Born-Haber cycle	lattice enthalpy	valence electron
crystalline solid	Lewis formula	variable valence
duet (duplet)	Lewis symbol	
ionic solid	octet	

COVALENT BONDS

8.4   FROM ATOMS TO MOLECULES

KEY CONCEPT   A covalent bond is a shared pair of electrons

Unlike the ionic bond, which involves the transfer of one or more valence electrons from a metal to a nonmetal, the covalent bond involves the sharing of a pair of electrons, generally (but not always) between two nonmetals. The sharing is facilitated by the overlap and merging together of orbitals that contain valence electrons. As usual, a bond forms because its formation lowers the energy of the atoms involved in the bond. Why does a covalent bond form instead of an ionic bond, or vice versa? There are complex underlying reasons for formation of one type of bond over the other, but the basic reason is that the bond that lowers the energy the most is the one that forms.

EXAMPLE   Understanding the covalent bond

What two orbitals merge to form a covalent bond in the F_2 molecule?

SOLUTION The electron configuration of fluorine is $1s^2 2s^2 2p^5$. Therefore, the only unpaired electron in fluorine is in a $2p$-orbital. The easiest way for two fluorines to covalently bond is for a $2p$-orbital on one fluorine (the orbital with a single electron) to merge with a similar $2p$-orbital on the other fluorine, thereby resulting in a shared pair of electrons between the two fluorines.

EXERCISE How many unpaired electrons are there in the fluorine molecule?

[*Answer:* None]

8.5   THE OCTET RULE AND LEWIS STRUCTURES

KEY CONCEPT A   The electron-pair bond

When a nonmetal atom bonds to another nonmetal atom, formation of an ionic bond is energetically unfavorable. Instead, the atoms bond to each other by sharing one or more pairs of electrons, to form what is called a **covalent bond.** Generally, the energetically most favorable bonding is achieved when each atom attains a noble-gas configuration of eight valence electrons (two valence electrons for hydrogen). For the purpose of counting valence electrons, shared electrons are counted as being in the valence shell of each atom sharing them and are therefore counted twice. The tendency for atoms to attain a noble-gas configuration is called the **octet rule** (eight valence electrons) or **duet rule** (two valence electrons for hydrogen).

EXAMPLE   Counting valence electrons

Water can be represented by the picture H:Ö:H. (a) For this picture, draw circles around the electrons that have formed covalent bonds by being shared by two atoms. (b) How many covalent bonds does water possess? (c) Show that all of the atoms in water obey the octet rule or duet rule.

SOLUTION (a) The electrons that are shared are those that are drawn between atoms. These are circled:

$$\text{H } \boxed{\vdots}\text{O}\boxed{\vdots}\text{ H}$$

(b) There are two covalent bonds because each shared pair of electrons is called a bond. The two electrons that make up a bond are often indicated by a line (—) rather than by a pair of dots (:), so water may also be written as H—Ö—H. (c) To count valence electrons for purposes of checking the octet rule, shared electrons are counted with both atoms that share them. The circles indicate the electrons counted with each atom:

| 2 valence electrons for hydrogen | 8 valence electrons for oxygen | 2 valence electrons for hydrogen |

Each hydrogen has achieved the He noble-gas configuration by having two electrons in its valence shell, and oxygen has achieved the neon noble-gas configuration by having eight valence electrons. The octet (duet for hydrogen) rule has been satisfied.

EXERCISE (a) How many valence electrons does the nitrogen atom in ammonia have?

$$\text{H}-\overset{\displaystyle ..}{\text{N}}-\text{H}$$
$$|$$
$$\text{H}$$

(b) How many covalent bonds are there in this molecule? (c) Do all of the atoms have a noble-gas configuration?

[*Answer:* (a) 8; (b) 3; (c) yes]

PITFALL   What atoms form covalent bonds?

The formation of covalent bonds always occurs when one nonmetal bonds to another nonmetal. However, a variety of metals, including Li, Be, Mg, Al, and Sn, form covalent bonds with some nonmetals, as in $BeCl_2$ and $AlCl_3$.

KEY CONCEPT B   Lone pairs

A pair of electrons that is shared by two atoms is called a covalent bond. Many compounds have pairs of electrons that are not shared but that belong exclusively to one atom. Such a pair is called a **lone pair.** Lone pairs of electrons have profound effects on the chemistry and structure of compounds, so it is essential to recognize them in the structure of a compound. When you count electrons to see if an atom obeys the octet rule, lone pairs are counted only with the single atom to which they belong.

EXAMPLE   Counting lone pairs

How many lone pairs of electrons does water possess?

SOLUTION Valence electrons that are not shared to form a covalent bond are lone pairs. Water has two lone pairs (four electrons), which are circled:

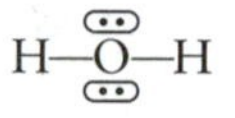

$$H—\overset{\cdot\cdot}{\underset{\cdot\cdot}{O}}—H$$

Two lone pairs

EXERCISE How many lone pairs of electrons are present in ammonia, NH_3? Which atom do they belong to?

[*Answer:* One lone pair; nitrogen]

KEY WORDS   Define or explain each term in a written sentence or two.

covalent bond   lone pair
duet rule   octet rule
Lewis structure   valence

THE STRUCTURES OF POLYATOMIC SPECIES

8.6   LEWIS STRUCTURES

KEY CONCEPT A   Multiple bonds

At times, more than one pair of electrons is shared by two atoms. When two pairs of electrons are shared by two atoms, the resulting bond is called a **double bond.** The Lewis structure for sulfur dioxide has a double bond:

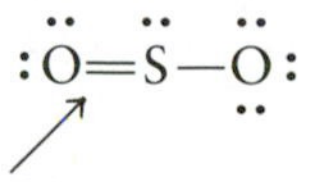

$$:\overset{\cdot\cdot}{O}=\overset{\cdot\cdot}{S}—\overset{\cdot\cdot}{\underset{\cdot\cdot}{O}}:$$

Double
bond

A triple bond consists of three pairs of electrons shared by two atoms. Acetylene (ethyne) has a **triple bond:**

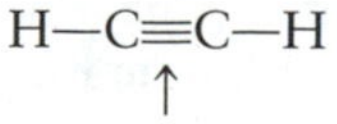

$$H—C\equiv C—H$$

Triple
bond

Double and triple bonds are counted in the usual way when you check to see if an atom obeys the octet rule. You should confirm that all of the atoms in SO_2, including the double-bonded atoms, have eight electrons in their valence shells. Similarly, both carbon atoms in acetylene are surrounded by eight electrons. It is useful to remember that C, O, N, and S often form multiple bonds.

EXAMPLE   Using the octet rule with double bonds

How many double bonds are present in carbon dioxide, $:\overset{..}{O}=C=\overset{..}{O}:$? Do all of the atoms in carbon dioxide obey the octet rule?

SOLUTION Carbon dioxide has two double bonds:

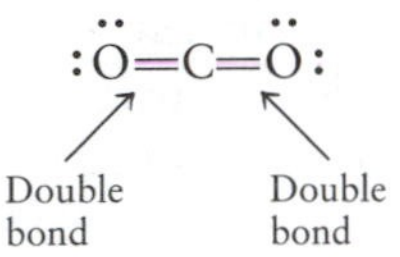

$$:\overset{..}{O}=C=\overset{..}{O}:$$

Double   Double
bond   bond

All of the atoms in carbon dioxide have eight valence electrons in their outer shell and, therefore, obey the octet rule:

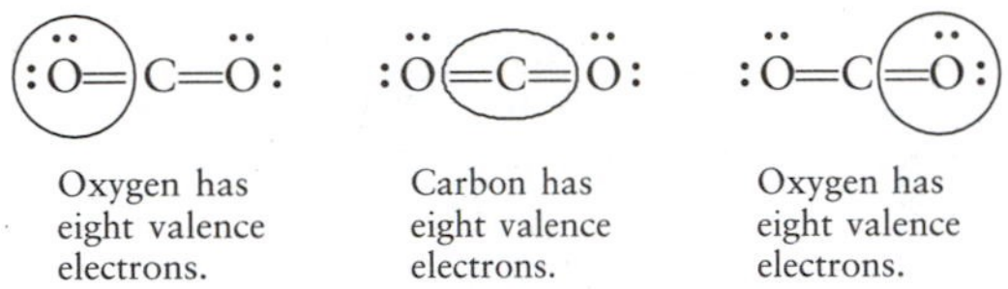

$$:\overset{..}{O}=C=\overset{..}{O}: \qquad :\overset{..}{O}=C=\overset{..}{O}: \qquad :\overset{..}{O}=C=\overset{..}{O}:$$

Oxygen has   Carbon has   Oxygen has
eight valence   eight valence   eight valence
electrons.   electrons.   electrons.

EXERCISE Carbon dioxide (CO_2) has three atoms, each of which obeys the octet rule. Does this mean that CO_2 has 24 (3×8) valence electrons?

[*Answer:* No, because shared electrons are counted twice. CO_2 has 16 valence electrons.]

KEY CONCEPT B   Writing Lewis structures I: Drawing atoms in the proper arrangement

We use three steps to draw the correct Lewis structure of a compound or ion:

Step 1. Draw the structure of the compound (or ion) with the correct arrangement of atoms, that is, so that each atom ends up bonded to the correct atom(s).
Step 2. Determine the number of valence electrons in the compound (or ion).
Step 3. Draw in bonds and, possibly, lone pairs, following the octet rule where possible. Multiple bonds may be required.

Here we focus on the first (and most difficult) step, drawing the atoms in their correct positions. It is helpful to note that symmetrical arrangements, especially those with a less electronegative atom surrounded by more electronegative atoms, are quite common. For instance,

For	CO_2,	O	C	O	is correct, not	O	O	C
For	SO_3,		O		is correct, not	S	O	O	O
		O	S	O	nor	O	O	S	O
					nor	S
						O	O	O

Also, certain elements tend to form a specific number of bonds. For example, H, one bond; C, four bonds; N, three bonds; O, two bonds; F, Cl, Br, and I, one bond. Even with these aids to drawing structures, getting the correct arrangement of atoms can be tricky. Practice and experience will make you proficient in this task.

EXAMPLE   Arranging atoms for a Lewis structure

What is the correct arrangement of atoms in the carbonate ion, CO_3^{2-}.

SOLUTION CO_3^{2-} contains three of the same atom (O) and one other atom (C). In addition, oxygen has a higher electronegativity than carbon. A symmetrical structure, with the less electronegative atom at the center and the more electronegative atoms surrounding it, gives the desired arrangement:

O

O   C   O

EXERCISE What is the arrangement of atoms in CCl_4?

Cl

Answer: Cl   C   Cl

Cl

PITFALL   The number of bonds formed by an element

The listing of the number of bonds formed by certain elements reflects a tendency, not a certainty. Bonding is a complex matter, and exceptions to these numbers occur.

KEY CONCEPT C   Writing Lewis structures II: Counting valence electrons

Here we focus on the second step for drawing Lewis structures, counting the number of valence electrons that belong in the structure. The total number of electrons in a Lewis structure is the sum of the valence electrons of the atoms in the structure, plus any adjustments in that sum resulting from the charge on ionic structures. As an example, let's work out the number of valence electrons in nitrogen trifluoride, NF_3. Nitrogen is a Group 15 element, so it possesses five valence electrons. Fluorine is in Group 17, so each fluorine possesses seven valence electrons, resulting in a total of 21 for the three fluorines. The sum of valence electrons from the atoms is 26:

$$
\begin{aligned}
\text{N:} \quad & 1 \times 5 = 5 \text{ valence electrons} \\
3\text{F:} \quad & 3 \times 7 = \underline{21} \text{ valence electrons} \\
\text{Total:} \quad & 26 \text{ valence electrons}
\end{aligned}
$$

Because there is no charge on NF_3, there is no need to go any further. NF_3 possesses 26 valence electrons. We recall that for a charged species, a negative charge results from the gain of electrons and a positive charge from the loss of electrons. The following table presents some examples of how to take into consideration the charge on a species.

Species	Electrons Gained or Lost by Original Atoms	Total Number of Valence Electrons
neutral molecule	0	sum of valence electrons of atoms
-1 ion	1 gained	(sum of valence electrons) $+1$
-2 ion	2 gained	(sum of valence electrons) $+2$
$+1$ ion	1 lost	(sum of valence electrons) -1
$+2$ ion	2 lost	(sum of valence electrons) -2

EXAMPLE   Counting valence electrons

How many valence electrons are present in the carbonate ion CO_3^{2-}?

SOLUTION We first determine the number of valence electrons that originate with the single carbon atom and three oxygen atoms, using the fact that carbon is a Group 14 element and oxygen a Group 16 element.

Then we take into account the -2 charge on the ion, which occurs because the carbonate ion has gained two electrons above and beyond those supplied by the original atoms:

$$
\begin{array}{rll}
\text{1C:} & 1 \times 4 = & 4 \text{ valence electrons} \\
\text{3O:} & 3 \times 6 = & 18 \text{ valence electrons} \\
-2 \text{ charge:} & & \underline{+2} \text{ valence electrons} \\
\text{Total:} & & 24 \text{ valence electrons}
\end{array}
$$

CO_3^{2-} has 24 valence electrons.

EXERCISE How many valence electrons are present in the methylammonium ion, $CH_3NH_3^+$?

[*Answer:* 14]

PITFALL   Count valence electrons, not all the electrons

Be certain you count only valence electrons for a Lewis structure. Do not count the total number of electrons. Only for hydrogen is the number of valence electrons the same as the total number of electrons for an atom.

KEY CONCEPT D   Drawing Lewis structures III: Drawing in bonds and lone pairs

Here we focus on the last step used in drawing Lewis structures, drawing in bonds, and possibly lone pairs, following the octet rule where possible. Multiple bonds may be required. In this step we first draw a single bond (—) between any two bonded atoms. Atoms surrounding a central atom are all bonded to the central atom, but not to each other. If any atom, after this step, does not satisfy the octet rule, use multiple bonds, if possible. Do not use a multiple bond if it results in too many electrons for an atom; remember, the *usual* numbers of bonds formed are H, one bond; C, four bonds; N, three bonds; O, two bonds; and halogens, one bond. Finally, add the remaining valence electrons as lone pairs. We should note that N, O, and the halogens usually have one or more lone pairs, whereas hydrogen never has any and carbon rarely does. Some of the electron pairs that were originally drawn in as multiple bonds may have to be changed into lone pairs to use the correct number of valence electrons. (*Hint:* Use a pencil so that you can move electrons easily on paper.) Finally, check the structure by (1) counting the valence electrons to make sure you have the correct number and (2) checking that those atoms that tend to follow the octet rule (C, O, N, F) or duet rule (H) have the correct number of valence electrons.

EXAMPLE   Drawing in bonds and lone pairs

We have ascertained the arrangement of atoms and number of valence electrons (24) in the carbonate ion, CO_3^{2-}. Complete the Lewis structure by drawing in bonds and lone pairs.

SOLUTION The arrangement of atoms is

$$
\begin{array}{ccc}
 & O & \\
O & C & O
\end{array}
$$

We first draw in single bonds:

$$
\begin{array}{c}
O \\
| \\
O{-}C{-}O
\end{array} \quad \text{(6 electrons used)}
$$

CHAPTER 8

Now, carbon "always" follows the octet rule, as does oxygen. Making one double bond gives carbon its eight valence-shell electrons:

$$O-\overset{\displaystyle O}{\underset{\displaystyle \|}{C}}-O \qquad \text{(8 electrons used)}$$

No more double bonds can be made because doing so would put ten electrons around carbon. Therefore, we add the remaining electrons as lone pairs on oxygen. (Carbon already has eight electrons and normally does not have lone pairs.)

$$\left[\ :\ddot{\overset{..}{O}}\atop{\displaystyle \|}\atop :\ddot{O}-C-\ddot{O}:\ \right]^{2-} \qquad \text{(24 electrons used)}$$

No further adjustment is needed because all of the electrons have been used and the octet rule has been satisfied. The charge is always indicated with a Lewis structure, generally by drawing large brackets around any ionic structure and writing the charge as a superscript.

EXERCISE Draw the Lewis structure of XeO_3.

Answer: $:\ddot{O}-Xe-\ddot{O}:$ with $:\ddot{O}:$ above Xe

8.7 RESONANCE

KEY CONCEPT   Resonance structures

Two problems with some Lewis structures are simultaneously resolved by the concept of resonance. Let's look at the problems first.

1. For some species, more than one Lewis structure with the same energy can be drawn, and there is no way to decide which structure is correct. For example, the gas sulfur trioxide (SO_3) has three equal-energy structures, which are called **equivalent structures**:

$$:\ddot{O}-\overset{\displaystyle :O}{\underset{\displaystyle \|}{S}}-\ddot{O}: \qquad :\ddot{O}=\overset{\displaystyle :\ddot{O}:}{S}-\ddot{O}: \qquad :\ddot{O}-\overset{\displaystyle :\ddot{O}:}{S}=\ddot{O}:$$

2. When more than one equivalent Lewis structure exists, the actual molecular properties do not agree with those of any one of the structures. For example, from any one of the Lewis structures shown, we would predict that SO_3 has two single bonds and one double bond. However, measurements show that all three bonds are equivalent, with properties intermediate between those of a single and those of a double bond.

We solve both of these problems by assuming that the actual SO_3 molecule found in nature is not represented by any one of the equivalent Lewis structures, but by a blending of all three. Because the actual SO_3 is a composite of the three equivalent Lewis structures and no single Lewis structure by itself is correct, we do not have to choose one of the Lewis structures as the correct one; all are used. This solves the first problem. Our second problem is solved because a blending of the three Lewis structures leads to the prediction that each S—O bond is intermediate between an S—O single bond and S=O double bond. This blending of structures is called **resonance,** and the composite structure that results is called a **resonance hybrid** of the individual Lewis structures that contribute to the composite. Thus, SO_3 is a resonance hybrid of the three structures shown.

EXAMPLE   Drawing contributing Lewis structures for a resonance hybrid molecule

Draw the two principal contributing Lewis structures for nitromethane, $H_3C—NO_2$.

SOLUTION Nitromethane has a total of 24 valence electrons. Its arrangement of atoms is suggested by the way it is written in the problem, with carbon bonded to nitrogen and the hydrogens and oxygens as terminal atoms. We arrive at two equivalent contributing Lewis structures:

$$
\begin{array}{cc}
\text{H \quad :\ddot{O}} & \text{H \quad :\ddot{O}:} \\
| \quad \parallel \quad\ddot{} & | \quad | \quad \ddot{} \\
\text{H—C—N—O:} & \text{H—C—N=O:} \\
| \quad \ddot{} & | \\
\text{H} & \text{H}
\end{array}
$$

In these structures, only the placement of electrons is different, as is required for Lewis structures that make major contributions to a resonance hybrid.

EXERCISE For which of the following species is a single Lewis structure a good representation of the actual species: CO_2, O_3, SO_2, CO_3^{2-}.

[*Answer:* CO_2, because it does not exhibit resonance]

8.8   FORMAL CHARGE

KEY CONCEPT A   Formal charge

The **formal charge** on an atom is the charge the atom would have if we arbitrarily assign valence electrons to it in the following way:

1. All of the lone pairs on the atom are assigned to the atom.
2. Half of the electrons in bonds on the atom are assigned to the atom.

The formal charge on the atom equals the number of valence electrons on the neutral atom (from the group number) minus the number of electrons assigned to the atom:

$$
\text{Formal charge} = \binom{\text{number of valence electrons}}{\text{on the neutral atom}} - \binom{\text{number assigned}}{\text{to the atom}}
$$

The sum of the formal charges on all of the atoms in a Lewis structure must equal the overall charge on the Lewis structure; if not, the formal charges have been assigned incorrectly.

EXAMPLE   Determining formal charge

Assign formal charges to all of the atoms in the two inequivalent Lewis structures shown for dinitrogen oxide (nitrous oxide, N_2O):

$$
\begin{array}{cc}
\text{:N≡N—\ddot{O}:} & \text{:\ddot{N}—N≡O:} \\
\text{Structure I} & \text{Structure II}
\end{array}
$$

SOLUTION For each atom, we determine the number of valence electrons for the neutral atom from the group number. Then we assign to each atom in the structure half the electrons in every bond to that atom and all the electrons in unshared pairs.

Structure I

Atom	A, Electrons in Bonds	B, Electrons in Unshared Pairs	C, Valence Electrons in Neutral Atom	D, Electrons Assigned = $\frac{1}{2}$A + B	Formal Charge = C − D
N (terminal)	6	2	5	5	0
N (central)	8	0	5	4	+1
O	2	6	6	7	−1

Structure II

Atom	A, Electrons in Bonds	B, Electrons in Unshared Pairs	C, Valence Electrons in Neutral Atom	D, Electrons Assigned $= \frac{1}{2}A + B$	Formal Charge $= C - D$
N (terminal)	2	6	5	7	−2
N (central)	8	0	5	4	+1
O	6	2	6	5	+1

The answers are in the last column of each table.

EXERCISE Assign formal charges to all the atoms in the two inequivalent Lewis structures shown for xenon difluoride:

$$:\!\ddot{F}\!\!=\!\!\ddot{X}\!e\!-\!\ddot{F}\!: \qquad\qquad :\!\ddot{F}\!-\!\dot{\ddot{X}}\!e\!-\!\ddot{F}\!:$$
$$\quad\text{I}\qquad\qquad\qquad\qquad\text{II}$$

Answer: I: F (double bonded), −1; Xe, +1; F (single bonded), 0
II: F, 0; Xe, 0

KEY CONCEPT B Formal charge and plausible structure

Formal charges can be valuable in judging the relative energies of different Lewis structures of the same molecule. A Lewis structure in which the atoms do not have much formal charge is usually lower in energy than a structure that has atoms with a lot of formal charge. If one plausible structure has about the same amount of formal charge as another, the structure that puts negative charge on the more electronegative atoms (Section 8.13) is likely to have the lower energy.

EXAMPLE Choosing between Lewis structures

Of the two structures given for N_2O in the previous example, which is preferable?

SOLUTION Structure I has one atom with formal charge +1 and one with formal charge −1. Structure II has one atom with formal charge −2 and two with formal charge +1. Because structure I has less formal charge, it is the lower energy structure and, therefore, the preferred one.

EXERCISE Of the two structures given for XeF_2 in the previous example, which is preferred?

[*Answer:* II]

KEY WORDS Define or explain each term in a written sentence or two.

double bond   multiple bond   tetravalent
equivalent structures   resonance   triple bond
formal charge   resonance hybrid
Kekulé structure   single bond

EXCEPTIONS TO THE OCTET RULE

8.9 RADICALS AND BIRADICALS

KEY CONCEPT Odd-electron species

Chemical entities belonging to an interesting and important class have an odd number of valence electrons. Any such entity must have an unpaired electron and must therefore have at least one atom that violates the octet rule; an odd number of electrons cannot all be paired. When drawing the Lewis structure of such a compound, there is usually a choice of where to place the unpaired electron. In general, it belongs on the least electronegative of the possible atoms. So, for ClO_2,

$$:\ddot{\text{O}}-\ddot{\text{C}}\text{l}-\ddot{\text{O}}: \qquad \text{is preferred over} \qquad \cdot\ddot{\text{O}}-\ddot{\text{C}}\text{l}-\ddot{\text{O}}:$$

Odd-electron species may arise when a bond is broken in a reaction in such a way that the electrons originally in the bond become unpaired. When this happens, pieces of molecules with unpaired electrons are formed; this type of highly reactive molecular fragment is called a **radical.** The methyl radical $CH_3\cdot$ and hydroxyl radical $\cdot OH$ are two important and commonly encountered radicals. (When writing the symbol of a radical, the unpaired electron is explicitly drawn as a dot.)

<hr>

EXAMPLE   Drawing the Lewis structure of an odd-electron species

Draw the Lewis structure of NO_2.

SOLUTION One nitrogen and two oxygens possess 17 valence electrons. The less electronegative nitrogen is in the center of the molecule, and the two oxygens surround it. We place the odd electron on the less electronegative nitrogen to get

$$:\ddot{\text{O}}-\dot{\text{N}}=\ddot{\text{O}}:$$

EXERCISE Draw the Lewis structure of the $\cdot CF_3$ radical.

$$\textit{Answer:} \quad \cdot\overset{\displaystyle :\ddot{\text{F}}:}{\underset{\displaystyle :\ddot{\text{F}}:}{\text{C}}}-\ddot{\text{F}}:$$

8.10   EXPANDED VALENCE SHELLS

KEY CONCEPT   Expanded octets

Because of the large size of the atom and the presence of low-energy empty d-orbitals, p-block elements in the third, fourth, and fifth periods can accommodate more than eight electrons in their valence shells. Therefore, these elements may violate the octet rule. An atom in a Lewis structure with more than eight valence electrons in its valence shell is said to possess an **expanded octet.** This behavior is especially prominent with P, S, Se, Te, Br, I, Kr, and Xe. How can we tell when an expanded octet is formed? In most cases, the only way to fit all of the valence electrons into the structure will be by using an expanded octet; the expanded octet is forced on us by the structure itself. Occasionally, accurate quantum mechanical calculations indicate that an expanded octet should be used even where one does not seem required, as in the sulfate ion (SO_4^{2-}), in which a structure with an expanded octet on sulfur has a lower energy than one in which sulfur obeys the octet rule.

<hr>

EXAMPLE   Drawing a Lewis structure with an expanded octet

Draw the Lewis structure of XeF_4.

SOLUTION This molecule has the expected symmetric structure, with the four more electronegative fluorines surrounding the (presumably) less electronegative xenon. There are a total of 36 valence electrons in the structure, eight from xenon (Group 18) and seven each from four fluorines, for a total of 28 from the fluorines. Any structure in which all the atoms obey the octet rule does not have the correct number of valence electrons; for instance,

$$\overset{\displaystyle :\ddot{\text{F}}:}{\underset{\displaystyle :\ddot{\text{F}}:}{:\ddot{\text{F}}-\text{Xe}-\ddot{\text{F}}:}} \qquad \text{(incorrect, only 32 valence electrons)}$$

To use 36 valence electrons, we put two lone pairs on xenon, giving xenon an expanded octet:

$$:\ddot{F}:$$
$$:\ddot{F}-\overset{\cdot\cdot}{\underset{\cdot\cdot}{Xe}}-\ddot{F}:$$
$$:\ddot{F}:$$

We do not form double bonds with these two additional pairs of electrons, because fluorine does not violate the octet rule nor form double bonds.

EXERCISE Draw the Lewis structure of SF_6.

Answer:

KEY WORDS   Define or explain each term in a written sentence or two.

antioxidant   radical
biradical   variable covalence
expanded octet

LEWIS ACIDS AND BASES

8.11   THE UNUSUAL STRUCTURES OF GROUP 13 HALIDES

KEY CONCEPT   Group 13 elements form unusual halides

Boron and aluminum form some halides with unusual structures. Boron trifluoride (BF_3), for instance, is an example of a compound in which boron has an incomplete octet. Boron has only six electrons in its valence shell in BF_3, instead of the octet expected for a second period element. BCl_3 is very similar in its bonding. However, $AlCl_3$ acts very differently. When it is heated to 180°C, it sublimes and forms a gas of Al_2Cl_6 molecules, with two chlorine atoms acting as a bridge between the aluminum atoms (see structure **41** in the text). Because boron in BF_3 has only six electrons in its outer shell, it can form a fourth bond with, for instance, a fluoride ion. In this case, a tetrafluoroborate ion, BF_4^-, results.

EXAMPLE   Understanding boron and aluminum hydrides

How does the valence of boron change when it converts from boron trifluoride to tetrafluoroborate ion?

SOLUTION The valence of an element is the number of covalent bonds it forms. In BF_3, boron has three covalent bonds and, therefore, a valence of 3. In the BF_4^- ion, boron has four covalent bonds, so it has a valence of 4.

EXERCISE Does aluminum follow the octet rule in Al_2Cl_6? Does chlorine?

[*Answer:* Both aluminum and chlorine have eight valence electrons in Al_2Cl_6.]

8.12 LEWIS ACID-BASE COMPLEXES

KEY CONCEPT   The definition of Lewis acids and bases

From the Lewis acid/base point of view, a base is a chemical species with an unshared pair of electrons and an acid is a species that can form a covalent bond by accepting an unshared pair of electrons. An acid-base reaction, therefore, results in formation of a covalent bond:

$$H_3N: + Ag^+ \longrightarrow Ag—NH_3^+$$

Complex

Species with unshared pair of electrons (base)

Species that can accept unshared pair to form a covalent bond (acid)

With this reaction in mind, we can make the following definitions: A **Lewis base** is an electron pair donor; a **Lewis acid** is an electron pair acceptor; and a **complex** is the species formed when a Lewis acid reacts with a Lewis base.

EXAMPLE 1   Recognizing a Lewis base from its structure

Which of the following can function as a Lewis base? (a) OH^-, (b) CO, (c) NH_4^+

SOLUTION Because a Lewis base is characterized by possession of at least one unshared pair of electrons, we must investigate the electron-dot structure of each species to see if unshared pairs of electrons are present. The Lewis structure of each is

$$\left[:\ddot{O}—H \right]^- \qquad :C\equiv O: \qquad \left[\begin{matrix} H \\ | \\ H—N—H \\ | \\ H \end{matrix} \right]^+$$

(a)   (b)   (c)

As we can see, both the hydroxide ion and carbon monoxide have unshared pairs of electrons, so both can function as Lewis bases. The ammonium ion does not have any unshared pairs of electrons, so it cannot function as a Lewis base.

EXERCISE Which of the following can function as a Lewis base? (a) H_2O, (b) CN^-, (c) BF_3

[*Answer:* H_2O, CN^-]

PITFALL   Does a Lewis base lose its electrons?

It should be noted that when a Lewis base "donates" its electrons, it does not lose them (as when an acid donates a proton). The electrons stay with the Lewis base as they are donated, and a covalent bond between the Lewis base and Lewis acid results.

EXAMPLE 2   Analyzing Lewis acid-base reactions

Adding acid to a solution of sodium cyanide can be very dangerous because the reaction of H^+ with CN^- results in formation of the very toxic gas HCN. Explain this reaction as a Lewis acid-base reaction.

SOLUTION We recognize that the cyanide ion, with two unshared pairs of electrons can function as a Lewis base; in fact, it is a strong Lewis base. The hydrogen ion has an empty 1s-orbital that can accept an electron pair from cyanide. Thus, the reaction proceeds by the donation of a pair of electrons by cyanide to hydrogen ion:

$$H^+ + :C\equiv N:^- \longrightarrow H—C\equiv N:$$

Lewis acid   Lewis base   Complex

EXERCISE Heating copper(II) sulfate ($CuSO_4$) to a temperature in excess of 600°C results in decomposition to copper(II) oxide and sulfur trioxide. Explain this reaction in terms of Lewis acids and bases.

[*Answer:* $CuSO_4 \longrightarrow SO_3 + CuO$]

KEY WORDS   Define or explain each term in a written sentence or two.

acid-base reaction	incomplete octet
complex	Lewis acid
coordinate covalent bond	Lewis base

IONIC VERSUS COVALENT BONDS

8.13   CORRECTING THE COVALENT MODEL

KEY CONCEPT   The variation of ionic character with electronegativity

Almost every bond is partially ionic and partially covalent. The covalent character results in the presence of electron density between the two atoms in the bond. The ionic character is a manifestation of the movement of electron density away from one atom and toward the other; this results in the development of a positive charge on one atom (the one that electron density moves away from) and development of a negative charge on the other atom (the one that electron density moves toward). The extent of the development of ionic character and its attendant positive and negative charges depends on the difference in electronegativities of the two bonded atoms. A large difference in electronegativity leads to a high ionic character; a small difference in electronegativity leads to a low ionic character. If the difference in electronegativities is above 2, the bond is considered ionic; if it is below 1.5, the bond is considered covalent. When the difference in electronegativities is between 1.5 and 2, the bond is neither clearly ionic nor clearly covalent.

Bond	Differences in Electronegativities	Partial Positive Charge On	Partial Negative Charge On	Type of Bond
Cs—F	3.2	Cs	F	ionic
Ca—O	2.1	Ca	O	ionic
Be—Cl	1.6	Be	Cl	uncertain
O—H	1.2	H	O	covalent
P—Cl	1.0	P	Cl	covalent
C—H	0.4	H	C	covalent
F—F	0	—	—	covalent

EXAMPLE   Using electronegativities

Which bond has the highest ionic character, P—O or C—Cl?

SOLUTION The ionic character of a bond is determined by the difference in the electronegativities of the atoms involved in the bond. The electronegativities of the atoms from Figure 8.15 are P, 2.2; C, 2.6; O, 3.4; Cl, 3.2. Thus, for the P—O bond, the differences in electronegativities is 1.2; for C—Cl, it is 0.6. Because the P—O bond has the larger difference in electronegativities, it has the higher ionic character.

EXERCISE Which bond is more covalent, F—Cl or S—Br?

[*Answer:* S—Br]

8.14 CORRECTING THE IONIC MODEL

KEY CONCEPT   Polarizability and polarizing power

Imagine that we have an ionic compound in which a small, compact cation is next to a large, diffuse anion. The small size and overall positive charge of the cation keep its electrons very tightly held in place. Conversely, the large size and overall negative charge of the anion allow its electrons more freedom to spread out and move away from the ion itself. In this situation, the cation attracts some of the electron charge cloud from the anion, resulting in the movement of some electron density into the space between the ions. The positioning of electrons between two atoms is precisely what constitutes a covalent bond, so the movement of electrons has added some covalent character to what was formerly an ionic bond. Atoms and ions that have charge clouds that can be easily distorted are said to be **polarizable**. Ions that can cause this electron charge cloud distortion have a high **polarizing power**. When an ion with high polarizing power is bonded to an easily polarizable ion, there will be substantial covalent character in the bond. Small highly charged cations have high polarizing power, and large highly charged anions are very polarizable.

Cations with High Polarizing Power	Highly Polarizable Anions
Li^+, Be^{2+}, Mg^{2+}, Al^{3+}, B^{3+}	I^-, Se^{2-}, Te^{2-}, As^{3-}, P^{3-}

EXAMPLE   Predicting the polarization of compounds

One of the compounds, KCl, BCl_3, and $BaCl_2$, is covalent. Which one?

SOLUTION The anion for each of these compounds is the same, so we must decide which of the cations has the highest polarizing power. Of the three cations, K^+, Ba^{2+}, and B^{3+}, the B^{3+} is the smallest because it combines being "high" and "to the right" in the periodic table. It is also the most highly charged. These factors make it the cation with the highest polarizing power, so we conclude that BCl_3 is the covalent compound. Even though Cl^- is not listed as a highly polarizable anion, it can be polarized by a strongly polarizing cation.

EXERCISE Explain why $SnCl_4$ is a covalent compound.

> [*Answer:* The highly charged Sn^{4+} ion is highly polarizing. (It is so highly polarizing that it never really exists as a simple ion.)]

KEY WORDS   Define or explain each term in a written sentence or two.

covalent character   polarizable
electronegativity   polarizing power
ionic character

DESCRIPTIVE CHEMISTRY TO REMEMBER

- **Calcium phosphate** [$Ca_3(PO_4)_2$] is a main component of teeth and bones.
- **Lattice enthalpies** are largest for small ions of high charge.
- **MgO** and **AgCl** are insoluble in water. **AgF** is soluble, however.
- **ClF** is one of many **interhalogen** compounds, compounds that contain halogen atoms only. Another example is ClF_4.
- **Benzene** (C_6H_6) exists as a hexagonal ring of six carbon atoms with one hydrogen attached to each carbon.

- Two important **radicals** are the methyl radical, $\cdot CH_3$, and the hydroxyl radical, $\cdot OH$. Radicals contain at least one unpaired electron, which makes them highly reactive and unstable. O_2 is a **biradical.**
- At room temperature, **phosphorus pentachloride** is actually an ionic solid consisting of PCl_4^+ cations and PCl_6^- anions. At 160°C, it sublimes to a gas of PCl_5 molecules.
- **Boron trichloride** is a colorless, reactive gas of BCl_3 molecules.
- **Aluminum trichloride** is a white solid that sublimes to Al_2Cl_6 molecules at 180°C.
- The **oxide ion** (O^{2-}) is a strong Lewis base.
- **BeCl₂** consists of long chains of covalently bonded $BeCl_2$ units.

CHEMICAL EQUATIONS TO KNOW

- In a limited supply of chlorine, phosphorus reacts to form **phosphorus trichloride:**

$$P_4(g) + 6Cl_2(g) \longrightarrow 4PCl_3(l)$$

- In a plentiful supply of chlorine, phosphorus reacts to form **phosphorus pentachloride:**

$$P_4(g) + 10Cl_2(g) \longrightarrow 4PCl_5(s)$$

- **Boron trifluoride** is produced in large amounts by heating boron oxide and calcium fluoride together:

$$B_2O_3(s) + 3CaF_2(s) \longrightarrow 2BF_3(s) + 3CaO(s)$$

- In a **Lewis acid-base reaction,** boron trifluoride (Lewis acid) and fluoride ion (Lewis base) react to form tetrafluoroborate ion (complex):

$$F^- + BF_3 \longrightarrow BF_4^-$$

- In a **Lewis acid-base reaction,** boron trifluoride (Lewis acid) and ammonia (Lewis base) react to form a complex of the two molecules:

$$NH_3 + BF_3 \longrightarrow NH_3BF_3$$

- In a **Lewis acid-base reaction,** oxide ion (Lewis acid) and water (Lewis base) react to form hydroxide ion (complex):

$$O^{2-}(aq) + H_2O(aq) \longrightarrow 2OH^-(aq)$$

- Dry fluorine, when diluted with nitrogen and passed over a film of sulfur at −75°C, reacts to form **SF₄:**

$$S_8(s) + 2F_2(g) \longrightarrow SF_4(g)$$

SELF-TEST EXERCISES

Ionic bonds

1. Which of the following is the Lewis symbol of nitrogen?
(a) $:\overset{..}{N}\cdot$ (b) $:\overset{..}{\underset{..}{N}}\cdot$ (c) $:N:$ (d) $:\overset{.}{N}\cdot$ (e) $:\overset{..}{N}:$

2. Which of the following is the Lewis symbol of the chloride ion, Cl^-?
(a) $:\overset{..}{Cl}\cdot{}^-$ (b) $:\overset{.}{Cl}\cdot{}^-$ (c) $:\overset{..}{\underset{..}{Cl}}:^-$ (d) $\cdot\overset{.}{Cl}\cdot{}^-$ (e) $:Cl^-$

3. Which of the following is the Lewis symbol of Ca?

(a) $\overset{\cdot}{Ca}$ (b) $\overset{\cdot\cdot}{Ca}\cdot$ (c) $:\overset{\cdot}{\underset{\cdot\cdot}{Ca}}:$ (d) $\overset{\cdot}{Ca}\cdot$ (e) $Ca:$

4. For which equation is the enthalpy of reaction equal to the lattice enthalpy of NaCl?

(a) $NaCl(s) \longrightarrow Na^+(g) + Cl^-(g)$ (b) $2NaCl(s) \longrightarrow 2Na(s) + Cl_2(g)$

(c) $NaCl(g) \longrightarrow Na^+(g) + Cl^-(g)$ (d) $NaCl(g) \longrightarrow Na(s) + Cl(g)$

(e) $NaCl(s) \longrightarrow Na^+(aq) + Cl^-(aq)$

5. What factor is most important in determining the type of bond formed between two atoms?

(a) size of the atoms

(b) energy changes that accompany bond formation

(c) physical state of the elements

(d) Lewis dot structure of the atoms

(e) charge on the nuclei of the atoms

6. Without consulting any references, predict which of the following has the highest lattice enthalpy.

(a) KF   (b) NaCl   (c) MgO   (d) BaTe   (e) MgS

7. Use the Born-Haber cycle to calculate the lattice enthalpy of NaBr. Use 108 kJ for the formation of 1 mol of Na(g) from Na(s) and 112 kJ for the formation of 1 mol of Br(g) from $Br_2(l)$.

(a) 678 kJ   (b) 1400 kJ   (c) 28 kJ   (d) 511 kJ   (e) 750 kJ

8. Use the Born-Haber cycle to calculate the lattice enthalpy of $MgCl_2$. Use 146 kJ for the formation of 1 mol of Mg(g) from Mg(s), 242 kJ for the formation of 2 mol of Cl(g) from $Cl_2(g)$, and -641 kJ for the enthalpy of formation of $MgCl_2$.

(a) 2517 kJ   (b) 1067 kJ   (c) 811 kJ   (d) 2168 kJ   (e) 1621 kJ

9. Use the Born-Haber cycle to calculate the enthalpy of formation of BaO. Use $143 \text{ kJ} \cdot \text{mol}^{-1}$ for the enthalpy of sublimation of Ba(s).

(a) $-211 \text{ kJ} \cdot \text{mol}^{-1}$ (b) $-1303 \text{ kJ} \cdot \text{mol}^{-1}$ (c) $-1162 \text{ kJ} \cdot \text{mol}^{-1}$

(d) $-553 \text{ kJ} \cdot \text{mol}^{-1}$ (e) $-1425 \text{ kJ} \cdot \text{mol}^{-1}$

10. Which of the following is likely to have the highest melting point?

(a) F_2 (b) CaO   (c) NaCl   (d) H_2O (e) $BeCl_2$

11. Only one of the following, when dissolved in water, forms a solution that conducts electricity. Which one?

(a) $C_6H_{12}O_6$ (b) Ne   (c) KF   (d) CH_3OH (e) O_2

12. Which of the following best describes ionic solids?

(a) random array of oppositely charged ions

(b) regular array of covalently bonded molecules

(c) regular array of free radicals

(d) regular array of oppositely charged ions

(e) random array of neutral atoms

Covalent bonds

13. Without refrence to any particular compound, which two atoms are most likely to form a covalent (rather than an ionic) bond?

(a) Na, Br   (b) Ca, Sc   (c) Fe, O   (d) C, H   (e) Zn, S

14. What two orbitals merge when a covalent bond forms between hydrogen and chlorine?

(a) H(2s), Cl(3p)   (b) H(1s), Cl(3p)   (c) H(1s), Cl(1s)

(d) H(1s), Cl(2p)   (e) H(3s), Cl(3p)

15. How many unpaired electrons are present in the compound CF_4, shown below?

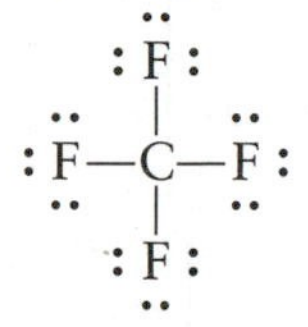

$$: \ddot{F} :$$
$$|$$
$$: \ddot{F} - C - \ddot{F} :$$
$$|$$
$$: \ddot{F} :$$

(a) 2   (b) 8   (c) 28   (d) 56   (e) 0

16. What is the valence of carbon in CF_4 (Exercise 15)?

(a) 8   (b) 0   (c) 4   (d) 2   (e) 32

17. Which of the following is the correct Lewis structure for Br_2?

(a) $: \dot{Br} - Br :$ (b) $: Br - Br :$ (c) $: \dot{Br} = \dot{Br} :$

(d) $: \ddot{Br} - \ddot{Br} :$ (e) $: Br = Br :$

18. How many lone pairs of electrons are present in hydrazine, N_2H_4?

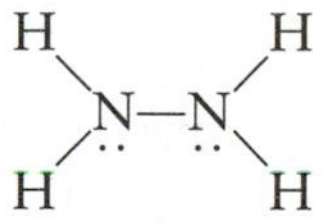

$$\begin{array}{ccc} H & & H \\ \diagdown & & \diagup \\ & N - N & \\ \diagup & & \diagdown \\ H & & H \end{array}$$

(a) 0   (b) 1   (c) 2   (d) 7   (e) 8

19. True or false, all the atoms in the structure of ethanol (below) obey the octet (or duet) rule?

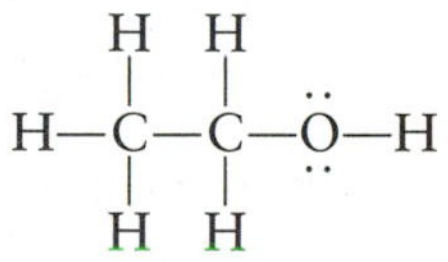

$$\begin{array}{cc} H & H \\ | & | \\ H - C - C - \ddot{O} - H \\ | & | \\ H & H \end{array}$$

(a) true   (b) false

20. How many lone pairs are present in CF_4? (See Exercise 15 for the Lewis structure of CF_4.)

(a) 0   (b) 4   (c) 8   (d) 12   (e) 24

The structures of polyatomic species

21. How many electrons are there in a double bond?

(a) 0   (b) 6   (c) 1   (d) 2   (e) 4

22. What is the correct arrangement of atoms in the Lewis structure of phosphorus trichloride, PCl_3?

(a) Cl P Cl Cl   (b)   P   (c)   Cl   (d) P Cl Cl Cl
                         Cl Cl Cl   Cl P Cl

23. What is the correct arrangement of atoms in the Lewis structure of NO_2Cl?

(a)   N   (b)   Cl   (c) N O O Cl   (d) Cl N O O
     Cl O O   O N O

24. How many valence electrons are there in the Lewis structure of difluoromethane, CH_2F_2?

(a) 20   (b) 32   (c) 8   (d) 26   (e) 18

25. How many valence electrons are there in the Lewis structure of the perchlorate ion, ClO_4^-?
(a) 32   (b) 31   (c) 30   (d) 17   (e) 16

26. Which of the following is the correct Lewis structure for the sulfite ion, SO_3^{2-}?

(a) $\left[:O\!=\!\overset{\overset{..}{O}}{\underset{}{S}}\!=\!O: \right]^{2-}$ (b) $\left[:\ddot{O}\!-\!\overset{\overset{..}{S}}{\underset{..}{O}}\!-\!\ddot{O}: \right]^{2-}$

(c) $\left[:\ddot{O}\!-\!\overset{\overset{..}{O}}{\underset{}{S}}\!-\!\ddot{O}: \right]^{2-}$ (d) $\left[:\ddot{O}\!-\!\overset{\overset{..}{O}}{\underset{}{S}}\!=\!O: \right]^{2-}$

27. Which of the following is the correct Lewis structure for ammonia, NH_3?

(a) $H\!-\!N\!=\!H\!-\!H$ (b) $H\!-\!\overset{\overset{:\ddot{N}:}{|}}{H}\!-\!H$ (c) $H\!-\!\overset{\overset{H}{||}}{N}\!-\!H$ (d) $H\!-\!\overset{\overset{H}{|}}{\underset{..}{N}}\!-\!H$

28. All but one of the following species exhibit the phenomenon of resonance. Which one does not?
(a) NO_3^- (b) SO_2 (c) CO_2 (d) SO_4^{2-} (e) C_6H_6

29. All but one of the statements regarding resonance is correct. Which one is incorrect?
(a) A species that undergoes resonance changes rapidly from one form to another.
(b) Resonance lowers the energy of a species.
(c) One Lewis structure will not properly describe a molecule that undergoes resonance.
(d) Resonance involves changing the positions of double bonds in Lewis structures.
(e) Resonance averages out certain bond lengths in a Lewis structure.

30. Which of the following structures represent resonance structures of a single species?

$$:\ddot{X}\!=\!\ddot{A}\!-\!\ddot{X}: \qquad :\ddot{X}\!-\!\ddot{A}\!=\!\ddot{X}: \qquad :\ddot{X}\!-\!\ddot{X}\!=\!\ddot{A}:$$
$$\text{I} \qquad\qquad\qquad \text{II} \qquad\qquad\qquad \text{III}$$

(a) I and III   (b) I and II   (c) II and III
(d) I, II, and III   (e) none of these

31. What is the formal charge on sulfur in the Lewis structure of SO_2 shown?

$$:\ddot{O}\!-\!\ddot{S}\!=\!\ddot{O}:$$

(a) +2   (b) −2   (c) 0   (d) −1   (e) +1

32. What is the formal charge on the single-bonded oxygens (either one will do) in the Lewis structure of NO_3^- shown?

$$\left[:\ddot{O}\!-\!\overset{\overset{:\ddot{O}:}{|}}{N}\!=\!\ddot{O}: \right]^-$$

(a) −1   (b) 0   (c) −2   (d) +1   (e) +2

33. Use formal charge to tell which of the following Lewis structures of the carbon dioxide is the best one.
(a) $:C\!\equiv\!O\!-\!\ddot{O}:$ (b) $:O\!\equiv\!O\!-\!\ddot{C}:$ (c) $:\ddot{O}\!=\!C\!=\!\ddot{O}:$ (d) $:O\!\equiv\!C\!-\!\ddot{O}:$

Exceptions to the octet rule

34. Which element displays variable valence?

(a) C (b) O (c) S (d) F (e) Al

35. Each of the elements shown except one is capable of achieving an expanded octet. Which one is not?

(a) I (b) S (c) O (d) Xe (e) P

36. Which of the following compounds has an atom with an expanded octet?

(a) SF_5 (b) H_2O (c) PCl_3 (d) HBr (e) NH_3

37. Which of the following is an odd-electron molecule?

(a) N_2O (b) N_2O_4 (c) NH_3 (d) NO_3 (e) N_2O_5

38. Which of the following is a radical?

(a) NO_2 (b) CH_4 (c) NO_3^- (d) OH^- (e) CH_3

39. Which molecule has a central atom with an expanded octet?

(a) CO_2 (b) XeF_4 (c) CF_4 (d) SO_3 (e) NCl_3

40. Which of the following species has a single unpaired electron on nitrogen?

(a) NO_3^- (b) N_2F_4 (c) NO_2^- (d) N_2O (e) NO_2

Lewis acids and bases

41. Which of the following can act as a Lewis base?

(a) Al^{3+} (b) CH_4 (c) H^+ (d) H_2 (e) OH^-

42. Which of the following can act as a Lewis acid?

(a) H^+ (b) CF_4 (c) H_2O (d) Cl_2 (e) Cl^-

43. A species with a _____ charge and _____ radius is likely to be a strong Lewis acid.

(a) high, large (b) low, small (c) low, large (d) high, small

44. In the reaction between H^+ and H_2O to form H_3O^+, _____ is the Lewis acid and _____ is the Lewis base.

(a) H_2O, H^+ (b) H^+, H_3O^+ (c) H_3O^+, H^+

(d) H_2O, H_3O^+ (e) H^+, H_2O

Ionic versus covalent bonds

45. Which bond is predominantly ionic? (Consult Figure 8.15 of the text.)

(a) Ge—F (b) Sn—Cl (c) As—Br (d) N—O (e) Si—O

46. Which bond has the highest ionic character?

(a) C—O (b) H—F (c) C—H (d) F—F (e) Si—Cl

47. Which ion has the highest polarizing power?

(a) Pb^{2+} (b) Al^{3+} (c) Se^{2-} (d) Ca^{2+} (e) Br^-

48. Which ion is most polarizable?

(a) Na^+ (b) O^{2-} (c) Cl^- (d) Mg^{2+} (e) Te^{2-}

Descriptive chemistry

49. Which of the following is the symbol for a methyl group?

(a) CO_3^{2-} (b) Mn   (c) CO_2H (d) CH_3 (e) OH

50. Which of the following is an interhalogen compound?

(a) HCl   (b) CCl_4 (c) OF_2 (d) F_2 (e) ClF

51. In a limited supply of Cl_2, phosphorus reacts to form

(a) PCl_3 (b) PCl_5 (c) PCl   (d) P_2Cl (e) P_4Cl_2

52. In the industrial production of BF_3, calcium fluoride is one of the reactants. What is the other?

(a) B_2O_3 (b) B   (c) B_2H_6

53. Which of the following is a strong Lewis base?

(a) O^{2-} (b) Cl^- (c) H^+ (d) $BeCl_2$ (e) Ag^+

54. What is the product formed when O^{2-} reacts with water?

(a) OH^- (b) H_2 (c) H_3O^+ (d) H_2O_2 (e) O_2

CHAPTER 9

MOLECULES: SHAPE, SIZE, AND BOND STRENGTH

The sizes and shapes of molecules are partially responsible for their physical and chemical properties, including their biochemical activity. The relative bond strengths of the different bonds in a molecule frequently determine the course of a chemical reaction. In this chapter, we explore these aspects of molecules.

THE SHAPES OF MOLECULES AND IONS

9.1 THE VSEPR (VALENCE-SHELL ELECTRON-PAIR REPULSION) MODEL

KEY CONCEPT Electron repulsions determine common molecular shapes

The shapes of molecules (and polyatomic ions) are determined by the requirement that electrons, which strongly repel one another remain as far from one another as possible. Because covalent bonds contain electrons, the electrons in the bonds adjust their positions to minimize electron-electron repulsions, and thereby determine the shape of the molecule. Lone pairs are particularly effective in repelling other electrons, so they, too, are important in determining the shape of a molecule.

EXAMPLE The common shapes of molecules

Use Figure 9.1 of the text to determine the possible shapes for a molecule that has four atoms attached to a central atom. Which of the possible shapes is the most symmetrical?

SOLUTION To answer this question, we must carefully consider the possible shapes given in Figure 9.1 of the text and find those that have four lines attached to a central sphere (indicating four atoms attached to a central atom). Based on this process, we find only the following possibilities: tetrahedral, seesaw, and square planar. Determining which is most symmetrical is somewhat, but not entirely, intuitive. In the tetrahedral structure, every angle from any one bond (a straight line attached to the central sphere) to any other is 109.5°; it is the most symmetrical.

EXERCISE How many atoms are attached to the central atom in a molecule with a square pyramidal shape?

[*Answer:* Five]

9.2 MOLECULES WITHOUT LONE PAIRS ON THE CENTRAL ATOM

KEY CONCEPT A   The arrangement of electron pairs in VSEPR theory

Probably the most important concept in **valence-shell electron-pair repulsion (VSEPR) theory** is that molecules take on the shapes they do because of electron-electron repulsions. We start out by concerning ourselves with how these repulsions cause regions of high electron concentration to arrange themselves around a central atom. It makes no difference at this point whether region of high electron concentration is a lone pair or a bonding pair. We just imagine that a certain number of such regions are attached to a central atom and that they arrange themselves so as to be as far apart as possible to minimize electron-electron repulsions. The arrangements that result are given in Figure 9.4 of the text and are repeated here.

Number of Regions of High Electron Concentration	Arrangement Around Central Atom
2	linear
3	trigonal planar
4	tetrahedral
5	trigonal bipyramidal
6	octahedral
7	pentagonal bipyramidal

It is imperative that you understand the three-dimensional arrangements to which each of these correspond. A good way of doing this is to build models of the shapes, using toothpicks to represent electron pairs and a styrofoam ball (or gumdrop) to represent the central atom. In your models, you should note that these shapes do result in electrons being as far apart as possible.

EXAMPLE   Understanding molecular shapes

Use a three-dimensional model of the trigonal planar arrangement to determine the angle between regions of high electron concentration in this arrangement.

SOLUTION In the trigonal planar arrangement of regions of high electron concentration, the three regions are all in the same plane. A representation of this is shown in the following figure, with M as the central atom and lines as the regions of high electron concentration.

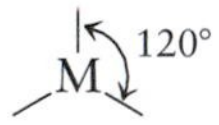

The angle between electron pairs is 120°.

EXERCISE In the trigonal bipyramidal arrangement, three of the regions of high electron concentration are in a plane and 120° apart; let's label these pairs a, b, and c. If we label the fourth and fifth regions d and e, what is the angle between region d and region a? What is the angle between region d and region e? (Hint: Use a model.)

[*Answer:* d and a, 90°; d and e, 180°]

PITFALL   Octa means "eight," doesn't it?

Don't be misled by the term *octahedral,* which refers to an eight-sided solid. A molecule with an octahedral shape has only six atoms on its periphery, not eight. Similarly, an octahedral arrangement of regions of high electron concentration refers to six electron pairs around the central atom. The *eight* in octahedral refers to the number of faces in the regular solid that has six vertices.

KEY CONCEPT B   Lewis structures with single bonds

The shapes of molecules and molecular ions that have Lewis structures with single bonds follow directly from the arrangement of regions of high electron concentration around the central atom. The regions arrange themselves as we have already noted, but now we recognize that regions of high electron concentration can be either bonding pairs or lone pairs. In structures with no lone pairs, all the regions are bonding pairs; and because each region has an atom attached to it, the shape of the molecule is the same as the arrangement of regions of high electron concentration around the central atom. To determine the shape, we need only draw the Lewis structure of the molecule or molecular ion, count the regions of high electron concentration around the central atom, and then deduce the shape from Figure 9.4 of the text. The following table gives some examples.

Molecule	Number of Regions of High Electron Concentration	Arrangement of Electrons	Number of Lone Pairs	Shape of Molecule
$BeCl_2$	2	linear	0	linear
BF_3	3	trigonal planar	0	trigonal planar
CH_4	4	tetrahedral	0	tetrahedral
PCl_5	5	trigonal bipyramidal	0	trigonal bipyramidal
SF_6	6	octahedral	0	octahedral

EXAMPLE   Using VSEPR theory

Predict the shape of silane, SiH_4.

SOLUTION The first step in using VSEPR theory to predict the structure of a compound is to draw the Lewis structure of the compound. For SiH_4, this is

$$
\begin{array}{c}
H \\
| \\
H-Si-H \\
| \\
H
\end{array}
$$

The number of regions of high electron concentration around the central atom is four. That is, each Si—H bond counts as one region. We use the preceding table to determine the arrangement of the regions and the shape of the molecule. Four regions adopt a tetrahedral arrangement, so we conclude that SiH_4 is tetrahedral.

EXERCISE Predict the shape of IF_6^+.

[*Answer:* Octahedral]

9.3   MULTIPLE BONDS IN THE VSEPR MODEL

KEY CONCEPT   In VSEPR theory, multiple bonds count as one region of high electron concentration

Although it may seem paradoxical, double and triple bonds are counted as only *one* region of high electron concentration for the purposes of VSEPR theory. Because both bonds in a double bond follow the molecular framework, if we count a multiple bond as a single region for the purposes of VSEPR theory, we do predict the correct shapes of molecules. This approach is illustrated in the following example.

EXAMPLE   Using VSEPR Theory with double bonds present

Predict the shape of CO_2.

SOLUTION The Lewis structure of CO_2 is

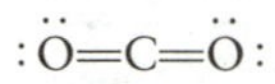

$$:\ddot{O}{=}C{=}\ddot{O}:$$

If we count each double bond as a single region of high electron concentration (as we must for VSEPR theory), there are two such regions around carbon. Hence, the electrons, and the attached oxygen atoms, take on a linear arrangement around the central carbon atom (Figure 9.4). Thus, carbon dioxide is a linear molecule.

EXERCISE What is the shape of SO_3?

[*Answer:* Trigonal planar]

9.4 MOLECULES WITH LONE PAIRS ON THE CENTRAL ATOM

KEY CONCEPT   VSEPR and the effect of lone pairs

Lone pairs are treated like any other pair of electrons when counting regions of high electron concentration around the central atom. However, because no atom is bonded to a lone pair, the shape of a molecule containing a lone pair is not the same as the arrangement of electron regions around the central atom. This is because the shape of a molecule is based on the location of the atoms in the molecule. Let's consider the electron-deficient compound GeF_2 as an example. Its Lewis structure is

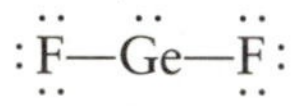

$$:\ddot{F}{-}Ge{-}\ddot{F}:$$

There are three regions of high electron concentration around Ge, the two bonds and the lone pair. These take on a trigonal planar shape around the Ge, as shown in the following figure.

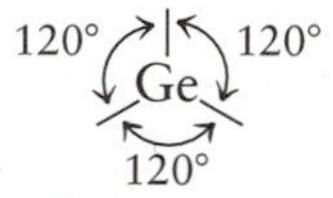

But only two of the three regions have atoms attached. So the three atoms in the molecule form an angular structure, not a trigonal planar structure. At this point, we would predict that the bond angle in GeF_2 is 120°, the normal trigonal planar bond angle. However, lone pairs have a predictable effect on the structure. A lone pair repels bonding pairs very strongly and pushes them away; in GeF_2, this repulsion results in a bond angle smaller than 120°.

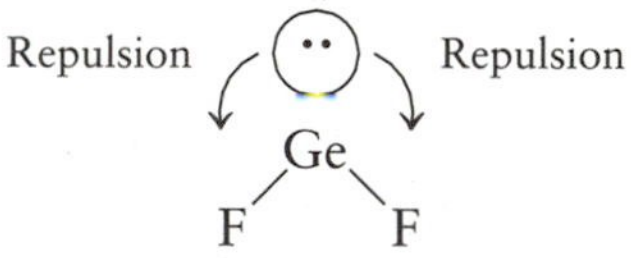

EXAMPLE   Using VSEPR theory

Predict the shape of $AsCl_3$.

SOLUTION We first draw the Lewis structure of the compound.

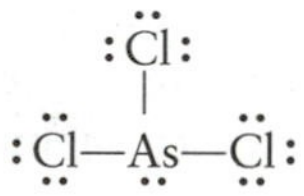

The four regions of high electron concentration take on a tetrahedral arrangement around the central As atom. Only three of the regions have atoms attached, thus resulting in a trigonal pyramidal structure. (A trigonal pyramid is a pyramid with a three-sided base.) The lone pair–bonding pair repulsions cause the bond angles to be smaller than the normal 109.5° tetrahedral bond angle.

9.5 THE DISTORTING EFFECT OF LONE PAIRS

KEY CONCEPT A   VSEPR and lone pairs on larger molecules

When we determine the shape of molecules with two, three, or four regions of high electron concentration, we do not have to decide where to spatially locate unshared pairs and bonding pairs; each position around the central atom is equally likely to have an unshared pair. However, when five regions are present, and the trigonal bipyramidal arrangement of electrons is obtained, we do have a choice to make. For this arrangement of electrons, lone pairs are found at equatorial positions and not at axial positions. (The equatorial positions are the three positions that point at the vertices of an equilateral triangle and are, therefore, 120° apart. The axial positions are the two positions that protrude from the top and bottom of the plane of the equilateral triangle and are 180° apart.) For the octahedral arrangement, we again have no decision to make when one lone pair is present; all six positions are equivalent. However, two lone pairs will always be positioned 180° apart from each other, that is, across from each other, with the central atom in between. The names given to the various shapes that arise when lone pairs are present on both small and large structures are summarized in the following table for the most common shapes found.

Number of Regions of High Electron Concentration	Arrangement of Regions of High Electron Concentration	Number of Lone Pairs	Shape of Molecule
3	trigonal planar	1	angular
		2	linear
4	tetrahedral	1	trigonal pyramidal
		2	angular
5	trigonal pyramidal	1	seesaw
		2	T-shaped
		3	linear
6	octahedral	1	square pyramidal
		2	square planar

EXAMPLE   Using VSEPR theory

Predict the shape of XeF$_4$.

SOLUTION We start with the Lewis structure:

$$:\overset{..}{F}:$$
$$|$$
$$:\overset{..}{F}\!-\!\overset{..}{\underset{..}{Xe}}\!-\!\overset{..}{F}:$$
$$|$$
$$:\overset{..}{F}:$$

The molecule has six regions of high electron concentration around xenon, so there is an octahedral arrangement of electrons. The two lone pairs are positioned across from each other, 180° apart. The four fluorine atoms are positioned at the remaining four positions, at the corners of a square. Thus, the molecule has a square planar shape.

EXERCISE Predict the shape of IF$_2^-$.

[*Answer:* Linear]

KEY CONCEPT B Lone pairs decrease bond angles through repulsion

As discussed in previous examples, the repulsion of lone pairs is so strong that they can change the normal bond angles in a molecule. In examples already used (GeF_2, $AsCl_3$, SF_2), the trigonal planar arrangement of the three electron pairs around the central atom would lead us to predict that the F—Ge—F bond angle is 120°. However, the lone pair of electrons exerts an extremely strong repulsive force on the electrons in the Ge—F bond and forces the bond angle to close up a little to something less than 120°. Although the presence of the lone pair allows us to predict that the bond angle is smaller, it doesn't allow us to predict how much smaller. An experiment to measure the bond angle is required if we desire to know the exact angle.

EXAMPLE 1 Predicting bond angles

Predict the value of the bond angle in PF_3.

SOLUTION The Lewis structure of PF_3 is

$$
\begin{array}{c}
\text{F} \\
| \\
\text{F—} \underset{\cdot\cdot}{\text{P}} \text{—F}
\end{array}
$$

There are four regions of high electron concentration around the central phosphorus atom. That is, 3 bonds + 1 lone pair = 4 regions. Therefore, we predict that the regions have a tetrahedal arrangement, with angles of 109.5° between them. However, only three atoms are attached to the central atom, so the shape of the molecule is trigonal pyramidal (not tetrahedral); remember, the shape of a molecule is given by the positions of the atoms in the molecule. Because the lone pair exerts a strong repulsive effect on the bonding electrons, it pushes the bonding electrons closer together than we might expect, and the actual bond angle is somewhat less than 109.5°.

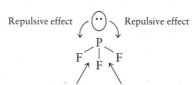

Bond angle smaller than 109.5°

EXERCISE What is the shape and bond angles (two types) in SF_4?

[*Answer:* Seesaw, less than 120° for one angle and less than 90° for four others.]

EXAMPLE 2 Using VSEPR theory with lone pairs present

Predict the shape of SO_2.

SOLUTION The Lewis structure of SO_2 is

$$:\overset{\cdot\cdot}{\text{O}}{=}\overset{\cdot\cdot}{\text{S}}{-}\overset{\cdot\cdot}{\underset{\cdot\cdot}{\text{O}}}: $$

If we count the double bond as one regions of high electron concentration (as we must for VSEPR theory), there are three such regions around sulfur. That is, one lone pair + one single-bond + one region for the double bond, which equals three regions of electrons. Thus, the electrons take on a trigonal planar arrangement around the central sulfur atom. Only two of the regions have atoms attached, so SO_2 is an angular molecule. The presence of the lone pair results in a bond angle smaller than the usual trigonal planar angle of 120°.

EXERCISE Predict the shape of HCN.

[*Answer:* Linear]

CHAPTER 9

axial lone pair   region of high electron concentration
equatorial lone pair   valence-shell electron-pair repulsion model

CHARGE DISTRIBUTION IN MOLECULES

9.6   POLAR BONDS

KEY CONCEPT   Dipole moments and polar bonds

The electronegativity of an atom is a measure of the atom's ability to attract electrons in a bond. As a result, if two bonded atoms have different electronegativities, they will engage in a tug-of-war for the electrons in the bond. The atom with the higher electronegativity tends to get an excess of electron density; the one with lower electronegativity loses some of its electron density. As the negatively charged electron-charge cloud undergoes this shift, the more electronegative atom acquires a partial negative charge and the less electronegative atom a partial positive charge. The ultimate outcome, a positive charge next to a negative charge of the same magnitude, is called an **electric dipole.**

When an electric dipole is formed on bonded atoms, the bond is called a **polar bond.** The magnitude of the polarity of the bond depends on the difference in electronegativities of the two atoms in the bond. A large difference in electronegativities results in a highly polar bond. If the difference in electronegativities is zero, the bond is usually completely nonpolar, which means that no electric dipole exists.

Bond	Difference in Electronegativities	Polarity of Bond	Partial + Charge On	Partial − Charge On
Si—F	2.1	highly polar	Si	F
O—H	1.2	very polar	H	O
P—Cl	1.0	polar	P	Cl
C—H	0.4	slightly polar	H	C
F—F	0	nonpolar	—	—

EXAMPLE   Predicting the polarity of bonds

Based on electronegativities, which of the following bonds would you predict to be most polar, S—Cl, N—F, or P—O?

SOLUTION The difference in electronegativities of the bonding atoms is usually a good measure of the polarity of a bond. For the bonds listed, we calculate the following differences, using the electronegativities in Figure 8.15 of the text: S—Cl, 0.6; N—F, 1.0; P—O, 1.2. The largest difference occurs for the P—O bond, so this bond is predicted to be the most polar.

EXERCISE Which of the following bonds is most polar, Sn—Cl, C—F, or Cl—Cl?

[*Answer:* C—F]

9.7   POLAR MOLECULES

KEY CONCEPT   The polarity of molecules

We learned earlier that a polar bond is a bond that possesses an electric dipole. A **polar molecule** is a molecule that as a whole possesses an electric dipole. The sources of an electric dipole on a molecule are electric dipoles on the bonds in the molecule. However, there is an important qualifier; the presence of bond dipoles in a molecule does not guarantee that a molecular dipole will exist. It is possible for the shape of the molecule to position bonds so that the bond dipoles will cancel to give a net zero electric dipole. Thus, it is possible for a molecule to have polar bonds but still be a nonpolar molecule. The canceling of individual bond dipoles occurs when a number of the same atoms are bonded to a central atom in a highly symmetrical way. A summary of some common molecular shapes and their polarity is given in Figure 9.10 of the text.

EXAMPLE   Predicting the polarity of a molecule

Is ClF_5 polar?

SOLUTION   At a first glance, we might imagine that because this looks like an AX_5 molecule (Figure 9.10 of the text), it is nonpolar. But beware! It is not only the formula but also the shape of a molecule that determines its polarity. The Lewis structure of ClF_5 is

$$F\!-\!\overset{\displaystyle F}{\underset{\displaystyle F\qquad F}{\overset{..}{Cl}}}\!-\!F$$

It possesses a lone pair and, therefore, has an octahedral arrangement of electrons and a square pyramidal shape. The square pyramidal shape results in a polar molecule. (It is actually an AX_5E molecule.)

EXERCISE   Is CS_2 polar?

[*Answer:* No]

KEY WORDS   Define or explain each term in a written sentence or two.

electric dipole

polar covalent bond

electric dipole moment

polar molecule

nonpolar molecule

STRENGTHS AND LENGTHS OF BONDS

9.8   BOND STRENGTHS

Imagine that you attempt to pull two bonded atoms in opposite directions until the bond breaks. The enthalpy that is required to completely separate the two atoms, in $kJ \cdot mol^{-1}$, is called the bond enthalpy (ΔH_B). It is relatively difficult to pull the atoms apart, the bond has a high bond enthalpy; if it is relatively easy to pull the atoms apart, a low bond enthalpy is involved. Table 9.2 of the text gives the bond enthalpies of a variety of diatomic molecules; Table 9.3 gives the average bond enthalpies of a number of bonds commonly found in polyatomic molecules.

EXAMPLE   Understanding bond strengths

You are probably aware that the oxygen in the atmosphere is quite reactive; many things burn. However, the nitrogen is fairly inert; in fact, when a fire is deprived of oxygen, it goes out even though there may be an abundance of nitrogen present. In what way might bond enthalpies explain this difference?

SOLUTION   From Table 9.2, we can see that the bond enthalpy of oxygen (O_2) is 496 $kJ \cdot mol^{-1}$ and that of nitrogen (N_2) is 944 $kJ \cdot mol^{-1}$. For O_2 (or N_2) to react, it typically must break apart into separate atoms; because of the lower bond enthalpy, it is much easier to break an oxygen molecule into individual atoms than to do the same to a N_2 molecule. Hence, we would expect the O_2 to be more reactive.

EXERCISE Which of the following halogens would you expect to be most reactive, F_2, Cl_2, or Br_2?

[*Answer:* F_2]

9.9 THE VARIATION OF BOND STRENGTH

KEY CONCEPT   The factors affecting bond strength

There are four major factors that affect the strength of a bond between the two atoms:

1. **Multiple bonds.** For any two atoms A and B, triple bonds are typically stronger than double bonds, and double bonds stronger than single bonds. (A and B may be the same.)

$$\text{Strongest} \quad A\equiv B > A=B > A-B \quad \text{Weakest}$$

$$C\equiv C > C=C > C-C$$

$$C\equiv O > C=O > C-O$$

2. **Lone pairs.** Lone pairs repel each other, so the presence of lone pairs on the bonding atoms makes bonds weaker than they would otherwise be, especially if atoms are small.
3. **Size of atoms.** All other things being equal, bonds between large atoms are weaker than similar bonds between small atoms.
4. **Resonance.** A bond that is part of a resonance structure has a different energy than would be predicted from a single Lewis structure. (A⋯B represents a resonance-affected bond, as in the text.)

$$A\text{⋯}B \quad \text{weaker than} \quad A=B$$

$$A\text{⋯}B \quad \text{stronger than} \quad A-B$$

Factors 1–3 are rules of thumb and cannot be relied on in all circumstances; there are many subtleties that affect bond strength.

EXAMPLE   Estimating relative bond strengths

Nitrogen and fluorine atoms are similar in size, but the N_2 molecule is far more stable than the F_2 molecule. Explain why.

SOLUTION We first draw the Lewis structures of each molecule in order to see what bonds and unshared pairs are present:

$$:\!\ddot{F}-\ddot{F}\!: \qquad :N\equiv N:$$

Each atom in F_2 has three unshared pairs; for small atoms such as fluorine, the unshared pairs on one atom repel those on the other atom quite strongly and tend to drive the atoms apart. In addition, F_2 has a single bond. Both factors result in a relatively weak bond in F_2, and the molecule is, therefore, relatively unstable. Conversely each nitrogen atom in N_2 has only one unshared pair, and the N_2 bond is a triple bond. Both these factors make the N_2 very strong, and the N_2 is, therefore, very stable. In fact, N_2 is one of the most stable diatomic species known.

EXERCISE Based on average bond enthalpies of $210 \text{ kJ} \cdot \text{mol}^{-1}$ for an N—O bond and $630 \text{ kJ} \cdot \text{mol}^{-1}$ for an N=O bond, and the Lewis structure of NO_3^-, we would predict that 1050 kJ ($2 \times 210 \text{ kJ} + 1 \times 630 \text{ kJ}$) are required to completely break all the bonds in NO_3^-. Is this prediction correct?

[*Answer:* No; NO_3^- undergoes resonance and, therefore, is more stable than predicted by any one Lewis structure. It will take more than 1050 kJ to break all the bonds in NO_3^-.]

KEY CONCEPT   Using bond enthalpies to predict enthalpies of reaction

As you might imagine, it takes energy to break a chemical bond. The enthalpy change that accompanies the breaking of a covalent bond in such a way that the two electrons in the bond become unpaired is called the **bond enthalpy**. Average bond enthalpies (Table 9.3 of the text) and the measured bond enthalpies of diatomic molecules (Table 9.2 of the text) are valuable for estimating the enthalpy of a reaction. For such a calculation, we must remember that breaking a bond requires energy (positive contribution to ΔH) and that forming a bond releases energy (negative contribution to ΔH). The reaction enthalpy is estimated by adding the bond enthalpies of all the bonds broken in the reaction and subtracting from this result the sum of the bond enthalpies for all the bonds formed. The result is only an estimate, not a precise calculation of the enthalpy of reaction; it works best when the reactants and products are in the gas phase.

EXAMPLE   Estimating the enthalpy of reaction by using bond enthalpies

Use bond enthalpies to estimate the enthalpy of reaction for the following reaction. The C—Cl bond enthalpy is 330 kJ·mol^{-1}.

$$CH_4(g) + Cl_2(g) \longrightarrow CH_3Cl(g) + HCl(g)$$

SOLUTION We must determine which bonds are broken and which are formed in order to use bond enthalpies to estimate the enthalpy of reaction. Writing the Lewis structures of the reactants and products makes this task a straightforward matter:

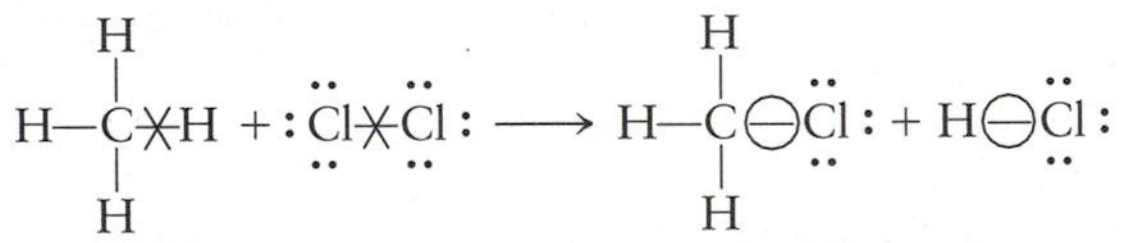

One C—H bond and one Cl—Cl bond are broken, and one C—Cl and one H—Cl bond are formed. (The bonds broken are crossed and the bonds made are circled.) Using the fact that breaking bonds takes energy and making bonds releases energy, we arrive at the following:

Break 1 mol C—H bonds:	+412 kJ
Break 1 mol of Cl—Cl bonds:	+242 kJ
Make 1 mol of C—Cl bonds:	−330 kJ
Make 1 mol of H—Cl bonds:	−431 kJ
Estimated reaction enthalpy:	−107 kJ

EXERCISE Estimate the enthalpy of reaction for the oxidation of ethane to ethanol in the vapor phase.

$$C_2H_6(g) + \tfrac{1}{2}O_2(g) \longrightarrow C_2H_5OH(g)$$

[*Answer:* −163 kJ]

9.11   BOND LENGTHS

KEY CONCEPT   The factors affecting bond length

The length of a bond is the distance between the nuclei of the two atoms joined by the bond. For instance, in the F_2 molecule, the bond length is 142 picometers (pm).

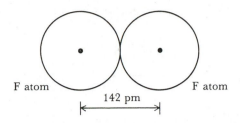

F atom   142 pm   F atom

A variety of factors affect bond length:

1. Bonds between heavier atoms tend to be longer than bonds between lighter atoms.
2. Multiple bonds are shorter than single bonds between the same atoms, for instance,

$$\text{Shortest} \quad A\equiv B < A=B < A-B \quad \text{Longest}$$
$$C\equiv C < C=C < C-C$$
$$C\equiv O < C=O < C-O$$

3. Resonance averages the lengths of the bonds participating in the resonance.

Bond lengths can be estimated by summing the covalent radii (Figure 9.18 of the text) of the atoms in the bond.

EXAMPLE   Estimating bond lengths

Use the covalent radii in Figure 9.18 of the text to show that a typical carbon-oxygen double bond is shorter than a typical carbon-oxygen single bond.

SOLUTION The pertinent covalent radii from Figure 9.18 are shown in the following table, along with the sums that give the desired approximate bond length:

	Carbon Radius, pm	Oxygen Radius, pm	Carbon + Oxygen, pm
C—O	77	74	151
C=O	67	60	127

The double bond length (127 pm) is predicted to be less than the single length (151 pm), as expected. It should be noted that the values predicted from this calculation are not the same as the average values given in Table 9.4; this difference emphasizes that summing covalent radii to get a bond length is only an approximation.

EXERCISE Without consulting any references, predict which bond is longer: S—Cl or S—F. Confirm your prediction by estimating the bond lengths by using covalent radii.

[*Answer:* S—Cl is predicted to be longer; estimated lengths are S—Cl, 200 pm, and S—F, 174 pm.]

KEY WORDS   Define or explain each term in a written sentence or two.

average bond enthalpy
bond dissociation
bond enthalpy

bond length
covalent radius

ORBITALS AND BONDING

9.12 SIGMA AND PI BONDS

KEY CONCEPT A σ-bonds in diatomic molecules

Three things occur when a σ-bond forms:

1. An unpaired electron on one atom pairs with an unpaired electron on another atom (for most σ-bonds).
2. An atomic orbital on one atom overlaps and merges with an atomic orbital on the other atom to form a new, larger orbital that encompasses both bonding atoms. For orbitals such as atomic p-orbitals, which have a long axis, the long axis of one orbital lines up with the long axis of the other, much like placing two hot dogs end-to-end.
3. Each bond results in the sharing of two electrons by the two bonding atoms.

A σ-bond concentrates a lot of electron density directly between the two nuclei of the bonded atoms; also, if you slice the orbital perpendicular to its long axis, the cross section is roughly circular (see Figures 9.19, 9.20 and 9.21 of the text). The σ-bonds in diatomic molecules can form from the overlap of various types of orbitals:

Molecule	Orbitals That Overlap
H_2	$H(1s) + H(1s)$
HCl	$H(1s) + Cl(3p)$
Cl_2	$Cl(3p) + Cl(3p)$

EXAMPLE   Understanding σ-bonds

Describe how a σ-bond forms in HBr. Use orbital diagrams.

SOLUTION The valence electron configurations and orbital diagrams of H and Br are

H: $1s^1$

Br: $[Ar]3d^{10}4s^24p^5$

Hydrogen has an unpaired electron in a $1s$-orbital, and bromine has an unpaired electron in a $2p$-orbital. Head-to-head overlap of the H($1s$) and Br($4p$) orbitals allows the electrons to pair; a new larger orbital that includes both the hydrogen atom and the bromine atom forms. Formation of this single new orbital with two electrons in it holds the two atoms together; thus, a σ-bond is formed.

EXERCISE Describe the bonding in F_2. Use orbital diagrams.

[*Answer:* A $2p$-orbital in one fluorine overlaps a $2p$-orbital in the other to form a σ-bond:

F : $1s^22s^22p^5$

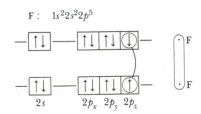

KEY CONCEPT B π-bonds

In certain situations, atomic orbitals with unpaired electrons cannot overlap in a head-to-head fashion to form a bond, but they can overlap in a side-by-side fashion. This situation generally occurs when two atoms, each having a p-orbital containing a single electron, are brought close to each other through σ-bonding. The side-by-side overlap results in formation of a π-bond; the two electrons (one from each atom) pair and are shared by both atoms, resulting in a covalent bond. A π-bond forms only after a σ-bond is formed. When a single π-bond exists with the σ-bond, a double bond results; when two π-bonds exist with the σ-bond, a triple bond results. A π-bond concentrates electron density above and below the internuclear axis rather than in between the two bonded atoms (see Figure 9.22 of the text); hence, it is weaker than the σ-bond. A single π-bond has two lobes, one above the internuclear axis and one below the internuclear axis.

EXAMPLE   Understanding π-bonds

Why are there no π-bonds in the Cl_2 molecule?

SOLUTION Chlorine has the electron configurations $1s^2 2s^2 2p^6 3s^2 3p^5$ and the orbital diagram (for valence electrons)

A σ-bond is formed by the overlap of the $3p$-orbital on one chlorine with the $3p$-orbital on the other, accompanied by the pairing and sharing of the electrons in the orbitals. At this point, all of the electrons on both chlorines are paired and no new bonds can form. Hence, no π-bond forms.

EXERCISE Account for the location of all the valence electrons in the cyanide ion, CN^-.

[*Answer:* CN^- has 10 valence electrons, 2 in a σ-bond, 4 in two π-bonds, and 4 in two unshared pairs.]

9.13   HYBRIDIZATION OF ORBITALS

KEY CONCEPT   Hybridization of atomic orbitals

Our bonding theory at this point has two problems. The first is that, in many cases, the shapes of molecules are inconsistent with the location of atomic orbitals that form bonds; the second is that the number of bonds to an atom frequently does not match the number of unpaired electrons in the unbonded atom. For example, according to our (so far, deficient) ideas, the smallest compound between hydrogen and carbon should be an angular CH_2 molecule, whereas , as we know, it is, a tetrahedral CH_4 molecule. We can patch up both problems by invoking orbital **hybridization,** in which new largely *equivalent* atomic orbitals form from old ones:

$$\text{Old atomic orbitals} \xrightarrow{\text{hybridization}} \text{new atomic orbitals}$$

For the carbon in methane, for instance,

$$2s + 2p_x + 2p_y + 2p_z \xrightarrow{\text{hybridization}} sp^3 + sp^3 + sp^3 + sp^3$$

(A way to imagine what occurs during hybridization is that the four electron clouds of the old atomic orbitals combine into a single cloud, which is then broken up into four new equivalent electron clouds.) A mathematical analysis indicates that the four sp^3 orbitals take on a tetrahedral arrangement, which agrees with the observed shape of the methane molecule. The four new sp^3 hybrid orbitals all have the same energy, and (following Hund's rule) each has one of the four valence electrons in it; thus, carbon can form four σ-bonds, again in agreement with the actual methane molecule.

EXAMPLE   Describing hybridized orbitals

In most (but not all) cases, when a boron atom undergoes hybridization, two $2p$-orbitals hybridize with one $2s$-orbital. How many new hybrid orbitals will form?

SOLUTION   One of the rules obeyed when hybridization occurs is that the number of new hybrid orbitals that form must be the same as the number of the original atomic orbitals that form the hybrids. Because the boron uses three atomic orbitals ($2s + 2p_x + 2p_y$) in its hybridization, three new hybrid orbitals form.

EXERCISE   With the type of hybridization described in the preceding example, will boron be able to form π-bonds?

[*Answer:* No; all three valence electrons are used in the hybrid orbitals. No valence electrons are left in the unhybridized p-orbitals.]

9.14   HYBRIDIZATION IN MORE COMPLEX MOLECULES

KEY CONCEPT   Formation of σ-bonds with hybridized orbitals

Hybrid atomic orbitals are very effective in forming σ-bonds because they can achieve significant orbital overlap. The steps involved in hybridization and bonding are as follows:

Step 1. Electrons are promoted from low-energy orbitals to higher energy orbitals to create enough unpaired electrons to form the required number of σ-bonds.

Step 2. Half-filled low-energy orbitals are hybridized to create enough hybrid orbitals to form the required number of σ-bonds. Remember that the number of hybridized orbitals that result must equal the number of orbitals that are used to create them.

Number of hybridized orbitals = number of orbitals used to create the hybridized orbitals

(Low-energy orbitals with pairs of electrons may also have to be hybridized, because unshared pairs often exist in hybridized orbitals.)

Step 3. Orbitals then overlap to form σ-bonds. Once the orbitals overlap, two electrons pair and become shared between two atoms; hence, a familiar covalent bond results.

Many of the σ-bonds in the familiar species we have already encountered involve overlap of hybridized orbitals, as the following table indicates.

σ-Bond	Atomic Orbital on First Atom	Atomic Orbital on Second Atom
C—H in CH_4	$C(sp^3)$	$H(1s)$
B—F in BF_3	$B(sp^2)$	$F(sp^3)$
S—F in SF_6	$S(sp^3d^2)$	$F(sp^3)$
C—C in C_2H_4	$C(sp^2)$	$C(sp^2)$

Table 9.5 of the text summarizes the different types of hybridization that are associated with different molecular shapes; it also indicates the correlation between the number of regions of high electron concentration around the central atom and the corresponding hybridization scheme.

EXAMPLE 1   Predicting hybridization

What is the hybridization of Xe in XeF_2?

SOLUTION We must determine the number of regions of high electron concentration around Xe in order to ascertain its hybridization. To do this, we first draw the Lewis structure of XeF_2:

$$:\!\ddot{F}\!-\!\dot{Xe}\!-\!\ddot{F}\!:$$

We then count the number of regions around Xe, five in this case (two bonding pairs and three lone pairs). Table 9.5 of the text indicates that five electron pairs result in sp^3d hybridization.

EXERCISE What is the hybridization of As in $AsCl_3$?

[*Answer:* sp^3]

EXAMPLE 2   Describing hybridization in a compound

Using orbital diagrams, describe the hybridization of boron in BF_3. (Consult the preceding table).

SOLUTION There are three single bonds to boron in BF_3. Boron's hybridization is therefore sp^2. The electron configuration of boron before hybridization is $1s^2 2s^2 2p^1$. Its orbital diagram is

$$\boxed{\uparrow\downarrow}\;\;\boxed{\uparrow\downarrow}\;\;\boxed{\uparrow\;|\;\;|\;\;}$$
$$1s\qquad 2s\qquad\quad 2p$$

To form three bonds, we need three unpaired electrons. Thus, we will promote one of the $2s$-electrons to a $2p$-orbital.

$$\boxed{\uparrow\downarrow}\;\;\boxed{\uparrow}\;\;\boxed{\uparrow\;|\;\uparrow\;|\;\;}$$
$$1s\qquad 2s\qquad\quad 2p$$

We finally hybridize the $2s$- and two of the $2p$-orbitals to properly account for the shape of BF_3.

$$\boxed{\uparrow\downarrow}\;\;\boxed{\uparrow\;|\;\uparrow\;|\;\uparrow\;}\;\;\boxed{\;\;}$$
$$1s\qquad\quad sp^2\qquad\quad 2p$$

To form the three σ-bonds to boron, a singly occupied sp^3-orbital on each of the three fluorines overlaps the three singly occupied sp^2 hybrid orbitals on boron.

EXERCISE What is the shape of a molecule in which the central atom is sp^3d hybridized and has two lone pairs?

[*Answer:* T-shaped]

9.15   HYBRIDS INVOLVING d-ORBITALS

KEY CONCEPT   Expanded valence shells utilize d-orbitals

When five, six, or seven regions of high electron concentration surround an atom to give the trigonal bipyramidal, octahedral, or pentagonal bipyramidal arrangement of regions, d-orbitals are used in the hybridization scheme. This pattern occurs because the use of a single valence s-orbital and the three accompanying p-orbitals in the same shell only yields four hybrid orbitals; four hybrid orbitals will not accommodate the trigonal bipyramidal, octahedral, or pentagonal bipyramidal arrangements that require five, six, and seven hybrid orbitals, respectively.

EXAMPLE   Using promotion and hybridization for bond formation

Describe how promotion and hybridization can be used to explain the bonding in AsF_5.

SOLUTION The Lewis structure of AsF_5 is

$$
\begin{array}{c}
\ddot{\textrm{:}}\ddot{\textrm{F}}\ddot{\textrm{:}} \\
\ddot{\textrm{:}}\ddot{\textrm{F}}\quad|\quad\ddot{\textrm{F}}\ddot{\textrm{:}} \\
\diagdown\textrm{As}\diagup \\
\ddot{\textrm{:}}\ddot{\textrm{F}}\quad\quad\ddot{\textrm{F}}\ddot{\textrm{:}} \\
\end{array}
$$

The central arsenic must form five σ-bonds to fluorine atoms. The electron configuration for arsenic is $[Ar]3d^{10}4s^2 4p^3$. Its valence electron orbital diagram is

—[↑↓|↑↓|↑↓|↑↓|↑↓]— —[↑↓]— —[↑|↑|↑]— —[| | | |]—
        3d   4s   4p   4d

To form five bonds, we need five unpaired electrons; promotion of a 4s-electron to a 4d-orbital gives us the required number:

—[↑↓|↑↓|↑↓|↑↓|↑↓]— —[↑]— —[↑|↑|↑]— —[↑| | |]—
        3d   4s   4p   4d

We finally hybridize the singly occupied 4s-, 4p-, and 4d-orbitals to generate five $sp^3 d$ hybrid orbitals. This results in the correct shape for AsF_5 as well as maximal overlap of atomic orbitals as bonds form.

—[↑↓|↑↓|↑↓|↑↓|↑↓]— —[↑|↑|↑|↑|↑]— —[| | |]—
        3d $sp^3 d$ 4d

To form the compound, we overlap a singly occupied 2p-orbital on each of five fluorine atoms with each of the singly occupied $sp^3 d$ hybrid orbitals on arsenic. The electrons pair, and each pair is shared by a fluorine and the arsenic to form a covalent bond. As the bond is formed, it is likely that the valence orbitals on each fluorine hybridize to form four sp^3 hybrid orbitals; this gives maximal orbital overlap. Each As—F bond is a (As $sp^3 d$, F sp^3) bond.

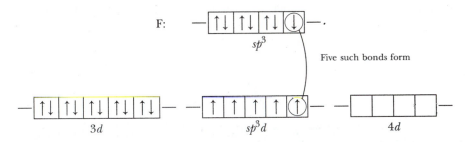

EXERCISE How many d-orbitals are used to hybridize xenon in XeF_4? Which d-orbitals are used?

[*Answer:* Two 5d-orbitals are used. (The 4d-orbitals are filled.)]

9.16   MULTIPLE CARBON-CARBON BONDS

KEY CONCEPT   Formation of π-bonds with hybridized orbitals

A π-bond is part of the makeup of any double or triple bond. We will use the double bond in ethylene to illustrate formation of a π-bond with hybridized orbitals:

H   H
 \ /
  C=C
 /   \
H   H

From the point of view of VSEPR theory, each carbon atom has three regions of high electron concentration around it, and each, therefore, has a trigonal planar structure. Thus, each carbon must also be sp^2 hybridized. The sp^2 hybridization uses a 2s-orbital and two 2p-orbitals on carbon, but leaves one 2p-orbital unaffected. This 2p-orbital protrudes above and below the carbon, perpendicular to the trigonal planar sp^2 hybrid orbitals, as shown in the next figure.

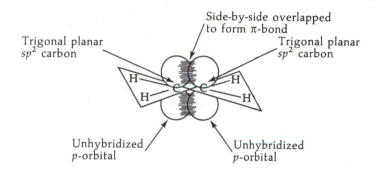

The two 2p-orbitals now overlap with each other in a side-by-side manner (rather than head-on, as in a σ-bond) to form a π-**bond**. The two electrons (one from each 2p-orbital) are paired and shared by the two carbon atoms, so a covalent bond forms. Hybridization does not occur in π-bond formation, because hybridization would result in less side-by-side overlap and, therefore, less effective bond formation.

EXAMPLE   Describing π-bonds

Describe how π-bonds may form in CO_2.

SOLUTION VSEPR theory predicts that CO_2 is linear; and, therefore, the central carbon is sp hybridized.

C: — | ↑↓ | — — | ↑ | ↑ | | — $\xrightarrow{\text{hybridize}}$ — | ↑ | ↑ | — — | ↑ | ↑ | —
          2s   2p   sp   2p
                                                                    (unhybridized)

The electrons in the sp hybrid orbitals are used for the two σ-bonds, one to each oxygen atom. The electrons in the unhybridized 2p-orbitals are used to form π-bonds, one to each oxygen atom.

EXERCISE Based on promotion and hybridization, predict how many π-bonds are in XeO_4?

[*Answer:* None]

9.17   CHARACTERISTICS OF DOUBLE BONDS

KEY CONCEPT   Double bonds have unique properties

We have already learned, and should recall, that a double bond is **shorter** and **stronger** than a single bond between the same atoms; in the same vein, a triple bond is stronger and shorter than a double bond between the same atoms. However, even though in Lewis structures we represent a multiple bond as two or three lines between atoms, e.g., C=C, or C≡C, we must keep in mind that a double bond is neither twice as strong nor half the length of a single bond, nor is a triple bond three times as strong nor one-third the length of a single bond. This is because π-bonds are weaker than σ-bonds. Another important characteristic of multiple bonds is that rotation around the bond is severely **restricted,** because rotation would destroy the overlap between the *p*-orbitals that participate in bond making and requires a large amount of energy. Rotation around single bonds requires very little energy and, at room temperature, usually occurs very easily.

EXAMPLE   Understanding the covalent bond

How many different forms of the compound $HClC{=}CHCl$ are there?

SOLUTION The key to answering this problem lies in the fact of restricted rotation around the double bond. We can actually draw two different Lewis structures for this compound,

These structures correspond to two different compounds. In I, the two chlorines are on the same side of the molecule; in II, the two chlorines are on opposite sides. If there were facile rotation around the double bond, one could easily interconvert into the other and I and II would then represent the same compound. But rotation around the double bond does not occur, so I and II are distinguishable, different compounds. Thus, there are two forms of $HClC{=}CHCl$.

EXERCISE How many forms of the compound $ClH_2C{-}CH_2Cl$ are there?

[*Answer:* One]

KEY WORDS   Define or explain each term in a written sentence or two.

electron promotion   restricted rotation
hybrid orbital σ-bond
nonbonding orbital   valence-bond theory

DESCRIPTIVE CHEMISTRY TO REMEMBER

- **Ethene** (ethylene) is $CH_2{=}CH_2$, ethyne (acetylene) is $HC{\equiv}CH$.
- The bond angle in **water** is 104.5°.
- **Teflon** is a polymer of polytetrafluoroethene.
- The **Group 14 hydrides** (CH_4, SiH_4, SnH_4, PbH_4) decrease in stability as the central atom increases in size.
- Because of **resonance** the bonds in benzene (C_6H_6) are intermediate in length between a single bond and double bond.
- Vision depends on the molecule **retinal,** which contains a double bond. When light hits the molecule, the π-bond in the double bond is broken, thus allowing the molecule to rotate around the remaining σ-bond. The π-bond then reforms, imparting a new shape to the molecule. The new shape imparts a signal along the optic nerve.

CHEMICAL EQUATIONS TO KNOW

- **Ethene** reacts with hydrogen chloride to form a compound with no double bond.

$$CH_2{=}CH_2(g) + HCl(g) \longrightarrow CH_3{-}CH_2Cl(g)$$

SELF-TEST EXERCISES

The shapes of molecules and ions

1. Within the framework of VSEPR theory, how many regions of high electron concentration surround the central atom in carbon tetrabromide, CBr_4?

(a) 0   (b) 4   (c) 2   (d) 3   (e) 32

2. Within the framework of VSEPR theory, how many regions of high electron concentration surround the central atom in sulfur tetrafluoride, SF_4?

(a) 5   (b) 4   (c) 34   (d) 6   (e) 3

3. Within the framework of VSEPR theory, how many regions of high electron concentration surround the central atom in carbonate ion, CO_3^{2-}?

(a) 4   (b) 6   (c) 5   (d) 3   (e) 8

4. Within the framework of VSEPR theory, how many regions of high electron concentration surround the central atom in SO_2?

(a) 2   (b) 5   (c) 4   (d) 6   (e) 3

5. What is the arrangement of four regions of high electron concentration (in the VSEPR sense) around a central atom?

(a) linear   (b) octahedral   (c) tetrahedral
(d) trigonal planar   (e) square planar

6. What is the arrangement of six regions of high electron concentration (in the VSEPR sense) around a central atom?

(a) octahedral   (b) hexagonal   (c) trigonal planar
(d) trigonal bipyramidal   (e) tetrahedral

7. What is the effect of lone pairs of electrons on bond angles?

(a) make angles smaller   (b) make angles larger   (c) have no effect

8. What is the shape of an AX_4E_2 molecule?

(a) tetrahedral   (b) octahedral   (c) linear
(d) trigonal planar   (e) square planar

9. What is the shape of an AX_3E_2 molecule?

(a) trigonal bipyramidal   (b) trigonal planar   (c) octahedral
(d) trigonal pyramidal   (e) T-shaped

10. What is the shape of $GaCl_3$?

(a) trigonal bipyramidal   (b) tetrahedral   (c) trigonal planar
(d) linear   (e) T-shaped

11. What is the shape of $TeCl_4$?

(a) square planar   (b) seesaw   (c) trigonal pyramidal
(d) tetrahedral   (e) square pyramidal

12. What is a reasonable value of the bond angle in ammonia (NH_3)?

(a) 109.5°   (b) 108°   (c) 90°   (d) 120°   (e) 111°

13. What is a reasonable value of the bond angle in bromine pentafluoride (BrF_5)?

(a) 180° (b) 93° (c) 90° (d) 120° (e) 88°

Charge distribution in molecules

14. Which bond is most polar?

(a) S—O (b) P—Cl (c) O—H (d) Cl—Br (e) C—H

15. Which of the following symbolizes a dipole moment? Each sign refers to a charge on a species.

(a) ⊕ ⓪ (b) ⊕ ⊖ (c) ⊖ ⊖
(d) ⊕ ⊕ (e) ⊕ ⊖ ⊕

16. Which of the following molecules is polar?

(a) CO_2 (b) NH_3 (c) CF_4 (d) BF_3 (e) SeO_3

17. A molecule consists of a central atom surrounded by some number of atoms of the same element. What shape molecule of this type is likely to be polar?

(a) linear (b) octahedral (c) square planar
(d) T-shaped (e) trigonal planar

18. Which of the following is nonpolar?

(a) SeO_2 (b) CH_3F (c) H_2O (d) ClF_3 (e) XeF_4

The strengths and lengths of bonds

19. What energy is required for the following process? Consult Table 9.2 of the text.

$$HCl(g) \longrightarrow H(g) + Cl(g)$$

(a) 426 kJ (b) 431 kJ (c) 862 kJ (d) 334 kJ (e) 224 kJ

20. Assume the bond energy of an X—X bond (for an unknown element X) is 125 kJ. What is a reasonable estimate of the X=X bond energy?

(a) 401 kJ (b) 212 kJ (c) 62.5 kJ (d) 250 kJ (e) 266 kJ

21. Use bond enthalpies to estimate the reaction enthalpy for the reaction of ethanol with oxygen to form acetic acid. (You will have to draw the Lewis structures of the reactants and products to answer this question.)

$$CH_3CH_2OH(g) + O_2(g) \longrightarrow HC_2H_3O_2(g) + H_2O(g)$$

(a) −612 kJ (b) −688 kJ (c) −298 kJ
(d) +350 kJ (e) −349 kJ

22. Which has no effect on bond energies?

(a) existence of multiple bonds (b) existence of unshared pairs
(c) number of neutrons in nuclei (d) size of atoms
(e) none of these; all affect bond energies

23. Use covalent radii to predict the approximate length of a C—S single bond.

(a) 90 pm (b) 72 pm (c) 102 pm (d) 179 pm (e) 204 pm

24. Which molecule has the largest bond length?

(a) O_2 (b) Te_2 (c) Se_2 (d) S_2

25. Which of the following bonds is shortest?

(a) $C \equiv C$ (b) $C = C$ (c) C—C

26. Which bond is weakest?

(a) HO—H   (b) HS—H   (c) HSe—H   (d) HTe—H

Orbitals and bonding

27. How many σ-bonds and π-bonds are there in an ethene (C_2H_4) molecule?

(a) 6 σ, 0 π (b) 6 σ, 1 π (c) 4 σ, 2 π

(d) 2 σ, 4 π (e) 5 σ, 1 π

28. What atomic orbitals overlap in the formation of an HCl molecule? Assume no hybridization occurs.

(a) H($1s$), Cl($3p$)   (b) H($1s$), Cl($1s$)   (c) H($2s$), Cl($2p$)

(d) H($1s$), Cl($2s$)   (e) H($2p$), Cl($3p$)

29. How many σ-bonds and π-bonds are there in an ethyne (C_2H_2) molecule?

(a) 5 σ, 0 π (b) 0 σ, 5 π (c) 3 σ, 2 π

(d) 2 σ, 3 π (e) 4 σ, 1 π

30. The hypothetical element Q has the valence-shell configuration shown. According to valence-bond theory, what is the Lewis structure of an Q_2 molecule?

(a) $:\ddot{Q}—\ddot{Q}:$ (b) $:\ddot{Q}=\ddot{Q}:$ (c) $:Q \equiv Q:$

31. One of the following does not apply to a covalent bond. Which one?

(a) electrons pair   (b) electrons are shared

(c) orbitals overlap   (d) may be polar

(e) electrons localized on one atom

32. What hybridization is associated with a trigonal planar arrangement of regions of high electron concentration?

(a) sp^3d (b) sp^2 (c) sp^3d^2 (d) sp (e) sp^3

33. What hybridization is associated with an octahedral arrangement of regions of high electron concentration?

(a) sp^3 (b) sp (c) sp^3d^2 (d) sp^3d (e) sp^2

34. What arrangement of regions of high electron concentration occurs with sp^3 hybridization?

(a) trigonal planar   (b) T-shaped   (c) octahedral

(d) tetrahedral   (e) trigonal bipyramidal

35. An octahedral arrangement of regions of high electron concentration can lead to all of the molecular shapes given except one. Which molecular shape cannot arise from an octahedral arrangement of regions?

(a) trigonal planar   (b) octahedral   (c) square planar

(d) square pyramidal   (e) linear

36. An AX_4E_2 molecule has two lone pairs. What is the hybridization of the central atom?

(a) sp (b) sp^3d (c) sp^3 (d) sp^2 (e) sp^3d^2

37. Head-to-head overlap of an sp^3 hybrid orbital with an sp^2 hybrid orbital creates a

(a) double bond   (b) σ-bond   (c) π-bond   (d) triple bond

38. Side-by-side overlap of a p-orbital with another p-orbital creates a

(a) single bond   (b) σ-bond   (c) π-bond   (d) triple bond

39. A triple bond consists of

(a) three σ-bonds   (b) one σ-bond and two π-bonds

(c) two σ-bonds and one π-bond   (d) three π-bonds

40. An atom with the valence electron configuration shown is expected to form four σ-bonds. What hybridization is expected?

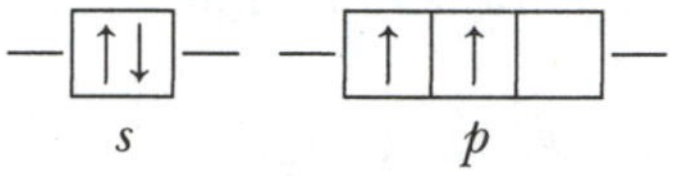

(a) sp (b) sp^3 (c) sp^3d (d) sp^2 (e) sp^3d^2

41. The orbital diagram that follows shows the valence-shell configuration of an atom after promotion of electrons but before hybridization. What hybridization is most likely?

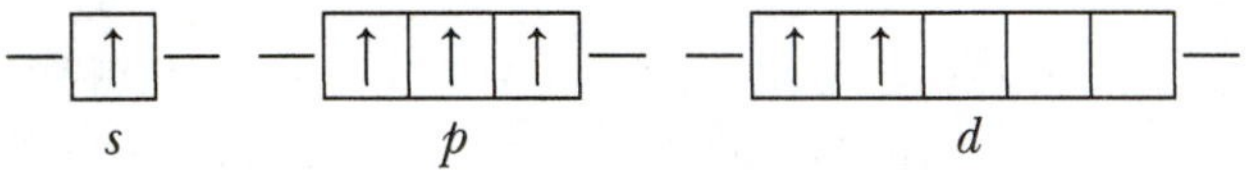

(a) sp (b) sp^3 (c) sp^3d (d) sp^2 (e) sp^3d^2

42. How many different forms of $BrFC{=}CFBr$ are there?

(a) 1   (b) 3   (c) 6   (d) 2   (e) 4

43. How many different forms of $H_2FC{-}CFH_2$ are there?

(a) 6   (b) 4   (c) 1   (d) 2   (e) 3

Descriptive chemistry

44. What bond in Teflon makes it (Teflon) so inert?

(a) $C{=}C$ (b) $C{-}F$ (c) $C{-}H$ (d) $H{-}H$ (e) $F{-}F$

45. What is the product(s) in the following reaction?

$$CH_2{=}CH_2(g) + HCl(g) \longrightarrow ?$$

(a) CH_3CH_3, Cl_2 (b) CH_3CH_2Cl (c) CH_4, Cl_2

(d) CH_2ClCH_2Cl (e) CH_2CHCl, H_2

46. Which Group 14 hydride is so unstable that its existence is in doubt?

(a) CH_4 (b) SiH_4 (c) PbH_4 (d) SnH_4

47. Which of the following is ethylene?

(a) CH_4 (b) $CH_2{=}CH_2$ (c) $CH{\equiv}CH$ (d) $CH_3{-}CH_3$ (e) NH_3

CHAPTER 10

LIQUID AND SOLID MATERIALS

Water is certainly the most important liquid on the Earth's surface; its many unusual properties actually make it a unique substance. Despite its uniqueness, water is an excellent material to examine when trying to explain the liquid state. There are many questions to ponder: For instance, why do liquids exist at all? And, a constant theme that reoccurs in our chemistry course, how are the macroscopic properties of a material explained by its underlying structure.

INTERMOLECULAR FORCES

10.1 LONDON FORCES

KEY CONCEPT   The London force

The London force arises because, for an instant, the electrons in a molecule can be slightly "out of position" relative to the nuclei of the atoms in the molecule. When such an event occurs, a slight positive charge appears in one part of the molecule (the part with slightly less electron charge cloud than it should have), and a slight negative charge appears elsewhere in the molecule (where there is slightly more electron charge cloud than there should be). Thus, a small transitory electric dipole appears in the molecule. This transitory electric dipole can encourage formation of a similar transitory electric dipole in a neighboring molecule so that the negative end of one of the transitory electric dipoles is next to the positive end of a neighboring one. Thus, a slight attractive interaction, called the **London force,** results. The London force increases as the electron polarizability increases because highly polarizable electrons move out of position most easily. Large molecules experience the greatest London forces; they have a large number of electrons and, therefore, a substantial electron charge cloud density far removed from the atomic nuclei. The London force is a very short-range interaction. The shape of a molecule also determines the strength of the London force. For two different molecules with the same number of electrons, the London force will be higher for the least spherical molecule. All molecules experience the London force. The large number of London force interactions makes it a potent intermolecular attractive force, despite its weakness.

> ### EXAMPLE   Predicting the relative effect of the London force
>
> Which element, Cl_2 or F_2, is predicted to have the highest boiling point? Why?
>
> **SOLUTION** Both Cl_2 and F_2 are nonpolar, so the major intermolecular attraction that holds the molecules of each in the liquid state is the London force. Because both molecules are diatomic molecules, both have the same shape; so it is safe to assume that the number of electrons in the molecule alone will determine the strength of the London force. Cl_2 with its 34 electrons is more polarizable than F_2 with its 18 electrons; therefore, Cl_2 experiences stronger London forces than F_2. The Cl_2 molecules are held more strongly together, and Cl_2 has the higher boiling point.

EXERCISE Which of the following has the highest boiling point: Ne, Ar, or Kr?

[*Answer:* Kr]

PITFALL   All molecules experience the London force

In polar molecules, the London force often is more important than dipole-dipole interactions.

10.2   DIPOLE-DIPOLE INTERACTIONS

KEY CONCEPT   Dipole-dipole interactions

Two polar molecules, that is two molecules with permanent dipole moments, experience an attractive interaction when the negative end of the dipole on one molecule approaches the positive end of the dipole on the other. The attractive interaction is called a **dipole-dipole interaction.** As with the other interactions discussed in this chapter, the attraction is due to the fundamental attraction of a positive charge for a negative charge. The dipole-dipole interaction is weaker in rotating molecules than in nonrotating molecules. Also, the more polar a molecule, the stronger is the dipole-dipole interaction. For the three interactions covered so far, the strength, on an individual molecule-molecule basis is

<table>
<tr><td>Strongest
interaction</td><td>ion-ion > dipole-dipole > London force</td><td>Weakest
interaction</td></tr>
</table>

EXAMPLE   Predicting the polarity of molecules and the dipole-dipole interaction

Which of the following molecules in the solid phase, HCl or HBr, is likely to experience a stronger dipole-dipole interaction?

SOLUTION The strength of the dipole-dipole interaction increases as the electric dipole on the molecule gets larger and as the molecules get closer to each other. Because Cl has a larger electronegativity (3.0) than Br (2.8), we expect, if all other things are the same, the HCl is more polar than HBr; this means it has a larger electric dipole. In addition, Cl is smaller than Br, so we expect that HCl molecules in solid HCl can get closer to each other than HBr molecules in solid HBr. Both the higher electric dipole on HCl and its smaller size indicate that HCl experiences a stronger dipole-dipole interaction than HBr in the solid phase. (The bond length of HCl is smaller than that of HBr; this difference has a slight effect on the dipole moment.)

EXERCISE Which will have the higher boiling point, *cis*-1,2-dichloroethene or *trans*-1,2-dichloroethene?

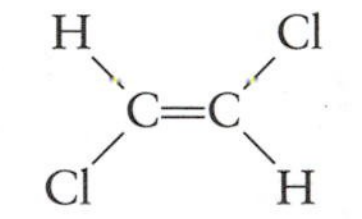

trans-1,2-dichloroethene 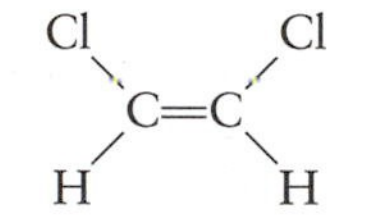   *cis*-1,2-dichloroethene

[*Answer:* *cis*-1,2-Dichloroethene, 60°C; *trans*-1,2-dichloroethene, 48°C]

10.3   HYDROGEN BONDING

KEY CONCEPT   Hydrogen bonding

Hydrogen bonding is an especially strong intermolecular attractive force that occurs when a hydrogen atom that is covalently bonded to F, O, or N moves into position close to a second F, O, or N atom. For example, water undergoes extensive hydrogen bonding, as illustrated in the following figure.

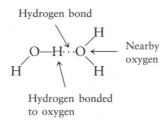

Hydrogen bond

H

O—H···O ← Nearby oxygen

H   H

Hydrogen bonded to oxygen

Three effects at play here make hydrogen bonding particularly strong. First, the electric dipoles on the molecules involved are very large because F, O, and N are among the most electronegative atoms in the periodic table. Thus, the partial positive charge on hydrogen is exceptionally high and the partial negative charge on F, O, or N (whichever is involved) is also very high; these high partial charges lead to a very strong attractive interaction. Second, the atoms involved (H, F, O, and N) are all quite small, so there is an excellent opportunity for the H atom to get very close to the F, O, or N; this close approach also increases the strength of the attractive interaction. Third, F, O, and N usually have lone pairs of electrons that can be distorted, and therefore attracted, to the positively charged hydrogen, further increasing the attractive interaction.

EXAMPLE   Evaluating hydrogen bonding

Explain why hydrogen bonding is important for HF and not for HCl.

SOLUTION To answer this question we will evaluate HF and HCl for the three factors that lead to hydrogen bonds.

1. Both HF and HCl are highly polar molecules with substantial electric dipoles. F, with electronegativity 4.0, and Cl, with electronegativity 3.0, both cause a large electric dipole to exist. However, the difference alone cannot account for the difference in hydrogen bonding because N, with electronegativity 3.0 (the same as Cl), does participate in hydrogen bonding.
2. F in HF and Cl in HCl each have three lone pairs, so the presence of lone pairs cannot account for the difference.
3. F has a covalent radius of 72 pm and Cl, 99 pm. The smaller size of the F allows the closer approach of a hydrogen and is the dominant factor that accounts for the different behavior of F and Cl with regard to hydrogen bonding.

EXERCISE Three of the following undergo hydrogen bonding. Which one doesn't?

$$H_2O \quad H_3C—NH_2 \quad CH_3F \quad CH_3OH$$

[*Answer:* CH_3F]

KEY WORDS   Define or explain each term in a written sentence or two.

dipole-dipole interaction
hydrogen bonding
instantaneous dipole moment

intermolecular forces
London force

LIQUID STRUCTURE

10.4   VISCOSITY

KEY CONCEPT   A viscous liquid (like honey) doesn't flow well

Viscosity can be thought of as a measure of the resistance of a substance to flow. Materials with a high resistance to flow, such as honey or toothpaste, have a high viscosity. Water, which flows much more easily, has a lower viscosity than they do. When intermolecular attractions between molecules are very strong, the molecules cannot move past one another very well, and the liquid does not flow easily. When the intermolecular attractions are weak, molecules can move and jostle past one another with little difficulty, and the liquid flows easily. Large, chainlike molecules that can intertwine and tangle together also cannot move past one another very easily. Flow is hindered for such molecules; they have high viscosities.

EXAMPLE   Understanding factors that affect viscosity

Large biological molecules such as those found in egg whites are often long and chainlike. Account for the fact that egg whites undergo an increase in viscosity when they are cooked.

SOLUTION We expect, from knowledge of the factors that affect viscosity, that cooking the egg increases intermolecular attractions and/or increases the chainlike character of the molecules. In fact, both occur in this case. An uncooked molecule forms a roughly spherical structure by intertwining with itself, much like a piece of string rolled into a ball. Cooking the egg causes the molecule to "unwind." It can then intertwine with other molecules, and resistance to flow is increased. In addition, the unwound molecule can undergo many more intermolecular attractions than a self-intertwined molecule; these attractions also contribute to the increase in viscosity that occurs when an egg is cooked. (Some chemical cross-linking also occurs in the cooking process.)

EXERCISE Of the three compounds acetone, propyl alcohol and ethyl methyl ether, which has the highest viscosity? These compounds are all approximately the same size.

Acetone   Propyl alcohol   Ethyl methyl ether

[*Answer:* Propyl alcohol (because of hydrogen bonding)]

10.5   SURFACE TENSION

KEY CONCEPT   The creation of surface area is an unfavorable process

A liquid has a surface tension because a molecule finds it slightly more favorable, energetically, to be buried in the body of a liquid sample, where it is surrounded on all sides by other molecules, than to be on the surface of the liquid, where it has fewer neighbors. It takes energy to force a molecule onto the surface of a liquid. This energy (which might better be called the "surface energy") is called the **surface tension**. The surface tension has units of joules per square meter ($J \cdot m^{-2}$); it is the energy needed to create a square meter of surface area. In any liquid sample, the surface area formed is the minimum possible under the circumstances; the creation of additional surface area requires energy and is an unfavorable process. The surface tension phenomenon results from intermolecular attractions. Substances that have the highest intermolecular attractions have the highest surface tension, because they require the most energy to move a molecule from the body of the liquid to the surface.

EXAMPLE   Using surface tension effects

Insects called water striders can walk on the surface of a pond because of surface tension effects. Explain how.

SOLUTION We will idealize the insect's foot as a block. A sketch of the foot on the surface of the pond (a) and how it would look if it sank (b) follows.

(a) Liquid surface   Foot

(b) Additional liquid surface

As the foot sinks, liquid surface area is created around the foot. Because it takes energy to create surface area, this is an unfavorable process. If the insect is light enough and has a large enough foot, the gravitational energy supplied (due to gravity on the insect's mass) will not be enough to create the new surface area; so the insect will not sink. (A fat water strider with small feet would sink, however.)

EXERCISE Considering surface tension effects only, would 0.05 mL of water exist as one large drop or as 5 drops of 0.01 mL each?

[*Answer:* As one larger drop]

KEY WORDS   Define or explain each term in a written sentence or two.

adhesion	meniscus
capillary action	surface tension
cohesion	viscosity

SOLID STRUCTURES

10.6   CLASSIFICATION OF SOLIDS

KEY CONCEPT   Two different classifications of solids

One way of classifying solids depends on the orderliness of the molecules that make up the solid. When the molecules are well ordered and regular (like soldiers marching in a parade), the solid is called **crystalline.** Conversely, when the molecules of the solid are more randomly organized (like a mob with no order), the solid is called **amorphous.** Amorphous solids are sometimes called glasses, and the material we call "glass" is actually an amorphous solid. Sodium chloride, with its regularly shaped cubic crystals, is typical of a crystalline material. Another classification of solids relies on two factors: the first is the type of units that makes up the solid, that is, whether the solid is composed of ions, molecules, or atoms; the second is the nature of the interactions that hold the solid together. Table 10.3 of the text summaries some of the features of the four different types of solids that arise from these considerations.

EXAMPLE   Predicting the class of a solid

A sample of a solid is extremely hard, is completely insoluble in water, has a very high melting point, and does not conduct electricity. What type of solid is it?

SOLUTION Table 10.3 of the text lists properties of solids. The properties of this sample match those of a network solid.

EXERCISE Solid CO_2 sublimes quite rapidly at a very low temperature. What type of solid is likely to do this?

[*Answer:* Molecular (the forces holding molecules together are very weak)]

KEY CONCEPT A   Close-packing in metals

A **close-packed structure** is one in which the atoms packed together occupy the smallest possible volume. In the following figure, (a) shows a structure that is not close-packed and (b) shows a close-packed structure.

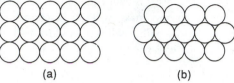

(a)   (b)

Two different close-packed structures are possible for packing spheres in three dimensions. To understand them, we imagine we have two layers of spheres, with the second layer fitting in the depressions made by the first. We now add a third layer. In the structure called **hexagonal close-packing (hcp)**, each sphere in the third layer is placed so that it lies directly above a sphere in the first layer; the first and third layers duplicate each other. In **cubic close-packing (ccp)**, each sphere in the third layer is placed in a depression that does not lie directly above a sphere in the first layer; the first and third layers are now not duplicates of each other. These two ways for placing the third layer are the only ones possible for close-packing, and each results in a complete third layer. Both close-packing schemes are equally efficient in packing spheres into the smallest possible volume. (Building these structures in three dimensions with Styrofoam balls should be a great aid in understanding packing.) Every packing arrangement of spheres has an associated **coordination number**, which is the number of nearest neighbors an atom has in a particular structure. The coordination numbers for a few common arrangements are given in the following table.

Structure	Coordination Number
hexagonal close-packing (hcp)	12
cubic close-packing (ccp); generates face-centered cubic (fcc) unit cell	12
body-centered cubic (bcc) unit cell	8

EXAMPLE   Counting coordination numbers

What is the coordination number of atom A in the two-dimensional structure shown below?

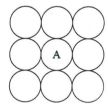

SOLUTION The coordination number of an atom is the number of "nearest neighbors." To find the nearest neighbors, we look for all atoms that are closer to A than any other atoms. In the next figure, the atoms marked NN (nearest neighbor) are closer to A than the other atoms and are its nearest neighbors. Atom A has four nearest neighbors and a coordination number of four:

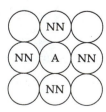

EXERCISE Use about 20 equal-sized Styrofoam spheres to build a model of a two-layer close-packed structure (because a third layer is not present, we cannot call the model hcp or ccp). By looking at your model, confirm that there are "holes" (air spaces) inside the structure. Looking from the top, you will see two different kinds of holes, one that lies directly above an atom in the lower layer and one that allows you to see through the two-layered structure to the table top. Now, take a very small sphere (a small marble will do) and place it in one of the holes that allows you to see to the table top. What is the coordination of the marble, as determined by the number of Styrofoam-sphere nearest neighbors?

[Answer: 6]

KEY CONCEPT B   Unit cells

A crystal is characterized by a repetitive, orderly arrangement of molecules, atoms, or ions. The smallest group of molecules, atoms, or ions that repeats in space in three dimensions to make up the crystal is called the **unit cell** for that crystal. A simple analogy is to imagine a rubber stamp that would be used over and over again to stamp out the pattern of a crystal; the information on the rubber stamp would be the unit cell. Consider the following two-dimensional tile pattern, and try to imagine what the unit cell, or repeat unit, for this pattern is

The repeat unit, or unit cell, is the pattern that when repeated in both dimensions reproduces the whole tile pattern. In this case, the repeat unit is

In crystals, unit cells often contain parts of atoms at corners or in faces, because this definition makes it convenient to describe three-dimensional structures with highly symmetric unit cells.

EXAMPLE   Counting the atoms in a unit cell

What is the total number of atoms contained in a face-centered cubic (fcc) unit cell?

SOLUTION Some square wood blocks, each representing a cubic unit cell, might prove helpful in thinking about this problem. But first, we note that a fcc unit cell contains *parts* of 14 atoms; an atom is located at each of the 8 corners of the unit cell, and an atom is located in each of the 6 faces. Let's consider what part of an atom in a face is in the unit cell. Putting two blocks together, face to face, should convince you that an atom in a face is shared by two unit cells; thus, each unit cell contains $\frac{1}{2}$ of an atom in the face. Now place eight of the blocks together in a cube-shaped structure. Buried inside the structure is a point at which the corners of all eight blocks come together; you can't see it, but it is there. The implication is that an atom in a corner is shared by eight unit cells; so each unit cell contains $\frac{1}{8}$ of an atom at the corner of the unit cell. The total number of atoms in the unit cell is calculated as

Total atoms in fcc unit cell = 6 faces $\times$ $\frac{1}{2}$ atom per face + 8 corners $\times$ $\frac{1}{8}$ atom per corner

$$= 3 + 1$$

$$= 4$$

EXERCISE How many atoms are contained in a body-centered cubic unit cell?

[Answer: 2]

10.8 PROPERTIES OF METALS

KEY CONCEPT Conductivity results from the movement of electrons

As indicated at the start of this chapter, one goal of our studies is to understand the properties of solids that are based on the underlying structure of the solid. Here are a few examples of this reasoning:

Many of the properties of metals are a result of the fact that metals have a large number of essentially **free electrons** that can easily move from place to place in the solid. Metals are lustrous because the electrons can easily absorb and reradiate incident light. The malleability and ductility of metals results from the fact that the high mobility of the electrons allows the positively charged nuclei to adjust to new positions as the metal is bent and shaped. Finally, the conductivity of metals is explained by the free movement of electrons from one place to another in the metal. A good conductor has a low resistance.

There are four different types of conductivity: (1) An **insulator** has a very high resistance, it does not conduct electricity. (2) A **metallic conductor** has a very low resistance that increases as the temperature is raised. (3) A **semiconductor** has a modest resistance that decreases as the temperature is raised. (4) A **superconductor** has zero resistance.

> **EXAMPLE** Predicting the resistance of a conductor
>
> Explain why the resistance of a metal increases as the temperature is raised.
>
> **SOLUTION** When the temperature is increased, atoms of the metal vibrate more and more violently. The vibrating atoms get in the way of the electrons trying to move through the solid and hinder their movement through the metal. Hence, the conductivity decreases, so the resistance increases.
>
> **EXERCISE** Why does the resistance of a semiconductor decrease as the temperature is raised?
>
> > [*Answer:* As the temperature is increased, electrons are liberated from their normal orbitals and become available for conduction.]

10.9 ALLOYS

KEY CONCEPT Alloy properties are different from those of the pure metal

It is possible to dissolve small amounts of a variety of elements in a molten metal; when the mixture cools and solidifies, the resulting mixture is called an alloy. Some typical alloys are listed in Table 10.4 of the text. One of the most interesting results of alloying a metal is that the properties of the alloy are often substantially different from that of the original metal. For example, pure gold is far too soft to use in jewelry; alloying the gold with a small amount of copper and silver makes it much harder and, therefore, suitable for jewelry. Structural alloys are designed to be harder and less brittle than the original metal. Solders, on the other hand, are designed to have a lower melting point than the original metal. Depending on how the alloying element fits into the metal lattice, a **substitutional alloy** (alloying element replaces a metal atom) or an **interstitial alloy** (alloying element fits into voids in the metal lattice) may result.

> **EXAMPLE** Predicting the structures of alloys
>
> Potassium has a radius of 227 pm and rubidium 248 pm. When they are alloyed, are they more likely to form a substitutional alloy or an interstitial alloy?

SOLUTION The text indicates that substitutional alloys can form only if the radii of the two elements involved are within about 15% different from each other. (If one of the elements is very much smaller than the other, an interstitial alloy will tend to form.) The radius of rubidium is only 9.3% larger than that of potassium; they will form a substitutional alloy.

EXERCISE Wolfram has a radius of 141 pm and carbon 91 pm. When they are alloyed, are they more likely to form a substitutional alloy or an interstitial alloy?

[*Answer:* Interstitial]

10.10 IONIC STRUCTURES

KEY CONCEPT   The rock-salt and cesium-chloride structures in ionic crystals

Unlike metals, ionic crystals are often made up of different-sized "spheres." That is, the size of the anion in the crystal might be very much different from that of the cation. The problem of how the ions pack together is much more complicated than in a metal. It turns out that the relative sizes of the anion and cation are very important in determining the actual packing arrangement. In the **rock-salt structure**, the larger anions pack in face-centered cubic unit cells. The smaller cations fit into holes between the anions in such a way that the cations also form a face-centered cubic unit cell; unlike the anions, however, the cations are not right next to each other. In this structure each anion has six cations as nearest neighbors and each cation six anions, so the structure has (6,6)-coordination. In the **cesium-chloride structure** (which is the structure of cesium chloride as well as many other salts), the very large cations occupy the corners of a cube and a smaller anion is at the center of the cube. The nearest neighbors of the centrally located anion are the eight cations at the corners of the cube, so the anion has a coordination number of 8. From another point of view, the anions in the cesium-chloride structure are also located at the corners of a cube, with the larger cation at the center of the cube. The cation therefore also has coordination number 8, so the cesium-chloride structure has (8,8)-coordination. The coordination number is expressed as two numbers: cation coordination number and anion coordination number. The two numbers are not always the same.

EXAMPLE   Recognizing the effect of size on the structure of an ionic solid

Is CaS more likely to crystallize in the rock-salt structure or the cesium-chloride structure?

SOLUTION The relative sizes of the ions in a crystal largely (but not solely!) determine the structure found in the crystal. Sodium chloride crystallizes in the rock-salt structure and cesium chloride in the cesium-chloride structure, so we compare the size of the ions in CaS to the size of the ions in NaCl and CsCl.

	Size of Ion, pm		
Ion	NaCl	CsCl	CaS
anion	181	181	184
cation	102	170	100

As we can see, the Cl^- ion is about the same size as the S^{2-} ion; but the Ca^{2+} ion with its 100-pm radius is much closer in size to the Na^+ ion with its 102-pm radius than to the Cs^+ ion with a 170-pm radius. Because the ions in CaS are much closer in actual size to those in NaCl than to those in CsCl, we conclude that CaS crystallizes in the NaCl structure. We should be careful to note that it is the sizes of the ions relative to each other that determine the structure, not their absolute sizes. LiF, for example, is composed of ions that are much smaller than those in NaCl, but it also crystallizes in the rock-salt structure because the size of F^- relative to Li^+ is the same as that of Cl^- relative to Na^+.

10.11   MOLECULAR SOLIDS

KEY CONCEPT   Molecular solids

In a molecular solid, individual molecules are held together in the solid phase by one or more of the weak intermolecular forces we discussed earlier (dipole-dipole attractions, London forces, and hydrogen bonds). The weakness of the force holding the molecules together means it is easy to pull the molecules away from one another. Therefore, molecular solids generally melt at low temperatures and are relatively soft. Candle wax and butter (both mixtures) are typical molecular solids. However, some molecular solids are hard. For example, when hydrogen bonds are present (as in sugar and water), strong interactions between molecules make the solid more rigid than would normally be expected of a molecular solid.

EXAMPLE   The properties of solids

Should a molecular solid be a good electrical conductor?

SOLUTION Electrical conduction occurs when an extensive molecular orbital extends throughout the solid. In a molecular solid, there are molecular orbitals within each molecule, but because the molecules do not covalently bond to one another, no molecular orbital extends throughout the solid. The lack of an extended molecular orbital leads us to predict that molecular solids are insulators.

EXERCISE Will methanol (CH_3OH) form a relatively hard or relatively soft solid phase? Explain your answer.

[*Answer:* Hard, because of hydrogen bonds.]

10.12   NETWORK SOLIDS

KEY CONCEPT   Network solids are extensively bonded

A network solid is characterized by an extensive network of covalent bonds that links all of the atoms in the solid. If we imagine that the covalent bonds are roads, it would be possible to drive from any atom to every other atom in the solid. Network solids are often very hard because the extensive network of bonds holds the atoms in place very well. However, if some of the bonds in the solid are weak, as in graphite and $AlCl_3$, the solid is not hard. The bonds in a network solid are usually highly localized, so they do not possess low-energy molecular orbitals that extend throughout the solid in a continuous fashion. Thus, they are usually insulators. (Graphite is an exception.) Graphite and diamond are both network solids made of linked carbon atoms; graphite and diamond are also **allotropes,** or different forms of an element in which the atoms are bonded differently.

EXAMPLE   Understanding bonding in network solids

Many texts claim that a diamond is actually one large molecule. In what way is this true?

SOLUTION A molecule is a group of atoms bonded together as a unit. A diamond consists of a very large number of carbon atoms bonded to one another, so in some sense it can be thought of as a molecule. However, most chemists would agree that a molecule should follow the law of constant composition; that is, a molecule always has the same number of atoms of each kind in it. Diamonds of different sizes have different numbers of carbon atoms. So in some ways a diamond is a huge molecule; in some ways it is not.

EXERCISE Is the fact that the network solid graphite is soft consistent with being a conductor?

[*Answer:* Yes, because both properties indicate that some of its bonds are not highly localized]

KEY WORDS   Define or explain each term in a written sentence or two.

allotrope	crystalline solid	semiconductor
alloy	cubic close-packing	superconductor
amorphous solid	hexagonal close-packing	unit cell
close-packed structure	insulator	
coordination number	metallic conductor	

PHASE CHANGES

10.13   VAPOR PRESSURE

KEY CONCEPT A   Vapor pressure

Imagine that we set up an experiment in which a beaker of water is placed on a laboratory bench and covered with a glass bell jar. The bell jar has a pressure gauge attached (which is set to 0 Torr at the start of the experiment) and is sealed with an airtight seal. Now, we observe the level of water in the beaker and the reading of the pressure gauge over a period of a few days.

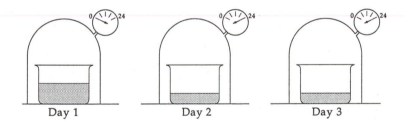

Because water molecules leave the liquid and enter the gas phase as the water in the beaker evaporates, the level of the water in the beaker on day 2 is lower than on day 1. Also, on day 2 additional gas-phase molecules are trapped in the volume enclosed by the bell jar; so the pressure inside the bell jar increases. On day 3, however, we observe that the level of water and the pressure are the same as on day 2. At this point, the number of gas-phase water molecules that collide with the surface of the liquid and rejoin the liquid equals the number that evaporate from the liquid; so the level of the liquid stays the same. In addition, because the number entering the gas phase equals the number leaving the gas phase, the number of gas-phase molecules stays the same, and the pressure does not change. At this point, a **dynamic equilibrium** exists. *Equilibrium* exists because there is no observable change over time for the system under observation; it is *dynamic* because there is an underlying motion or movement that is responsible for maintaining the equilibrium—that is, the continuous evaporation and condensation of the water. A good way of thinking about dynamic equilibria is, "Whatever is done is undone." The pressure recorded by the pressure gauge is called the **vapor pressure** of the liquid; it increases as the temperature of the liquid increases.

EXAMPLE   Understanding the vapor-pressure experiment

In the experiment just described, the level of water in the beaker stayed the same after day 2 and would, presumably, stay the same from then on. Given this fact, why does a puddle of water completely evaporate to dryness on a summer day?

SOLUTION The difference between the experiment and a puddle lies in the presence of the bell jar. The bell jar traps the gas-phase water molecules in a small volume and does not let them escape from the vicinity of the liquid. The random motion of these molecules then permits a large number of them to collide with the surface of the liquid and recondense into the liquid. The molecules that evaporate from a puddle are under no such constraints. They diffuse far away from the puddle after evaporation and do not recondense into the liquid. Eventually, all the liquid evaporates because the liquid phase is not being replenished to make up for the molecules lost by evaporation.

EXERCISE How will the vapor pressure of a liquid change if the surface area of the liquid is doubled?

[*Answer:* It will not change at all.]

KEY CONCEPT B   Vapor pressure and molecular structure

A liquid that evaporates easily has a higher vapor pressure than one that does not evaporate easily. The vapor pressure of a liquid is determined by the net number of molecules that evaporate into the enclosed volume above the liquid. If evaporation occurs easily, a large number of molecules will evaporate before the rate of condensation catches up to and equals the rate of evaporation. The large number of vapor-phase molecules then results in a high vapor pressure. The ease of evaporation itself depends on the intermolecular attractive forces present in the liquid. If the attractive forces are large, molecules are held back in the liquid phase and evaporation does not occur easily; the vapor pressure of the liquid is low. If the attractive forces are weak, molecules can leave the liquid with ease; evaporation occurs easily, and the vapor pressure is high.

EXAMPLE   Predicting the relative vapor pressures of different substances

Methane (CH_4) and carbon tetrafluoride (CF_4) are both liquids at 125 K. Predict which has the highest vapor pressure at this temperature.

SOLUTION Both CH_4 and CF_4 are tetrahedral and are therefore nonpolar. The intermolecular attractions in both are due to London forces. Because both molecules have the same shape, the strength of the London forces depends largely on the number of electrons in each compound. The compound with fewer electrons is less polarizable, will experience the weaker London forces, and will have the higher vapor pressure. CH_4 has fewer electrons than CF_4, so CH_4 has the higher vapor pressure.

EXERCISE Methanol (CH_3OH) and fluoromethane (CH_3F) are both liquids at 0°C. Both have approximately the same molecular weight. Which has the higher vapor pressure and why?

[*Answer:* CH_3F; because it does not hydrogen bond as does CH_3OH]

10.14   BOILING

KEY CONCEPT   Vapor pressure and boiling point

Boiling is characterized by formation of large bubbles in a liquid. These bubbles are not air bubbles; they are filled with molecules from the liquid that have formed a small pocket of gas-phase molecules. This occurs because the vapor pressure of the liquid is equal to the external pressure on the liquid, and molecules entering the gas phase can successfully push against the applied external pressure to form the bubble. The bubbles in a boiling liquid are small "balloons" filled with molecules from the liquid. The key idea is that a liquid boils when its vapor pressure equals the external pressure (usually atmospheric pressure) on the liquid. Because the boiling point (actually the boiling temperature) of a liquid depends on the external pressure, it is convenient to define some type of standard boiling point; the **normal boiling point** of a liquid is such a standard and is defined as the temperature at which the vapor pressure of the liquid equals exactly 1 atm. Remember, when you boil water in the laboratory or at home, the applied atmospheric pressure is usually *not* exactly 1 atm, and the temperature of the boiling water is not exactly 100.00°C.

EXAMPLE   Predicting the boiling point of water at a given pressure

Assume we double the pressure on a sample of water from 0.4 atm to 0.8 atm. Does the boiling point (in kelvins) also double?

SOLUTION   To answer this question, we must determine the actual boiling point of water at the two given pressures. The actual boiling point is the temperature at which the vapor pressure equals the external pressure. From Figure 10.50 of the text, the vapor pressure of water is 0.4 atm at 76°C, or 349 K. Thus, the boiling point of water at 0.4 atm is 76°C (349 K). Similarly, at 0.8 atm, the boiling point is 94°C (367 K). 367 K is not twice 349 K, so the boiling temperature does not double when the pressure doubles.

EXERCISE   Instructions for canning many nonacid foods call for cooking the food in a pressure cooker filled with water at a total pressure of 1.7 atm. The reason for the high pressure is to increase the boiling temperature of the water so that it is high enough to assure killing the bacteria that cause botulism. What is the boiling point of water at 1.7 atm? (Use Figure 10.50 of the text.)

[*Answer:* 112°C]

10.15   FREEZING AND MELTING

KEY CONCEPT   Solidification

For most substances, the molecules in the solid are closer together than in the liquid. An increase in pressure pushes the molecules together and makes it easier for the solid to form; the liquid freezes at a higher temperature when the pressure on it is increased. The effect is small but measurable. Water is unusual in this respect. Water molecules are farther apart in the solid than in the liquid, so applying pressure to water tends to favor the liquid state. That is, increasing the pressure makes it easier to form the liquid, and liquid water must be cooled to a lower temperature than expected to get it to freeze when pressure is applied; the freezing temperature of water drops as pressure is applied.

EXAMPLE   Predicting the change in freezing temperature as a function of pressure

A science fiction story about the life of an android living on the surface of Jupiter (where the atmospheric pressure is enormous) contains a scene in which the android makes an ax by first melting some water over a fire and then letting it freeze in an ax-shaped mold. Even though the writer assumed the temperature was very cold, could the high pressure make freezing impossible?

SOLUTION   Yes. Increasing the pressure on water lowers its freezing point. At the high pressure described in the story, the freezing point of water might be so low that it would not freeze, even at the temperature used by the writer.

EXERCISE   The text, in Section 10.15, describes how the weight of a glacier causes a film of water to form under the glacier; the glacier then slides downhill on the lubricating film of water. Could the same mechanism work if the glacier was made of ethyl alcohol instead of water? Explain your answer.

[*Answer:* No; water is the only common substance for which the melting point decreases as pressure increases. The melting point of ethyl alcohol increases as pressure increases, so it would not melt under the pressure of a glacier.]

10.16   PHASE DIAGRAMS AND COOLING CURVES

KEY CONCEPT A   Phase diagrams

A **phase diagram** is a graph that shows the physical state of a substance as a function of the pressure and temperature of the substance. A sketch of a phase diagram is shown in the following figure.

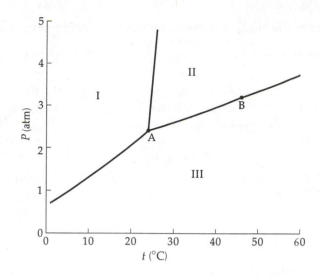

The x-axis is a temperature axis and refers to the actual temperature of the sample. The y-axis is a pressure axis. When a vapor in equilibrium with some other phase or a pure vapor is present, the y-axis refers to the pressure of the vapor; when no vapor is present, it refers to the applied pressure on the sample. The three separate areas on the graph show where the substance exists as a solid (I), as a liquid (II), and as a vapor (III). As we expect, the solid is present at high pressures and low temperatures, and the vapor at high temperatures and low pressures. Each of the three lines represents the temperatures and pressures at which two phases are at equilibrium. The word *equilibrium* is important here; two (or more) phases may be present at almost any temperature and pressure, but the only temperatures and pressures at which two phases are at equilibrium fall on a line. The point at which the three lines meet is called the **triple point** (A). It is a unique point, representing the only temperature and pressure at which the solid, liquid, and vapor can coexist at equilibrium. The triple point for the diagram above is at $t = 24°C$ and $P = 2.4$ atm.

EXAMPLE   Interpreting a phase diagram

For the phase diagram shown above, determine the lowest pressure at which the substance can exist as a liquid.

SOLUTION The liquid region of the curve defines all the temperatures and pressures at which the substance can exist as a liquid. The lowest pressure for existence of the liquid is at the triple point; at lower pressures, the graph indicates that only the solid and vapor may exist. The pressure at the triple point, 2.4 atm, is the lowest pressure at which the liquid can exist.

EXERCISE What phase(s) of the substance is (are) present at $t = 46°C$ and $P = 3.2$ atm, point B?

[*Answer:* Liquid and vapor are present at equilibrium.]

KEY CONCEPT B   Cooling and heating curves

A cooling curve is a graph that shows the temperature of a substance (on the y-axis) as heat is withdrawn (the amount of heat withdrawn is on the x-axis); a heating curve shows the temperature as heat is added to a substance. Important points to notice regarding such a curve are the following:

1. Some parts of the curve correspond to changes in temperature. During these changes, no phase change occurs; added (or withdrawn) heat causes a change in molecular motion and, therefore, a change in temperature. To calculate the temperature change that accompanies the withdrawal (or addition) of a certain amount of heat, you must use the specific heat of the substance (for the correct physical state).

2. Other parts of the curve (the flat parts) correspond to a constant temperature. At these points, a phase change occurs; added (or withdrawn) heat, instead of changing molecular motion, causes the phase change to occur. You should consider this

carefully; heat can be added or withdrawn with no accompanying temperature change. To calculate the amount of heat required to accomplish a particular phase change, you must use the enthalpy of fusion (for freezing or melting) or the enthalpy of vaporization (for boiling or condensation).

3. The flat parts of the curve correspond to the melting temperature and boiling temperature of the substance.

EXAMPLE **Calculating a portion of a heating curve**

We add 1000 J to a 15.0-g sample of liquid carbon tetrachloride at 55.0°C. Describe what happens.

Substance	Freezing Point, °C	Enthalpy of Fusion, $J \cdot g^{-1}$	Specific Heat Capacity (Liquid), $J \cdot g^{-1} \cdot K^{-1}$	Boiling Point, °C	Heat of Vaporization, $J \cdot g^{-1}$	Specific Heat Capacity (Vapor), $J \cdot g^{-1} \cdot K^{-1}$
CCl_4	−23.0	21.3	0.862	76.7	194	0.540

SOLUTION We note that at 55.0°C carbon tetrachloride is a liquid. Adding heat increases the temperature of the liquid up to the boiling temperature; at this point, a phase change occurs. We first calculate the heat required to warm the liquid to its boiling point, which involves the specific heat capacity of the liquid:

Heat = mass × specific heat capacity × change in temperature

$$= 15.0 \text{ g} \times 0.862 \text{ J} \cdot g^{-1} \cdot K^{-1} \times (76.7°C - 55.0°C)$$

$$= 15.0 \text{ g} \times 0.682 \text{ J} \cdot g^{-1} \cdot K^{-1} \times (21.7 \text{ K}) \quad [\text{remember } \Delta T \text{ (°C)} = \Delta T \text{ (K)}]$$

$$= 281 \text{ J}$$

At this point, the liquid is at its boiling temperature with 719 J (1000 J − 281 J) of heat left to add. As this heat is added, the liquid vaporizes. The heat of vaporization is 194 $J \cdot g^{-1}$, so it is clear that not all the carbon tetrachloride can be vaporized; we must calculate what mass of carbon tetrachloride can be vaporized with the remaining 719 J.

$$\text{Heat} = \text{mass} \times \text{heat of vaporization (in } J \cdot g^{-1})$$

$$\text{Mass} = \frac{\text{heat}}{\text{heat of vaporization}}$$

$$= \frac{719 \text{ J}}{194 \text{ J} \cdot g^{-1}}$$

$$= 3.71 \text{ g}$$

Thus, we are left with 11.3 g of liquid carbon tetrachloride at its boiling temperature of 76.7°C and 3.71 g of gaseous carbon tetrachloride at 76.7°C.

EXERCISE How much heat is required to vaporize the remaining carbon tetrachloride and warm the whole sample to 85.0°C?

[*Answer:* 2.26 kJ]

10.17   CRITICAL PROPERTIES

KEY CONCEPT   The critical temperature

The temperature of a substance can be increased to the point where it is impossible to liquefy the substance, no matter how high the applied pressure. The temperature above which it is impossible to liquefy a substance is called the **critical temperature** of the substance and is denoted by the symbol T_c.

EXAMPLE   Interpreting the critical temperature

The critical temperature of methane is 191 K. What pressure must be applied to methane at 298 K in order to liquefy it?

SOLUTION The critical temperature of a substance is the maximum temperature at which it can exist in the liquid form. Above the critical temperature, it is impossible for the substance to be liquefied. Because 298 K is higher than methane's critical temperature of 191 K, it is impossible to liquefy the methane at 298 K, no matter what pressure is applied.

EXERCISE Solid carbon dioxide is called "dry ice" and, as you may have observed, does not form a liquid at 25°C and 1 atm. The critical temperature of carbon dioxide is 31°C. Is the failure of dry ice to form a liquid at 25°C and 1 atm due to the relative values of its actual temperature and critical temperature?

[*Answer:* No]

KEY WORDS   Define or explain each term in a written sentence or two.

critical temperature	phase	supercritical fluid
critical pressure	phase boundary	triple point
dynamic equilibrium	phase diagram	vapor pressure
normal boiling point	phase transition	volatile

DESCRIPTIVE CHEMISTRY TO REMEMBER

- **Water** is unusual because it expands when it freezes.
- Magnesium and zinc crystallize in a **hcp** structure; aluminum, copper, silver, and gold in a **ccp** structure; iron, sodium, and potassium in a **bcc** structure.
- **Titanium chloride** ($TiCl_4$) is a liquid that boils at 136°C and freezes at −25°C to a molecular solid.
- **Rhombic sulfur,** a molecular solid of S_8 rings, melts at 113°C to form a mobile straw-colored liquid.
- **Mercury** forms a meniscus that bulges upward; water forms a meniscus that bulges downward.
- Common glass is an **amorphous** solid.
- Compressing graphite at over 80,000 atm and 1500°C in the presence of a trace amount of a metal such as chromium or iron results in the formation of natural **diamond.**
- **Graphite** is a black, lustrous, electrically conducting, slippery solid which sublimes at 3700°C.
- **Ice** is an open network of water molecules held together by hydrogen bonds.
- F_2 and Cl_2 are gases, Br_2 a liquid, and I_2 a solid.
- As a result of **hydrogen bonding,** liquid HF contains zigzag chains of HF molecules; gaseous HF contains short chains and $(HF)_6$ rings.
- As a result of **hydrogen bonding,** gaseous acetic acid contains dimers.
- **Steel** is an alloy of carbon in iron.

SELF-TEST EXERCISES

Intermolecular forces

1. Which molecule experiences London forces only?

(a) HCl   (b) H_2Se (c) H_2O (d) P_4 (e) HF

2. Which molecule experiences London forces only?

(a) NH_3 (b) H_2O (c) CO_2 (d) CH_3OH (e) SO_2

3. Which of the atoms shown experiences the largest London forces?

(a) Kr   (b) He   (c) Xe   (d) Ne   (e) Ar

4. Predict which of the following has the highest boiling point.

(a) C_4H_{10} (*n*-butane, cylindrical)

(b) C_4Cl_{10} (perchoro-*n*-butane, cylindrical)

(c) C_4H_{10} (isobutane, roughly spherical)

(d) C_4Cl_{10} (perchloro-isobutane, roughly spherical)

5. Only one of the following molecules experiences dipole-dipole interactions. Which one?

(a) $BeCl_2$ (b) CO_2 (c) SO_2 (d) XeF_2 (e) F_2

6. Only one of the following does not experience dipole-dipole interactions. Which one?

(a) $CHCl_3$ (b) CF_4 (c) ClF_3 (d) IF_5 (e) PF_3

7. Which of the following is predicted to have the highest boiling point?

(a) 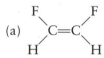   (b) 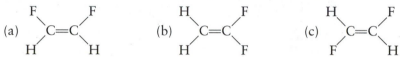   (c)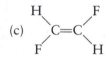

8. Which of the following is predicted to have the highest boiling point?

(a) 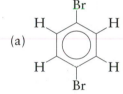   (b)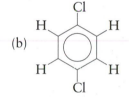

(c) 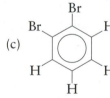   (d)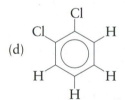

9. Which of the following undergoes hydrogen bonding?

(a) HCl   (b) H_2 (c) CH_4 (d) CFH_3 (e) CH_3OH

10. A hydrogen bond is

(a) the covalent bond between two hydrogen atoms in H_2.

(b) an especially strong intermolecular attractive force.

(c) a covalent bond between hydrogen and oxygen.

(d) an ionic bond between H^- and a Group 1 metal cation.

11. All but one of the following contribute to the hydrogen bonding. which one doesn't?

(a) small size of hydrogen

(b) polarizability of hydrogen

(c) small size of the more electronegative atom (O, F, N)

(d) high electronegativity of the more electronegative atom (O, F, N)

(e) low electronegativity of hydrogen

12. Which of the following Group 16 hydrides has an unexpectedly high boiling point?
(a) H_2O (b) H_2S (c) H_2Se (d) H_2Te

Liquid structure

13. Which has the highest viscosity?
(a) water   (b) gasoline   (c) rubbing alcohol   (d) honey

14. For most liquids, the viscosity _____ as temperature _____.
(a) decreases, increases   (b) increases, increases

15. On the basis of its structure, which of the following would be predicted to have the highest viscosity?

(a)
```
     H  H  H
     |  |  |
 H — C— C— C—OH
     |  |  |
     H  H  H
```
(b)
```
    OH OH OH
     |  |  |
 H — C— C— C—H
     |  |  |
     H  H  H
```
(c)
```
     H  H  H
     |  |  |
 H — C— C— C—H
     |  |  |
     H  H  H
```

16. A molecule at the surface of a liquid has _____ energy than a molecule in the bulk of the liquid.
(a) higher   (b) lower   (c) the same

17. If a liquid in a glass tube forms a meniscus that is essentially flat (it bulges neither upward nor downward), the forces between liquid molecules are _____ the forces between the liquid molecules and the glass.
(a) greater than   (b) equal to   (c) less than

18. Which type of solid has the smallest surface area for a given volume?
(a) cylinder   (b) pyramid   (c) cube   (d) sphere   (e) cone

19. Assume a cube-shaped drop of liquid has a surface area of 0.54 cm^2. What is the surface area of a spherical drop with the same volume? All answers are in cm^2.
(a) 0.081   (b) 0.64   (c) 0.027   (d) 0.54   (e) 0.044

20. Capillary action refers to a phenomenon in which a liquid
(a) flows through a pipe because of a pressure differential.
(b) is pushed up a tube by a pressure differential.
(c) forms a sphere because of cohesive forces.
(d) forms a meniscus.
(e) rises up a narrow tube because of adhesive forces.

Solid structures

21. Which of the following forms an amorphous solid?
(a) sodium fluoride   (b) copper   (c) water   (d) glass   (e) sucrose

22. Which type of solid has a disordered molecular structure?
(a) amorphous   (b) crystalline   (c) neither

23. Four of the following phrases apply to a crystalline solid. Which one does not?

(a) near-random arrangement of molecules   (b) contains unit cell

(c) repetitive pattern   (d) orderly array

(e) well-defined forces

24. What is the coordination number of a metal atom in a hexagonal close-packed structure?

(a) 6   (b) 12   (c) 8   (d) 10   (e) 4

25. Predict the density of gold. Gold has an atomic radius of 144 pm and crystallizes in a cubic close-packed arrangement (ccp). Remember that the ccp structure has a face-centered cubic cell. All answers are in $g \cdot cm^{-3}$.

(a) 1.00   (b) 0.0763   (c) 19.4   (d) 13.1   (e) 0.0515

26. What percentage of the volume of a face-centered cubic (fcc) unit cell is occupied by atoms in the cell. (The fcc unit cell is a cubic close-packed arrangement; the correct answer corresponds to the most efficient way to pack equal-sized spheres.)

(a) 66.7%   (b) 44.4%   (c) 81.5%   (d) 100%   (e) 74.0%

27. A representation of a brick wall is shown below.

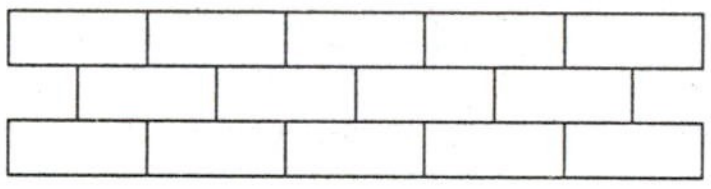

What is the number of bricks in the unit cell for this two-dimensional pattern?

(a) 2   (b) 1   (c) 3   (d) $1\frac{1}{2}$ (e) 4

28. Which mixture is an alloy?

(a) $NaCl + H_2O$ (dissolved)   (b) $NaCl + KNO_3$ (ground together)

(c) $H_2O + CH_3OH$ (dissolved together)   (d) $Au + Sn$ (dissolved together)

29. Which is a semiconductor?

(a) Cu   (b) Co doped in Au   (c) NaCl

(d) graphite   (e) As doped in Si

30. Which type of solids is best described as "cations held together by a sea of electrons"?

(a) ionic   (b) molecular   (c) network   (e) metal

31. The following shows a top view of two layers in a close-packed arrangement (darker atoms below, lighter ones above). Where does the next atom go to create hexagonal close-packing?

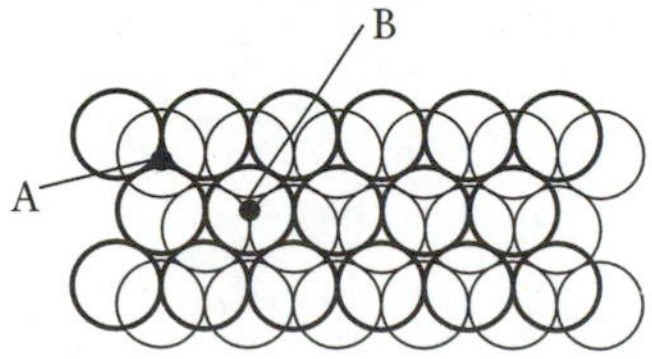

(a) position A   (b) position B   (c) elsewhere

32. In a copper wire, an electric current is carried by which species?

(a) Cu^+ (b) Cu^{2+} (c) Cu   (d) p^+ (e) e^-

33. A substance with a very high resistance is called a(n)

(a) conductor   (b) superconductor   (c) insulator   (d) semiconductor

34. Which of the following are most likely to form a substitutional alloy? Use data from Chapter 8 of the text.

(a) Ni, Cu   (b) H, Fe   (c) C, W   (d) Bi, Fe   (e) N, Pt

35. Which element is most likely to form an interstitial alloy with titanium? Use data from Chapter 8 of the text.

(a) Bi   (b) Au   (c) Hg   (d) H   (e) Fe

36. The properties of an alloy are likely to be _____ the properties of the original metals.

(a) different from   (b) the same as

37. Which of the following alkali halides is predicted to exist in a rock-salt structure? Use data from Chapter 7 of the text.

(a) LiI   (b) NaF   (c) CsCl   (d) LiF   (e) RbCl

38. Cesium chloride crystallizes in a body-centered cubic unit cell with a Cs^+ ion at the center of the unit cell and Cl^- ions at the corners. What is the coordination number of Cs^+ in this structure?

(a) 8   (b) 6   (c) 4   (d) 12   (e) 10

39. What happens to the resistance of a conductor when its temperature is increased?

(a) decreases   (b) increases   (c) stays the same

40. What is a typical resistance for a superconductor?

(a) zero   (b) very high   (c) very low   (d) depends on the superconductor

41. What type of coordination is present in the rock-salt structure?

(a) (8, 8)   (b) (6, 6)   (c) (12, 12)   (d) (8, 4)   (e) (6, 3)

42. Which describes a typical network solid?

(a) atoms all connected to one another   (b) very small band gap

(c) soft, and low melting   (d) contains a vast array of ions

(e) held together by intermolecular attractive forces

43. Which of the following is a pair of allotropes?

(a) ^{12}C, ^{14}C (b) H_2O, H_2O_2 (c) ice, water

(d) S_8, S_2 (e) Na, K

44. Which of the following forms a molecular solid?

(a) C (graphite)   (b) H_2O (c) NaCl   (d) C (diamond)   (e) $AlCl_3$

Phase changes

45. Which liquid, on the basis of its structure, has the lowest vapor pressure?

          H H H   H H H   OH OH OH

          | | | | | | | | |

(a) H—C—C—C—F   (b) H—C—C—C—H   (c) H—C—C—C—H

          | | | | | | | | |

          H H H   H H H   H H H

46. What is the normal boiling point of the liquid that has the vapor pressure curve shown in the figure?

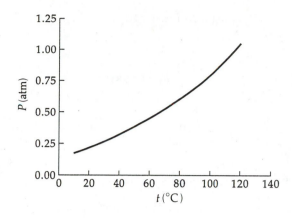

(a) 18°C   (b) 126°C   (c) 83°C   (d) 100°C   (e) 118°C

47. Which of the liquids in Exercise 45 has the lowest boiling point?

Use the following phase diagram for Exercises 48–50.

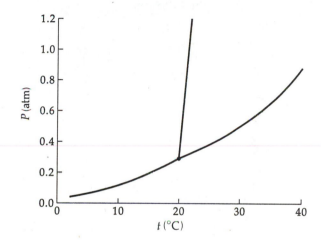

48. What is the physical state of the substance at 10°C and 1.0 atm?

(a) solid   (b) liquid

(c) vapor   (d) vapor in equilibrium with liquid

(e) solid in equilibrium with liquid

49. What is the minimum pressure that must be applied to liquefy the substance?

(a) 1.2 atm   (b) 0.2 atm   (c) 0.1 atm   (d) 0.3 atm   (e) 0.5 atm

50. At what temperature and pressure are only the solid and vapor in equilibrium?

(a) 10°C, 0.2 atm   (b) 20°C, 0.5 atm   (c) 18°C, 0.3 atm

(d) 30°C, 0.5 atm   (e) 15°C, 0.2 atm

51. How much heat must be withdrawn from 12.0 g of liquid carbon tetrachloride (CCl_4) at 25.0°C to completely solidify it?

Substance	Freezing Point, °C	Enthalpy of Fusion, $J \cdot g^{-1}$	Specific Heat Capacity (Liquid), $J \cdot g^{-1} \cdot K^{-1}$	Boiling Point, °C	Heat of Vaporization, $J \cdot g^{-1}$	Specific Heat Capacity (Vapor), $J \cdot g^{-1} \cdot K^{-1}$
CCl_4	−23.0	21.3	0.862	76.7	194	0.540

(a) 863 J   (b) 256 J   (c) 752 J   (d) 497 J   (e) 514 J

52. The critical temperature of oxygen is $-118°C$. What pressure is required to liquefy oxygen at room temperature, about $25°C$?

(a) 1 atm   (b) 10 atm   (c) 25 atm

(d) 1000 atm   (e) It is impossible to liquefy oxygen at $25°C$.

Descriptive chemistry

53. Which substance is noted for the fact that it expands when it freezes?

(a) copper   (b) silicon   (c) iron(III) oxide

(d) water   (e) all substances expand when they freeze

54. Which of the following crystallizes in an hexagonal close-packed structure?

(a) Al   (b) Cu   (c) Ag   (d) Au   (e) Zn

55. Which of the following halides forms a molecular solid?

(a) $TiCl_4$ (b) $CaCl_2$ (c) NaCl   (d) $MgCl_2$ (e) $FeCl_3$

56. Steel is an alloy of iron and which of the following?

(a) Zn   (b) H   (c) C   (d) Sn   (e) Bi

57. Which of the following elements is a liquid at room temperature?

(a) Cl_2 (b) K   (c) Ne   (d) Br_2 (e) N_2

CHAPTER 11

CARBON-BASED MATERIALS

In this chapter we begin an exploration of organic chemistry, the chemistry of carbon compounds. Carbon compounds are used as foods and fuels. Cotton, wool, and silk are composed of carbon compounds and, as you might know, life on Earth is carbon based. Although the breadth of organic chemistry is so enormous that it might seem impossible to learn, it is greatly simplified by the fact that the chemistry of carbon compounds is largely determined by small groups of atoms in the compound rather than by the whole compound itself.

HYDROCARBONS

11.1   TYPES OF HYDROCARBONS

KEY CONCEPT   There are many types of hydrocarbons

Hydrocarbons are compounds that contain only carbon and hydrogen. For example, C_6H_{12} and C_4H_8 are hydrocarbons; CH_4O and C_3H_9N are not. Two major classes of hydrocarbons are **saturated** hydrocarbons, which contain single bonds only, and **unsaturated** hydrocarbons, which contain one or more multiple bonds. A second, and different, classification system distinguishes between **aromatic** hydrocarbons, which contain one or more benzene rings, and **aliphatic** hydrocarbons, which do not contain a benzene ring. The stick structures used to represent hydrocarbons do not show any of the atoms present in the structure; only the bonds are shown.

EXAMPLE   Understanding hydrocarbon classifications and stick structures

What is the molecular formula of the compound represented by the following stick structure? Is it saturated or unsaturated? Is it aromatic or aliphatic?

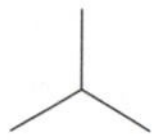

SOLUTION Each line in the structure represents a single bond with a carbon attached at each end. If any of the carbon atoms in the stick structure does not satisfy its valence of four through bonding to other carbon atoms, we must add hydrogen atoms to satisfy the carbon valence. For instance, from the lines in the stick structure, we can see that the carbon at the center of the structure has only three bonds; one hydrogen is added to complete the carbon valence. The carbons at the end of each line have only one bond; three additional hydrogen atoms are required for each of these. The following structure illustrates the meaning of the stick structure.

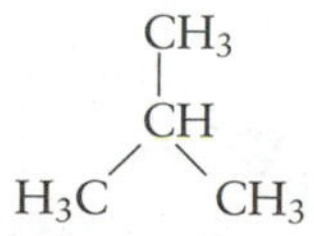

Thus, the formula of the compound is C_4H_{10}. Because there are only single bonds in the compound, it is unsaturated; because there are no benzene rings, it is aliphatic.

EXERCISE What is the molecular formula of the compound represented by the following stick structure? Is it saturated or unsaturated? Is it aromatic or aliphatic?

[*Answer:* C_8H_{10}, unsaturated, aromatic]

11.2 ALKANES

KEY CONCEPT The properties of alkanes

Because the only important intermolecular attractions in the alkanes are London forces, the alkanes show a smooth variation in many physical properties as molecular size increases. In branched alkanes, the side chains prevent the molecules from getting as close to one another as unbranched molecules of the same size might, so the branched molecules experience lower molecular attractions and have physical properties that reflect these smaller attractions. Chemically, the strong C—H and C—C bonds of the alkanes make them relatively unreactive. Two important types of reactions that the alkanes do undergo are oxidation and substitution. A typical oxidation is the complete combustion to carbon dioxide and water in excess oxygen; this is always a highly exothermic reaction and is the basis for using the alkanes as fuels. In a **substitution reaction,** a group or atom is substituted for another group or atom in the original molecule. The substitution of a chlorine atom for a hydrogen atom through a radical chain reaction is an example of an alkane substitution reaction. For example, in methane the reaction is

$$CH_4(g) \ + \ Cl_2(g) \ \longrightarrow \ CH_3Cl(g) \ + \ HCl(g)$$

Many such reactions are difficult to control, and multiple substitutions may occur. For the chlorination of methane, it is not unusual to get CH_2Cl_2, $CHCl_3$, and CCl_4 as products as well as CH_3Cl.

EXAMPLE Predicting the physical and chemical properties of the alkanes

Arrange the following alkanes in order of increasing boiling points.

$$CH_3CH_2CH_3 \qquad CH_3CH_2CH_2CH_3 \qquad CH_3CH(CH_3)CH_3 \qquad CH_3CH_2CH_2CH_2CH_3$$

Propane   Butane   Methylpropane   Pentane

SOLUTION For the alkanes, the boiling points increase as the size (molar mass) of the compound increases. The smallest is propane, so it has the lowest boiling point; the largest is pentane, so it has the highest boiling point. Butane and methylpropane, the intermediate compounds, are both the same size (4 carbons); but branched compounds experience smaller intermolecular attractions than do unbranched compounds of the same size, so methylpropane has a lower boiling point than butane. The order is

Lowest boiling point   propane < methylpropane < butane < pentane   highest boiling point

EXERCISE Write a balanced chemical equation for the combustion of gaseous butane, C_4H_{10}.

[*Answer:* $2C_4H_{10}(g) \ + \ 13O_2(g) \rightarrow 8CO_2(g) \ + \ 10H_2O(g)$]

11.3 ALKENES AND ALKYNES

KEY CONCEPT A Alkenes

The **alkenes** are a homologous series that contain a carbon-carbon double bond and have a molecular formula of the type C_nH_{2n}. C_2H_4 (ethene, also called ethylene) and CH_2=$CHCH_3$ (propene)

are the two simplest alkenes. Alkenes can be produced from alkanes or substituted alkanes by an **elimination reaction;** in this reaction a small molecule such as water, HCl, HBr, or H_2 is driven out of (eliminated from) the reactant molecule, leaving a double bond. The atoms or groups removed are usually on adjacent carbon atoms in the reactant molecule. Catalytic **dehydrogenation** eliminates H_2:

$$CH_3CH_3 \longrightarrow CH_2{=}CH_2 + H_2$$

The characteristic reaction of the double bond is an **addition reaction,** a reaction in which a small molecule such as H_2O, HCl, HBr, Cl_2, Br_2, or H_2 adds to the double bond. In the addition, one part of the small molecule ends up on one carbon associated with the double bond and the rest of the small molecule ends up on the other double-bonded carbon. In the process, the double bond becomes a single bond. For example,

> **Hydrogenation** (addition of H_2): $CH_2{=}CH_2 + H_2 \longrightarrow CH_3CH_3$
>
> **Halogenation** (addition of a halogen molecule): $CH_2{=}CH_2 + Cl_2 \longrightarrow CH_2ClCH_2Cl$
>
> **Hydrohalogenation** (addition of a hydrohalide): $CH_2{=}CH_2 + HBr \longrightarrow CH_3CH_2Br$
>
> **Hydration** (addition of water): $CH_2{=}CH_2 + H_2O \longrightarrow CH_3CH_2OH$

EXAMPLE   Identifying the products of addition reactions

What product is formed when HBr adds to the double bond in ethene, $CH_2{=}CH_2$.

SOLUTION In this addition reaction, H adds to one of the double-bonded carbon atoms and Br adds to the other. The product is bromoethane:

$$CH_2{=}CH_2 + HBr \longrightarrow CH_2BrCH_3$$

EXERCISE What product is formed when Br_2 adds to propene, $CH_2{=}CH{-}CH_3$?

[*Answer:* $CH_2Br{-}CHBr{-}CH_3$]

KEY CONCEPT B   Alkynes

Alkynes are hydrocarbons that contain a carbon-carbon triple bond. They are a homologous series with formulas of the type C_nH_{2n-2}. The two simplest alkynes are $CH{\equiv}CH$ (ethyne) and $CH{\equiv}CCH_3$ (propyne). The reactions of alkynes are similar to those of alkenes, with addition across the triple bond to form a double-bonded compound followed by continued addition being the common mode of reaction.

EXAMPLE   Understanding alkynes

What is the product formed when Cl_2 is added to propyne, $CH{\equiv}CCH_3$?

SOLUTION Using the alkenes as a model, we can conclude that when one Cl_2 adds to a multiple bond, one of the π-bonds is destroyed. Because there are two π-bonds in the triple bond of propyne, two Cl_2 molecules can add across the triple bond. Therefore, the resulting product is $CHCl_2CCl_2CH_3$.

EXERCISE Using bond energies, estimate the enthalpy of reaction for the reaction discussed in the preceding example (addition of 2 mol Cl_2 to 1 mol propyne)? Use 352 kJ·mol^{-1} for the C—Cl bond enthalpy.

[*Answer:* -435 kJ]

11.4   AROMATIC COMPOUNDS

KEY CONCEPT A   Aromatic hydrocarbons

Aromatic compounds are compounds that are based on or contain a benzene (C_6H_6) ring:

Such compounds are also called **arenes**. Even though they are unsaturated, they are quite unreactive; and when they react, they do so more by substitution than by addition. All carbon-carbon bonds in the benzene ring are equivalent. There are no isolated double bonds in benzene but rather a π system with delocalized orbitals that extend over all six carbon atoms. The delocalized π orbitals contain six electrons. There are also **polycyclic** arenes, which contain two or more benzene-type rings fused together. The conventional symbols for benzene and the polycyclic naphthalene and anthracene are

Benzene   Naphthalene ($C_{10}H_8$)   Anthracene ($C_{14}H_{10}$)

EXAMPLE   Recognizing aromatic hydrocarbons

What is the molecular formula of the compound depicted below?

SOLUTION Only two chlorine atoms are explicitly shown, but many other atoms are present. A hexagon with a circle represents a benzene molecule with six carbons in the ring (the hexagon) and one hydrogen only attached to each carbon. When two chlorines are attached to the benzene, two hydrogens must be removed (because each carbon has four bonds), so four hydrogen atoms are present. The formula is $C_6H_4Cl_2$. A more complete picture of the molecule follows.

EXERCISE What is the molecular formula of naphthalene, shown below?

[*Answer:* $C_{10}H_8$]

KEY CONCEPT B   Reactions of aromatic hydrocarbons

Reactions involving benzene rings are generally substitution reactions in which some atom or group replaces a hydrogen; the π system is not disrupted. For instance, in the presence of the catalyst

FeBr$_3$, the reaction of benzene with Br$_2$ results in the formation of bromobenzene:

$$C_6H_6 + Br_2 \xrightarrow{FeBr_3} C_6H_5Br + HBr$$

When benzene, minus one hydrogen, is a substituent, it is called a **phenyl group** (C$_6$H$_5$). Thus, the compound CH≡C(C$_6$H$_5$) is given the common name phenylacetylene. The benzene ring can be made to undergo addition under relatively harsh conditions.

EXAMPLE **Understanding the reactions of aromatic hydrocarbons**

What is the product formed when Br$_2$ is mixed with benzene and gently heated, with no catalyst present?

SOLUTION No reaction occurs. The benzene ring is very resistant to attack under normal conditions; specifically, the Br$_2$ will not add across the bonds in benzene. Because no catalyst is present, no reaction will occur.

EXERCISE Classify the reaction of Br$_2$ with benzene (in the presence of a catalyst) as one of the following: elimination, substitution, addition.

[*Answer:* Substitution]

KEY WORDS   Define or explain each term in a written sentence or two.

aliphatic	aliphatic hydrocarbon	hydrocarbon
alkane	aromatic hydrocarbon	isomer
alkene	condensed structural formula	saturated hydrocarbon
alkyne	cycloalkane	unsaturated hydrocarbon

TOOLBOX 11.1   How to name hydrocarbons

KEY CONCEPT A   Alkane nomenclature

If the carbon atoms in an alkane are bonded in a row, so that the formula is of the form CH$_3$(CH$_2$)$_m$CH$_3$, the compound is named according to the number of carbon atoms present. Compounds of this type are called **unbranched alkanes.** Table 11.1 of the text gives examples of such compounds. All are named with a stem name and the suffix *-ane.* A **branched alkane** has one or more carbons that are not in a row of carbon atoms and thus form **side chains.** Butane is an unbranched alkane, and methylpropane is a branched alkane with a CH$_3$ side chain. When naming the branched alkanes, we treat side chains as **substituents,** which are atoms or groups of atoms that have been substituted for a hydrogen atom. To name a branched alkane, we follow these steps:

Step 1. Identify the longest chain of carbon atoms in the molecule.
Step 2. Number the carbon atoms on the longest chain so that the carbon atoms with substituents have the lowest possible numbers.
Step 3. Identify each substituent and the number of the carbon atom it is located on.
Step 4. Name the compound by listing the substituents in alphabetical order, with the numbered location of each substituent preceding its name; the name(s) of the substituent(s) are followed by the parent name of the alkane, which is determined by the number of carbon atoms in the longest chain. For instance, the octane with a methyl group at the number 3 carbon would be 3-methyloctane.

Step 5. If two or more of the same substituents are present (such as two methyl groups), a Greek prefix, such as *di-*, *tri-*, or *tetra-* is attached to the name of the group; and the numbers of the carbon atoms to which the groups are attached, separated by commas, are included in the name. For instance, if our octane had a methyl group at carbon 4 as well as at carbon 3, its name would be 3,4-dimethyloctane.

Step 6. Numbers in a name are always separated from letters by a hyphen; numbers are separated from each other by commas.

EXAMPLE 1   Naming unbranched alkanes

What is the name of the alkane $CH_3(CH_2)_6CH_3$?

SOLUTION The alkane contains a total of eight carbon atoms, so the stem of the name is obtained by combining the prefix for eight (*oct-*) with the alkane suffix *-ane* to get the name *octane*.

EXERCISE What is the name of the alkane $CH_3(CH_2)_7CH_3$?

[*Answer:* Nonane]

EXAMPLE 2   Naming a branched alkane

Name the alkane

$$
\begin{array}{c}
CH_3 \\
| \\
CH_3-CH_2-CH_2-C-CH_3 \\
| \\
CH_3
\end{array}
$$

SOLUTION The first step is to number the carbons of the longest carbon chain so that the substituents appear on the lowest numbered carbons:

$$
\begin{array}{c}
\qquad\qquad\qquad CH_3 \\
\qquad\qquad\qquad | \\
CH_3-CH_2-CH_2-C-CH_3 \\
\;\;5\qquad 4\qquad 3\quad 2|\quad 1 \\
\qquad\qquad\qquad CH_3
\end{array}
$$

The longest carbon chain contains five carbons, so the compound is a pentane. There are two methyl groups on carbon number 2. Combining all this information gives the name 2,2-dimethylpentane.

EXERCISE Name the alkane

$$
\begin{array}{c}
\qquad\quad CH_3\;\; CH_2CH_3 \\
\qquad\quad |\qquad | \\
CH_3-CH_2-CH-CH-CH-CH_3 \\
\qquad\qquad\qquad\qquad | \\
\qquad\qquad\qquad\quad CH_3
\end{array}
$$

[*Answer:* 3-Ethyl-2,4-dimethylhexane]

KEY CONCEPT B   Alkene nomenclature

Alkenes are named by using the stem name of the corresponding alkane with a number that specifies the location of the double bond. The number is obtained by numbering the longest carbon chain so that the double bond is associated with the lowest numbered carbon possible and assigning to the double bond the lower number of the two double-bonded carbons:

$$
\overset{5\quad\;\; 4\quad\;\; 3\qquad 2\;\; 1}{CH_3CH_2CH{=}CHCH_3}
$$

2-Pentene (not 3-pentene)

Other substituents are named as explained earlier for the alkanes, with a number specifying the location of the group and an appropriate prefix denoting how many groups of each kind are present.

Numbering the longest chain so that the double bond has the lowest number takes precedence over keeping the substituent numbers low.

EXAMPLE   Naming an alkene

Name the alkene

$$CH_3-CH_2-\underset{\underset{CH_3}{|}}{C}=CH_2$$

SOLUTION The carbon chain is numbered so that the carbons associated with the double bond have the lowest possible numbers. The double bond is assigned the number 1 because this is the lower of the numbers of the two double-bonded carbons, and the methyl group is at position 2. The name is 2-methyl-1-butene.

$$\underset{4}{CH_3}-\underset{3}{CH_2}-\underset{2}{\underset{\underset{CH_3}{|}}{C}}=\underset{1}{CH_2}$$

EXERCISE Name the following alkene.

$$CHCl_2-CH_2-\underset{\underset{CH_3}{|}}{C}=CH-CH_3$$

[*Answer:* 5,5-Dichloro-3-methyl-2-pentene. (Note that a low number for the double bond takes precedence over low numbers for the substituents.)]

KEY CONCEPT C   Aromatic nomenclature

When a single substituent is present on a benzene ring, the compound is named by using the substituent name as a prefix in front of the word *benzene*. Thus, C_6H_5Cl is called chlorobenzene. If two of the same substituents are present, the common method of naming uses the prefix *ortho-* (abbreviated *o-*) to denote that the two substituents are on adjacent carbons, the prefix *meta-* (abbreviated *m-*) to denote that the two substituents are separated by one carbon, and the prefix *para-* (abbreviated *p-*) to denote that the two substituents are separated by two carbons (i.e., they are across the ring from each other). For example,

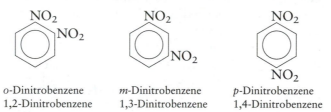

| *o*-Dinitrobenzene | *m*-Dinitrobenzene | *p*-Dinitrobenzene |
| 1,2-Dinitrobenzene | 1,3-Dinitrobenzene | 1,4-Dinitrobenzene |

Systematic names are obtained by numbering the carbon atoms on the ring, starting at one of the groups and naming the compound as a substituted benzene. Numbering is done so as to place substituents on the lowest numbered carbon atoms possible. The ortho, meta, and para labels are also used in some special circumstances, such as in naming substituted toluenes ($C_6H_5CH_3$ is toluene). Thus, 1-chloro-4-methylbenzene is also called *p*-chlorotoluene.

EXAMPLE   Naming aromatic compounds

What is the name of the following compound?

SOLUTION The two chlorine atoms are on adjacent carbons. If we number the carbons of the ring, starting with one of the chlorines (which one is immaterial) and numbering so as to keep the numbers of the substituents as low as possible, the two chlorines are on carbons 1 and 2, and the compound is 1,2-dichlorobenzene. Using the ortho, meta, and para system and noting that the two chlorines are on adjacent carbon atoms, we name the compound *o*-dichlorobenzene.

EXERCISE Name the following compound.

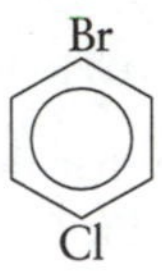

[*Answer:* 1-Bromo-3-chlorobenzene or *p*-bromochlorobenzene]

KEY WORDS Define or explain each term in a written sentence or two.

addition reaction
elimination reaction
phenyl group
substituent
substitution reaction

FUNCTIONAL GROUPS

11.5 ALCOHOLS

KEY CONCEPT There are three types of alcohols

Alcohols are compounds with a formula such as $CH_3CH_2CH_2OH$, in which an —OH group is attached to an alkane-like group ($CH_3CH_2CH_2$ in this case). To be an alcohol, the —OH cannot be attached to a benzene ring or to a C═O group. Alcohols are neither acids nor bases; the —OH group in an alcohol is not ionic as it is in NaOH, for instance. Alcohols are classified according to the number of R (organic) groups attached to the carbon that holds the —OH group. If there is one such group, a primary alcohol is obtained. If there are two, it is a secondary alcohol, and three, a tertiary alcohol. Examples of each follow:

CH_3CH_2OH	CH_3—C(CH_3)(H)—OH	CH_3CH_2—C(CH_3)(CH_2CH_3)—OH
Primary alcohol	Secondary alcohol	Tertiary alcohol

Alcohols are able to hydrogen bond both to water and to themselves very well. Thus, small alcohols are quite soluble in water; also, alcohols have unusually high boiling points.

EXAMPLE Recognizing the nature of alcohols

Only one of the following compounds is an alcohol. Which one?

I: CH_3—C(CH_3CHCH_3)(H)—OH

II:

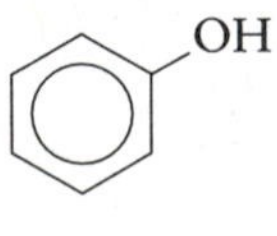

III: CH_3CH_2—C(═O)—OH

IV: KOH

SOLUTION Compound I has two R groups (one CH_3, one CH_3CHCH_3) attached to a carbon to which the —OH is bound; it is a secondary alcohol. Compound II has the —OH directly attached to a benzene ring; it is not an alcohol (it is a phenol). Compound III has the —OH attached to a C=O; it is not an alcohol (it is a carboxylic acid). Compound IV is the ionic compound potassium hydroxide; it is not an alcohol.

EXERCISE Which of the following compounds is a tertiary alcohol?

$$CH_3CH_2-\underset{\underset{H}{|}}{\overset{\overset{CH_3}{|}}{C}}-OH \qquad CH_3-\underset{\underset{H}{|}}{\overset{\overset{CH_3\overset{\overset{CH_3}{|}}{C}CH_3}{|}}{C}}-OH \qquad CH_3CH_2-\underset{\underset{CH_2CH_3}{|}}{\overset{\overset{CH_2CH_3}{|}}{C}}-OH \qquad CH_3-\underset{\underset{H}{|}}{\overset{\overset{H}{|}}{C}}-OH$$

$$\text{I} \qquad\qquad\qquad \text{II} \qquad\qquad\qquad\quad \text{III} \qquad\qquad\qquad \text{IV}$$

[*Answer:* III]

11.6 ETHERS

KEY CONCEPT   Ethers have unique properties

Ethers are compounds of the form R—O—R, in which the symbol R denotes an alkyl group such as CH_3 or $(CH_3)_2CH$. Although the R group in this case can be a benzene ring, the oxygen atom in R—O—R cannot be bonded to a C=O group in order to have an ether. Ethers cannot hydrogen bond, so they have high vapor pressures, that is, they are volatile and evaporate easily. For the same reason, they have low boiling points. Even though ethers do not undergo a wide variety of reactions, they are often highly inflammable and must be handled with care.

EXAMPLE   Recognizing ethers and their properties

Only one of the following compounds is an ether. Which one?

$$CH_3OH \qquad \underset{\text{II}}{\overset{\overset{O}{\parallel}}{O-CCH_3}} \qquad \underset{\text{III}}{CH_3-\overset{\overset{O}{\parallel}}{C}-OCH_2CH_3} \qquad \underset{\text{IV}}{\overset{\overset{CH_3}{|}}{\underset{\underset{CH_3}{|}}{\underset{CH_2}{|}}{CH-O-CH_3}}}$$

$$\text{I} \qquad\qquad\qquad \text{II}$$

SOLUTION Compound I has an H and an R group attached to the central oxygen; it is an alcohol. In both II and III, the central oxygen is attached to a C=O; these are not ethers. (In fact, they are called esters). Compound IV has two R groups, a $CH_3CHCH_2CH_3$ and a CH_3 attached to a central oxygen; it is the ether.

EXERCISE The compound dimethyl ether (CH_3OCH_3) has a higher molar mass than methanol (CH_3OH), yet it has a lower boiling point than methanol. Explain why.

[*Answer:* Methanol can hydrogen bond and dimethyl ether cannot.]

11.7 PHENOLS

KEY CONCEPT   An —OH attached to a benzene ring forms a phenol

The compound formed when an —OH group is attached to a benzene ring is called **phenol**. The formula for phenol is C_6H_5OH, and it is the prototype for a class of compounds called phenols, which are compounds with an —OH group attached to an aromatic ring. The structure of phenol itself and two phenols are shown below. Unlike alcohols, phenols are weak acids. The hydrogen attached to the oxygen can ionize in typical acid like fashion.

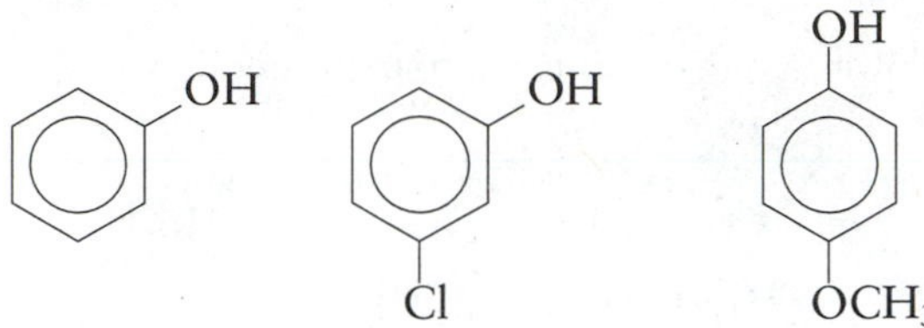

Phenol   *m*-Chlorophenol   *p*-Cresol (a phenol)

EXAMPLE   Distinguishing phenols from alcohols

Because both alcohols and phenols are organic compounds that contain an —OH group, do they both belong to the same class of compounds?

SOLUTION No. Alcohols and phenols are considerably different in their chemical properties. Although we cannot delve into the details here, the presence of the aromatic ring has a dramatic effect on the chemical properties of phenols. One important difference we do know about is that although alcohols are neither acids nor bases, phenols are acids.

EXERCISE What is the molecular formula of *p*-cresol?

[*Answer:* $C_7H_8O_2$]

11.8   ALDEHYDES AND KETONES

KEY CONCEPT   Aldehydes and ketones contain a C=O group

Aldehydes and ketones are built around the **carbonyl group**, C=O. If two R groups (alkyl and/or aryl) are attached to the carbon, a **ketone** results. If one or two hydrogen atoms are attached to the carbon, an **aldehyde** results. Examples of both are shown below.

Propanone   Benzophenone   Ethanal   Benzaldehyde
(a ketone)   (a ketone)   (an aldehyde) (an aldehyde)

Ketones can be prepared by oxidation (with $Na_2Cr_2O_7$) of a secondary alcohol. Aldehydes can be prepared by gentle oxidation of a primary alcohol.

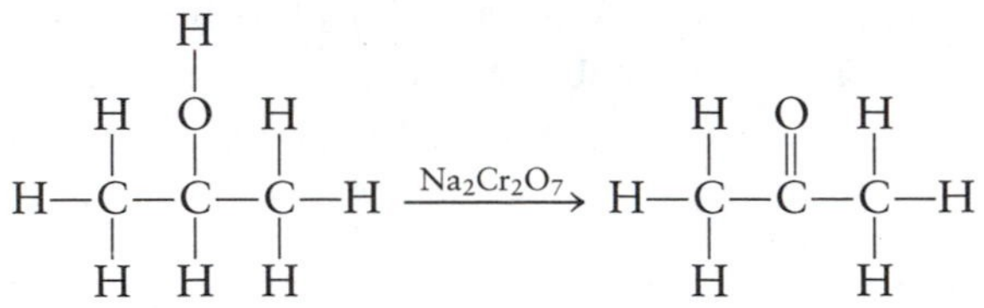

Secondary alcohol   Ketone

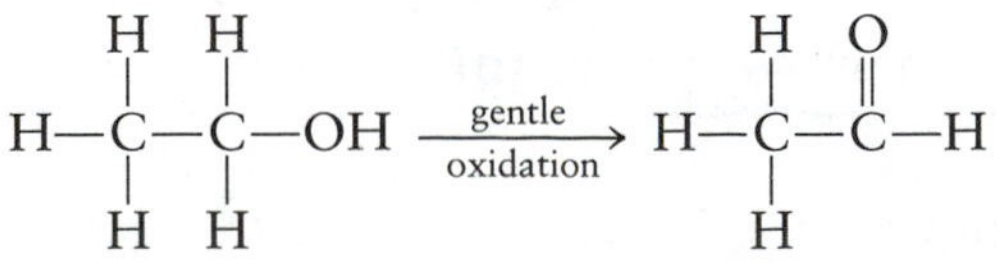

Primary alcohol   Aldehyde

In both cases, oxidation involves removal of two hydrogen atoms from the alcohol. As is often the case in reactions involving organic compounds, the chemical equation is not necessarily balanced. It simply shows the path of reaction for the major organic compound.

EXAMPLE **Writing the reactions of aldehydes and ketones**

What reaction occurs when the aldehyde $CH_3CH(CH_3)CHO$ is reacted with Ag^+ from Tollens reagent? Write down the reaction.

SOLUTION Tollens reagent is used to identify aldehydes, because the silver ion in it reacts to form a beautiful silver metal mirror on the inside of the glassware in which the reaction is performed. In this reaction the silver ion is reduced (to silver metal) and the aldehyde oxidized (to a carboxylic acid). The formation of the mirror indicates that an aldehyde is present.

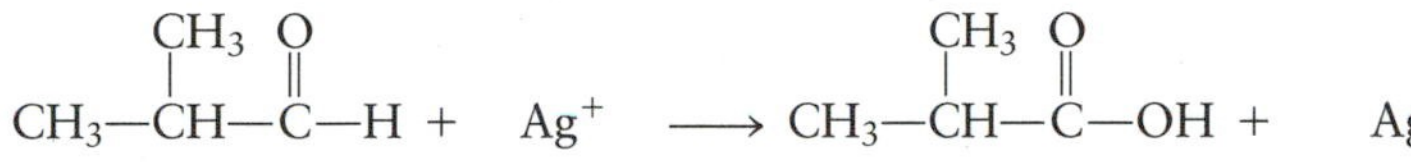

$$CH_3\text{—}\underset{\underset{\text{Aldehyde}}{}}{CH}\text{—}\overset{O}{\overset{\|}{C}}\text{—H} + \underset{\text{Silver ion}}{Ag^+} \longrightarrow CH_3\text{—}\underset{\underset{\text{Carboxylic acid}}{}}{CH}\text{—}\overset{O}{\overset{\|}{C}}\text{—OH} + \underset{\text{Silver mirror}}{Ag}$$

EXERCISE Will a silver mirror form when the ketone $CH_3C(O)CH_2CH_3$ is placed in Tollens reagent?

[*Answer:* No. Ketones cannot be oxidized by Ag^+.]

11.9 CARBOXYLIC ACIDS

KEY CONCEPT   Carboxylic acids contain a —COOH group

The —COOH group (carboxyl group) that characterizes **carboxylic acids** has a double-bonded oxygen and an —OH bonded to the carbon. Another group must be attached to the carbon in order to have four bonds to the carbon; this could be an alkyl group, an aryl group, or even a hydrogen atom. Examples of some carboxylic acids are shown below.

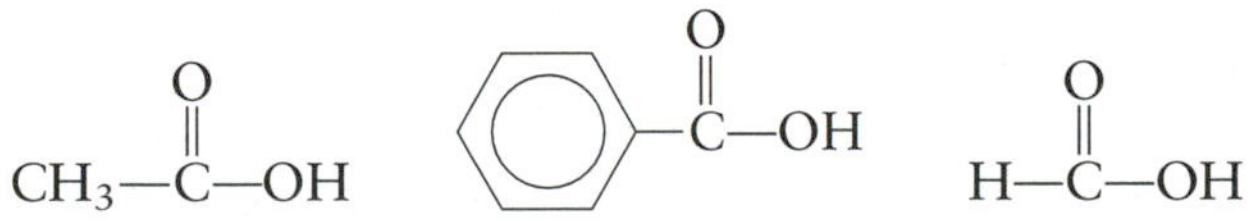

$$CH_3\text{—}\overset{O}{\overset{\|}{C}}\text{—OH} \qquad \text{Benzoic acid} \qquad H\text{—}\overset{O}{\overset{\|}{C}}\text{—OH}$$

Acetic acid   Benzoic acid   Formic acid

Carboxylic acids can be prepared by oxidation (with $Na_2Cr_2O_7$) of an aldehyde or of a primary alcohol; when the primary alcohol is oxidized, it first forms an aldehyde that then further reacts to form a carboxylic acid.

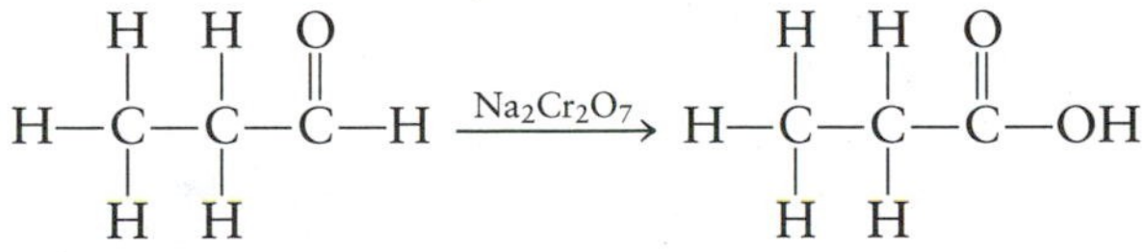

Aldehyde   Carboxylic acid

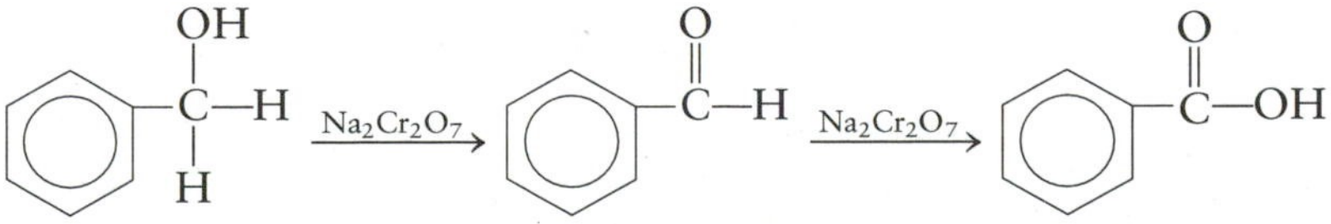

Primary alcohol   Aldehyde   Carboxylic acid

When a carboxylic acid is reacted with an alcohol, a molecule of water is split out and an **ester** forms. Esters often have a pleasant smell and are found in many foods and medications.

$$\underset{\text{Carboxylic acid}}{\overset{\displaystyle H \quad H \quad O}{\underset{\displaystyle H \quad H}{H-C-C-C-\boxed{OH}}}} + \boxed{H}-OCH_3 \longrightarrow \underset{\text{Ester}}{\overset{\displaystyle H \quad H \quad O}{\underset{\displaystyle H \quad H}{H-C-C-C-OCH_3}}}$$

$$\underset{\text{Alcohol}}{} \qquad H_2O \longleftarrow \text{Water split out}$$

EXAMPLE **Writing the reactions that produce carboxylic acids and esters**

Write the structure of the product formed when the aldehyde $CH_3CH_2CH_2CHO$ is oxidized with $Na_2Cr_2O_7$.

SOLUTION When an aldehyde is oxidized, a carboxylic acid is formed. The carboxylic acid structure that results can be drawn by changing the —CHO group in the aldehyde to a —COOH group to get the carboxylic acid.

$$CH_3CH_2CH_2-\overset{\displaystyle O}{\overset{\|}{C}}-H \xrightarrow{Na_2Cr_2O_7} CH_3CH_2CH_2-\overset{\displaystyle O}{\overset{\|}{C}}-OH$$

EXERCISE What is the structure of the ester formed when acetic acid (CH_3COOH) reacts with ethanol (CH_3CH_2OH)?

Answer: $CH_3-\overset{\displaystyle O}{\overset{\|}{C}}-OCH_2CH_3$

11.10 AMINES AND AMIDES

KEY CONCEPT   Amines are derivatives of ammonia

If we replace one or more of the hydrogen atoms in ammonia with an alkyl group, an **amine** results.

$$\underset{\text{Ethylamine}}{\overset{\displaystyle H}{CH_3CH_2-\overset{|}{N}-H}} \qquad \underset{\text{Diethylamine}}{\overset{\displaystyle H}{CH_3CH_2-\overset{|}{N}-CH_2CH_3}} \qquad \underset{\text{Triethylamine}}{\overset{\displaystyle CH_2CH_3}{CH_3CH_2-\overset{|}{N}-CH_2CH_3}}$$

A primary or secondary amine can react with a carboxylic acid to form an **amide;** in this reaction, a water molecule is split out. Special conditions are needed for this particular condensation reaction. In addition, tertiary amines will not work because a tertiary amine does not possess a hydrogen bonded to the nitrogen (to form the water molecule). The hydrogens bonded to carbon will not serve to form the water molecule.

$$\text{Water splits out}$$
$$H_2O \nwarrow$$

$$\underset{\text{Acid}}{CH_3CH_2-\overset{\displaystyle O}{\overset{\|}{C}}-\boxed{OH}} + \underset{\text{Secondary amine}}{\boxed{H}-\overset{\displaystyle CH_2CH_3}{\overset{|}{N}}-CH_3} \longrightarrow \underset{\text{Amide}}{CH_3CH_2-\overset{\displaystyle O}{\overset{\|}{C}}-\overset{\displaystyle CH_2CH_3}{\overset{|}{N}}-CH_3}$$

EXAMPLE **Identifying condensation reactions that produce amines and amides**

Only one of the following amines will *not* undergo a condensation with an acid to form an amide. Which one?

$$\underset{I}{\overset{\displaystyle CH_2CH_3}{H-\overset{|}{N}-CH_3}} \qquad \underset{II}{\overset{\displaystyle CH_2CH_3}{CH_3CH_2-\overset{|}{N}-CH_2CH_3}} \qquad \underset{III}{\overset{\displaystyle H}{CH_3CH_2-\overset{|}{N}-CH_2CH_3}}$$

SOLUTION When an amine and a carboxylic acid condense to form an amide, an H atom from the amine and the —OH from the acid combine to form a water molecule. However, a tertiary amine has no hydrogen (bonded to N) to form the water, so it cannot undergo the condensation. Amine II is a tertiary amine; it does not possess the requisite hydrogen and will not react with an acid to form an amide.

EXERCISE A quaternary ammonium ion is formed when four groups bond to a central nitrogen. Draw the structure of the quaternary ammonium ion formed when two hydrogen atoms and two methyl groups bond to the nitrogen. Show *all* atoms and *all* bonds, and don't forget the charge!

$$\text{Answer:}\quad \begin{array}{ccccccc} & H & & H & & H & \\ & | & & | & & | & \\ H\!-\!C\!-\!N^{\pm}\!-\!C\!-\!H \\ & | & & | & & | & \\ & H & & H & & H & \end{array}$$

KEY WORDS   Define or explain each term in a written sentence or two.

alcohol (primary, secondary, tertiary)	carbonyl group	functional group
aldehyde	condensation reaction	hydroxyl group
amide	diol	phenol
amine	ether	quaternary ammonium ion

ISOMERS

11.11   STRUCTURAL ISOMERS

KEY CONCEPT   Different structures may have the same molecular formula

Isomerism is the occurrence of two or more different compounds with the same molecular formula. **Structural isomerism** is a type of isomerism in which at least one atom in one isomer is bonded to neighbors different from those with which it bonds in another isomer. For instance, there are two compounds with the formula C_4H_{10}:

$$\begin{array}{cccccccc} & H & H & H & H & & & \\ & | & | & | & | & & & \\ H\!-\!C\!-\!C\!-\!C\!-\!C\!-\!H & & & & \\ & | & | & | & | & & & \\ & H & H & H & H & & & \end{array}$$

Butane

$$\begin{array}{ccccc} & & H & & \\ & & | & & \\ & & H\!-\!C\!-\!H & & \\ & H & | & H & \\ & | & | & | & \\ H\!-\!C\!-\!C\!-\!C\!-\!H \\ & | & | & | & \\ & H & H & H & \end{array}$$

Isobutane or methylpropane

The central carbon in methylpropane is bonded to three other carbons and one hydrogen, as neighbors; there is no carbon with these same neighbors in butane, so methylpropane and butane are structural isomers. Another way of indicating these structures is

$$CH_3CH_2CH_2CH_3 \qquad CH_3CH(CH_3)CH_3$$

Butane   Methylpropane

The parentheses around the CH_3 in the formula for methylpropane indicate that the —CH_3 group is bonded to the preceding carbon atom. A third way of writing the formula for butane is $CH_3(CH_2)_2CH_3$. In this case, the parentheses are used to indicate how many methylene groups in a row are in the formula.

EXAMPLE Understanding isomerism in the alkanes

Without referring to the text, draw all the structural isomers that have the formula C_5H_{12}.

SOLUTION A good method for drawing structural isomers is to first draw the carbon "skeletons" and then add the required hydrogen atoms. We start with the five carbons bonded in a straight chain and then move one or more of the carbons to get all the other structural isomers:

$$
C-C-C-C-C \qquad C-C-\underset{\displaystyle C}{\overset{\displaystyle C}{\underset{|}{\overset{|}{C}}}}-C \qquad C-\underset{\displaystyle C}{\overset{\displaystyle C}{\underset{|}{\overset{|}{C}}}}-C
$$

You may wonder why we do not include

$$
C-\underset{}{\overset{\displaystyle C}{\overset{|}{C}}}-C-C \qquad \text{and} \qquad C-C-\underset{\displaystyle C}{\underset{|}{C}}-C
$$

But if you make a list of the carbon atoms and the neighbors they have, you will find that these structures represent the same isomer as the center structure in the first figure. Molecular models (constructed from Styrofoam balls and toothpicks, or gum drops and toothpicks, or a commercial-model set) also will show that the three skeletons match up atom-to-atom. Thus, the only possible arrangements for the five carbons of an alkane are those in the first figure. Any others you can draw are simply replicas of one of these three, oriented differently on the paper. To complete the drawings, we add in the hydrogen atoms, remembering that there must be a total of 12 hydrogens and that every carbon atom is bonded to 4 other atoms:

EXERCISE The formulas for four alkanes with formula C_6H_{14} are shown in the figure. Which are distinct isomers and which represent the same compound?

$$
\underset{\text{I}}{CH_3CH_2CH_2\overset{\displaystyle CH_3}{\overset{|}{C}}HCH_3} \qquad \underset{\text{II}}{CH_3CH_2\overset{\displaystyle CH_3}{\underset{\displaystyle CH_3}{\overset{|}{\underset{|}{C}}}}-CH_3} \qquad \underset{\text{III}}{CH_3-\overset{\displaystyle CH_3}{\overset{|}{C}}HCH_2CH_2CH_3}
$$

$$
\underset{\text{IV}}{CH_3CH_2\overset{\displaystyle}{\underset{\displaystyle CH_3}{\underset{|}{C}}}HCH_2CH_3}
$$

[*Answer:* I, II, and IV are distinct; I and III are the same.]

11.12   GEOMETRICAL AND OPTICAL ISOMERS

KEY CONCEPT A   Geometrical isomers can be cis- or trans-

The effective lack of rotation around the double bond makes possible a type of isomerism called **geometrical isomerism.** Geometrical isomers are compounds with the same molecular formulas and with atoms bonded to the same neighbors, but with different arrangements of atoms in space. As an example, for 2-butene, two geometrical isomers are possible:

$$CH_3 \diagdown \quad \diagup CH_3$$
$$C=C$$
$$H \diagup \quad \diagdown H$$

cis-2-Butene

$$CH_3 \diagdown \quad \diagup H$$
$$C=C$$
$$H \diagup \quad \diagdown CH_3$$

trans-2-Butene

In the trans isomer, the two methyl groups are across the double bond from each other (hence the designation *trans*). In the cis isomer, they are on the same side of the double bond. Because rotation around the double bond does not occur under normal circumstances, the two isomers are distinct compounds with different chemical and physical properties. If enough energy is added to a pure sample of either compound, a **cis-trans isomerization** reaction may occur in which the energy excites rotation around the double bond and part of the sample is converted to the other isomer.

EXAMPLE   Understanding cis-trans isomerism

Is cis-trans isomerism possible for 1-butene?

SOLUTION We first draw the compound in a way that would show any possible cis-trans isomerism:

$$CH_3CH_2 \diagdown \quad \diagup H$$
$$C=C$$
$$H \diagup \quad \diagdown H$$

As we can see, no difference results if we interchange the two hydrogens on the right side of the structure or if we interchange the ethyl group (CH_3CH_2) with the hydrogen beneath it. Thus, cis-trans isomerism is not possible, and 1-butene exists in only one form.

EXERCISE What is the name of the following compound?

$$CH_3CH_2 \diagdown \quad \diagup CH_3$$
$$C=C$$
$$H \diagup \quad \diagdown H$$

[*Answer:* cis-2-Pentene]

KEY CONCEPT B   Optical isomers are nonidentical mirror images

If two molecules are mirror images of each other and are also different from each other, they are **optical isomers.** By analogy, your left and right hand are mirror images, but they are not the same; if you doubt this, try putting your right hand in a left glove or your left hand in a right glove. Remember also, it is possible for two mirror images to be identical; for instance, two pencils are mirror images, but they are the same. For molecules to be isomers, they must be different. A molecule that has a nonidentical mirror image is also called a **chiral** molecule. A chiral molecule and its mirror image molecule are called a pair of **enantiomers.** By analogy, again, your left hand is chiral and your right hand is chiral. If they were molecules, your left hand would be one enantiomer of a pair of enantiomers and your right hand would be the other enantiomer (enantiomers always come in pairs). The pencils mentioned above, however, would be **achiral.** If we have a mixture of two enantiomers in identical proportions, the mixture is called a **racemic mixture.** A pair of enantiomers have (almost) identical chemical and physical properties, except when they interact with other chiral molecules. In addition, they rotate plane-polarized light in equal and opposite directions. One way to tell whether a molecule is chiral is to look for a carbon atom with four different groups attached; if such a carbon is present, the molecule is almost certain to be chiral.

EXAMPLE Identifying chiral molecules

Which of the following molecules is chiral?

$$\begin{array}{cccc} H & F & H & H \\ | & | & | & | \\ H-C-C-C-C-H \\ | & | & | & | \\ H & Cl & H & H \end{array} \qquad \begin{array}{cccc} F & H & Cl & H \\ | & | & | & | \\ H-C-C-C-C-H \\ | & | & | & | \\ H & H & Cl & H \end{array} \qquad \begin{array}{cccc} H & F & H & H \\ | & | & | & | \\ H-C-C-C-C-H \\ | & | & | & | \\ H & F & H & H \end{array}$$

I   II   III

SOLUTION For a molecule to be chiral, it must have four different groups attached to one carbon. In molecule I, the second carbon from the left has four different groups (F, Cl, CH_3, CH_2CH_3) attached, so it is a chiral molecule. In molecule II, each carbon has at least two of the same atoms or groups attached, so it is not chiral. The same is true of molecule III. The only chiral molecule is I.

EXERCISE One of the enantiomers of glutamine ($C_4H_{10}N_2O_3$) has a melting point of 185°C. What is the melting point of the other enantiomer?

[*Answer:* 185°C]

KEY WORDS   Define or explain each term in a written sentence or two.

achiral molecule	isomers	optical isomers
chiral molecule	geometrical isomers	racemic mixture
enantiomer	optical activity	structural isomers

POLYMERS

11.13   ADDITION POLYMERIZATION

KEY CONCEPT   A polymer is a huge molecule with many units

Alkenes undergo a reaction called **addition polymerization,** in which one alkene adds to another alkene, then a third adds to the first two, then a fourth, then a fifth, and so on. This series of reactions results in a long chain of linked units. The single alkene is called the **monomer;** the linked long-chain species formed from it is called a **polymer.** In this **polymerization reaction** the double bond is converted to a single bond (as in the previous addition reactions). The polymerization may occur through a free-radical chain mechanism in which an initiator starts the chain reaction and produces a free radical. (The smell we all associate with a new car is largely due to initiators evaporating from the plastic in the passenger compartment.)

$$\text{R-O}\cdot \ + \ \underset{\underset{X}{|}}{\overset{\overset{H}{|}}{CH_2=C}} \ \longrightarrow \ \text{R-O-CH}_2-\underset{\underset{X}{|}}{\overset{\overset{H}{|}}{C}}\cdot$$

Initiator   Monomer   Free radical

The newly produced free radical now reacts with a second monomer unit to form a larger free radical, which reacts with a third monomer unit and so on, until the massive polymer is produced.

$$\text{R-O-CH}_2-\underset{\underset{X}{|}}{\overset{\overset{H}{|}}{C}}\cdot \ + \ \underset{\underset{X}{|}}{\overset{\overset{H}{|}}{CH_2=C}} \ \longrightarrow \ \text{R-O-CH}_2-\underset{\underset{X}{|}}{\overset{\overset{H}{|}}{C}}-\text{CH}_2-\underset{\underset{X}{|}}{\overset{\overset{H}{|}}{C}}\cdot$$

Free radical   Monomer   New free radical

Another important polymerization procedure uses a **Ziegler-Natta catalyst,** such as triethylaluminum, $Al(CH_2CH_3)_3$. This catalyst causes side groups to be spatially arranged in a regular fashion along the polymer chain in either an **isotactic polymer** or **syndiotactic polymer** (see Figure 11.19 of the text). This leads to a high-density material because the regular arrangement allows the polymer chains to pack closely together. In an **atactic** polymer, the side groups are randomly arranged and the polymer chains cannot pack as closely together; the material is of lower density. Rubber is a natural polymer made of isoprene $[CH_2{=}C(CH_3){-}C(CH_3){=}CH_2]$ monomers. Vulcanization adds S—S cross-links to the natural polymer to strengthen it. A Ziegler-Natta catalyst is used to produce synthetic rubber; without this type of catalyst, a useless product results.

EXAMPLE   Understanding polymerization

What side chain results when propene is polymerized by using a free-radical chain reaction?

S O L U T I O N We imagine that we start with a propene unit that has been converted to a free radical, and we allow it to add to a second propene. We treat the methyl group as the X in the reaction written previously. Thus, the side group is a methyl group:

$$CH_3{-}\underset{\underset{CH_3}{|}}{CH}\cdot \;+\; CH_2{=}CH_2CH_3 \;\longrightarrow\; CH_3{-}\underset{\underset{CH_3}{|}}{CH}{-}CH_2{-}\underset{\underset{CH_3}{|}}{CH}{-}CH_2\cdot$$

E X E R C I S E Does rotation around carbon-carbon single bonds convert an isotactic polymer into a syndioactic polymer?

[*Answer:* No]

11.14   CONDENSATION POLYMERIZATION

KEY CONCEPT   Condensation polymers form from condensation reactions

It is possible to produce a polymer through a series of condensation reactions such as esterification or formation of amides. An example is the formation of Dacron, which is produced from monomer units that are a diacid and a diol. The schematic that follows indicates how the units fit together to form a polymer.

H₂O splits out   H₂O splits out   H₂O splits out

This type of polymer is called a **polyester.** If a diamine is used instead of a diol, and amide bonds formed instead of ester bonds, the resulting polymer is called a **polyamide.**

EXAMPLE   Drawing the structures of condensation polymers

Nylon is a polyamide produced by the polymerization of 1,6-hexanedioic acid with 1,6-hexanediamine. Write the structure of the polymer that results.

$$\underset{\text{1,6-Hexanedioic acid}}{HO{-}\overset{\overset{O}{\|}}{C}{-}(CH_2)_4{-}\overset{\overset{O}{\|}}{C}{-}OH} \qquad \underset{\text{1,6-Hexanediamine}}{H_2N{-}(CH_2)_6{-}NH_2}$$

SOLUTION Formation of the polymer starts by formation of an amide bond between the carboxylic acid group on the right of the 1,6-hexanedicarboxylic acid and the amine group to the left on the 1,6-hexanedi-amine. Then the other amine group on the 1,6-hexanediamine forms an amide bond with a second 1,6-hexanedioic acid. This acid molecule bonds to a third 1,6-hexanediamine and so on.

$$HO—\overset{\overset{O}{\|}}{C}—(CH_2)_4—\overset{\overset{O}{\|}}{C}—\boxed{OH\ H}—\overset{\overset{H}{|}}{N}—(CH_2)_6—\overset{\overset{H}{|}}{N}—\boxed{H\ HO}—\overset{\overset{O}{\|}}{C}—(CH_2)_4—\overset{\overset{O}{\|}}{C}—\boxed{OH\ H}—\overset{\overset{H}{|}}{N}—(CH_2)_6—\overset{\overset{H}{|}}{N}—etc$$

Water splits out   Water splits out   Water splits out

The resulting polymer is

$$HO—\overset{\overset{O}{\|}}{C}—(CH_2)_4—\overset{\overset{O}{\|}}{C}—\overset{\overset{H}{|}}{N}—(CH_2)_6—\overset{\overset{H}{|}}{N}—\overset{\overset{O}{\|}}{C}—(CH_2)_4—\overset{\overset{O}{\|}}{C}—\overset{\overset{H}{|}}{N}—(CH_2)_6—\overset{\overset{H}{|}}{N}—etc$$

EXERCISE Is the following polymer a polyester or a polyamide? What monomers are used to produce it?

$$HO—(CH_2)_4—O—\overset{\overset{O}{\|}}{C}—\bigcirc—\overset{\overset{O}{\|}}{C}—O—(CH_2)_4—O—\overset{\overset{O}{\|}}{C}—\bigcirc—\overset{\overset{O}{\|}}{C}—O—etc$$

Answer: Polyester, $HO—(CH_2)_4—OH$ and

$$HO—\overset{\overset{O}{\|}}{C}—\bigcirc—\overset{\overset{O}{\|}}{C}—OH$$

11.15 COPOLYMERS AND COMPOSITES

KEY CONCEPT   There are many types of copolymers

Copolymers are polymers that are produced with more than one monomer. Thus, a polymer of the structure —A—A—A—A—A— is not a copolymer; it has only one repeating unit because it is produced from one monomer. There are four different types of copolymers:

An **alternating copolymer** is of the form —A—B—A—B—A—B—A—B— in which the monomer units alternate.

A **block copolymer** is of the form —A—A—A—A—A—B—B—B—B—B— in which long segments of one monomer are connected to long segments of the other.

A **random copolymer** is of the form —A—B—B—A—B—A—A—B—B—A— in which the monomers are randomly arranged in the polymer.

A **graft copolymer** is the form —A(B)—A(B)—A(B)—A(B)—A(B)—A(B)— in which there is a long chain of one monomer (A in this case), with a side chain (B in this case) attached to each A monomer.

EXAMPLE   Understanding the nature of copolymers

What type of copolymer is the following?

$$—CH_2—CH_2—CF_2—CF_2—CH_2—CH_2—CF_2—CF_2—$$

SOLUTION This a polymer of two monomers ($CH_2{=}CH_2$ and $CF_2{=}CF_2$) and is of the form —A—B—A—B—A—B—A—B—. It is an alternating copolymer.

EXERCISE What type of copolymer is the following?

CHAPTER 11

$$\begin{array}{ccccccc} & \overset{O}{\underset{\|}{}} & & \overset{O}{\underset{\|}{}} & & \overset{O}{\underset{\|}{}} & \\ O-C-CH_3 & & O-C-CH_3 & & O-C-CH_3 & \\ | & & | & & | & \\ -CH_2-CH-CH_2 & - & CH-CH_2 & - & CH-CH_2- & \end{array}$$

[*Answer:* Graft copolymer]

11.16   PHYSICAL PROPERTIES

KEY CONCEPT   The properties of a polymer depend on its structure

The properties of a polymer depend on the length of the polymer chains, the polarity of the side groups, and the extent of crystalline regions (formed by unbranched polymers). Long chains with polar side groups lead to a strong polymer. Crystalline regions are accommodated by the lack of branching side chains because this permits the main polymer chains to line up in a regular crystallike fashion; crystalline regions allow for favorable interactions between chains and add strength to the polymer.

EXAMPLE   Predicting the physical properties of polymers

Which polymer would you expect to be less flexible, polyethylene or polypropylene? (Remember the effect of size on intermolecular interactions.)

$$\begin{array}{cc} & \overset{CH_3}{\underset{|}{}} \\ -(CH_2-CH_2)_n- & -(CH_2-CH)_n- \\ \text{Polyethylene} & \text{Polypropylene} \end{array}$$

SOLUTION The methyl group on polypropylene adds to the London forces between chains and makes the polypropylene less flexible than the polyethylene.

EXERCISE Which of the following would you expect to have the highest softening point (when heated), polyethylene (above) or polyacrylate?

$$\begin{array}{c} \overset{CN}{\underset{|}{}} \\ -(CH_2-CH)_n- \\ \text{Polyacrylate} \end{array}$$

[*Answer:* Polyacrylate]

KEY WORDS   Define or explain each term in a written sentence or two.

addition polymerization	elasticity	polyester
atactic polymer	isotactic polymer	radical polymerization
copolymer (alternating, block, graft, random)	monomer	syndiotactic polymer
condensation polymerization	polyamide	Ziegler-Natta catalyst

EDIBLE POLYMERS

11.17   PROTEINS

KEY CONCEPT   A protein is an amino acid copolymer

An amino acid is a compound containing both a carboxylic acid group and an amine group. The biologically important amino acids have structures of the form

$$NH_2-\underset{\underset{H}{|}}{\overset{\overset{R}{|}}{C}}-CO_2H$$

Amino acids are the building blocks of proteins. In proteins, the amino acids are linked by peptide bonds. Proteins are **polypeptides**, which are copolymers containing a mixture of approximately 20 amino acids incorporated into the chain. The component amino acid building blocks of a protein are often called amino acid **residues**. (Remember that part of the amino acid structure is lost when it is incorporated into the polypeptide; what is left is called the residue.)

The **primary structure** of a protein is the sequence of amino acids in the peptide chain. Two peptides with different sequences of amino acids have different primary structures. The three-dimensional shape that is assumed by the chain of amino acids is called the **secondary structure.** One of the two common secondary structures is the **α helix,** a structure in which the long chain twists to form a helical structure (like a spiral staircase). A second common secondary structure involves the cross-linking of chains by hydrogen bonding to form a **β-pleated sheet.** Hydrogen bonds between the N—H and C=O groups of the amino acid residues give rise to these structures. The **tertiary structure** of a polypeptide refers to how the α helices or β-pleated sheets twist and turn to determine the overall shape of the polypeptide. A good analogy is provided by taking a helical toy Slinky and twisting it back onto itself to give a specifically shaped result. Both hydrogen bonding and **disulfide links** (S—S links between sulfur atoms) help to determine the tertiary structure of polypeptides. In many proteins, polypeptide units with the appropriate tertiary structure come together to form a more massive unit that is the functioning protein. The way in which the units join determines the **quaternary structure** of the protein. Any modification of any of the structures (primary, secondary, tertiary, or quaternary) usually results in the malfunction or dysfunction of the protein involved. The loss of one or more of the structural features of a protein is called **denaturation** of the protein. A denatured protein cannot function as it should.

EXAMPLE   Understanding the structure and properties of polypeptides

Draw the structure of Ala-Gly-Gly, and locate the peptide linkages.

SOLUTION A peptide is formed through a condensation of the carboxylic-acid end of an amino acid with the amine end of another amino acid, accompanied by the elimination of a water molecule. The three amino acids we have are

The elimination of water (as indicated by the rectangles) and the formation of C—N bonds results in the formation of peptide linkages, which are shown in boldface in the following.

EXERCISE Let us assume that a protein is produced in which the primary structure is not what it should be. Which of the other three structures of the protein is likely to be affected? Will the protein function properly?

[*Answer:* If the primary structure of a protein chain is wrong, all the other structures (secondary, tertiary, and quaternary) are likely to be affected to some extent. In many (but not all) cases, a protein with the improper primary structure will not function properly.]

11.18 CARBOHYDRATES

KEY CONCEPT   A polysaccharide is a glucose polymer

Carbohydrates are a class of compounds that includes sugars and starches. Sugars contain many hydroxyl (—OH) groups and may be either aldehydes or ketones:

$$\begin{array}{cccccccc} & OH & OH & OH & OH & OH & O \\ & | & | & | & | & | & \| \\ H-C & -C & -C & -C & -C & -C-H \\ & | & | & | & | & | \\ & H & H & H & H & H \end{array} \qquad \begin{array}{cccccccc} & OH & OH & OH & OH & O & OH \\ & | & | & | & | & \| & | \\ H-C & -C & -C & -C & -C & -C-H \\ & | & | & | & | & & | \\ & H & H & H & H & & H \end{array}$$

Glucose (an aldehyde)   Fructose (a ketone)

Because of the many —OH groups, sugars (and their polymers) hydrogen bond to one another extremely well. The **polysaccharides** ("many sugars") are polymers of glucose; before forming a polymer, the glucose changes into a cyclic form (see Figures 11.35 to 11.37 in the text). Starch and cellulose are two well-known polysaccharides. Because of the nature of the bond that holds the monomer units together in starch, it is digestible by humans. In cellulose, the bond that holds the glucose units together is subtly different; this slight change makes cellulose indigestible (try eating a piece of wood if you doubt this!). Starch is made up of amylose and amylopectin, which differ only in the fact that the amylopectin has more branching. In cellulose, individual polymers are straight and rodlike; one polymer hydrogen bonds to others to form a very strong bundle of molecules that can then become the strong structural component of plants (such as wood).

EXAMPLE   Understanding the properties of carbohydrates

The formula of glucose is $C_6H_{12}O_6$. When two glucose units bond together to form maltose, the resulting dimer has the formula $C_{12}H_{22}O_{11}$. The bond used to make maltose is important because it is the same kind of bond that is used to link the thousands of glucose units together to make amylose and to make the basic structure of amylopectin. What do the formulas of glucose and maltose suggest about the splitting out of a small molecule to form the bond?

SOLUTION If we added up the atoms that are contained in the two glucose units that make up maltose, we would get the formula $C_{12}H_{24}O_{12}$. The actual formula of maltose, however, has 2 fewer hydrogen atoms and 1 less oxygen. This suggests, correctly, that a molecule of water splits out when the bond between the two glucose units in maltose is formed:

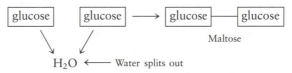

EXERCISE Which overall shape is more likely for amylose, rodlike (like cellulose), or helical?

[Answer: Helical; starch is not rigid like wood.]

11.19   DNA AND RNA

KEY CONCEPT   DNA, like RNA, is a copolymer of four nucleotides

The biological ribonucleic acids (RNA) and deoxyribonucleic acids (DNA) are polymers with structures (for both) that can be schematically represented as

The bases involved are adenine, guanine, cytosine, and thymine for DNA, and the first three and uracil for RNA. The sugars in both DNA and RNA are derived from ribose in its furanose form. A base-sugar unit is called a **nucleoside;** these are formed by a condensation between the base and ribose, with the elimination of a water molecule. There are four nucleosides in DNA and four in RNA, corresponding to the four bases plus one sugar for each. A base-sugar-PO_4H_2 unit, formed by the condensation of a nucleoside with phosphoric acid, again with the elimination of a water molecule, is called a **nucleotide.** As noted earlier, four such nucleotides are used in the construction of RNA and four in DNA. Nucleotides are the monomer units that make up DNA and RNA, and these two **polynucleotides** are copolymers of the different nucleotides. DNA itself is usually found as two chains intertwined to form a double helix. The double helix is held together by hydrogen bonding between adenine and thymine and between guanine and cytosine. Adenine and thymine fit together spatially to form hydrogen bonds, as do guanine and cytosine. Thus, they are called the AT (adenine-thymine) and GC (guanine-cytosine) **base pairs.** Other pairings are not as favorable as these and do not (or, at least, should not) occur.

EXAMPLE   Understanding the structure of DNA and RNA

How does the polymer bonding change if a thymine is substituted for a cytosine in DNA?

SOLUTION Because of the DNA structure, the substitution of any base for another has no effect on the bonding in the polymer. The bases, although critically important to the function of DNA do not affect the way in which one nucleotide bonds to the others to make the polymer. This can be seen in the following schematic of DNA; the bases are not part of the polymer backbone.

EXERCISE What is the effect on the double helical structure of DNA if a thymine is substituted for a cytosine?

[*Answer:* Base pairing and, therefore, the helical structure would be disrupted. (See Figure 11.41 of the text.)]

CHAPTER 11

KEY WORDS   Define or explain each term in a written sentence or two.

α helix	essential amino acid	polysaccharide
β-pleated sheet	nucleic acid	primary structure
carbohydrate	oligopeptide	quaternary structure
denaturation	peptide	secondary structure
disulfide link	peptide bond	tertiary structure

DESCRIPTIVE CHEMISTRY TO REMEMBER

- All **alkanes** are insoluble in water.
- The **alkanes** are unaffected by concentrated H_2SO_4, boiling HNO_3, boiling NaOH, and strong oxidizing agents.
- **Ethylene glycol** ($HOCH_2CH_2OH$) is used as an antifreeze.
- **Phenols** are weak acids.
- The **vulcanization** of rubber involves the formation of sulfur bridges between the polyisoprene chains.
- The two principal **secondary structures** of proteins are the α helix and the β-pleated sheet.

CHEMICAL EQUATIONS TO KNOW

- Complete combustion of **alkanes** results in the formation of water and carbon dioxide:

$$CH_4(g) + 2O_2(g) \longrightarrow CO_2(g) + 2H_2O(g)$$

- In a **substitution reaction,** a chlorine can substitute for a hydrogen on an alkane:

$$CH_4(g) + Cl_2(g) \longrightarrow CH_3Cl(g) + HCl(g)$$

- In an **addition reaction,** chlorine (or bromine) can add to an alkene to form a chlorinated (brominated) alkane:

$$CH_2{=}CH_2(g) + Cl_2(g) \longrightarrow CH_2Cl-CH_2Cl(g)$$

- In a **substitution reaction,** a bromine can substitute for a hydrogen on a benzene ring:

- Dichromate oxidation of **secondary alcohols** produces a ketone:

- Vigorous oxidation of **primary alcohols** produces a carboxylic acid:

- An **ester** is formed by the condensation reaction between an alcohol and an acid:

$$CH_3CH_2-\overset{\overset{\displaystyle O}{\|}}{C}-OH + HOCH_3 \longrightarrow CH_3CH_2-\overset{\overset{\displaystyle O}{\|}}{C}-OCH_3 + H_2O$$

- An **amide** is formed by the condensation reaction between a carboxylic alcohol and an amine:

$$CH_3CH_2-\overset{\overset{\displaystyle O}{\|}}{C}-OH + H_2NCH_3 \longrightarrow CH_3CH_2-\overset{\overset{\displaystyle O}{\|}}{C}-\overset{\overset{\displaystyle H}{|}}{N}-CH_3 + H_2O$$

SELF-TEST EXERCISES

Hydrocarbons

1. What is the molecular formula of the compound represented by the following stick structure?

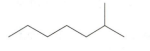

(a) $C_{10}H_{18}$ (b) C_8H_{18} (c) C_8H_{16} (d) C_9H_{17} (e) $C_{10}H_{20}$

2. What is the molecular formula of the compound represented by the following structure?

(a) C_5H_{12} (b) C_5H_5 (c) C_5H_{10} (d) C_5H_8 (e) C_5H_{14}

3. Which alkane has the highest boiling point?
(a) 2,2-dimethylpentane   (b) pentane   (c) hexane
(d) 3-methylpentane   (e) 3,3-dimethylbutane

4. What product is formed when CH_3CH_3 is mixed with chlorine and flashed with light?
(a) no reaction occurs   (b) CH_3CH_2Cl (c) CH_2ClCH_2Cl
(d) CH_3CHCl_2 (c) compounds (b), (c), and (d)

5. Which of the following could be the formula of a noncyclic alkene?
(a) CH_4 (b) C_3H_8 (c) C_4H_6 (d) C_5H_{10} (e) C_6H_{14}

6. Which of the following is a product of the reaction of Cl_2 with $CH_3CH{=}CH_2$?
(a) $CH_2Cl-CH_2-CH_2Cl$ (b) $CH_3-CHCl-CH_2Cl$ (c) $CH_3-CH_2-CH_2Cl$
(d) $CH_3-CCl{=}CHCl$ (e) $CH_2Cl-CH{=}CH_2$

7. Use the following bond enthalpies to estimate the enthalpy of reaction for the following reaction. C=C, 612 kJ·mol^{-1}; C—C, 348 kJ·mol^{-1}; Br—Br, 193 kJ·mol^{-1}; C—Br, 280 kJ·mol^{-1}. All answers are in kJ.

$$CH_3-CH{=}CH_2 + Br_2 \longrightarrow CH_3CHBr-CH_2Br$$

(a) -296 (b) -631 (c) $+444$ (d) $+177$ (e) -103

8. What product is formed when Br_2 is mixed with benzene in the presence of $FeBr_3$?
(a) 2,2-dibromohexane   (b) 2,2-dibromohexene   (c) bromobenzene
(d) cyclohexane   (e) bromocyclohexane

9. What is the name of $CH_3(CH_2)_2CH_3$?
(a) pentane   (b) 1-methylpentane   (c) cyclopentane
(d) hexane   (e) butane

10. What is the name of $CH_3C(CH_3)_2CH_2CH_2CH_3$?
(a) dimethylpentane   (b) 4,4-dimethylpentane   (c) 4-dimethylpentane
(d) 2,2-dimethylpentane   (e) heptane

11. Name $CH_3CH(CH_3)CH_2CH=CH_2$.
(a) 4-methyl-2-pentene   (b) 2-methyl-4-pentene   (c) hexene
(d) 4-methyl-1-pentene   (e) 2-methyl-5-pentene

12. Name the following compound.

$$\underset{H}{\overset{CH_3}{\diagdown}}C=C\underset{CH_2CH_3}{\overset{H}{\diagup}}$$

(a) *cis*-2-pentene   (b) *trans*-2-pentene   (c) *cis*-3-pentene
(d) *trans*-3-pentene   (e) pentene

13. Name the following compound:

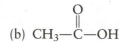

(a) *m*-chlorotoulene   (b) chloroethylbenzene   (c) xylene
(d) 1-chloro-3-ethylbenzene   (e) 4-chloro-6-ethylbenzene

Functional groups

14. Which of the following is an alcohol?

(a)   (b) $CH_3-\overset{O}{\overset{\|}{C}}-OH$ (c) $CH_3\overset{CH_3}{\overset{|}{C}}HCH_2OH$ (d) CsOH

15. Which of the following is a secondary alcohol?
(a) CH_3OH (b) $CH_3CH(OH)CH_3$ (c) $(CH_3)_3COH$
(d) CH_3CH_2OH (e) $CH_3CH_2CH_2OH$

16. Which of the following has the highest boiling point?
(a) CH_3CH_2 (b) $CH_3OCH_2CH_3$ (c) $CH_3CH_2CH_2OH$
(d) $CH_3CH_2CH_2CH_3$ (e) $CH_3CH_2CH_2F$

17. Which of the following is an ether?

(a) $CH_3CH_2OCH_2CH_3$ (b) $CH_3CH_2CH_2CH_2Br$ (c) $CH_3-\overset{\overset{\displaystyle O}{\|}}{C}-OCH_3$

(d) CH_3NH_2 (e)

18. Which of the following is a phenol?

(a)   (b) $CH_3-\overset{\overset{\displaystyle O}{\|}}{C}-OH$ (c) $CH_3\overset{\overset{\displaystyle CH_3}{|}}{C}HCH_2OH$ (d) CsOH

19. Which compound is the strongest acid?
(a) ethanol   (b) 1-butanol   (c) phenol
(d) 2-butanol   (e) 2-methyl-2-propanol

20. Which of the following is an aldehyde?

(a) $CH_3CH_2-\overset{\overset{\displaystyle O}{\|}}{C}-CH_3$ (b) $\overset{\overset{\displaystyle O}{\|}}{C}$ with NH_2 (c) $CH_3CH_2OCH_3$

(d) $CH_3CH_2-\overset{\overset{\displaystyle O}{\|}}{C}-H$ (e) $CH_3-\overset{\overset{\displaystyle O}{\|}}{C}-OH$

21. What is formed by the mild oxidation of the following compound?

$$CH_3\overset{\overset{\displaystyle CH_3}{|}}{C}HCH_2OH$$

(a) $CH_3\overset{\overset{\displaystyle CH_2OH}{|}}{C}HCH_2OH$ (b) $CH_3\overset{\overset{\displaystyle CH_3}{|}}{C}H-\overset{\overset{\displaystyle O}{\|}}{C}-OH$ (c) $CH_3\overset{\overset{\displaystyle CH_3}{|}}{C}H-OCH_3$

(d) CO_2 and H_2O (e) $CH_3\overset{\overset{\displaystyle CH_3}{|}}{C}H-\overset{\overset{\displaystyle O}{\|}}{C}-H$

22. What is formed by the oxidation of the following compound with sodium dichromate?

$$CH_3CH_2\overset{\overset{\displaystyle CH_3}{|}}{\underset{\underset{\displaystyle CH_3}{|}}{C}}HCHOH$$

(a) $CH_3CH_2\overset{\overset{\displaystyle CH_3}{|}}{C}H-\overset{\overset{\displaystyle O}{\|}}{C}-H$ (b) $CH_3CH_2\overset{\overset{\displaystyle CH_3}{|}}{C}H-\overset{\overset{\displaystyle O}{\|}}{C}-OH$

(c) $CH_3CH_2\overset{\overset{\displaystyle CH_3}{|}}{C}H-\overset{\overset{\displaystyle O}{\|}}{C}-OCH_3$ (d) $CH_3CH_2\overset{\overset{\displaystyle CH_3}{|}}{C}H-\overset{\overset{\displaystyle O}{\|}}{C}-CH_3$

(e) $CH_3CH_2\overset{\overset{\displaystyle CH_3}{|}}{C}HO-\overset{\overset{\displaystyle O}{\|}}{C}-OCH_3$

23. Which of the following is a carboxylic acid?

(a) $CH_3CH_2-\overset{\overset{\textstyle O}{\|}}{C}-H$ (b) $CH_3CH_2-\overset{\overset{\textstyle O}{\|}}{C}-OH$ (c) $CH_3-\overset{\overset{\textstyle O}{\|}}{C}-CH_3$

(d) $CH_3-\overset{\overset{\textstyle O}{\|}}{C}-OCH_3$ (e) $CH_3-\overset{\overset{\textstyle O}{\|}}{C}-NH_2$

24. What compound is formed by the following reaction?

$$CH_3CH_2-\overset{\overset{\textstyle O}{\|}}{C}-OH \; + \; CH_3\overset{\overset{\textstyle OH}{|}}{C}HCH_3 \longrightarrow \; ?$$

(a) $CH_3CH_2-\overset{\overset{\textstyle O}{\|}}{C}-O-CH_3CH_3CH_2$ (b) $CH_3CH_2-\overset{\overset{\textstyle O}{\|}}{C}-O-CH_2CH_2CH_3$

(c) $\overset{\overset{\textstyle CH_3}{|}}{\underset{\underset{\textstyle CH_3}{|}}{C}}H-\overset{\overset{\textstyle O}{\|}}{C}-O-CH_2CH_3$ (d) $CH_3CH_2-\overset{\overset{\textstyle O}{\|}}{C}-O-\overset{\overset{\textstyle CH_3}{|}}{\underset{\underset{\textstyle CH_3}{|}}{C}}H$

(e) $CH_3CH_2CH_2-\overset{\overset{\textstyle O}{\|}}{C}-O-CH_2CH_3$

25. Which of the following is a secondary amine?

(a) $(CH_3)_2NH$ (b) $CH_3CH_2N(CH_3)_2$ (c) NH_3

(d) CH_3NH_2 (e) $CH_3NH_3^+$

26. What is the name of $(CH_3)_2NCH_2CH_2CH_3$?

(a) tetramethylamine   (b) dimethylpropylamine

(c) methylmethylpropylamine   (d) butylamine

27. What compound is formed by the following reaction?

$$CH_3-\overset{\overset{\textstyle O}{\|}}{C}-OH \; + \; CH_3NH_2 \longrightarrow \; ?$$

(a) $CH_3-\overset{\overset{\textstyle O}{\|}}{C}-\overset{\overset{\textstyle H}{|}}{N}CH_3$ (b) $CH_3-\overset{\overset{\textstyle O}{\|}}{C}-\overset{\overset{\textstyle NH_2}{|}}{C}H_2$ (c) $CH_3-\overset{\overset{\textstyle O}{\|}}{C}-\overset{\overset{\textstyle NH_2}{|}}{O}CH_2$

(d) $CH_3-\overset{\overset{\textstyle O}{\|}}{C}-O\overset{\overset{\textstyle H}{|}}{N}CH_3$ (e) $CH_3-\overset{\overset{\textstyle OH}{|}}{C}H-\overset{\overset{\textstyle H}{|}}{N}CH_3$

28. The name of $CH_3CH(OH)CH(CH_3)CH_3$ is

(a) 2-methyl-3-butanol   (b) 1-methyl-2-butanol   (c) 3-methyl-2-butanol

(d) 2-methyl-1-butanol   (e) methylbutanol

29. What compound is formed by the gentle oxidation of 1-propanol?

(a) propanal   (b) propanone   (c) propanoic acid

(d) propane   (e) 2-propanol

30. What is the name of the following compound?

$$\underset{}{O}$$

$$CH_3\overset{\|}{C}HCH_2CH_3$$

(a) butanol   (b) butanal   (c) butanone   (d) 2-butanol   (e) butene

31. What is the name of the following compound?

$$\underset{}{O}$$

$$CH_3CH(CH_3)\overset{\|}{C}H$$

(a) 3-methyl-4-propanol   (b) 2-methylpropanone   (c) 2-methyl-1-propanone

(d) 3-methyl-4-propanone   (e) 2-methylpropanal

32. The oxidation of 2-pentanol produces

(a) 2-pentanone   (b) pentanal   (c) 3-pentanone

(d) 2-butanone   (e) butanal

33. The gentle oxidation of an alcohol results in a compound that reduces Tollens reagent. Which of the following may be the alcohol?

(a) 2-pentanol   (b) 2-butanol   (c) 1-propanol

(d) 2,2-dimethyl-2-pentanol   (e) 3-pentanol

34. What is the name of $CH_3CH(CH_3)COOH$?

(a) 2-methylbutanoic acid   (b) 2-methylpropanoic acid

(c) 2-methyl-3-propanoic acid   (d) 1-methylpropanoic acid

35. Which of the following represents an ester?

(a) R—OH   (b) R—O—R   (c) $R-\underset{\underset{R}{|}}{\overset{\overset{OR}{|}}{C}}-OR$ (d) $R-\overset{\overset{O}{\|}}{C}-OR$

Isomers

36. All but one of the following is a structural isomer of $CH_3CH_2CH(CH_3)CH_3$. Which one is not?

(a) $CH_3CH_2CH_2CH_3$ (b) $CH_3CH_2CH_2CH_2CH_3$ (c) $C(CH_3)_4$

37. Which of the following amino acids is chiral?

(a) $H-\underset{\underset{NH_2}{|}}{\overset{\overset{CH_3}{|}}{C}}-CO_2H$ (b) $H-\underset{\underset{NH_2}{|}}{\overset{\overset{H}{|}}{C}}-CO_2H$ (c) $CH_3-\underset{\underset{NH_2}{|}}{\overset{\overset{CH_3}{|}}{C}}-CO_2H$

38. Which of the following are a pair of enantiomers?

(a) a pair of *cis-trans* isomers

(b) any molecule and its mirror image

(c) any two structural isomers

(d) a chiral molecule and its mirror image

(e) an alkene and a cycloalkane with the same formula

39. Which property is different for each of a pair of enantiomers?

(a) melting point   (b) boiling point   (c) solubility in water

(d) color   (e) rotation of plane-polarized light

Polymers

40. Which of the following could serve as a monomer in an addition polymerization process?

(a) ethane   (b) propene   (c) hydrogen peroxide

(d) terephthalic acid   (e) ethylene glycol

41. Which of the following represents isotactic polypropylene? (Solid bonds face toward you, dashed bonds face away).

42. What is the structure of the polymer that forms if the following monomer undergoes addition polymerization?

43. In condensation polymerization, what kind of bond is formed when polymers form?

(a) carbon-carbon   (b) ester   (c) amide

(d) (a) and (b) both   (e) (b) and (c) both

44. What is the structure of the condensation polymer formed from the following compounds?

$$HOOC-CH_2-CH_2-COOH \quad \text{and} \quad H_2N-CH_2-NH_2$$

(c) $(-NH-CH_2-NH-O-\overset{\overset{\displaystyle O}{\|}}{C}-CH_2-CH_2-\overset{\overset{\displaystyle O}{\|}}{C}-O-)_n$

(d) $(-\overset{\overset{\displaystyle NH_2}{|}}{\underset{\underset{\displaystyle NH_2}{|}}{C}}H-\overset{\overset{\displaystyle O}{\|}}{C}-CH_2-CH_2-\overset{\overset{\displaystyle O}{\|}}{C}-)_n$

45. Which of the following represents a graft copolymer?

(a) —A—A—A—A—A—

(b) —A—B—A—B—A—B—

(c) —A—A—A—B—B—B—

(d) —A—A—B—A—B—B—A—

(e) $-A-\underset{\displaystyle B}{\overset{\displaystyle |}{A}}-A-\underset{\displaystyle B}{\overset{\displaystyle |}{A}}-A-\underset{\displaystyle B}{\overset{\displaystyle |}{A}}-$

Edible polymers

46. Proteins are copolymers of

(a) amino acids   (b) nucleotides   (c) esters

(d) alkenes   (e) glucose

47. The spatial arranging of a polypeptide into an α helix or β-pleated sheet refers to which structure of a protein?

(a) primary   (b) secondary   (c) tertiary   (d) quaternary

48. Which of the following represents a peptide linkage?

(a) $-\overset{\overset{\displaystyle O}{\|}}{C}-\overset{\overset{\displaystyle H}{|}}{N}-$ (b) $-\overset{\overset{\displaystyle O}{\|}}{C}-O-\overset{\overset{\displaystyle }{|}}{\underset{\underset{\displaystyle }{|}}{C}}-$ (c) $-\overset{\overset{\displaystyle }{|}}{\underset{\underset{\displaystyle }{|}}{C}}-O-\overset{\overset{\displaystyle }{|}}{\underset{\underset{\displaystyle }{|}}{C}}-$ (d) $-\overset{\overset{\displaystyle O}{\|}}{C}-O-\overset{\overset{\displaystyle O}{\|}}{C}-$

49. The common polysaccharides are polymers of

(a) amino acids   (b) nucleotides   (c) esters

(d) alkenes   (e) glucose

50. Only one of the following functional groups is not commonly found in sugars. Which one?

(a) aldehyde   (b) ketone   (c) carboxylic acid   (d) hydroxy

51. A nucleotide is a

(a) DNA base

(b) DNA base + sugar

(c) DNA base + phosphate fragment

(d) DNA base + sugar + phosphate fragment

52. The double helix of DNA is held together by

(a) S—S bridges   (b) AC and GT hydrogen bonds

(c) peptide bonds   (d) AT and GC hydrogen bonds

Descriptive chemistry

53. Which of the following will react with a typical alkane?

(a) Cl_2 with ultraviolet light   (b) conc. H_2SO_4 (c) boiling HNO_3

(d) boiling NaOH   (e) cold N_2

54. Which of the following is a weak acid?

(a) HCl   (b) ethanol   (c) diethyl ether

(d) phenol   (e) trimethylamine

55. Which of the following is insoluble in water?

(a) ethanol   (b) ethane   (c) acetic acid

(d) methylamine   (e) phenol

CHAPTER 12

THE PROPERTIES OF SOLUTIONS

Solutions are mixtures in which the mixing occurs on a molecular level. Although there are many different kinds of solutions, we shall concentrate on those that have a liquid as the solvent; in fact, most of our work will be devoted to the most common type of solution, that with water as the solvent. It's interesting to note that most of the Earth's surface is covered by a solution—the oceans!

SOLUTES AND SOLVENTS

12.1   THE MOLECULAR NATURE OF DISSOLVING

KEY CONCEPT   Solutions form on a molecular level

Imagine a mixture of salt crystals and sugar crystals. In this mixture, very few of the sugar molecules come into close contact with the sodium and chloride ions. The mixing occurs in a way such that there is no opportunity for close contact; even when two crystals are touching each other, they are not really in close contact on an atomic level. However, when a solution forms, the molecules (or ions) that serve as the solute are in intimate contact with molecules of the solvent. The dissolving process reflects this; in an aqueous solution, for instance, the water molecules pull the solute species (molecules or ions) away from the pure solute one by one. Thus, at the end of the dissolving process, each solute molecule (or ion) is in close contact with and surrounded by a number of water molecules.

EXAMPLE   Describing how solutes dissolve in water

When glucose dissolves in water, it does so with the cyclic structure shown below. How do the water molecules interact with the sugar molecules as the sugar dissolves?

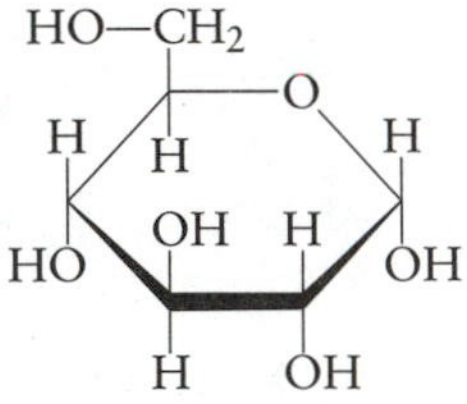

SOLUTION   The oxygen atoms in the sugar have a partial negative charge (due to the O—H bond polarity) and the hydrogen atoms a partial positive charge. The same is true for the hydrogen and oxygen atoms in water. We expect the positive hydrogens to interact with the negative oxygens (because opposite charges attract). Examples of this interaction are shown below with dotted lines. Remember, only the hydrogens attached to oxygens have the partial positive charge required for the interaction. (The interactions shown by the dotted lines are schematic—some of the details of the interactions are more complicated than indicated).

EXERCISE How do the ions in $BaCl_2$ interact with water molecules when $BaCl_2$ is dissolved in water?

Answer: Each ion is surrounded by many waters, with interactions as schematically illustrated below.

12.2   SOLUBILITY

KEY CONCEPT   Often only a limited amount of solute will dissolve

For most solutes, there is a limit to the amount of solute that can be dissolved in a given amount of a particular solvent. For instance, we can imagine what would happen if we tried to dissolve a 5-lb bag of sugar in a cup of water. At some point, the water would contain as much sugar as it can possibly hold and no more would dissolve; any more added sugar would just sink to the bottom of the cup. At this point, the solution is called a **saturated solution**. The amount of sugar that we dissolved, which is the maximum amount the water can hold, is called the **solubility** of the sugar. The solubility of any liquid or solid depends on the solvent and the temperature. When a mass of undissolved solid is in contact with a saturated solution, like our undissolved sugar on the bottom of a cup, a dynamic equilibrium is set up between the undissolved solute and the solute in the solution:

$$Sugar(s) \rightleftharpoons sugar(aq)$$

The mass of dissolved solute and of undissolved solute stays the same, but there is a constant interchange of dissolved solute molecules for undissolved solute molecules and vice versa. For every solute molecule that dissolves from the solid mass of solute, a solute molecule rejoins the solid. "Whatever is done is undone."

EXAMPLE   Determining whether a solution is saturated

If we add 75.0 g of solid $CaCl_2$ to 100 g of water in a beaker at 0°C, does a saturated solution result? If so, what is the mass of the undissolved $CaCl_2$ at the bottom of the beaker? Does a dynamic equilibrium exist? The solubility of $CaCl_2$ in water at 0°C is 59.5 g/100 g H_2O.

SOLUTION The solubility of 59.5 g/100 g H_2O means that the maximum amount of $CaCl_2$ that can be dissolved in 100 g of water at 0°C is 59.5 g. If we add 75.0 g $CaCl_2$ to 100 g water at 0°C, 59.5 g dissolve and 15.5 g remain undissolved. The solution would now contain as much $CaCl_2$ as it could hold, so it is saturated. In this situation, the undissolved solid becomes involved in a dynamic equilibrium with solute in the solution.

EXERCISE What is the molarity of a saturated aqueous solution of CaF_2 at 0°C? Because CaF_2 is only sparingly soluble, assume that dissolving CaF_2 in 1000 mL of water results in 1000 mL of solution; that is, the addition of a small amount of CaF_2 to 1000 mL of water results in no appreciable change in the volume of the water. The solubility of CaF_2 is 1.7×10^{-3} g/100 g H_2O.

[*Answer:* 2.2×10^{-4} mol·L^{-1}]

KEY WORDS   Define or explain each term in a written sentence or two.

molar solubility
saturated solution

FACTORS AFFECTING SOLUBILITY

12.3   SOLUBILITIES OF IONIC COMPOUNDS

KEY CONCEPT   Ionic solubility depends on the anion and the cation

The solubility of an ionic compound depends on the nature of both the anion and cation involved. A large anion with a low charge such as NO_3^- is not strongly held in the ionic crystal lattice, so it is relatively easy for water molecules to pull it away from the crystal (and dissolve the crystal). An ion with a large charge such as phosphate PO_4^{3-}, is strongly held in the crystal lattice, so it is difficult for the water molecules to pull it into solution. If the phosphate were smaller, it would be held even more strongly in the crystal lattice. You should note that most phosphates are insoluble and virtually all nitrates are soluble. A similar analysis could be done for cations.

EXAMPLE   Predicting the solubility of ionic compounds

Only one of the three compounds $Cd(CN)_2$, CdO, $CdBr_2$, is insoluble in water. Which one?

SOLUTION   All three compounds are ionic with Cd^{2+} as the ion, so the cadmium cannot be the reason for any difference in solubility. Of the three ions, the cyanide is large with a small (-1) charge, the oxide is small with a larger (-2) charge, and the bromide is large with a -1 charge. On the basis of the observation that a large charge on the anion tends to make compounds insoluble, we predict that the CdO is insoluble in water.

EXERCISE   Only one of the compounds $Pb(ClO_3)_2$, $PbBr_2$, and $PbCO_3$ is insoluble in water. Which one?

[*Answer:* $PbCO_3$]

12.4   THE LIKE-DISSOLVES-LIKE RULE

KEY CONCEPT   The dependence of solubility on the solvent

A rough rule of thumb used to predict whether a solute is soluble in a specific solvent is "like dissolves like." This colloquialism translates to the following:

1. Nonpolar solutes are relatively soluble in nonpolar solvents but are relatively insoluble in polar and hydrogen-bonded solvents.
2. Ionic, polar, and hydrogen-bonded solutes are relatively soluble in polar and hydrogen-bonded solvents but are relatively insoluble in nonpolar solutes.

These rules are only approximations. In exacting work, where solubilities must be accurately known, they should be looked up.

EXAMPLE Predicting solubilities

Which of the following is least soluble in H_2O: benzene (C_6H_6), hydrogen peroxide (H_2O_2), or dinitrogen trioxide (N_2O_3)?

SOLUTION Water is a hydrogen-bonded solvent, so we expect that polar, ionic, and hydrogen-bonded substances will be relatively soluble in it and that nonpolar substances will be relatively insoluble. H_2O_2 forms hydrogen bonds; it is extremely soluble in water. N_2O_3 is polar; it is soluble in water. C_6H_6 is nonpolar; it is relatively insoluble in water and is the least soluble of the three compounds given.

EXERCISE Is CCl_4 more soluble in benzene or in water?

[*Answer:* Benzene]

12.5 PRESSURE AND SOLUBILITY: HENRY'S LAW

KEY CONCEPT Gas solubility increases with gas partial pressure

Henry's law states that the solubility of a gas increases linearly with the partial pressure of the gas:

$$\text{Solubility of gas} = \text{constant} \times \text{partial pressure of gas}$$

$$S = k_H \times P_{gas}$$

The Henry's law constant k_H depends on the identity of the gas, the solvent, and the temperature.

EXAMPLE Using Henry's law

How many grams of CO_2 are trapped in a 500-mL bottle of soda water under a pressure of 5.0 atm at 25°C? (Assume that the partial pressure of CO_2 is 3.0 atm.)

SOLUTION Because the soda is mostly water, we can use the Henry's law constants given in the text (Table 12.2). For CO_2, $k_H = 2.3 \times 10^{-2}$ mol·L^{-1}·atm^{-1}. Henry's law gives, for the solubility of CO_2,

$$S = k_H \times P_{gas}$$

$$S(CO_2) = 2.3 \times 10^{-2} \text{ mol·L}^{-1}\cdot\text{atm} \times 3.0 \text{ atm}$$

$$= 6.9 \times 10^{-2} \text{ mol·L}^{-1}$$

Thus, 6.9×10^{-2} mol of CO_2 dissolve per liter of soda. This is equivalent to 3.5×10^{-2} mol (69/2) per 500 mL (500 mL = 0.500 L), and

$$3.5 \times 10^{-2} \text{ mol} \times \frac{44.0 \text{ g}}{1 \text{ mol } CO_2} = 1.5 \text{ g } CO_2$$

EXERCISE What mass of N_2 (in mg) is dissolved in 1 L of pond water at 25°C at a total atmospheric pressure of 1.0 atm? Assume the atmosphere is 79% N_2.

[*Answer:* 2×10 mg]

12.6 TEMPERATURE AND SOLUBILITY: THERMAL POLLUTION

KEY CONCEPT Gas solubility increases with gas partial pressure

The solubility of a substance is generally dependent on temperature. Even though many ionic solids become more soluble as the temperature is increased, there are some that don't follow this behavior (see Figure 12.9 of the text). For gases, we can reliably predict that the gas becomes less soluble as the temperature of the solvent (water, for instance) rises.

EXAMPLE   **Understanding solubility changes with temperature**

For which of the ionic compounds shown in Figure 12.9 does the solubility increase the most as the temperature increases?

SOLUTION Only three of the compounds shown in the graph undergo a consistent increase in solubility as temperature increases: $AgNO_3$, KI, and $NaNO_3$. The one that has the steepest slope is the one for which the solubility increases the most with temperature; this is $AgNO_3$.

EXERCISE For which of the ionic compounds shown in Figure 12.9 does the solubility consistently decrease as the temperature increases?

[*Answer:* Li_2SO_4]

PITFALL   **Be certain to distinguish the rate of dissolving from solubility**

All substances dissolve faster when the temperature of the solvent increases. However, the amount of material that dissolves (the solubility) may become lower, stay the same, or become higher as the temperature increases. Be certain to distinguish between how fast something happens (the rate of dissolving) and the ultimate end point (how much dissolves).

KEY WORDS   Define or explain each term in a written sentence or two.

the bends	Henry's law constant	surfactant
colloid	hydrophilic	thermal pollution
effervesce	hydrophobic	
Henry's law	micelle	

WHY DOES ANYTHING DISSOLVE?

12.7   THE ENTHALPY OF SOLUTION

KEY CONCEPT   The dissolving process involves two steps

We would like to know what factors contribute to the enthalpy of solution, what makes it exothermic or endothermic. We can do this by envisioning that a two-step process occurs when an ionic compound dissolves. First, the ionic compound breaks up into gas-phase ions; second, the ions dissolve in the solvent. Breaking the solid into the gas-phase ions is always endothermic and requires an amount of energy that we earlier called the lattice enthalpy. When the ions dissolve, the solvent molecules usually arrange themselves around the ions in a low-energy configuration (a process called **solvation**), so energy is released during this process. In water, the process is called **hydration** and the energy released called the **enthalpy of hydration**. For NaCl dissolving in water, the two-step process is as follows:

Solid breaks up into gas phase ions:

$$NaCl(s) \longrightarrow Na^+(g) + Cl^-(g) \qquad \Delta H_L^\circ = \text{lattice enthalpy}$$

Water molecules hydrate ions as they dissolve:

$$Na^+(g) + Cl^-(g) \longrightarrow Na^+(aq) + Cl^-(aq) \qquad \Delta H_H^\circ = \text{hydration enthalpy}$$

The (aq) after an ion indicates it is hydrated. Summing these two reactions gives the reaction for the dissolving of the solid. According to Hess's law, the enthalpy of solution equals the sum of the two reactions, that is, the sum of the lattice enthalpy plus the hydration enthalpy:

$$NaCl(s) \longrightarrow Na^+(aq) + Cl^-(aq) \qquad \Delta H^\circ_{sol} = \Delta H^\circ_L + \Delta H^\circ_H$$

EXAMPLE Estimating the hydration enthalpy

The lattice enthalpy of AgCl is $+905$ kJ$\cdot$mol^{-1}, and its heat of solution in water is $+65.5$ kJ$\cdot$mol^{-1}. Calculate the hydration enthalpy of AgCl.

SOLUTION The sum of the lattice enthalpy and hydration energy equals the enthalpy of solution:

$$\Delta H^\circ_{sol} = \Delta H^\circ_L + \Delta H^\circ_H$$

We are given $\Delta H^\circ_{sol} = +65.5$ kJ$\cdot$mol^{-1} and $\Delta H^\circ_L = +905$ kJ$\cdot$mol^{-1}. Substitution of these into the given equation permits calculation of ΔH°_H:

$$65.5 \text{ kJ}\cdot\text{mol}^{-1} = 905 \text{ kJ}\cdot\text{mol}^{-1} + \Delta H^\circ_H$$

$$\Delta H^\circ_H = 65.5 \text{ kJ}\cdot\text{mol}^{-1} - 905 \text{ kJ}\cdot\text{mol}^{-1}$$

$$= -840 \text{ kJ}\cdot\text{mol}^{-1}$$

EXERCISE Calculate the enthalpy of hydration of AgF. Considering the difference between the Cl^- ion and F^- ion, does this result seem reasonable, compared with the result just obtained for AgCl?

[*Answer:* -993 kJ$\cdot$mol^{-1}; the difference is reasonable because the smaller F^- should hydrate more strongly than Cl^- and result in a more negative ΔH°_H for AgF than for AgCl.]

12.8 INDIVIDUAL ION HYDRATION ENTHALPIES

KEY CONCEPT The hydration energy of individual ions can be estimated

An individual ion hydration enthalpy (ΔH°_H) is the enthalpy change that occurs when the gas-phase ion dissolves and becomes hydrated. For instance, for sodium ion, ΔH°_H is -444 kJ; thus,

$$Na^+(g) \longrightarrow Na^+(aq) \qquad \Delta H^\circ_H = -444 \text{ kJ}$$

A large number of individual ion enthalpies have been determined; some are shown in Table 12.5 of the text. From this table, we should note that a small, highly charged ion hydrates very strongly, resulting in a large negative enthalpy change for the hydration process. Conversely, a large ion with low charge does not hydrate very strongly; its hydration energy is not as large, and hydration of such an ion is not as energetically favorable a process as it is for a smaller, more highly charged ion.

EXAMPLE Using and interpreting individual ion enthalpies

Use individual ion enthalpies to calculate the enthalpy of hydration of MgBr$_2$.

SOLUTION We want the enthalpy change for the overall process

$$Mg^{2+}(g) + 2Br^-(g) \longrightarrow Mg^{2+}(aq) + 2Br^-(aq) \qquad \Delta H^\circ_H = ?$$

Using Hess's law, we can determine this value by summing the equations that represent hydration of the individual ions. The individual ion enthalpies are read from Table 12.5 of the text.

$Mg^{2+}(g) \longrightarrow Mg^{2+}(aq)$	$\Delta H^\circ_H = 1 \text{ mol} \times (-2003 \text{ kJ}\cdot\text{mol}^{-1}) =$	-2003 kJ
$2Br^-(g) \longrightarrow 2Br^-(aq)$	$\Delta H^\circ_H = 2 \text{ mol} \times (-309 \text{ kJ}\cdot\text{mol}^{-1}) =$	-618 kJ
$Mg^{2+}(g) + 2Br^-(g) \longrightarrow Mg^{2+}(aq) + 2Br^-(aq)$	$\Delta H^\circ_H =$	-2621 kJ

EXERCISE Use individual ion enthalpies to calculate the enthalpy of hydration of SrI_2. Comment on why the value differs from that for $MgBr_2$, calculated in the example.

[*Answer:* -2116 kJ; it is more positive because Sr^{2+} is larger than Mg^{2+} and I^- is larger than Br^-.]

12.9 SOLUBILITY AND DISORDER

KEY CONCEPT Enthalpy, solubility, and disorder

Although it is not always obvious, nature favors disorder and randomness over order. When given a chance, molecules tend to spread throughout as large a volume as possible. In a similar fashion, the available energy in a system tends to spread out over as many quantum levels as possible, as chaotic thermal motion. The spread-out state is a more disordered state. When a solid dissolves, the molecules change from a highly ordered, confined state (the compact solid) to a disordered, spread-out state (the molecules spread throughout the solution); so the dissolving process is a favorable process insofar as the disorder of molecules is concerned. What happens energetically depends on whether the enthalpy of solution is exothermic or endothermic. If the solute dissolves exothermically, energy that was confined to the relatively few molecules of the dissolving system spreads to all the molecules of the surroundings. As far as the energy is concerned, this is a change from a more ordered to a more disordered state, and the process is favorable. However, if the solute dissolves endothermically, energy that was spread among all the molecules of the surroundings becomes confined to the relatively few molecules of the system. Energetically, the change is from a more disordered to a more ordered state and is unfavorable.

	Exothermic Dissolving	Endothermic Dissolving
molecules	ordered $\rightarrow$ disordered	ordered $\rightarrow$ disordered
energy	ordered $\rightarrow$ disordered	disordered $\rightarrow$ ordered

Exothermic dissolving is always a favorable process because the changes in the disorder of molecules and energy are both favorable. Endothermic dissolving occurs only if the favorable ordered $\rightarrow$ disordered change for the molecules overwhelms the unfavorable disordered $\rightarrow$ ordered change for the energy.

EXAMPLE Understanding disorder and order in the solution process

When a nonpolar solute dissolves in a nonpolar solvent, the enthalpy of solution is often close to zero. Because the energy change is unimportant, why does the solute dissolve?

SOLUTION The solute dissolves because the solute molecules change from a more ordered to a more disordered state during the dissolving process, and such a change is favorable.

EXERCISE When a helium balloon pops in a room, why do the helium molecules spread throughout the room rather than stay in place where they were before the balloon broke?

[*Answer:* If we assume helium and air to be ideal gases, a zero enthalpy change will occur when the helium molecules disperse throughout the room (ideal gas molecules have zero energy of interaction). The spreading out of the molecules is a favorable process because of the increase in disorder (entropy) and because there are no opposing unfavorable energetic processes.]

KEY WORDS Define or explain each term in a written sentence or two.

enthalpy of hydration
enthalpy of solution
hydration

individual in hydration therapy
ion-dipole interaction
solvation

COLLIGATIVE PROPERTIES

KEY CONCEPT   A survey of colligative properties

Addition of a nonvolatile solute to a solvent affects the solvent in such a way that many of the physical properties of the resulting solution are different from those of the pure solvent. The properties thus affected are called *colligative properties*. The properties we will consider and the difference between the pure solvent and a solution are summarized in the table.

Colligative Property	Difference between Solution and Pure Solvent
vapor pressure	solution vapor pressure lower than that of pure solvent
boiling point	solution boiling point higher than that of pure solvent
freezing point	solution freezing point lower than that of pure solvent
osmotic pressure	exists only when solution is in contact with pure solvent, with semipermeable membrane present

Colligative properties depend only on the relative number of solute particles (molecules or ions) present and not on the specific identity of the solute. For instance, NaCl ionizes in water to produce two ions per dissolved formula unit, whereas glucose, which does not ionize, produces one molecule per dissolved molecule. For equal concentrations of glucose and NaCl, the NaCl produces twice as many solute particles and has approximately twice the effect on the four colligative properties just listed. For example, if a certain concentration of glucose raises the boiling point of water by 0.5°C, the same concentration of NaCl raises it by approximately twice as much, or 1.0°C.

EXAMPLE   Predicting the effect of the number of solute particles on colligative properties

A 0.50 m aqueous glucose solution boils at 100.3°C; that is, the boiling point of the solution is 0.3°C above that of pure water. What is the boiling point of a 0.50 m aqueous $Ca(NO_3)_2$ solution?

SOLUTION Glucose does not ionize in solution, whereas $Ca(NO_3)_2$ ionizes into three ions ($1Ca^{2+}$ + $2NO_3^-$ = 3 ions) when it is dissolved. There are three times as many solute particles present in a 0.50 m $Ca(NO_3)_2$ solution as there are in a 0.50 m glucose solution, so we expect the increase in the boiling point to be three times larger in 0.50 m $Ca(NO_3)_2$ than in 0.50 m glucose. Because 3 × 0.3°C = 0.9°C, the boiling point of the $Ca(NO_3)_2$ is predicted to be 100.9°C (100.0°C + 0.9°C = 100.9°C).

EXERCISE Addition of 1.5 mol NaCl to 1.0 kg H_2O lowers the freezing point of the solution to −5.6°C. What is the freezing point of the solution made by adding 1.5 mol Na_2SO_4 to 1.0 kg water?

[*Answer:* −8.4°C]

PITFALL   Nonvolatile solutes only!

In this section we consider nonvolatile solutes only. These are solutes that do not evaporate easily and, for the purposes of our discussion, are considered to have zero vapor pressure. Ionic compounds and sugars (such as glucose, fructose, and sucrose) are examples of nonvolatile solutes; in contrast, the alcohols that easily dissolve in water are generally quite volatile.

12.10   MEASURES OF CONCENTRATION

KEY CONCEPT A   Mole fraction

We now discuss a slightly different kind of concentration unit, one that gives the number of moles of one substance relative to the total number of moles of all of the substances in the solution. The volume of solution or solvent is not involved in the mole fraction, only the moles of substance present. Even though a solute and solvent are present, the mole fraction concentration unit treats all substances equally. So, if we have a solute mixed with a solvent,

$$\text{Mole fraction solute} = \frac{\text{mole solute}}{\text{mole solute} + \text{mole solvent}}$$

$$\text{Mole fraction solvent} = \frac{\text{mole solvent}}{\text{mole solute} + \text{mole solvent}}$$

Using the symbols

$$n_A = \text{mol solvent} \qquad n_B = \text{mol solute}$$

$$x_A = \text{mole fraction solvent} \qquad x_B = \text{mole fraction solute}$$

we get

$$x_A = \frac{n_A}{n_A + n_B} \qquad x_B = \frac{n_B}{n_A + n_B}$$

It is always true that the sum of the mole fractions of all of the substances in a solution equals 1. For the solution just discussed, $x_A + x_B = 1$. The mole fraction is dimensionless.

EXAMPLE   Calculating mole fractions

If we add 100 g C_2H_5OH to 100 g H_2O, what is the mole fraction of each component of the solution? Do the mole fractions sum to 1?

SOLUTION The mole fraction of each component is given by

$$x_{C_2H_5OH} = \frac{n_{C_2H_5OH}}{n_{C_2H_5OH} + n_{H_2O}} \qquad x_{H_2O} = \frac{n_{H_2O}}{n_{C_2H_5OH} + n_{H_2O}}$$

where $n_{C_2H_5OH}$ = mol C_2H_5OH and n_{H_2O} = mol H_2O. We calculate the moles of each component, using their respective molar mass:

$$n_{C_2H_5OH} = 100 \text{ g} \times \frac{1 \text{ mol } C_2H_5OH}{46.1 \text{ g}} \qquad n_{H_2O} = 100 \text{ g} \times \frac{1 \text{ mol } H_2O}{18.0 \text{ g}}$$

$$= 2.17 \text{ mol } C_2H_5OH \qquad = 5.56 \text{ mol } H_2O$$

Substituting these values into the defining equations for mole fraction gives

$$x_{C_2H_5OH} = \frac{n_{C_2H_5OH}}{n_{C_2H_5OH} + n_{H_2O}} \qquad x_{H_2O} = \frac{n_{H_2O}}{n_{C_2H_5OH} + n_{H_2O}}$$

$$= \frac{2.17 \text{ mol}}{2.17 \text{ mol} + 5.56 \text{ mol}} \qquad = \frac{5.56 \text{ mol}}{2.17 \text{ mol} + 5.56 \text{ mol}}$$

$$= 0.281 \qquad = 0.719$$

For the second part of the question, the mole fractions sum to 1, as they should:

$$x_{C_2H_5OH} + x_{H_2O} = 0.281 + 0.719 = 1.000$$

EXERCISE What is the mole fraction of each component of the solution made by dissolving 25.0 g CCl_4 in 50.0 g C_5H_{12}? Do the mole fractions sum to 1?

[*Answer:* $x_{CCl_4} = 0.190$, $x_{C_5H_{12}} = 0.810$; yes]

KEY CONCEPT B   Molality

The molality m of a solution is obtained by dividing the number of moles of solute by the kilograms of solvent. For this unit of concentration, the volume of solution is irrelevant.

$$\text{Molality} = \frac{\text{moles of solute}}{\text{kilograms of solvent}}$$

This unit is convenient because the concentration expressed in molality does not change as the temperature of the solution changes. The units of molality are moles per kilogram ($\text{mol} \cdot \text{kg}^{-1}$).

EXAMPLE Preparing a solution of known molality

What is the molality of a solution made by adding 125 g Na_2SO_4 to 500 g H_2O?

SOLUTION The molality is defined as

$$\text{Molality} = \frac{\text{moles of solute}}{\text{kilograms of solvent}}$$

For this problem, we have

$$\text{Molality } Na_2SO_4 = \frac{\text{mol } Na_2SO_4}{\text{kg } H_2O}$$

The moles of Na_2SO_4 are calculated with the molar mass of Na_2SO_4, which is $142.1 \text{ g} \cdot \text{mol}^{-1}$:

$$125 \text{ g} \times \frac{1 \text{ mol } Na_2SO_4}{142.1 \text{ g}} = 0.880 \text{ mol } Na_2SO_4$$

We note that 500 g = 0.500 kg and calculate the molality by substituting into the equation for molality:

$$\text{Molality} = \frac{0.880 \text{ mol } Na_2SO_4}{0.500 \text{ kg } H_2O}$$

$$= 1.76 \text{ mol} \cdot \text{kg}^{-1}$$

EXERCISE What is the molality of the solution made by dissolving 10.0 g ethylene glycol ($C_2H_6O_2$) in 50.0 g water?

[*Answer:* $3.22 \text{ mol} \cdot \text{kg}^{-1}$]

PITFALL   Distinguishing molality from molarity

Care must be taken to distinguish the concentration unit molality from the concentration unit molarity. Molarity is defined as moles solute per liter solution and is symbolized by an uppercase Roman M. Molality is defined as moles solute per kilogram solvent and is symbolized by a lowercase italic *m*.

$$\text{Molarity (M)} = \frac{\text{moles solute}}{\text{liters solution}} \qquad \text{Molality } (m) = \frac{\text{moles solute}}{\text{kilograms solvent}}$$

Because the density of water is close to $1 \text{ kg} \cdot \text{L}^{-1}$, the molarity and molality of a dilute aqueous solution are very close to the same number. For a nonaqueous solution in which the density of solvent is not 1 or for a concentrated aqueous solution, the molarity and molality can be very different.

12.11   VAPOR-PRESSURE LOWERING

KEY CONCEPT   The lowering of vapor pressure and Raoult's law

Addition of a nonvolatile solute to a solvent lowers the vapor pressure, so the vapor pressure of the resulting solution is less than that of the pure solvent. The effect is quantitatively described by Raoult's law:

Vapor pressure of solution = mole fraction solvent × vapor pressure of pure solvent

$$P_A \qquad = \qquad x_A \qquad \times \qquad P_A^*$$

where P_A = vapor pressure of solution, x_A = mole fraction of solvent and P_A^* = vapor pressure of pure solvent.

EXAMPLE   Calculating vapor-pressure lowering

What is the vapor pressure, at 25°C, of the solution that results from adding 60.0 g glucose ($C_6H_{12}O_6$) to 500 g water? The vapor pressure of water at 25°C is 23.8 Torr. Assume ideal solution behavior.

SOLUTION Because glucose is a nonvolatile solute, Raoult's law can be used to calculate the vapor pressure of the solution:

$$P_A = x_A \times P_A^*$$

For our problem, in which water is the solvent, this translates to

$$P_{\text{solution}} = x_{H_2O} \times P_{H_2O}^*$$

The problem states that $P_{H_2O}^* = 23.8$ Torr. The mole fraction of water in the solution x_{H_2O} is given by

$$x_{H_2O} = \frac{n_{H_2O}}{n_{H_2O} + n_{C_6H_{12}O_6}}$$

We use the molar mass of water (18.0 g/mol) and glucose (180.2 g/mol) to calculate the moles of each in the solution:

$$n_{H_2O} = 500 \text{ g} \times \frac{1 \text{ mol } H_2O}{18.0 \text{ g}} \qquad n_{C_6H_{12}O_6} = 60.0 \text{ g} \times \frac{1 \text{ mol } C_6H_{12}O_6}{180.2 \text{ g}}$$

$$= 27.8 \text{ mol} \qquad\qquad\qquad = 0.333 \text{ mol}$$

Substituting these values into the equation for the mole fraction of water gives

$$x_{H_2O} = \frac{27.8}{27.8 + 0.333}$$

$$= \frac{27.8}{28.1}$$

$$= 0.988$$

Substituting this value and the vapor pressure of pure water into Raoult's law gives the desired answer:

$$P_{\text{solution}} = 0.988 \times 23.8 \text{ Torr}$$

$$= 23.5 \text{ Torr}$$

As expected, the vapor pressure of the solution is lower than that of the pure solvent.

EXERCISE What is the vapor pressure, at 23°C, of the solution that results from adding 75.0 g I_2 to 500 g carbon tetrachloride, CCl_4. The vapor pressure of CCl_4 at 23°C is 100 Torr. Assume I_2 is nonvolatile, and assume ideal solution behavior.

[*Answer:* 91.5 Torr]

12.12   BOILING-POINT ELEVATION AND FREEZING-POINT DEPRESSION

KEY CONCEPT   Solutes change the boiling and freezing points

Addition of a nonvolatile solute to a solvent **increases** the boiling point of the resulting solution above that of the pure solvent. In a similar fashion, addition of a nonvolatile solute to a solvent

decreases the freezing point of the resulting solution below that of the pure solvent. For the freezing phenomenon, the solution must be cooled to a lower temperature than the pure solvent in order to get it to freeze. Quantitatively, the effect can be calculated with the relation

$$\text{Freezing-point depression} = i \times k_f \times \text{molality of solute}$$

The constant k_f (the **freezing-point constant**) depends on the identity of the solvent. The factor i is the van't Hoff factor and is equal to the number of moles of solute particles that result from 1 mol of solute. This relation is illustrated in the table.

Compound	Solute Particles Formed upon Dissolving	Number of Moles of Solute Particles per Mole Solute	i
$C_6H_{12}O_6$	$C_6H_{12}O_6$	1	1
NaCl	Na^+, Cl^-	2	2
Na_2SO_4	$2Na^+$, SO_4^{2-}	3	3
$Ca(NO_3)_2$	Ca^{2+}, $2NO_3^-$	3	3

The van't Hoff factor i is dimensionless; the dimensions of k_f are $K \cdot kg \cdot mol^{-1}$. Molecular solutes such as glucose do not ionize, and $i = 1$. With such solutes, we get

$$\text{Freezing-point depression} = k_f \times \text{molality of solute}$$

(In this context, the term depression means "lowering".)

EXAMPLE 1 Calculating a freezing-point depression

What is the freezing point of a mixture consisting of 0.300 g S_8 dissolved in 20.0 g camphor? The freezing point of pure camphor is 179.8°C; S_8 is a nonvolatile molecular solute.

SOLUTION The equation used to calculate the lowering of the freezing point for a solution is

$$\text{Freezing-point depression} = i \times k_f \times \text{molality of solute}$$

For a molecular solute $i = 1$; k_f for camphor is $39.7\ K \cdot kg \cdot mol^{-1}$ (Table 12.7 of the text). The molar mass of S_8 is $256\ g \cdot mol^{-1}$, so its molality is calculated as follows.

$$\text{Molality of } S_8 = \frac{\text{mol } S_8}{\text{kg camphor}}$$

$$\text{Mol } S_8 = 0.300\ g \times \frac{1\ \text{mol } S_8}{256\ g} = 1.17 \times 10^{-3}\ \text{mol } S_8$$

$$\text{kg camphor} = 20.0\ g \times \frac{1\ kg}{1000\ g} = 0.0200\ kg$$

$$\text{Molality of } S_8 = \frac{1.17 \times 10^{-3}\ \text{mol } S_8}{0.0200\ kg} = 0.0585\ mol \cdot kg^{-1}$$

Substituting into the equation for the freezing-point depression gives

$$\text{Freezing-point depression} = i \times k_f \times \text{molality of solute}$$

$$= 1 \times 39.7\ K \cdot kg \cdot mol^{-1} \times 0.0585\ mol \cdot kg^{-1}$$

$$= 2.3\ K = 2.3°C$$

Because the freezing point of pure camphor is 179.8°C and the freezing-point depression (lowering) is 2.3°C, the freezing point of the solution is 177.5°C.

EXERCISE What is the freezing point (in °C) of the solution made by adding 5.0 g $Ca(NO_3)_2$ to 400 g water?

[*Answer:* −0.43°C]

EXAMPLE 2   Using a freezing-point depression to measure a molar mass

The addition of 0.226 g of a nonionizing unknown compound to 100 g camphor lowers the freezing point of the camphor by 0.42 K. What is the molar mass of the unknown? $k_f = 39.7 \ K \cdot kg \cdot mol^{-1}$

SOLUTION The equation relating molality to freezing-point depression can be rearranged to

$$\text{Molality} = \frac{1}{k_f} \times \text{freezing-point depression}$$

We have $k_f = 39.7 \ K \cdot kg \cdot mol^{-1}$ and freezing-point depression = 0.42 K. Substituting into the equation gives

$$\text{Molality} = \frac{1}{39.7 \ \cancel{K} \cdot kg \cdot mol^{-1}} \times 0.42 \ \cancel{K}$$

$$= 0.011 \ mol \cdot kg^{-1}$$

This molality means that there is 0.011 mol unknown dissolved in 1000 g (1 kg) camphor. We can now calculate the moles of unknown in 100 g camphor.

$$100 \ g \ camphor \times \frac{0.011 \ mol \ unknown}{1000 \ g \ camphor} = 0.0011 \ mol$$

The problem states that there is 0.226 g unknown in 100 g camphor; therefore, 0.226 g unknown corresponds to 0.0011 mol unknown and

$$\text{Molar mass} = \frac{0.226 \ g}{0.0011 \ mol}$$

$$= 2.1 \times 10^2 \ g \cdot mol^{-1}$$

EXERCISE The addition of 0.243 g of a nonionizing unknown to 250 g carbon tetrachloride lowers its freezing point by 0.12 K. What is the molar mass of the unknown? $k_f = 29.8 \ K \cdot kg \cdot mol^{-1}$ for carbon tetrachloride.

[*Answer:* $2.4 \times 10^2 \ g \cdot mol^{-1}$]

PITFALL   Choosing the correct boiling-point and freezing-point constant

It is important to remember that the freezing-point constant k_f used in calculations of freezing-point depressions depends on the identity of the solvent and not that of the solute.

PITFALL   Is the change in temperature subtracted or added?

Boiling points are elevated (increased) by the presence of a solute. Freezing points are depressed (lowered) by the presence of a solute. After calculating the freezing-point depression, subtract the depression from the freezing point of the pure solvent so that the freezing point of the solution is less than the freezing point of the pure solvent. A good way to remember these points is to memorize the statement, "Boiling points start higher and go higher; freezing points start lower and go lower."

PITFALL   The van't Hoff constant

The ideal values of i we have given are valid only for very dilute solutions. At concentrations at which ions or molecules begin to interact with one another, experimental values of i must be used.

12.13   OSMOSIS

KEY CONCEPT   Osmosis and osmotic pressure

The osmosis process depends on the existence of a **semipermeable membrane,** which allows some species to pass through but is a barrier to the passage of other species. Many important semipermeable membranes allow water molecules to pass through the membrane but do not allow passage of larger biological molecules or ions. Osmosis occurs when a pure solvent is placed on one side of a semipermeable membrane and a solution on the other; the solvent molecules pass through the membrane but the solute molecules do not. The solvent molecules have a natural tendency to flow from where they are more concentrated (the pure solvent, which is 100% solvent) to where they are less concentrated (the solution, which is less than 100% solvent).

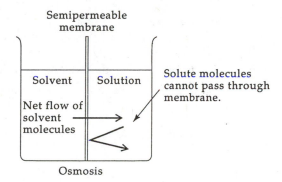

As osmosis occurs, solvent moves to the solution side of the membrane, so the level of the solution in the container increases and that of the solvent decreases. The pressure that must be applied to the top of the solution side to stop its level from rising is called the **osmotic pressure (Π)** and is given by

$$\Pi = i \times RT \times \text{molarity of solute}$$

In this equation, i = van't Hoff factor, R = ideal gas constant, and T = temperature in kelvins.

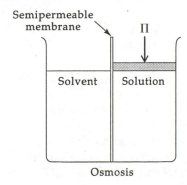

EXAMPLE   Calculating the osmotic pressure

We dissolve 3.00 g of a nonionizing biomolecule with molar mass $5.0 \times 10^3 \, \text{g} \cdot \text{mol}^{-1}$ in enough water to make 250 mL of solution. What is the osmotic pressure of the solution at 25°C?

SOLUTION   The osmotic pressure of a solution is given by

$$\Pi = i \times RT \times \text{molarity}$$

The molarity of the solute is given by

$$\text{Molarity} = \frac{\text{mol solute}}{\text{liter solution}}$$

$$\text{Mol solute} = 3.00 \text{ g} \times \frac{1 \text{ mol}}{5.0 \times 10^3 \text{ g}} = 6.0 \times 10^{-4} \text{ mol}$$

$$\text{Liter solution} = 250 \text{ mL} \times \frac{1 \text{ L}}{1000 \text{ mL}} = 0.250 \text{ L}$$

$$\text{Molarity} = \frac{6.0 \times 10^{-4} \text{ mol}}{0.250 \text{ L}} = 2.4 \times 10^{-3} \text{ mol} \cdot \text{L}^{-1}$$

$i = 1$ because the solute is nonionizing; $T = 25 + 273 = 298$ K, and $R = 0.0821$ L atm·mol^{-1}·K^{-1}. Substituting these values into the original equation gives

$$\Pi = i \times RT \times \text{molarity}$$

$$= 1 \times (0.0821 \text{ L atm} \cdot \text{mol}^{-1} \cdot \text{K}^{-1}) \times (298 \text{ K}) \times (2.4 \times 10^{-3} \text{ mol} \cdot \text{L}^{-1})$$

$$= 0.059 \text{ atm}$$

EXERCISE What is the osmotic pressure at 50°C of the solution made by dissolving 1.55 g sucrose ($C_{12}H_{22}O_{11}$) in enough water to make 100 mL of solution?

[*Answer:* 1.20 atm]

KEY WORDS   Define or explain each term in a written sentence or two.

colligative property	molality	Raoult's law
freezing-point constant	mole fraction	semipermeable membrane
freezing-point depression	nonideal solution	van't Hoff *i* factor
ideal solution	osmotic pressure	vapor-pressure lowering

DESCRIPTIVE CHEMISTRY TO REMEMBER

- The high solubility of **nitrates** (ionic compounds containing NO_3^-) in water results in their rarely being found in mineral deposits. Rainwater dissolves and washes away any potential nitrate mineral deposits.
- Bone is largely **calcium phosphate** [$Ca_3(PO_4)_2$]. Like most phosphates it is insoluble, a desirable feature for the skeletons of creatures that are mostly water.
- **Calcium hydrogen phosphate** [$Ca(HPO_4)_2$] is more soluble than calcium phosphate and is included in many commercial fertilizers.
- Hard water contains dissolved **magnesium** and **calcium** salts.
- Soaps react with the **calcium ions** in hard water to form scum. Pretreatment of the water with washing soda (Na_2CO_3) precipitates the calcium as $CaCO_3$, so the formation of scum is avoided.
- Most **ionic** and **molecular solids** are more soluble in hot water than in cold water. All **gases** are less soluble in hot water than in cold water.
- **Thermal pollution** is the damage caused to the environment by waste heat. One consequence of thermal pollution is that O_2 is driven out of solution. This loss of O_2 results in the death of the bacteria that degrade dead organic matter.
- Biological cell walls act like **semipermeable membranes**. They allow water and small ions to pass but not the large biomolecules synthesized and required inside the cell.
- **Salting** preserves meat because the flow of water from bacteria to the concentrated salt solution formed on the meat by salting dehydrates the bacteria and kills them.

- A form of **reverse osmosis** is the forcing of water molecules across a semipermeable membrane from a solution (seawater) to a pure-water collector by the application of pressure to the solution in contact with the membrane.
- **Salt** is placed on snow-covered roads to lower the melting point of the frozen water (snow or ice) below the outside temperature; the snow will then melt.

CHEMICAL EQUATIONS TO KNOW

- Treating **phosphate rock** with sulfuric acid results in the formation of calcium sulfate:

$$Ca_5(PO_4)_3OH(s) + 5H_2SO_4(aq) \longrightarrow 3H_3PO_4(aq) + 5CaSO_4(s) + H_2O(l)$$

- Carbon dioxide reacts with water to form **carbonic acid:**

$$CO_2(g) + H_2O(l) \longrightarrow H_2CO_3(aq)$$

- Calcium carbonate ($CaCO_3$) reacts with dissolved carbonic acid (H_2CO_3) to form the more soluble **calcium hydrogen carbonate** [$Ca(HCO_3)_2$]:

$$CaCO_3(s) + H_2CO_3(aq) \longrightarrow Ca^{2+}(aq) + 2HCO_3^-(aq)$$

- When aqueous calcium hydrogen carbonate [$Ca(HCO_3)_2$] is heated in hot-water pipes and kettles, carbon dioxide is driven off and calcium carbonate (**boiler scale**) precipitates:

$$Ca^{2+}(aq) + 2HCO_3^-(aq) \longrightarrow CaCO_3(s) + H_2O(l) + CO_2(g)$$

- Soap scum is a precipitate of **calcium stearate** formed by the reaction of Ca^{2+} in hard water with stearate ions $C_{17}H_{35}CO_2^-$ from soap (sodium stearate):

$$Ca^{2+}(aq) + 2C_{17}H_{35}CO_2^-(aq) \longrightarrow Ca(C_{17}H_{35}CO_2)_2(s)$$

MATHEMATICAL EQUATIONS TO KNOW AND UNDERSTAND

Mass percentage solute $= \dfrac{\text{mass of solute}}{\text{mass of solution}} \times 100$ definition of mass percentage

Molar concentration = moles of solute/liters of solution   definition of molarity

Mass concentration = mass of solute/liters of solution   definition of mass concentration

$x_A = \dfrac{n_A}{n_A + n_B}$ definition of mole fraction

Molality = moles of solute/kilograms of solvent   definition of molality

Solubility of gas $= k_H \times$ partial pressure of gas   Henry's law

$P_A = x_A \times P_A^*$ Raoult's law

Boiling point elevation $= i \times k_b \times$ molality of solute   boiling point elevation

Freezing point depression $= i \times k_f \times$ molarity of solute   freezing point depression

$\Pi = i \times RT \times$ molarity of solute   osmotic pressure

SELF-TEST EXERCISES

Solutes and solvents

1. How might a water molecule interact with methanol (CH_3OH) when the methanol dissolves?

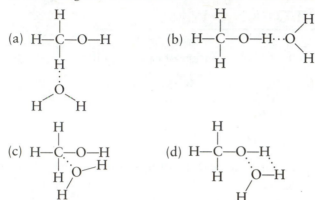

2. Suppose that a mixture is prepared by taking pieces of egg white, each containing 10^3 molecules and stirring them thoroughly with water. The small pieces are invisible to the naked eye and do not break up into smaller pieces when the mixing is done. Is the resulting mixture a solution?

(a) yes   (b) no   (c) need more information to tell

3. Use Figure 12.9 of the text to estimate the molar solubility of $NaNO_3$ at 40°C. All answers are in moles per liter.

(a) 100   (b) 20   (c) 0.85   (d) 1.2   (e) 0.010

4. Is a solution containing 23 g of NaCl in 100 g H_2O at 65°C a saturated solution? Use Figure 12.9 of the text.

(a) yes   (b) no   (c) need more information to tell

5. Imagine an experiment in which, at the start, a sample of solid sugar composed entirely of radioactive sugar molecules is in contact with a saturated aqueous solution of sugar. The sugar molecules in the saturated solution are all nonradioactive. After a week or two, which of the following is true?

(a) All the molecules in the solid are radioactive.

(b) All the molecules in the solution are radioactive.

(c) All the molecules in the solid are nonradioactive.

(d) All the molecules in the solution are nonradioactive.

(e) Some of the molecules in the solution and some in the solid are radioactive.

Factors affecting solubility

6. Without using any references, decide which of the following is likely to be the most insoluble in water.

(a) $NiHPO_4$ (b) $Ni(H_2PO_4)_2$ (c) $Ni_3(PO_4)_2$

7. An ionic compound with a _____ anion of _____ charge is most likely to be soluble.

(a) small, high   (b) small, low   (c) large, low   (d) large, high

8. Which of the following is most likely to be soluble in CCl_4?

(a) P_4 (b) KOH   (c) sucrose   (d) HCl   (e) H_2O

9. All of the following statements about soap except one are true. Which one is untrue?

(a) The hydrophobic end of a soap molecule is soluble in water.

(b) Soap molecules entrap dirt in a micelle.

(c) A soap molecule is a long chainlike molecule.

(d) One end of a soap molecule is soluble in nonpolar substances.

(e) Soap molecules react with Ca^{2+} to form a solid precipitate.

10. The solubility of a gas in water at 25°C is 6.0 mM. What is the solubility if the partial pressure of the gas above the water is doubled?

(a) 36 mM   (b) 3.0 mM   (c) 12 mM   (d) 6.0 mM   (e) 2.4 mM

11. What mass of argon gas is dissolved at 20°C in 100 mL of solution if the partial pressure of argon above the solution is 1.0×10^{-3} atm?

(a) 4.0×10^{-3} g   (b) 0.060 g   (c) 6.0 g

(d) 6.0×10^{-6} g   (e) 0.27 g

12. Imagine that three gases are trapped in a space above water, with partial pressures 24 Torr for gas A, 31 Torr for gas B, and 55 Torr for gas C. The solubility of C in the water (at 25°C) is 1.2×10^{-3} mol·L^{-1}. Gas D is added to the space above the water at a partial pressure of 110 Torr (so the total pressure doubles). What is the solubility of gas C after the addition of D?

(a) 2.4×10^{-3} mol·L^{-1} (b) 6.0×10^{-4} mol·L^{-1} (c) 3.0×10^{-4} mol·L^{-1}

(d) 1.2×10^{-3} mol·L^{-1} (e) 4.8×10^{-3} mol·L^{-1}

13. The solubility of a(n) _____ always decreases when the temperature is _____ .

(a) ionic solid, decreased   (b) molecular solid, decreased   (c) gas, decreased

(d) ionic solid, increased   (e) gas, increased

Why does anything dissolve?

14. For which of the following will an aqueous solution cool as the solute is dissolved? You may consult Table 12.3 of the text for data.

(a) NaCl   (b) $Mg(NO_3)_2$ (c) $Ca(OH)_2$ (d) $MgCl_2$ (e) Li_2CO_3

15. Use the lattice enthalpy of KBr (689 kJ·mol^{-1}) and its enthalpy of hydration (Table 12.4) to calculate the enthalpy of solution of KBr. All answers are in kJ·mol^{-1}.

(a) +1359   (b) −1359   (c) −19   (d) +19   (e) 0

16. What is the lattice enthalpy of AgF, calculated from its enthalpy of solution and enthalpy of hydration? All answers are in kJ·mol^{-1}.

(a) +1295   (b) −1016   (c) +1016   (d) +970   (e) −970

17. What is the enthalpy of solution of $MgCl_2$, as calculated from the lattice enthalpy (2524 kJ·mol^{-1}) and individual ion hydration enthalpies (Mg^{2+}, −2003 kJ·mol^{-1}; Cl^-, −340 kJ·mol^{-1})? All answers are in kJ·mol^{-1}.

(a) +181   (b) −181   (c) +4867   (d) −159   (e) −4867

18. Calculate the individual ion hydration enthalpy of Li^+ from the lattice enthalpy of LiCl (861 kJ·mol^{-1}), the enthalpy of solution of LiCl (−37.0 kJ·mol^{-1}) and the individual ion enthalpy of Cl^- (−340 kJ·mol^{-1}). All answers are in kJ·mol^{-1}.

(a) +484   (b) −484   (c) +558   (d) −558   (e) +1164

19. Which ion is expected to have the highest (most negative) individual ion hydration enthalpy?

(a) Na^+ (b) Al^{3+} (c) Se^{2-} (d) Br^- (e) Au^+

20. An ionic solid with a _____ lattice energy and _____ hydration energy is likely to have an exothermic enthalpy of solution.

(a) large, large   (b) small, small   (c) large, small   (d) small, large

21. Which of the following describes a solution process that is certain to occur naturally?

(a) Energy becomes disordered; molecules become disordered.

(b) Energy becomes disordered; molecules become ordered.

(c) Energy becomes ordered; molecules become disordered.

(d) Energy becomes ordered; molecules become ordered.

22. When a solid solute dissolves in a solvent, the molecules become more

(a) ordered   (b) disordered

Colligative properties

23. How many grams of $Ca(NO_3)_2$ are required to prepare 250 mL of a 0.12 m solution? Assume the density of the solution is $1.00 \text{ g} \cdot \text{mL}^{-1}$.

(a) 4.9 g   (b) 41 g   (c) 0.18 g   (d) 5.5 g   (e) 20 g

24. What is the molality of a solution of KF in water if the mole fraction of KF is 0.011? All answers are in molal units (m, $\text{mol} \cdot \text{kg}^{-1}$).

(a) 0.33   (b) 0.84   (c) 0.61   (d) 2.2   (e) 0.11

25. What is the molarity of a 4.38 m solution of $CaCl_2$ in water? The density of the solution is $1.35 \text{ g} \cdot \text{mL}^{-1}$. All answers are in molarity units (M, $\text{mol} \cdot \text{L}^{-1}$).

(a) 4.04   (b) 3.24   (c) 5.91   (d) 3.97   (e) 4.38

26. Which of the following solutions is likely to show the largest effect on colligative properties?

(a) 0.1 m $C_6H_{12}O_6$ (b) 0.1 m NaCl

(c) 0.1 m $CaCl_2$ (d) 0.1 m $(NH_4)_3PO_4$

(e) none of these—all have the same effect

27. When 2.0 mol of ethylene glycol, a molecular solvent, is dissolved in a certain mass of water, the freezing point of the resulting solution is $-0.77°C$. What is the freezing point of the solution made by adding 2.0 mol $Mg(NO_3)_2$ to the same mass of water?

(a) 0.0°C   (b) $-0.77°C$ (c) $-0.51°C$ (d) $-1.5°C$ (e) $-2.3°C$

28. What is the vapor pressure, at 30°C, of an aqueous solution of hexose in water in which $x_{\text{hexose}} = 0.16$? The vapor pressure of pure water at 30°C is 31.8 Torr.

(a) 5.1 Torr   (b) 17 Torr   (c) 38 Torr   (d) 32 Torr   (e) 27 Torr

29. What is the molarity of the solution made by adding enough water to 85.0 g $HC_2H_3O_2$ to make 600 mL solution?

(a) 0.236 M   (b) 2.36 M   (c) 1.11 M   (d) 0.142 M   (e) 0.813 M

30. How many grams of Na_2S_2 are present in 10.0 mL of a 0.20 M solution?

(a) 110 g   (b) 2.0×10^{-3} g   (c) 0.22 g

(d) 220 g   (e) 2.2×10^{-3} g

31. How many milliliters of 3.0 M HCl are required to obtain 50 mmol of HCl?

(a) 60   (b) 15   (c) 170   (d) 150   (e) 17

32. 200 g glycerol ($C_3H_8O_3$) is added to 100 g water. What is the mole fraction of glycerol?

(a) 1.00   (b) 0.521   (c) 0.718   (d) 3.56   (e) 0.281

33. What is the mole fraction of iodine (I_2) in a solution of I_2 in CCl_4 in which the mole fraction of CCl_4 is 0.28.

(a) 1.28   (b) 0.28   (c) 0.56   (d) 0.72   (e) 1.72

34. What is the molality of the solution made by adding 30.0 g $Ca(NO_3)_2$ to 250 g water?

(a) 0.732 m (b) 0.183 m (c) 0.0458 m

(d) 0.120 m (e) 0.618 m

35. What is the molality of the solution made by dissolving 125 g Br_2 in 750 g $CHCl_3$?

(a) 0.0938 m (b) 0.167 m (c) 0.666 m (d) 0.782 m (e) 1.04 m

36. How many grams of KBr should be added to 100 g water to prepare a 0.250 m solution?

(a) 4.00   (b) 29.8   (c) 0.336   (d) 2.98   (e) 25.0

37. What is the molality of a solution of cysteamine (C_2H_7NS) in propanol (C_3H_8O) if the mole fraction cysteamine is 0.110?

(a) 1.10 m (b) 3.21 m (c) 2.57 m (d) 0.0840 m (e) 2.06 m

38. What is the molarity of a 2.00 m aqueous KOH solution? The density of the solution is 1.18 $g \cdot mL^{-1}$. Use 1.00 $g \cdot mL^{-1}$ for the density of water.

(a) 1.91 M   (b) 1.52 M   (c) 1.63 M   (d) 2.00 M   (e) 2.12 M

39. What is the vapor pressure, at 25°C, of the solution made by adding 100 g P_4 to 500 g $CHCl_3$. The vapor pressure of pure $CHCl_3$ at 25°C is 182 Torr. Assume ideal solution behavior.

(a) 153 Torr   (b) 29.4 Torr   (c) 146 Torr

(d) 36.4 Torr   (e) 182 Torr

40. What is the freezing point of a 0.050 m solution of LiCl in H_2O?

(a) −0.19°C   (b) −0.093°C   (c) +0.093°C

(d) +0.19°C   (e) 0.00°C

41. What is the freezing point of a 5.00-g sample of camphor ($C_{10}H_{16}O$) that has been contaminated with 0.025 g naphthalene ($C_{10}H_8$)? Pure camphor freezes at 179.8°C; $k_f = 39.7$ $K \cdot kg \cdot mol^{-1}$ for camphor.

(a) 178.3°C   (b) 179.8°C   (c) 181.4°C   (d) 180.8°C   (e) 178.8°C

42. What is the osmotic pressure, at 25°C, of a 4.4×10^{-3} M solution of sodium stearate in water? Sodium stearate ionizes into sodium ion and stearate ion when it dissolves.

(a) 0.018 atm   (b) 0.0090 atm   (c) 0.22 atm

(d) 0.11 atm   (e) 0.16 atm

43. What is the osmotic pressure at 37°C of the solution prepared by dissolving 310 mg of a large nonionizing biomolecule (molar mass = 4.0×10^5 $g \cdot mol^{-1}$) in enough water to make 5.0 mL of solution?

(a) 0.36 Torr   (b) 5.6 Torr   (c) 3.0 Torr   (d) 2.8 Torr   (e) 6.0 Torr

44. An aqueous solution of a nonionizing solute with molar mass 161 $g \cdot mol^{-1}$ has a vapor pressure of 22.6 Torr at 25°C. What is the freezing point of the solution? The vapor pressure of pure water at 25°C is 23.8 Torr.

(a) +5.4°C   (b) −5.4°C   (c) +10.8°C   (d) −10.8°C   (e) 0.0°C

45. When 50.0 mg of an unknown compound is dissolved in 12.0 g of melted camphor, the camphor freezes at 177.5°C. What is the molar mass of the unknown? The freezing point of pure camphor is 179.8°C; $k_f = 39.7 \text{ K} \cdot (\text{kg} \cdot \text{mol})^{-1}$.

(a) $60 \text{ g} \cdot \text{mol}^{-1}$ (b) $5.5 \times 10^2 \text{ g} \cdot \text{mol}^{-1}$ (c) $1.7 \times 10^2 \text{ g} \cdot \text{mol}^{-1}$
(d) $1.5 \times 10^2 \text{ g} \cdot \text{mol}^{-1}$ (e) $72 \text{ g} \cdot \text{mol}^{-1}$

46. When 21.2 mg of a nonionizing unknown is dissolved in enough water to make 10.0 mL of solution at 25°C, the solution develops an osmotic pressure of 8.20 Torr. What is the molar mass of the unknown?

(a) $4.03 \times 10^3 \text{ g} \cdot \text{mol}^{-1}$ (b) $813 \text{ g} \cdot \text{mol}^{-1}$ (c) $4.80 \times 10^3 \text{ g} \cdot \text{mol}^{-1}$
(d) $3.61 \times 10^3 \text{ g} \cdot \text{mol}^{-1}$ (e) $1.55 \times 10^3 \text{ g} \cdot \text{mol}^{-1}$

47. 150 g of two nonionizing compounds, one with molar mass $161 \text{ g} \cdot \text{mol}^{-1}$ and one with molar mass $1052 \text{ g} \cdot \text{mol}^{-1}$ are each separately dissolved in enough water to make 1.0 L of solution. Which one has the largest effect on colligative properties, that is, causes the greatest vapor-pressure lowering, boiling-point elevation, freezing-point depression, and osmotic pressure?

(a) compound with molar mass = $161 \text{ g} \cdot \text{mol}^{-1}$

(b) compound with molar mass = $1052 \text{ g} \cdot \text{mol}^{-1}$

(c) neither, because equal masses of each are used

Descriptive chemistry

48. Which of the following is the form in which phosphorus is included in most commercial fertilizers?

(a) $Ca_3(PO_4)_2$ (b) P_4 (c) P_2O_5 (d) Na_3PO_4 (e) $Ca(HPO_4)_2$

49. The presence of which two ions makes water "hard"?

(a) Na^+, K^+ (b) Ca^{2+}, Mg^{2+} (c) Cl^-, HCO_3^-
(d) F^-, Cl^- (e) Na^+, Cl^-

50. What is the chemical composition of boiler scale?

(a) $Ca(HCO_3)_2$ (b) Na_2CO_3 (c) $CaCO_3$
(d) $NaCl$ (e) $Na(HCO_3)$

51. What is the formula of the scum formed by soap in hard water?

(a) $Ca_3(PO_4)_2$ (b) $Ca(C_{17}H_{35}CO_2)_2$ (c) $CaCO_3$
(d) $Ca(NO_3)_2$ (e) $CaHPO_4$

52. Why is salt placed on snowy roads in the winter? Remember that snow is frozen water.

(a) lower the melting point of the water (b) increase the boiling point of the water
(c) increase the outside temperature (d) increase the osmotic pressure of the water
(e) increase the vapor pressure of the water

53. What are the product(s) from the reaction of carbon dioxide with water?

$$CO_2(g) + H_2O(l) \longrightarrow \, ?$$

(a) $O_2 + H_2 + C$ (b) $CO + H_2O_2$ (c) $O_2 + CH_4$
(d) H_2CO_3 (e) CH_3OH

CHAPTER 13
CHEMICAL EQUILIBRIUM

A chemical equilibrium is a dynamic equilibrium. It is, first, an equilibrium because there are no observable changes in concentrations as time evolves. It is, second, dynamic because underlying the equilibrium there is a process, a "dynamic motion," in which the forward rate and reverse rate cancel out each other's effect. One insightful way to describe a dynamic equilibrium is, "Whatever is done is undone."

EQUILIBRIUM AND COMPOSITION

13.1 THE REVERSIBILITY OF CHEMICAL REACTIONS

KEY CONCEPT   Reactions at equilibrium

Imagine a chemical reaction in which the reactants are placed in a reaction vessel and allowed to form products:

$$\text{Reactants} \longrightarrow \text{products}$$

For many reactions, the products can also back-react to form reactants:

$$\text{Products} \longrightarrow \text{reactants}$$

At some point, the rate of formation of products equals the rate of formation of reactants, and a **dynamic equilibrium** is attained. At this point, there is no *net* formation of reactants or products, so the concentration of each is constant. Such an equilibrium is represented by double arrows $\rightleftharpoons$.

$$\text{Reactants} \rightleftharpoons \text{products}$$

EXAMPLE   Understanding equilibrium

Equilibrium exists at a certain time for the following reaction, with $[SO_2] = 0.50$, $[O_2] = 0.60$, and $[SO_3] = 4.11 \times 10^{-2}$. If no changes are made in the experimental conditions, how do the concentrations of each substance change with time?

$$2SO_2(g) + O_2(g) \rightleftharpoons 2SO_3(g)$$

SOLUTION The concentrations do not change with time once equilibrium has been established.

EXERCISE In the preceding equilibrium, assume that we set up an experiment in which, initially, there is only radioactive oxygen in the O_2 molecules. After a relatively long time, will all the radioactive oxygen still be only in O_2 molecules?

[*Answer:* No. The equilibrium is dynamic, with a constant exchange of atoms among the substances in the equilibrium; so the radioactive oxygen atoms will be distributed among all three oxygen-containing species, SO_2, O_2, and SO_3.]

13.2 THE EQUILIBRIUM CONSTANT

KEY CONCEPT A   An equilibrium constant can be written for any reaction

Experimentation has shown that, at equilibrium, the ratio of products to reactants is a constant at any temperature. More specifically, for the equilibrium

$$a\text{A} + b\text{B} \rightleftharpoons c\text{C} + d\text{D}$$

the concentrations of reactants and products obey the relationship

$$K_c = \frac{[\text{C}]^c [\text{D}]^d}{[\text{A}]^a [\text{B}]^b}$$

K_c is a constant at any temperature and is called the **equilibrium constant**. The concentrations used in the calculation of K_c must be the equilibrium concentrations,

EXAMPLE 1   Calculating an equilibrium constant

For the reaction of an alcohol with an acid at 50°C, the concentration (in $\text{mol} \cdot \text{L}^{-1}$) of each substance at equilibrium is [acid] = 2.2, [alcohol] 3.6, [ester] = 5.2, and [H_2O] = 3.8. What is K_c for this equilibrium?

$$\text{Acid} + \text{alcohol} \rightleftharpoons \text{ester} + H_2O$$

SOLUTION The equilibrium constant for this equilibrium is

$$K_c = \frac{[\text{ester}]^1 [H_2O]^1}{[\text{acid}]^1 [\text{alcohol}]^1}$$

We have been careful to remember that the concentration of each substance in the equilibrium must be raised to the power given by its coefficient in the chemical equation. In this case, each coefficient is 1. (In the future, powers of 1 will not be written.) Substituting each of the equilibrium values into the equation gives

$$K_c = \frac{5.2 \times 3.8}{2.2 \times 3.6}$$

$$= 2.5$$

EXERCISE Calculate K_c for the equilibrium shown if the equilibrium concentrations of gases are [CO_2] = 0.40, [H_2] = 0.18, [CO] = 0.30, and [H_2O] = 0.42.

$$CO_2(g) + H_2(g) \rightleftharpoons CO(g) + H_2O(g)$$

[*Answer:* 1.8]

EXAMPLE 2   Calculating an equilibrium constant

What is K_c for the equilibrium shown if the equilibrium concentrations of each substance are [SO_2] = 0.68, [O_2] = 0.88, and [SO_3] = 2.3.

$$2SO_2(g) + O_2(g) \rightleftharpoons 2SO_3(g)$$

SOLUTION For this equilibrium,

$$K_c = \frac{[SO_3]^2}{[SO_2]^2 [O_2]}$$

Substituting in the equilibrium concentrations gives

$$K_c = \frac{(2.3)^2}{(0.68)^2(0.88)}$$

$$= 13$$

EXERCISE What is K_c for the equilibrium shown if the equilibrium concentrations of each of the substances in the equilibrium are $[NO_2] = 0.016$, $[O_2] = 0.13$, and $[NO] = 0.26$.

$$2NO_2(g) \rightleftharpoons O_2(g) + 2NO(g)$$

[*Answer:* 34]

KEY CONCEPT B   How you write K_c depends on the chemical equation

The value of K_c for an equilibrium depends on how we choose to write the chemical equation for the equilibrium. If we represent an equilibrium as $aA \rightleftharpoons bB$, we get the following:

Form of Equilibrium Used	Equilibrium Constant
$aA \rightleftharpoons bB$	equilibrium constant $= K_c$
$bB \rightleftharpoons aA$	equilibrium constant $= \dfrac{1}{K_c}$
$naA \rightleftharpoons nbB$	equilibrium constant $= (K_c)^n$

The last equilibrium in the table represents the original equilibrium multiplied by a number n. This number may be a fraction such as $\frac{1}{2}$. In using the relationship, equilibrium constant $= (K_c)^n$, with fractional values of n, we should remember that

$$(K_c)^{1/z} = \sqrt[z]{K_c}$$

EXAMPLE   Determining the values of the equilibrium constant for different forms of a chemical equation

An equilibrium equation and the associated equilibrium constant are

$$2ICl(g) \rightleftharpoons I_2(g) + Cl_2(g) \qquad K_c = 0.10$$

What is the equilibrium constant for the reverse equilibrium.

$$I_2(g) + Cl_2(g) \rightleftharpoons 2ICl(g) \qquad K_c = ?$$

SOLUTION The relationship between the equilibrium constant for a forward reaction, $K_c(\text{forward})$, and a reverse reaction, $K_c(\text{reverse})$, is

$$K_c(\text{reverse}) = \frac{1}{K_c(\text{forward})}$$

Thus, for our problem,

$$K_c(\text{forward}) = \frac{1}{0.10}$$

$$= 10$$

EXERCISE What is the equilibrium constant for the equation

$$\frac{1}{2}I_2(g) + \frac{1}{2}Cl_2(g) \rightleftharpoons ICl(g)$$

[*Answer:* 3.2]

13.3   HETEROGENEOUS EQUILIBRIA

KEY CONCEPT   The concentration of a pure solid or liquid is not in K_c

Because the concentration of a **pure solid** or **pure liquid** is constant, the equilibrium expression for an equilibrium involving a pure liquid or pure solid does not explicitly contain the concentration of either. The constant concentration instead is collapsed into the equilibrium constant. Thus, the equilibrium constants reported in tables and books are based on the assumption that the concentrations of pure liquids and pure solids do not appear in the equilibrium expression. For instance, the equilibrium expression for the following equilibrium is

$$CuCl_2 \cdot 5H_2O(s) \rightleftharpoons CuCl_2(s) + 5H_2O(g)$$

$$K_c = [H_2O]^5$$

> **EXAMPLE   Writing the equilibrium constant for a heterogeneous equilibrium**
>
> Write the equilibrium expression for K_c for the equilibrium
>
> $$S(s) + O_2(g) \rightleftharpoons SO_2(g)$$
>
> **SOLUTION** This is a heterogeneous equilibrium because both a solid and gases are present. The concentration of the sulfur, which is a pure solid, does not appear in the equilibrium expression. Thus,
>
> $$K_c = \frac{[SO_2]}{[O_2]}$$
>
> **EXERCISE** Write the equilibrium expression for K_c for the equilibrium
>
> $$Al_2O_3(s) + 3C(s) + 3Cl_2(g) \rightleftharpoons 2AlCl_3(s) + 3CO(g)$$
>
> *Answer:* $K_c = \dfrac{[CO]^3}{[Cl_2]^3}$

13.4   GASEOUS EQUILIBRIA

KEY CONCEPT   Gas concentrations in K_p are expressed as partial pressures

For equilibria that involve gases but not dissolved solutes, it is possible to write an equilibrium constant K_p that uses the partial pressures of the gases rather than their concentrations. For instance, for the equilibrium

$$2NO_2(g) \rightleftharpoons 2NO(g) + O_2(g)$$

$$K_p = \frac{(P_{NO})^2 P_{O_2}}{(P_{NO_2})^2}$$

The equilibrium constant K_c for this equilibrium is

$$K_c = \frac{[NO]^2[O_2]}{[NO_2]^2}$$

It is important to understand that K_p and K_c describe the same equilibrium and that the amounts of reactants and products present do not depend on which equilibrium constant we choose to use. The two equilibrium constants are related to each other by the relation

$$K_p = K_c(RT)^{\Delta n}$$

Δn = moles gaseous product in the equation − moles gaseous reactant in the equation

In our example, there are 3 mol of products and 2 mol of reactants, so $\Delta n = +1$. Δn may be negative. If the moles of product are the same as the moles of reactant, then $\Delta n = 0$ and $K_c = K_p$. When we use $R = 0.0821\ \text{L}\cdot\text{atm}\cdot\text{mol}^{-1}\cdot\text{K}^{-1}$, the partial pressures must be expressed in atmospheres.

EXAMPLE 1   Relating K_p to K_c

For the equilibrium shown, at 650 K, $K_c = 4.0 \times 10^5$. What is K_p?

$$2NO_2(g) \rightleftharpoons 2NO(g) + O_2(g)$$

SOLUTION $\Delta n = -1$ (2 − 3 = −1) for this equilibrium. Substituting this result, $R = 0.0821\ \text{L}\cdot\text{atm}\cdot\text{mol}^{-1}\cdot\text{K}^{-1}$, and $T = 650$ K into the relationship between K_p and K_c gives

$$K_p = K_c(RT)^{\Delta n}$$
$$= 4.0 \times 10^5 \times (0.0821 \times 650)^1$$
$$= 2.1 \times 10^7$$

EXERCISE At 250°C, $K_c = 3.2 \times 10^3$ for the equilibrium shown in the example. What is K_p?

[*Answer:* 75]

EXAMPLE 2   Relating K_p to K_c

At 500 K, $K_p = 2.5 \times 10^{10}$ for the equilibrium shown. What is K_c at this temperature?

$$2SO_2(g) + O_2(g) \rightleftharpoons 2SO_3(g)$$

SOLUTION $\Delta n = -1$ (2 − 3 = −1) for this equilibrium. Substituting this result, $R = 0.0821\ \text{L}\cdot\text{atm}\cdot\text{mol}^{-1}\cdot\text{K}^{-1}$, and $T = 500$ K into the relationship between K_p and K_c gives

$$K_p = K_c(RT)^{\Delta n}$$
$$2.5 \times 10^{10} = K_c(0.0821 \times 500)^{-1}$$
$$= \frac{K_c}{(0.0821 \times 500)}$$
$$= \frac{K_c}{41.1}$$
$$K_c = 2.5 \times 10^{10} \times 41.1$$
$$= 1.0 \times 10^{12}$$

EXERCISE At 700 K, $K_p = 54$ for the equilibrium shown. What is K_c?

$$H_2(g) + I_2(g) \rightleftharpoons 2HI(g)$$

[*Answer:* 54]

KEY WORDS   Define or explain each term in a written sentence or two.

chemical equilibrium   heterogeneous equilibria
dynamic equilibrium   homogeneous equilibria
equilibrium constant   law of mass action

USING EQUILIBRIUM CONSTANTS

13.5 THE EXTENT OF REACTION

KEY CONCEPT A   The size of *K* determines the amount of products and reactants

The size of K_c tells us whether an equilibrium favors reactants, or products, or neither. Because

$$K_c = \frac{\text{concentration of products}}{\text{concentration of reactant}}$$

a large K_c indicates that the concentration of products is larger than the concentration of reactants at equilibrium. On the other hand, a small K_c indicates that the concentration of reactants is larger than that of products at equilibrium. If K_c is close to one, the concentrations of products and reactants are similar at equilibrium.

$K_c > 1000$	concentration of products > concentration of reactants
$1000 > K_c > 0.001$	concentration of products ≈ concentration of reactants
$K_c < 0.001$	concentration of products < concentration of reactants

EXAMPLE   Interpreting an equilibrium constant qualitatively

$K_c = 0.10$ for the equilibrium $2ICl(g) \rightleftharpoons I_2(g) + Cl_2(g)$. Without doing any calculations, decide which set of equilibrium values is correct for the equilibrium:

Reagent	Set 1	Set 2	Set 3
ICl	0.14 M	0.88 M	6.3×10^{-5} M
Cl_2	6.3×10^{-5} M	0.23 M	0.24 M
I_2	7.3×10^{-5} M	0.34 M	0.44 M

SOLUTION Because the value of K_c is less than 1000 and more than 0.001, we predict that the concentration of products and reactants will be similar to each other at equilibrium. Set 2 is the set of concentrations that most closely meets this criterion.

EXERCISE $K_c = 4.7 \times 10^4$ for the equilibrium $CO(g) + H_2O(g) \rightleftharpoons CO_2(g) + H_2(g)$. Would you expect the concentration of CO or CO_2 to be higher at equilibrium?

[*Answer:* CO_2]

PITFALL   Using K_c to estimate the relative amounts of products and reactants

Estimates of the relative amounts of product and reactant based on the size of K_c assume that if more than one product or more than one reactant is present, the concentrations of all the products are close to each other, and/or the concentrations of all the reactants are similar. If this is not true, such estimates can be wrong.

KEY CONCEPT B   Calculations with one unknown concentration

For some calculations, the equilibrium constant and all but one of the concentrations are known. Substitution of the data into the equilibrium expression leads to one equation with one unknown, so we can solve for the unknown concentration in a straightforward manner.

EXAMPLE 1   Solving for an unknown concentration or partial pressure

$K_p = 160$ for the equilibrium shown. What is the partial pressure of H_2 if the partial pressure of I_2 is 0.35 atm and that of HI is 6.3 atm?

$$H_2(g) + I_2(g) \rightleftharpoons 2HI(g)$$

SOLUTION The equilibrium constant expression for this equilibrium is

$$K_p = \frac{(P_{HI})^2}{(P_{H_2})(P_{I_2})}$$

Substituting the values $P_{HI} = 6.3$, $P_{I_2} = 0.35$, and $K_p = 160$ gives

$$160 = \frac{(6.3)^2}{P_{H_2}(0.35)}$$

$$P_{H_2} = \frac{39.7}{(0.35)(160)}$$

$$= 0.71 \text{ atm}$$

It is prudent to always check your results by substituting the equilibrium values obtained into the equilibrium constant expression to see if you get the correct value of the equilibrium constant:

$$\frac{(6.3)^2}{(0.71)(0.35)} = 160$$

EXERCISE $K_c = 0.10$ for the equilibrium shown. What is the concentration of I_2 when [ICl] = 0.50 and [Cl_2] = 0.063.

$$2ICl(g) \rightleftharpoons Cl_2(g) + I_2(g)$$

[*Answer:* 0.40 M]

EXAMPLE 2   Solving for an unknown concentration

$K_c = 4.4 \times 10^4$ for the equilibrium given. What is the molarity of H_2 at equilibrium when the molarity of N_2 is 0.11 M and that of NH_3 is 1.5 M?

$$N_2(g) + 3H_2(g) \rightleftharpoons 2NH_3(g)$$

SOLUTION The equilibrium expression for the equilibrium is

$$K_c = \frac{[NH_3]^2}{[N_2][H_2]^3}$$

Substitution of the given quantities $K_c = 4.4 \times 10^4$, [N_2] = 0.11, and [NH_3] = 1.5 gives

$$4.4 \times 10^4 = \frac{(1.5)^2}{(0.11)[H_2]^3}$$

$$[H_2]^3 = \frac{2.3}{(4.4 \times 10^4)(0.11)}$$

$$= 4.8 \times 10^{-4}$$

$$[H_2] = (4.8 \times 10^{-4})^{1/3}$$

Remember that raising a quantity to the $\frac{1}{3}$ power is the same as taking the cube root of the quantity. To accomplish this on a calculator that uses algebraic notation,

Enter 4.8×10^{-4}

Press Y^x (this may be X^y on some calculators)

CHEMICAL EQUILIBRIUM

Enter 0.33333333

Press =

Display shows 7.8297 − 02

This is your calculator's way of indicating 7.8297×10^{-2}. The correct answer to the problem is 7.8×10^{-2} M. Substitution of these values into the equilibrium expression gives the correct value for the equilibrium constant:

$$\frac{(1.5)^2}{(0.11)(7.8 \times 10^{-2})^3} = 4.3 \times 10^4$$

EXERCISE For the equilibrium in the example, what is $[NH_3]$ at equilibrium when $[H_2] = 6.1 \times 10^{-3}$ and $[N_2] = 0.15$?

[*Answer:* 0.039]

13.6 THE DIRECTION OF REACTION

KEY CONCEPT The size of Q determines the direction of reaction

Imagine we have a flask, containing NH_3, N_2, and H_2, in which equilibrium has not yet been attained:

$$2NH_3(g) \rightleftharpoons N_2(g) + 3H_2(g) \qquad K_c = 7.4 \times 10^{-4}$$

For example, assume $[NH_3] = 0.51$, $[N_2] = 0.040$, and $[H_2] = 0.12$. It is possible to calculate a quantity Q_c that symbolically resembles the equilibrium constant but is not numerically equal to the equilibrium constant because equilibrium concentrations are not present:

$$Q_c = \frac{[N_2][H_2]^3}{[NH_3]^2}$$

$$= \frac{(0.040)(0.12)^3}{(0.51)^2}$$

$$= 2.7 \times 10^{-4}$$

Note that Q_c (2.7×10^{-4}) is not equal to K_c (7.4×10^{-4}), a confirmation that equilibrium concentrations are not present. Q_c is called the **reaction quotient**. It tells us in which direction the reaction must occur in order to reach equilibrium.

When $Q_c > K_c$, reactants must form from products in order to reach equilibrium.

When $Q_c = K_c$, the reaction is already at equilibrium.

When $Q_c < K_c$, products must form from reactants to reach equilibrium.

The value of the reaction quotient tells us only what must occur for equilibrium to be reached. It does not give information regarding how fast the equilibrium will be attained, but it does tell us whether reactants or products have a tendency to form. For our example, $Q_c < K_c$, so some NH_3 must decompose into N_2 and H_2 to attain equilibrium.

EXAMPLE Predicting the direction of reaction

$K_c = 160$ at 500 K for the following equilibrium. If 0.20 M HI, 0.10 M H_2, and 0.10 M I_2 are introduced into a flask and the temperature increased to 500 K, in what direction must the reaction proceed to reach equilibrium?

$$H_2(g) + I_2(g) \rightleftharpoons 2HI(g)$$

SOLUTION We must calculate Q_c in order to answer this problem.

$$Q_c = \frac{[HI]^2}{[H_2][I_2]}$$

$$= \frac{(0.20)^2}{(0.10)(0.10)}$$

$$= 4.0$$

Because $Q_c < K_c$, some H_2 and I_2 must react to form HI to attain equilibrium.

EXERCISE $K_c = 0.050$ at 2200°C for the following equilibrium; 0.50 M of each gas in the equilibrium is introduced into a flask and the temperature increased to 2200°C. Calculate Q_c and state what must occur for equilibrium to be attained.

$$N_2(g) + O_2(g) \rightleftharpoons 2NO(g)$$

[*Answer:* $Q_c = 1$, and some NO must decompose into N_2 and O_2 to attain equilibrium.]

13.7   EQUILIBRIUM TABLES

KEY CONCEPT A   Calculating equilibrium constants from concentrations

When the concentrations of products and reactants at equilibrium are known or can be calculated in a straightforward way, the equilibrium constant can be calculated. In the two examples that follow, we determine the equilibrium concentrations in two ways:

1. The equilibrium concentrations of all substances are measured directly.
2. The equilibrium concentrations are calculated from initial known concentrations and the measured equilibrium concentration of one substance.

EXAMPLE 1   Calculating the equilibrium constant, method 1

The concentrations of the substances in the following reaction are measured at equilibrium and give $[Br_2] = 0.17$, $[Cl_2] = 0.17$, and $[BrCl] = 0.030$.

$$2BrCl(g) \rightleftharpoons Cl_2(g) + Br_2(g)$$

What is K_c for the equilibrium?

SOLUTION We first write the expression for the equilibrium constant to determine what information is required.

$$K_c = \frac{[Cl_2][Br_2]}{[BrCl]^2}$$

Because we have the equilibrium concentration of each substance in the equilibrium expression, we can substitute the given values into it to calculate K_c:

$$K_c = \frac{(0.17)(0.17)}{(0.030)^2}$$

$$= 32$$

EXERCISE The concentrations of the substances in the following reaction are measured at equilibrium and give $[SO_2] = 0.025$, $[Cl_2] = 0.025$, and $[SO_2Cl_2] = 4.1 \times 10^{-4}$. What is the value of K_c?

$$SO_2(g) + Cl_2(g) \rightleftharpoons SO_2Cl_2(g)$$

[*Answer:* 0.66]

EXAMPLE 2 Calculating the equilibrium constant, method 2

A sample of HI(g) is placed in a flask at a pressure of 0.10 atm. After the system comes to equilibrium, the partial pressure of HI(g) is 0.050 atm. What is K_p for the equilibrium?

$$2HI(g) \rightleftharpoons H_2(g) + I_2(g) \qquad K_p = ?$$

SOLUTION We solve this by means of a table showing the partial pressures of each substance before and after equilibrium. Also included in the table is the change in partial pressure of each substance expressed as the variable x.

	2HI(g)	$\rightleftharpoons$ H$_2$(g)	+ I$_2$(g)
before equilibrium	0.10 atm	0 atm	0 atm
change to reach equilibrium	$-2x$	$+x$	$+x$
equilibrium values	$0.10 - 2x$	x	x

Note that the values of x used in the row "change to reach equilibrium" must agree with the stoichiometry of the equilibrium; that is, the 2 in front of HI requires the change in the partial pressure of HI to be $2x$. In this problem, we know the equilibrium partial pressure of HI is 0.050 atm. This allows us to solve for x.

$$0.10 \text{ atm} - 2x = 0.050 \text{ atm}$$

$$-2x = 0.050 \text{ atm} - 0.10 \text{ atm}$$

$$x = 0.025 \text{ atm}$$

Because x is the equilibrium partial pressure of H$_2$ and I$_2$, we can calculate K_p. That is, $P_{H_2} = P_{I_2} = x = 0.025$ atm, and from the problem, $P_{HI} = 0.050$ atm; thus,

$$K_p = \frac{P_{H_2} \times P_{I_2}}{(P_{HI})^2}$$

$$= \frac{0.025 \times 0.025}{(0.050)^2}$$

$$= 0.25$$

EXERCISE A sample of CO(g) is placed in a flask at a pressure of 0.200 atm. After the system comes to equilibrium, the partial pressure of CO(g) is 0.190 atm. What is K_p for the equilibrium?

$$2CO(g) \rightleftharpoons C(s) + CO_2(g)$$

[*Answer:* 0.139]

KEY CONCEPT B   Equilibrium calculation for the decomposition of a single substance

When a single substance decomposes, the concentrations (or partial pressures) of each of the substances in the equilibrium can be determined from the stoichiometry of the chemical equation. The concentrations are expressed in terms of an unknown x, and x is solved for by using the equilibrium expression.

EXAMPLE 1 Calculating equilibrium concentrations when a single substance decomposes

$K_c = 0.10$ for the equilibrium shown. What are the equilibrium concentrations of each of the substances in the equilibrium when ICl, initially at 0.75 M, is allowed to come to equilibrium?

$$2ICl(g) \rightleftharpoons I_2(g) + Cl_2(g)$$

SOLUTION The first task is to set up a table showing the initial concentrations, the changes necessary to attain equilibrium, and the equilibrium concentrations. The changes and the equilibrium concentrations are expressed in terms of an unknown x, with x defined so that $2x$ is the amount of ICl that reacts to reach equilibrium.

	2ICl	$\rightleftharpoons$ I$_2$	+ Cl$_2$
initial concentration	0.75 M	0 M	0 M
change to reach equilibrium	$-2x$	$+x$	$+x$
equilibrium concentration	0.75 M $- 2x$	x	x

Note that the changes follow the stoichiometry of the balanced equation. That is, where $2x$ M ICl decomposes, x M I$_2$ and x M Cl$_2$ form. We next substitute the equilibrium concentrations into the equilibrium expression:

$$K_c = \frac{[I_2][Cl_2]}{[ICl]^2}$$

$$0.10 = \frac{(x)(x)}{(0.75 - 2x)^2}$$

$$= \frac{x^2}{(0.75 - 2x)^2}$$

We now take the square root of both sides of the equation:

$$0.32 = \frac{x}{0.75 - 2x}$$

$$0.32(0.75 - 2x) = x$$

$$x = 0.24 - 0.64x$$

$$1.64x = 0.24$$

$$x = 0.15$$

From the value of x, we can calculate the equilibrium concentrations of each of the substances in the equilibrium:

$$[ICl] = 0.75 - 2x = 0.45$$

$$[Cl_2] = [I_2] = x = 0.15$$

Substitution of these values into the equilibrium expression gives the correct value for the equilibrium constant:

$$\frac{(0.15)(0.15)}{(0.45)^2} = 0.11$$

EXERCISE $K_c = 1.9 \times 10^{-2}$ for the equilibrium shown. What are the equilibrium concentrations (in $mol \cdot L^{-1}$) of each of the substances in the equilibrium if HI, initially at 0.35 M, is allowed to come to equilibrium?

$$2HI(g) \rightleftharpoons H_2(g) + I_2(g)$$

[*Answer:* HI, 0.31 M; H$_2$, 0.042 M; I$_2$, 0.042 M]

EXAMPLE 2   Calculating equilibrium concentrations when a single substance decomposes

$K_p = 0.050$ at 900 K for the equilibrium shown. What is the equilibrium concentration of C$_2$H$_6$ when 2.0 atm of C$_2$H$_6$ is placed in a flask and allowed to come to equilibrium?

$$C_2H_6(g) \rightleftharpoons C_2H_4(g) + H_2(g)$$

SOLUTION We first set up a table showing the partial pressures of all the substances involved before and after equilibrium.

	$C_2H_6(g)$ $\rightleftharpoons$ $C_2H_4(g)$ + $H_2(g)$		
before equilibrium	2.0 atm	0 atm	0 atm
change to reach equilibrium	$-x$	$+x$	$+x$
equilibrium values	$2.0 - x$	x	x

We next set up the equilibrium expression and substitute the equilibrium partial pressures, written in terms of the unknown x, into the expression.

$$K_p = \frac{P_{C_2H_4} \times P_{H_2}}{P_{C_2H_6}}$$

$$= \frac{x^2}{2.0 - x}$$

We now solve for x by using the quadratic equation. We first must get the equation into the form $ax^2 + bx + c = 0$.

$$\frac{x^2}{2.0 - x} = 0.050$$

$$x^2 = (0.050)(2.0 - x)$$

$$x^2 + 0.050x - 0.10 = 0$$

Thus, $a = 1$, $b = 0.050$, and $c = -0.10$. Substituting these into the quadratic formula gives

$$x = \frac{-b \pm \sqrt{b^2 - 4ac}}{2a}$$

$$= \frac{-0.050 \pm \sqrt{0.0025 - 4(1)(-0.10)}}{2}$$

$$= \frac{-0.050 \pm \sqrt{0.4025}}{2}$$

$$= \frac{-0.050 \pm 0.634}{2}$$

$$x = \frac{-0.050 + 0.634}{2} \qquad x = \frac{-0.050 - 0.634}{2}$$

$$= 0.29 \qquad\qquad = -0.34$$

x represents the partial pressure of two gases; and because a partial pressure cannot be negative, we reject the negative answer as physically meaningless. We next substitute the equilibrium partial pressures of each component into the equilibrium expression to check our answer:

$$\frac{x^2}{2.0 - x} = \frac{(0.29)^2}{2.0 - x}$$

$$= \frac{0.084}{1.71}$$

$$= 0.049$$

This result (0.049) is the value of K_p (if we realize that some rounding error occurs) and shows that we have calculated the correct equilibrium concentrations. To answer the question: $P_{C_2H_6} = 1.71$ atm.

EXERCISE $K_p = 25$ at 500 K for the equilibrium shown. What is the equilibrium partial pressure of PCl_5 if 6.0 atm of PCl_5 are placed in a flask and allowed to come to equilibrium?

$$PCl_5(g) \rightleftharpoons PCl_3(g) + Cl_2(g)$$

[*Answer:* 1.0 atm]

KEY CONCEPT C Equilibrium calculations with arbitrary initial conditions

Solving an equilibrium problem often involves four distinct steps:

Step 1. Write the equilibrium equation and expression for K_c or K_p. If necessary, calculate Q_c or Q_p to decide in which direction the reaction must shift to reach equilibrium.
Step 2. Construct a table, showing the concentrations of components before and after equilibrium. Express the change in concentrations with a single variable x.
Step 3. Use the required mathematical tools to solve for x.
Step 4. Calculate the equilibrium amounts of each component with the value of x obtained.

Step 3 is often the most time consuming because the equation to solve often contains x^2, x^3, or x to a higher power. Thus, more than one root exists for the equation. In most cases the reasonableness of the answer allows us to choose which root to use; for instance, the value of x used cannot result in a negative concentration. In some cases, the different values of x that you find must be checked to see which one results in equilibrium concentrations that give the correct equilibrium constant when substituted in the equilibrium expression.

EXAMPLE 1 Calculating equilibrium concentrations by using the quadratic equation

$K_c = 6.5 \times 10^{-3}$ for the equilibrium given. What are the equilibrium concentrations of the substances in the equilibrium if N_2O_4, initially at 0.40 M, is allowed to come to equilibrium?

$$N_2O_4(g) \rightleftharpoons 2NO_2(g)$$

SOLUTION The first step, as usual, is to set up a table showing the initial concentrations, changes to attain equilibrium, and equilibrium concentrations.

	N_2O_4	$\rightleftharpoons$ $2NO_2$
initial concentration	0.40 M	0 M
change to attain equilibrium	$-x$	$+2x$
equilibrium concentration	0.40 M $- x$	$2x$

These values are substituted into the equilibrium expression to solve for x:

$$K_c = \frac{[NO_2]^2}{[N_2O_4]}$$

$$6.5 \times 10^{-3} = \frac{(2x)^2}{0.40 - x}$$

$$= \frac{4x^2}{0.40 - x}$$

To solve for x, we rearrange the equation into the form $ax^2 + bx + c = 0$ and use the quadratic equation:

$$(6.5 \times 10^{-3})(0.40 - x) = 4x^2$$

$$2.6 \times 10^{-3} - 6.5 \times 10^{-3}x = 4x^2$$

$$4x^2 + 6.5 \times 10^{-3}x - 2.6 \times 10^{-3} = 0$$

This equation shows the desired form, with $a = 4$, $b = 6.5 \times 10^{-3}$, and $c = -2.6 \times 10^{-3}$. We can now use the quadratic equation to solve for x:

$$x = \frac{-b \pm \sqrt{b^2 - 4ac}}{2a}$$

$$= \frac{-6.5 \times 10^{-3} \pm \sqrt{(6.5 \times 10^{-3})^2 - 4(4)(-2.6 \times 10^{-3})}}{2(4)}$$

$$= \frac{-6.5 \times 10^{-3} \pm \sqrt{4.3 \times 10^{-5} + 4.16 \times 10^{-2}}}{8}$$

$$= \frac{-6.5 \times 10^{-3} \pm \sqrt{4.2 \times 10^{-2}}}{8}$$

$$= \frac{-6.5 \times 10^{-3} \pm 0.20}{8}$$

Because $2x$ is the concentration of NO_2, x cannot be a negative number; so we ignore the negative root of the equation:

$$x = \frac{-6.5 \times 10^{-3} + 0.20}{8}$$

$$= 0.025$$

With this result, it is possible to calculate the equilibrium concentrations of NO_2 and N_2O_4:

$$[NO_2] = 2x = 0.050$$

$$[N_2O_4] = 0.40 - x = 0.38$$

Substitution of these values into the equilibrium expression gives the correct value for the equilibrium constant:

$$\frac{(0.050)^2}{(0.38)} = 6.6 \times 10^{-3}$$

EXERCISE $K_c = 6.8 \times 10^{-2}$ for the equilibrium given. What are the equilibrium concentrations of I_2 and I when I_2, initially at 3.5 M, is allowed to come to equilibrium?

$$I_2(g) \rightleftharpoons 2I(g)$$

[*Answer:* I_2, 3.3 M; I, 0.47 M]

EXAMPLE 2 Calculating equilibrium concentrations by using approximations

$K_c = 7.7 \times 10^{-11}$ for the equilibrium shown. What are the equilibrium concentrations of all the substances in the equilibrium when HBr, initially at 0.20 M, and H_2, initially at 0.15 M, are allowed to come to equilibrium?

$$2HBr(g) \rightleftharpoons H_2(g) + Br_2(g)$$

SOLUTION The first step is to set up a table showing the initial concentrations, the changes required to attain equilibrium and the equilibrium concentrations.

	2HBr	$\rightleftharpoons$ H$_2$	+ Br$_2$
initial concentration	0.20 M	0.15 M	0 M
change to attain equilibrium	$-2x$	$+x$	$+x$
equilibrium concentration	0.20 M $- 2x$	0.15 M $+ x$	x

We now substitute these values and the value of K_c into the equilibrium expression to solve for x.

$$K_c = \frac{[H_2][Br_2]}{[HBr]^2}$$

$$7.7 \times 10^{-11} = \frac{(0.15 + x)(x)}{(0.20 - x)^2}$$

This equation could be solved by putting it into the form $ax^2 + bx + c = 0$ and using the quadratic formula, but a simpler approach is possible. Because of the small size of K_c, we make the assumption that x is very small—so small, in fact, that adding x to 0.15 results in 0.15 and subtracting x from 0.20 results in 0.20.

$$0.15 + x \approx 0.15$$

$$0.20 - x \approx 0.20$$

With this assumption, the equation to solve now becomes

$$7.7 \times 10^{-11} = \frac{(0.15)x}{(0.20)^2}$$

$$x = \frac{(7.7 \times 10^{-11})(0.20)^2}{(0.15)}$$

$$= 2.1 \times 10^{-11}$$

Before proceeding, we check to see whether the assumption regarding the smallness of x is appropriate. A general rule of thumb is that x should be less than 5% of the quantity relative to which it is assumed to be small. In our example,

$$\frac{x}{0.15} \times 100 = \frac{2.1 \times 10^{-11}}{0.15} \times 100 = 1.4 \times 10^{-8}\%$$

x is even a smaller percentage of 0.20. Thus, x is less than 5% of both quantities relative to which it is assumed to be small, and the assumption that x is small is justified. The desired equilibrium concentrations can now be calculated.

$$[Br_2] = x = 2.1 \times 10^{-11}$$

$$[H_2] = 0.15 + x = 0.15$$

$$[HBr] = 0.20 - x = 0.20$$

Substitution of these values into the equilibrium expression gives the correct value for the equilibrium constant:

$$\frac{(2.1 \times 10^{-11})(0.15)}{(0.20)^2} = 7.9 \times 10^{-11}$$

EXERCISE $K_c = 1.6 \times 10^{-9}$ for the equilibrium shown. What are the equilibrium concentrations of the substances in the equilibrium when SO_3, initially at 0.65 M, and O_2, initially at 0.010 M, are allowed to come to equilibrium?

$$2SO_3(g) \rightleftharpoons 2SO_2(g) + O_2(g)$$

[*Answer:* O_2, 0.010 M; SO_2, 2.6×10^{-4} M; SO_3, 0.65 M]

KEY WORDS   Define or explain each term in a written sentence or two.

reaction quotient

THE RESPONSE OF
EQUILIBRIA TO CHANGE

13.8   ADDING AND REMOVING REAGENTS

KEY CONCEPT A   Le Chatelier's principle

Imagine that we have a system in dynamic equilibrium and that we disturb the equilibrium state in some manner. For example, let's assume we have a dissolved gas in equilibrium with undissolved gas above the liquid, and we suddenly increase the partial pressure of the gas above the liquid by introducing some of the gas into the experimental apparatus. According to Henry's law, increasing the partial pressure makes the gas more soluble; so some of the introduced gas dissolves in the liquid and the partial pressure of the gas decreases, as the following figure illustrates.

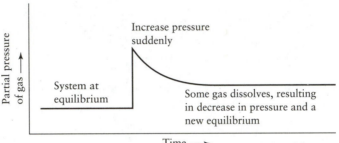

Le Chatelier's principle states that when a system at equilibrium is disturbed, it adjusts so as to partially cancel the effect of the disturbance. For the system shown, the disturbance is the sudden increase in pressure, and we can see that the way the system responds results in a partial canceling of the increase and a new state of equilibrium. This is a manifestation of Le Chatelier's principle.

> **EXAMPLE   Using Le Chatelier's principle**
>
> Assume we have a system in a dynamic equilibrium and that the molarity of a compound is determined by the equilibrium. The equilibrium is disturbed by adding some of the compound, increasing the molarity from the equilibrium value of 1.0 M to 2.0 M. Use Le Chatelier's principle to predict how the system responds to the sudden increase in the molarity.
>
> **SOLUTION** Le Chatelier's principle predicts that a system at equilibrium responds to a disturbance in the equilibrium by canceling some of the effect of the disturbance. It does so by establishing a new equilibrium state. In this case, we predict that, by establishment of a new equilibrium state, the molarity of the compound will fall below the 2.0 M achieved by the disturbance but above the starting molarity of 1.0 M.
>
> **EXERCISE** Assume a dynamic equilibrium is responsible for determining the volume of a system. The system has volume V_1; we suddenly increase the volume to V_2. How does the system respond to the increase?
>
> > [*Answer:* It establishes a new equilibrium with volume larger than V_1 but smaller than V_2.]

KEY CONCEPT B   The effect of adding or removing a reagent from an equilibrium mixture

When a reagent is added or removed from a system at equilibrium, the equilibrium is disturbed and adjusts to attain a new equilibrium state. As an example, consider the equilibrium shown, with [HI] = 0.21 and [H_2] = [I_2] = 0.017.

$$2HI(g) \rightleftharpoons H_2(g) + I_2(g) \qquad K_c = 6.3 \times 10^{-3}$$

Imagine that HI is suddenly added to the equilibrium mixture to make its concentration 0.46 M. There is now too much HI for equilibrium to exist; some of it has to decompose into H_2 and I_2 to establish a new equilibrium. The following graph portrays the concentrations of the substances in the equilibrium before and after the addition of the extra HI.

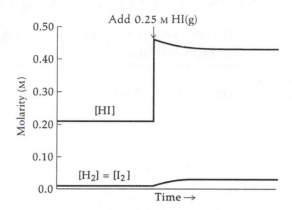

The new equilibrium concentrations after addition of the HI are [HI] = 0.43 and $[H_2]$ = $[I_2]$ = 0.034. As can be seen from the graph, addition of HI results in formation of additional products. This result can be generalized:

• Addition of reactants or removal of products in an equilibrium results in loss of reactant and formation of additional products.
• Addition of products or removal of reactants in an equilibrium results in loss of product and formation of additional reactants.

These effects are in accord with Le Chatelier's principle.

EXAMPLE 1   Shifting an equilibrium by adding reagents

Describe the effect on the equilibrium shown of (a) addition of SO_2, (b) addition of SO_3, (c) removal of O_2, and (d) removal of SO_3.

$$2SO_2(g) + O_2(g) \rightleftharpoons 2SO_3(g)$$

SOLUTION (a) Addition of a reactant to an equilibrium causes formation of additional product, so addition of SO_2 results in the reaction of some SO_2 with O_2 to form SO_3. (b) Addition of a product to an equilibrium causes formation of additional reactant, so addition of SO_3 results in the decomposition of some SO_3 into SO_2 and O_2. (c) Removal of reactant results in formation of additional reactants, so removal of O_2 results in the decomposition of some SO_3 into SO_2 and O_2. (d) Removal of product results in formation of additional product, so removal of SO_3 results in the reaction of some SO_2 with some O_2 to form SO_3.

EXERCISE Describe the effect on the equilibrium shown of (a) addition of HI; (b) addition of H_2; (c) removal of I_2; (d) removal of HI.

$$H_2(g) + I_2(g) \rightleftharpoons 2HI(g)$$

[*Answer:* (a) Reactants form; (b) products form; (c) reactants form; (d) products form]

EXAMPLE 2   Shifting an equilibrium by adding a pure liquid or solid

Describe the effect of the addition of C(s) to the equilibrium shown.

$$C(s) + O_2(g) \rightleftharpoons CO_2(g)$$

SOLUTION Addition of a pure liquid or pure solid to an equilibrium has no effect on the equilibrium. This is why the concentration of a pure liquid or pure solid does not appear in the equilibrium expression for an equilibrium.

EXERCISE Describe the effect of removal of exactly half of the $CuSO_4 \cdot 5H_2O$ from the equilibrium shown.

$$CuSO_4 \cdot 5H_2O(s) \rightleftharpoons CuSO_4(s) + 5H_2O(g)$$

[*Answer:* No effect]

13.9   COMPRESSING A REACTION MIXTURE

KEY CONCEPT   Compression of reacting gases changes the equilibrium

Imagine we have a gas-phase equilibrium that undergoes a change to a smaller volume, that is, the reaction vessel is somehow compressed. Le Chatelier's principle predicts that the gas molecules respond to a decrease in volume by reacting to form fewer molecules, thereby "fitting" into the smaller volume better. For the equilibrium

$$2A(g) \rightleftharpoons B(g)$$

the A molecules would react to form B molecules, resulting in fewer gas molecules overall. On the other hand, a decrease in pressure that is accomplished by an increase in volume causes the gas to react in such a way as to produce more gas molecules, thereby filling the additional volume created. For the preceding equilibrium, some B molecules would decompose into A molecules, thereby resulting in more gas-phase molecules overall. The following graphs show the results of an increase in pressure (accomplished by a decrease in volume) for different types of equilibria. The increase in pressure occurs at the arrow.

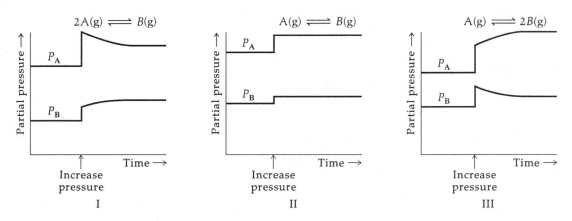

These results are derived rigorously in the text and can be summarized as follows:

An increase in pressure on an equilibrium at constant temperature, accomplished by a decrease in volume, results in a shift of the equilibrium so as to form fewer gas molecules.

A decrease in pressure on an equilibrium at constant temperature, accomplished by an increase in volume, results in a shift of the equilibrium so as to form more gas molecules.

EXAMPLE 1   Determining the effect of pressure changes on an equilibrium

What is the effect of increasing the pressure on the equilibrium shown? Assume there is no change in temperature.

$$2H_2(g) + CO(g) \rightleftharpoons CH_3OH(g)$$

SOLUTION An increase in pressure on an equilibrium at constant temperature, accomplished by a decrease in volume, results in a shift of the equilibrium so as to form fewer gas molecules. For this equilibrium, an increase in pressure causes some H_2 to react with some CO to form CH_3OH, resulting in fewer gas molecules overall.

EXERCISE What is the effect of decreasing the pressure at constant temperature on the equilibrium shown?

$$COCl_2(g) \rightleftharpoons CO(g) + Cl_2(g)$$

[*Answer:* Some $COCl_2$ decomposes to form CO and Cl_2.]

EXAMPLE 2   Determining the effect of pressure changes on an equilibrium

Describe the effect of increasing the pressure at constant temperature on the equilibrium shown.

$$H_2(g) + I_2(g) \rightleftharpoons 2HI(g)$$

SOLUTION Because the equilibrium has the same number of moles of gas on both sides, changing the pressure has no effect.

EXERCISE Describe the effect of decreasing the pressure at constant temperature on the following equilibrium.

$$H_2(g) + CO_2(g) \rightleftharpoons H_2O(g) + CO(g)$$

[*Answer:* No effect]

EXAMPLE 3   Determining the effect of pressure changes on an equilibrium

What is the effect of decreasing the pressure on the following equilibrium? Assume the temperature is constant.

$$2C(s) + O_2(g) \rightleftharpoons 2CO(g)$$

SOLUTION Decreasing the pressure on an equilibrium causes the equilibrium to shift in such a way as to form more *gas* molecules. In this case, decreasing the pressure causes some C and some O_2 to react to form CO.

EXERCISE Describe the effect of increasing the pressure on the following equilibrium at constant temperature.

$$CaCO_3(s) \rightleftharpoons CaO(s) + CO_2(g)$$

[*Answer:* Some CO_2 will combine with CaO to form $CaCO_3$.]

PITFALL   The effect of addition of an inert gas

When an inert gas is added to a gaseous reaction mixture, the total pressure of the gaseous mixture increases; however, there is no effect on the equilibrium. We have considered changing pressure only by changing the volume of the reaction vessel at constant temperature.

13.10   TEMPERATURE AND EQUILIBRIUM

KEY CONCEPT   The change in equilibrium as temperature is changed

Changing the temperature affects the forward and reverse rates of reaction in an equilibrium and therefore affects the equilibrium. The overall effect of a temperature change on an equilibrium is given in the following table.

Temperature Change	Exothermic Reaction	Endothermic Reaction
increase	more reactants form	more products form
decrease	more products form	more reactants form

EXAMPLE   Determining the effect of temperature changes on an equilibrium

What is the effect of increasing the temperature on the equilibrium shown? Assume constant pressure.

$$2HI(g) \rightleftharpoons H_2(g) + I_2(g) \qquad \Delta H = -52.96 \text{ kJ}$$

SOLUTION Because the reaction is exothermic, increasing the temperature results in the formation of reactants. Hence, increasing the temperature causes some I_2 to react with some H_2 to form HI.

EXERCISE What is the effect of increasing the temperature on the following equilibrium? Assume the pressure is constant.

$$CaCO_3(s) \rightleftharpoons CaO(s) + CO_2(g) \qquad \Delta H = +178.5 \text{ kJ}$$

[*Answer:* Some $CaCO_3$ decomposes to form CaO and CO_2.]

13.11   CATALYSTS AND HABER'S ACHIEVEMENT

KEY CONCEPT   A catalyst does not change the position of an equilibrium

We have noted previously that a catalyst speeds up a reaction. Because a catalyst has equivalent effects on the speed of the forward and backward reactions, addition of a catalyst to an equilibrium has no effect on the concentrations of the reactants and products in the equilibrium. However, a catalyst will increase the rate at which the equilibrium is attained. One of the great achievements of Haber and Carl Bosch was to find a catalyst that would speed up the attainment of the equilibrium that produces ammonia, thereby making it practical to produce the ammonia in a factory.

EXAMPLE   Understanding the effect of a catalyst on equilibrium

At 1000 K, the concentrations of the substances in the following equilibrium are SO_2, 0.950 M; O_2, 0.880 M; SO_3, 0.180 M. A catalyst is added that increases the forward rate of reaction by a factor of 1.88. What is the concentration of SO_3 after addition of the catalyst and reattainment of equilibrium?

$$2SO_2(g) + O_2(g) \rightleftharpoons 2SO_3(g)$$

SOLUTION Addition of a catalyst does not change the position of an equilibrium. By this we mean that the concentrations of the products and reactants do not change because of the addition of the catalyst. Because the forward reaction was increased in rate by a factor of 1.88, the reverse reaction must have been increased by the same rate. The new concentration of SO_3 is 0.180 M, the same as the original concentration.

EXERCISE The equilibrium constant, K_c, for the reaction in the preceding example is 0.0413 at 1000 K. What is the value of K_c after addition of the catalyst?

[*Answer:* 0.0413]

KEY WORDS   Define or explain each term in a written sentence or two.

Le Chatelier's principle
catalyst

DESCRIPTIVE CHEMISTRY TO REMEMBER

• Nitrogen has a strong triple bond and a correspondingly high bond enthalpy (944 kJ·mol^{-1}), so it is very unreactive.

CHEMICAL EQUATIONS TO KNOW

- An important reaction in the **gasification** of coal is

$$2CO(g) + 2H_2(g) \rightleftharpoons CH_4(g) + CO_2(g)$$

- Solid **calcium hydroxide** exists in equilibrium with its ions in aqueous solution:

$$Ca(OH)_2(s) \rightleftharpoons Ca^{2+}(aq) + 2OH^-(aq)$$

- Solid **calcium carbonate** exists in equilibrium with calcium oxide and carbon dioxide:

$$CaCO_3(s) \rightleftharpoons CaO(s) + CO_2(g)$$

- In the **Mond process,** nickel and carbon monoxide are in equilibrium with nickel tetracarbonyl:

$$Ni(s) + 4CO(g) \rightleftharpoons Ni(CO)_4(g)$$

- **Dinitrogen tetroxide** can exist in equilibrium with nitrogen dioxide:

$$N_2O_4(g) \rightleftharpoons 2NO_2(g)$$

- **Nitric oxide** (nitrogen monoxide) can exist in equilibrium with nitrogen and oxygen:

$$N_2(g) + O_2(g) \rightleftharpoons 2NO(g)$$

- **Nitrous oxide** (dinitrogen monoxide) can also exist in equilibrium with nitrogen and oxygen:

$$2N_2(g) + O_2(g) \rightleftharpoons 2N_2O(g)$$

- In the **Haber process,** nitrogen and hydrogen are in equilibrium with ammonia:

$$3H_2(g) + N_2(g) \rightleftharpoons 2NH_3(g)$$

- The production of **sulfur trioxide** (used in the production of sulfuric acid) involves an equilibrium with sulfur dioxide and oxygen:

$$2SO_2(g) + O_2(g) \rightleftharpoons 2SO_3(g)$$

- **Ozone** can exist in equilibrium with oxygen:

$$2O_3(g) \rightleftharpoons 3O_2(g)$$

- Gaseous **phosphorus pentachloride** can exist in equilibrium with phosphorus trichloride and chlorine:

$$PCl_5(g) \rightleftharpoons PCl_3(g) + Cl_2(g)$$

MATHEMATICAL EQUATIONS TO KNOW AND UNDERSTAND

$K_p = K_c(RT)^{\Delta n}$	relationship between K_c and K_p
when $Q_c > K_c$	reactants have a tendency to form
when $Q_c = K_c$	the reaction is at equilibrium
when $Q_c < K_c$	products have a tendency to form
$x = \dfrac{-b \pm \sqrt{b^2 - 4ac}}{2a}$	quadratic formula

SELF-TEST EXERCISES

Equilibrium and composition

1. A 0.751-g sample of $BaCO_3$ containing radioactive Ba is poured into a saturated solution of $BaCO_3$ (which contains Ba^{2+} and CO_3^{2-} ions) in which there are no radioactive Ba atoms. After thorough stirring for 30 min, there is still 0.751 g of $BaCO_3$ at the bottom of the solution. Which of the following is correct about the Ba atoms?

(a) All of the radioactive Ba remains in the 0.751 g of solid $BaCO_3$

(b) All of the radioactive Ba has gone into solution as Ba^{2+} ions.

(c) Some of the radioactive Ba remains in the solid $BaCO_3$ and some has gone into solution as Ba^{2+} ions.

2. Assume equilibrium exists for the reaction shown below. One of the statements regarding the equilibrium is untrue. Which one?

$$H_2(g) + Cl_2(g) \rightleftharpoons 2HCl(g)$$

(a) The rate of the reaction $H_2 + Cl_2 \rightarrow 2HCl$ equals the rate of reaction $2HCl \rightarrow H_2 + Cl_2$.

(b) The concentrations of H_2, Cl_2, and HCl stay constant.

(c) The rate of reaction $H_2 + Cl_2 \rightarrow 2HCl$ is zero; that is, the reaction does not occur.

(d) There is no observable change in the system as time evolves.

3. What is the correct form of the equilibrium constant K_c for the equilibrium shown?

$$4NH_3(g) + 3O_2(g) \rightleftharpoons 2N_2(g) + 6H_2O(g)$$

(a) $\dfrac{[N_2]^2[H_2O]^6}{[NH_3]^4[O_2]^3}$ (b) $\dfrac{[NH_3]^4[O_2]^3}{[N_2]^2[H_2O]^6}$ (c) $\dfrac{2[N_2] \times 6[H_2O]}{4[NH_3] \times 3[O_2]}$

(d) $\dfrac{4[NH_3] \times 3[O_2]}{2[N_2] \times 6[H_2O]}$ (e) $\dfrac{[N_2]^2 + [H_2O]^6}{[NH_3]^4 + [O_2]^3}$

4. What is the correct form of the equilibrium constant K_c for the equilibrium shown?

$$CO(g) + 2H_2(g) \rightleftharpoons CH_3OH(g)$$

(a) $\dfrac{[CO][H_2]^2}{[CH_3OH]}$ (b) $\dfrac{[CH_3OH]}{[CO][H_2]^2}$ (c) $[CH_3OH][CO][H_2]^2$

(d) $\dfrac{1}{[CH_3OH][CO][H_2]^2}$ (e) $\dfrac{[CH_3OH]}{[CO] + [H_2]^2}$

5. What is K_c for the equilibrium shown if the concentrations at equilibrium are $[NO_2] = 0.041$, $[CO] = 0.033$, $[NO] = 1.8$, and $[CO_2] = 2.0$?

$$NO_2(g) + CO(g) \rightleftharpoons NO(g) + CO_2(g)$$

(a) 3.8×10^{-4} (b) 2.9 (c) 8.2×10^3 (d) 2.7×10^3 (e) 3.7

6. What is K_c for the equilibrium shown if the concentrations at equilibrium are $[CO] = 0.045$, $[Cl_2] = 0.045$, and $[COCl_2] = 0.50$?

$$COCl_2(g) \rightleftharpoons CO(g) + Cl_2(g)$$

(a) 9.0×10^{-2} (b) 1.1×10^2 (c) 4.1×10^{-3}
(d) 2.5×10^2 (e) 8.1×10^{-3}

7. The equilibrium constant for equilibrium I is 22.5 at 380 K. What is the equilibrium constant for II at the same temperature?

$$SO_2Cl_2(g) \rightleftharpoons SO_2(g) + Cl_2(g) \qquad (I)$$

$$SO_2(g) + Cl_2(g) \rightleftharpoons SO_2Cl_2(g) \qquad (II)$$

(a) 507   (b) 22.5   (c) 4.74   (d) 0.0444   (e) 66.9

8. The equilibrium constant for equilibrium I is 4.0×10^{-4} at 1483°C. What is the equilibrium constant for II at the same temperature?

$$Br_2(g) \rightleftharpoons 2Br(g) \qquad (I)$$

$$2Br(g) \rightleftharpoons Br_2(g) \qquad (II)$$

(a) 50   (b) 6.3×10^6 (c) 5.0×10^3 (d) 0.0444   (e) 2.5×10^3

9. Given the equilibrium constants for equilibriums I and II, what is the equilibrium constant for III at the same temperature?

$$2NO(g) + O_2(g) \rightleftharpoons 2NO_2(g) \qquad K_c = 2.5 \times 10^2 \qquad (I)$$

$$2NO_2(g) \rightleftharpoons N_2O_4(g) \qquad K_c = 3.0 \times 10^3 \qquad (II)$$

$$2NO(g) + O_2(g) \rightleftharpoons N_2O_4(g) \qquad K_c = ? \qquad (III)$$

(a) 7.5×10^5 (b) 1.3×10^{-6} (c) 3.3×10^3
(d) 5.5×10^5 (e) 1.2×10^1

10. The equilibrium constant for equilibrium I is 47.9 at 400 K. What is the equilibrium constant for II at the same temperature?

$$N_2O_4(g) \rightleftharpoons 2NO_2(g) \qquad (I)$$

$$\tfrac{1}{2}N_2O_4(g) \rightleftharpoons NO_2(g) \qquad (II)$$

(a) 0.0209   (b) 6.92   (c) 2.29×10^3 (d) 47.9   (e) 24.0

11. What is the equilibrium expression for the equilibrium shown?

$$2HgO(s) \rightleftharpoons 2Hg(s) + O_2(g)$$

(a) $[Hg]^2[O_2]$ (b) $[O_2]$ (c) $[O_2][Hg]^2/[HgO]^2$
(d) $[O_2]/[HgO]^2$ (e) $[Hg]^2/[HgO]^2$

12. What is the correct form of the equilibrium constant K_c for the equilibrium shown?

$$Fe_2S_3(s) \rightleftharpoons 2Fe^{3+}(aq) + 3S^{2-}(aq)$$

(a) $\dfrac{1}{[Fe^{3+}]^2[S^{2-}]^3}$ (b) $[Fe^{3+}]^2[S^{2-}]^3$ (c) $\dfrac{[Fe^{3+}]^2[S^{2-}]^3}{[Fe_2S_3]}$

(d) $\dfrac{[Fe_2S_3]}{[Fe^{3+}]^2[S^{2-}]^3}$ (e) $\dfrac{1}{[Fe_2S_3]}$

13. Equilibrium is attained for the reaction below when 0.10 mol $CaSO_4 \cdot 2H_2O(s)$, 0.30 mol $CaSO_4(s)$, and 0.50 mol $H_2O(g)$ are all placed in a 1.0-L flask. What is K_c for this equilibrium?

$$CaSO_4 \cdot 2H_2O(s) \rightleftharpoons CaSO_4(s) + 2H_2O(g)$$

(a) 1.0   (b) 0.075   (c) 0.75   (d) 1.5   (e) 0.25

14. Equilibrium is attained for the reaction below when 0.50 mol $BaF_2(s)$ is in equilibrium with Ba^{2+} at a concentration of 0.017 M and F^- at a concentration of 0.010 M. What is K_c for this equilibrium?

$$BaF_2(s) \rightleftharpoons Ba^{2+}(aq) + 2F^-(aq)$$

(a) 3.4×10^{-6} (b) 8.5×10^{-3} (c) 3.4×10^{-2}

(d) 1.7×10^{-6} (e) 1.7×10^{-4}

15. What is K_p for the equilibrium shown if the equilibrium partial pressures are P_{Cl_2} = 2.2 atm, P_{H_2O} = 0.85 atm, P_{HCl} = 6.4 atm, and P_{O_2} = 1.6 atm?

$$2Cl_2(g) + 2H_2O(g) \rightleftharpoons 4HCl(g) + O_2(g)$$

(a) 19   (b) 5.3×10^{-2} (c) 6.4×10^{-2}

(d) 1.3×10^{-3} (e) 7.7×10^2

16. What is the correct form of the equilibrium constant K_p for the equilibrium shown?

$$2ICl(g) \rightleftharpoons I_2(g) + Cl_2(g)$$

(a) $\dfrac{P_{ICl}^2}{P_{I_2}P_{Cl_2}}$ (b) $\dfrac{P_{I_2} + P_{Cl_2}}{P_{ICl}^2}$ (c) $\dfrac{P_{I_2} + P_{Cl_2}}{2P_{ICl}}$ (d) $\dfrac{P_{I_2}P_{Cl_2}}{P_{ICl}^2}$ (e) $P_{I_2}P_{Cl_2}$

17. What is P_{O_2} in the equilibrium shown if P_{SO_3} = 2.5 atm and P_{SO_2} = 8.2×10^{-3} atm?

$$2SO_2(g) + O_2(g) \rightleftharpoons 2SO_3(g) \qquad K_p = 3.0 \times 10^4$$

(a) 9.1×10^6 atm   (b) 5.2×10^5 atm   (c) 3.1 atm

(d) 98 atm   (e) 4.2 atm

18. $K_c = 4.1 \times 10^8$ at 298 K for the equilibrium $N_2(g) + 3H_2(g) \rightleftharpoons 2NH_3(g)$. What is K_p?

(a) 1.7×10^7 (b) 1.0×10^{10} (c) 6.9×10^5

(d) 2.5×10^{11} (e) 5.2×10^{11}

19. K_p = 47.9 at 127°C for the equilibrium shown. What is K_c at this temperature?

$$N_2O_4(g) \rightleftharpoons 2NO_2(g)$$

(a) 1.57×10^3 (b) 5.17×10^4 (c) 6.18×10^3

(d) 4.44×10^{-2} (e) 1.46

Using equilibrium constants

20. For which K_c for the hypothetical equilibrium shown does the equilibrium most strongly tend toward products?

$$A(g) + B(g) \rightleftharpoons C(g) + D(g)$$

(a) 0.015   (b) 8.1×10^{-3} (c) 1.0   (d) 290   (e) 3.2×10^3

21. What is Q_c for the equilibrium if $[Br_2]$ = 0.26, [NO] = 0.88, and [NOBr] = 0.33?

$$2NOBr(g) \rightleftharpoons 2NO(g) + Br_2(g) \qquad K_c = 4.1 \times 10^{-4}$$

(a) 1.8   (b) 0.69   (c) 1.4   (d) 2.1   (e) 0.34

22. For Exercise 21, in what direction is there a tendency for the reaction to shift?

(a) to the left   (b) to the right   (c) neither direction

23. When N_2O_4 initially at 1.0 atm is placed in a flask at 298 K and allowed to come to equilibrium, the equilibrium partial pressure of N_2O_4 is 0.82 atm. What is the equilibrium partial pressure of NO_2?

$$N_2O_4(g) \rightleftharpoons 2NO_2(g)$$

(a) 0.18 atm   (b) 0.36 atm   (c) 0.82 atm

(d) 0.41 atm   (e) 1.64 atm

24. When NOCl, initially at 0.500 atm, is placed in a flask and allowed to come to equilibrium, the equilibrium partial pressure of Cl_2 is 0.201 atm. What is the equilibrium partial pressure of NOCl?

$$2NOCl(g) \rightleftharpoons 2NO(g) + Cl_2(g)$$

(a) 0.402 atm   (b) 0.299 atm   (c) 0.101 atm

(d) 0.098 atm   (e) 0.201 atm

25. For a hypothetical equilibrium, what is [B] and [A] = 2.5 and [C] = 0.50?

$$A(g) + 3B(g) \rightleftharpoons 2C(g) \qquad K_c = 6.4 \times 10^2$$

(a) 0.013   (b) 0.054   (c) 0.043   (d) 1.6×10^{-4} (e) 0.18

26. NO, initially at 5.0 M, is allowed to come to equilibrium. What is the equilibrium concentration of NO? (The answer is very sensitive to how and where you round off.)

$$2NO(g) \rightleftharpoons N_2(g) + O_2(g) \qquad K_c = 20$$

(a) 0.90 M   (b) 0.50 M   (c) 1.6 M   (d) 2.2 M   (e) 2.8 M

27. $K_c = 0.082$ for the equilibrium shown. What is the equilibrium concentration of Cl_2 if $COCl_2$, initially at 0.50 M, is allowed to come to equilibrium?

$$COCl_2(g) \rightleftharpoons CO(g) + Cl_2(g)$$

(a) 0.17 M   (b) 0.34 M   (c) 0.20 M   (d) 0.26 M   (e) 0.44 M

28. Assume the temperature is such that $K_p = 0.54$ for the equilibrium shown. What is the equilibrium partial pressure of PCl_5 if PCl_5, initially at 1.20 atm, is allowed to come to equilibrium at this temperature?

$$PCl_5(g) \rightleftharpoons PCl_3(g) + Cl_2(g)$$

(a) 0.58 atm   (b) 0.80 atm   (c) 0.54 atm

(d) 0.62 atm   (e) 0.40 atm

29. $K_c = 2.5 \times 10^{-9}$ at 298 K for the equilibrium shown. What is the equilibrium concentration of H_2 if NH_3, initially at 25 mM, is allowed to come to equilibrium at 298 K?

$$2NH_3(g) \rightleftharpoons N_2(g) + 3H_2(g)$$

(a) 6.3×10^{-8} mM   (b) 0.49 mM   (c) 1.5 mM

(d) 1.9 mM   (e) 2.1×10^{-3} mM

30. Assume the temperature is such that $K_p = 3.1 \times 10^{-8}$ for the equilibrium shown. What is the equilibrium partial pressure of S_2 if H_2S, initially at 0.50 atm, and H_2, initially at 0.010 atm, are allowed to come to equilibrium at this temperature?

$$2H_2S(g) \rightleftharpoons 2H_2(g) + S_2(g)$$

(a) 8.8×10^{-3} atm   (b) 1.3×10^{-3} atm   (c) 7.8×10^{-5} atm

(d) 1.6×10^{-6} atm   (e) 2.8×10^{-4} atm

The response of equilibria to change

31. Which of the following causes the equilibrium shown to shift to the right?

$$3H_2(g) + N_2(g) \rightleftharpoons 2NH_3(g)$$

(a) addition of NH_3 (b) addition of N_2
(c) removal of H_2 (d) removal of N_2

32. Removal of H_2 from the equilibrium shown shifts the equilibrium

$$H_2(g) + I_2(g) \rightleftharpoons 2HI(g)$$

(a) to the left   (b) to the right   (c) neither to the left nor right

33. Addition of Fe(s) to the equilibrium shown shifts the equilibrium

$$2Fe_2O_3(s) + 3C(s) \rightleftharpoons 4Fe(s) + 3CO_2(g)$$

(a) to the left   (b) to the right   (c) neither to the left nor right

34. Decreasing the volume of the equilibrium shown shifts the equilibrium

$$Br_2(g) + 3F_2(g) \rightleftharpoons 2BrF_3(g)$$

(a) to the left   (b) to the right   (c) neither to the left nor right

35. Decreasing the pressure on the equilibrium shown shifts the equilibrium

$$Xe(g) + F_2(g) \rightleftharpoons XeF_2(g)$$

(a) to the left   (b) to the right   (c) neither to the left nor right

36. Decreasing the pressure on the equilibrium shown shifts the equilibrium

$$H_2(g) + Br_2(g) \rightleftharpoons 2HBr(g)$$

(a) to the left   (b) to the right   (c) neither to the left nor right

37. For which of the following changes does K_c also change?
(a) change in pressure   (b) addition of a reagent   (c) change in temperature
(d) addition of a catalyst   (e) none of these

Descriptive chemistry

38. Gaseous methanol (CH_3OH) can exist in equilibrium with carbon monoxide (CO) and
(a) H_2 (b) H_2O (c) CH_4 (d) H_2O_2 (e) OH^-

39. PCl_5 can exist in equilibrium with
(a) P_4 and Cl_2 (b) PCl_3 and P   (c) PCl_3 (d) PCl_3 and Cl_2

40. The Haber process is used for the synthesis of
(a) NH_3 (b) esters   (c) CH_3OH (d) NO   (e) carbohydrates

41. Only one of the following gases is very unreactive. Which one?
(a) O_2 (b) N_2 (c) H_2 (d) Cl_2 (e) F_2

42. What compound exists in equilibrium with CO(g) and Ni(s)?
(a) $NiCO_2$ (b) Ni_2CO_3 (c) $Ni(CO_2)_3$ (d) $NiCO_3$ (e) $Ni(CO)_4$

CHAPTER 14

PROTONS IN TRANSITION: ACIDS AND BASES

Acids and bases are both generated as by-products of our technology. To understand the environmental consequences of these by-products, we must understand their chemical structure, their strengths as acids and bases, and the reactions they undergo. These topics are the concern of this chapter.

WHAT ARE ACIDS AND BASES?

14.1 BRØNSTED-LOWRY ACIDS AND BASES

KEY CONCEPT   Acids and bases are defined by proton transfers

The Brønsted-Lowry theory of acids and bases focuses on the transfer of an H^+ ion (a proton). In any reaction involving Brønsted acids and bases, we look for what the proton does in order to understand what chemistry is happening. With this in mind, we can make the following definitions. A **Brønsted acid** is a compound that donates a proton (H^+) in a chemical reaction; a **Brønsted base** is a compound that accepts a proton (H^+) in a chemical reaction. When a Brønsted acid donates a proton, a Brønsted base accepts it, so any chemical reaction involving Brønsted acids and bases must contain both; you cannot have one without the other. As an example, consider the **ionization** of HNO_3 in water:

$$HNO_3(aq) + H_2O(l) \longrightarrow H_3O^+(aq) + NO_3^-(aq)$$

In this reaction, HNO_3 donates a proton and H_2O accepts it. HNO_3 is a Brønsted acid (it donates a proton), and H_2O is a Brønsted base (it accepts a proton). The fact that water is the solvent does not affect our view of the reaction; it is simply a reactant.

EXAMPLE 1   Recognizing a Brønsted acid and Brønsted base in a chemical reaction

Identify the Brønsted acid and Brønsted base in the reaction

$$NH_3 + CH_3^- \rightleftharpoons NH_2^- + CH_4$$

SOLUTION The substance that donates a proton is the Brønsted acid, the one that accepts it is the Brønsted base. In this reaction, NH_3 donates a proton (and becomes NH_2^-) and CH_3^- accepts a proton (and becomes CH_4). NH_3 acts as a Brønsted acid, and CH_3^- acts as a Brønsted base.

EXERCISE Identify the Brønsted acid and Brønsted base in the reaction

$$HS^-(aq) + H_2O(l) \rightleftharpoons H_2S(aq) + OH^-(aq)$$

[*Answer:* HS^-, base; H_2O, acid]

EXAMPLE 2   Predicting the products of a Brønsted acid-base reaction

Predict the products of the following Brønsted acid-base reaction:

$$HCO_3^-(aq) + OH^-(aq) \rightleftharpoons ?$$

SOLUTION A Brønsted acid-base reaction involves proton transfer, and the first hurdle here is to determine which of the two hydrogens in the reactants transfers as a proton. We must call on our knowledge that OH^- often accepts a proton to form water to conclude that it is most likely the Brønsted base in this reaction. Thus, we conclude that HCO_3^- is the Brønsted acid and donates a proton (forming CO_3^{2-}) and that OH^- is the Brønsted base and accepts the proton (forming H_2O). The correct equation is

$$HCO_3^-(aq) + OH^-(aq) \rightleftharpoons CO_3^{2-}(aq) + H_2O(l)$$

EXERCISE Predict the products of the Brønsted acid-base reaction

$$HSO_4^-(aq) + OH^-(aq) \rightleftharpoons ?$$

[*Answer:* $SO_4^{2-}(aq) + H_2O(l)$]

14.2   CONJUGATE ACIDS AND BASES

KEY CONCEPT   A conjugate acid and base differ only by the presence of an H^+

Proton transfer reactions are extremely fast and facile. It is not surprising, therefore, to find that most Brønsted acid-base reactions actually exist as dynamic equilibria in which the proton rapidly moves back and forth between two Brønsted bases. For example, consider the ionization of aqueous hydrocyanic acid (HCN):

$$HCN(aq) + H_2O(l) \rightleftharpoons H_3O^+(aq) + CN^-(aq)$$

In this equilibrium, products form by transfer of a proton from HCN to H_2O; reactants reform by proton transfer from H_3O^+ to CN^-. For the forward reaction, HCN is a Brønsted acid and H_2O a Brønsted base; for the reverse reaction, H_3O^+ is a Brønsted acid and CN^- a Brønsted base. When a Brønsted acid loses its proton, the base formed is called the conjugate base of the acid; when a base accepts a proton to become an acid, the acid is called the conjugate acid of the base. CN^- is the conjugate base of HCN, and HCN is the conjugate acid of CN^-. Also, H_3O^+ is the conjugate acid of H_2O, and H_2O the conjugate base of H_3O^+. If we consider the reverse reaction, it becomes evident that H_3O^+ is an acid, as suggested, because it loses a proton; similarly, CN^- is a base because it gains a proton. As the following examples indicate, a conjugate acid and base differ only in the presence of an H^+.

Forward Reaction	Reverse Reaction	Conjugate Acid-Base Pair
HCN (acid)	CN^- (base)	HCN, CN^-
H_2O (base)	H_3O^+ (acid)	H_3O^+, H_2O

EXAMPLE   Recognizing conjugate acid-base pairs

Identify the conjugate acid-base pairs in the equilibrium

$$H_2S(aq) + NH_3(aq) \rightleftharpoons HS^-(aq) + NH_4^+(aq)$$

SOLUTION Members of a conjugate acid-base pair differ only in the presence of an H^+. In this reaction, H_2S donates a proton and becomes HS^-. H_2S and HS^- differ only in the presence of an H^+ and are a conjugate acid-base pair. The proton is accepted by NH_3, which then forms NH_4^+. NH_3 and NH_4^+ are a conjugate acid-base pair; they differ only in the presence of a proton.

EXERCISE Identify the conjugate acid-base pairs in the equilibrium

$$CN^-(aq) + H_2O(l) \rightleftharpoons HCN(aq) + OH^-(aq)$$

[*Answer:* CN^- and HCN; H_2O and OH^-]

14.3   PROTON EXCHANGE BETWEEN WATER MOLECULES

KEY CONCEPT A   An amphiprotic compound can undergo autoprotolysis

An **amphiprotic** compound is one that can donate a proton (to a proton acceptor) or accept a proton (from a proton donor). Such a compound can act as a Brønsted acid (when it donates a proton) or as a Brønsted base (when it accepts a proton). Amphiprotic solvents are common and undergo autoprotolysis. In the autoprotolysis process, one molecule of the solvent acts as a Brønsted acid and donates a proton and another molecule of the solvent acts as a Brønsted base and accepts the proton. As with most Brønsted acid-base reactions, autoprotolysis results in an equilibrium. For water, the autoprotolysis equilibrium is

$$H_2O(l) + H_2O(l) \rightleftharpoons OH^-(aq) + H_3O^+(aq)$$

EXAMPLE   Writing an autoprotolysis equilibrium

Write the autoprotolysis equilibrium for the amphiprotic solvent sulfuric acid, H_2SO_4.

SOLUTION Autoprotolysis of an amphiprotic solvent occurs when one solvent molecule donates a proton to a second solvent molecule. For H_2SO_4, this is

$$H_2SO_4 + H_2SO_4 \rightleftharpoons H_3SO_4^+ + HSO_4^-$$

EXERCISE Write the autoprotolysis equilibrium for the amphiprotic solvent HF.

[*Answer:* $HF + HF \rightleftharpoons H_2F^+ + F^-$]

KEY CONCEPT B   The water autoprotolysis constant

The autoprotolysis of water is a dynamic equilibrium and is described by an equilibrium constant, K_w. The concentration of water does not appear in K_w; that is, we get

$$H_2O(l) + H_2O(l) \rightleftharpoons H_3O^+(aq) + OH^-(aq)$$

$$K_w = [H_3O^+][OH^-]$$

At 25°C, the experimentally determined value of K_w is 1.0×10^{-14}. In other words, at 25°C, the concentration of H_3O^+ multiplied by the concentration of OH^- *must* equal 1.0×10^{-14}, no matter what the source of the ions:

$$[H_3O^+][OH^-] = 1.0 \times 10^{-14} \quad (25°C)$$

In pure water, the concentration of OH^- must be the same as that of H_3O^+, so we get

$$\text{Pure water: } [H_3O^+] = 1.0 \times 10^{-7} \quad [OH^-] = 1.0 \times 10^{-7}$$

EXAMPLE   Using K_w

An acid solution at 25°C is made up so that $[H_3O^+] = 3.4 \times 10^{-5}$ M. What is the hydroxide ion concentration in this solution? Is the hydroxide ion concentration higher or lower than that in pure water?

SOLUTION In any aqueous solution at 25°C, $[H_3O^+][OH^-] = 1.0 \times 10^{-14}$. We are given an H_3O^+ concentration of 3.4×10^{-5} M; thus,

$$[H_3O^+][OH^-] = 1.0 \times 10^{-14}$$

$$(3.4 \times 10^{-5})[OH^-] = 1.0 \times 10^{-14}$$

$$[OH^-] = \frac{1.0 \times 10^{-14}}{3.4 \times 10^{-5}}$$

$$= 2.9 \times 10^{-10} \text{ mol} \cdot L^{-1}$$

This OH^- concentration is lower than that of pure water because 2.9×10^{-10} M is less than 1.0×10^{-7} M. This result is expected because an increase in H_3O^+ concentration forces a decrease in the OH^- concentration.

EXERCISE What is the H_3O^+ concentration in a solution of a base in which $[OH^-] = 6.6 \times 10^{-5}$ M? Is the H_3O^+ concentration higher or lower than that in pure water?

[*Answer:* 1.5×10^{-10} M; lower]

14.4 THE pH SCALE

KEY CONCEPT H_3O^+ concentrations are expressed as pH

The acidity of a solution is often reported by using the **pH** of the solution:

$$pH = -\log[H_3O^+]$$

This definition leads to the following at 25°C:

Acid solution: $[H_3O^+] > 1.0 \times 10^{-7}$ M   pH < 7.00
Neutral solution: $[H_3O^+] = 1.0 \times 10^{-7}$ M   pH = 7.00
Basic solution: $[H_3O^+] < 1.0 \times 10^{-7}$ M   pH > 7.00

The pH of a solution can be approximately measured with **universal indicator paper,** which turns different colors when different pH solutions are placed on it. More accurate measurements are made with an instrument called a pH meter. Use of the pH is entirely a matter of convenience and conveys no more information than the hydronium ion concentration $[H_3O^+]$ itself.

EXAMPLE 1 Interpreting the pH

A typical hydronium ion concentration of the juices from a dill pickle is 3.3×10^{-2} M. To what pH does this correspond? Is it acidic, basic, or neutral?

SOLUTION The definition of pH is

$$pH = -\log[H_3O^+]$$

We are given $[H_3O^+] = 3.3 \times 10^{-2}$. Substituting into the definition of pH gives

$$pH = -\log(3.3 \times 10^{-2})$$

$$= -(-1.48)$$

$$= 1.48$$

Because the pH is less than 7.00, the solution is acidic.

EXERCISE What is the pH of a sample of human milk that has a hydronium ion concentration of 7.9×10^{-8} M? Is it acidic, basic, or neutral?

[*Answer:* 7.10; barely basic]

EXAMPLE 2   Calculating [OH⁻] from the pH

What is the hydroxide ion concentration of a sample of wine at 25°C with pH = 2.80? Is the wine acidic, basic, or neutral?

SOLUTION First, because the pH is less than 7.00, the wine is acidic. For the other part of the question, we can solve for [OH⁻] by calculating [H_3O^+] from the pH and using the relationship

$$[H_3O^+][OH^-] = 1.0 \times 10^{-14}$$

With the definitions antilog $z = 10^z$ for any z and antilog(log v) = v for any v, we take the antilog of both sides of the equation for pH and get

$$pH = -\log[H_3O^+]$$

$$[H_3O^+] = 10^{-pH}$$

$$= 10^{-2.80}$$

$$= 1.6 \times 10^{-3} \, M$$

Now, we substitute this value of [H_3O^+] into the equation $[H_3O^+][OH^-] = 1.0 \times 10^{-14}$:

$$(1.6 \times 10^{-3})[OH^-] = 1.0 \times 10^{-14}$$

$$[OH^-] = 6.3 \times 10^{-12} \, M$$

EXERCISE What is the hydroxide ion concentration of a sample of milk of magnesia, with pH = 10.47. Is it acidic, basic, or neutral?

[*Answer:* 2.95 × 10⁻⁴ M; basic]

14.5   THE pOH OF SOLUTIONS

KEY CONCEPT   OH⁻ concentrations are expressed as pOH

The molarity of hydroxide ions (OH⁻) in an aqueous solution is often reported as the pOH.

$$pOH = -\log[OH^-]$$

Because pure water has [OH⁻] = 1×10^{-7} M, the pOH of pure water is 7.0. There are a variety of p functions in chemistry, all defined in a similar fashion:

$$pX = -\log(X)$$

For instance, because $K_w = 1.0 \times 10^{-14}$, $pK_w = 14.00$. With this definition, and because $[H_3O^+][OH^-] = 1.0 \times 10^{-14}$ (at 25°C), we get

$$pH + pOH = pK_w = 14.00$$

EXAMPLE   Using the pOH

In the preceding example 2, you were asked to calculate the hydroxide ion concentration of a sample of wine with pH = 2.80. Redo the calculation, using the pOH.

SOLUTION The answer illustrates how using the pOH can greatly simplify a calculation. In this case, we first calculate the pOH and then use the definition of pOH to get [OH⁻].

$$pH + pOH = 14.00$$

$$pOH = 14.00 - pH$$

$$= 14.00 - 2.80$$

$$= 11.20$$

$$[OH^-] = 10^{-pOH}$$
$$= 10^{-11.20}$$
$$= 6.3 \times 10^{-12} \, M$$

This is the same answer we got earlier, but with a much simpler calculation!

EXERCISE Use the pOH to calculate the pH of a solution that is 6.7×10^{-8} M in OH^-.

[*Answer:* 1.5×10^{-7} M]

KEY WORDS   Define or explain each term in a written sentence or two.

acid	autoprotolysis constant	conjugate base
amphiprotic	base	pH
autoprotolysis	conjugate acid	pOH

WEAK ACIDS AND BASES

14.6 PROTON TRANSFER EQUILIBRIA

KEY CONCEPT A K_a indicates acid strength; K_b indicates base strength

The ionization of an acid in water and of a base in water can both be described by equilibrium constants. The symbol K_a is used for acid ionization constants, and K_b for base ionization constants. With the rule that the concentration of water is never used in the ionization equilibrium expression when water is the solvent, the following can be written:

Ionization of the acid HBrO:

$$HBrO(aq) + H_2O(l) \rightleftharpoons H_3O^+(aq) + BrO^-(aq) \qquad K_a = \frac{[H_3O^+][BrO^-]}{[HBrO]}$$

Ionization of the base triethylamine:

$$(C_2H_5)_3N(aq) + H_2O(l) \rightleftharpoons (C_2H_5)_3NH^+(aq) + OH^-(aq) \qquad K_b = \frac{[(C_2H_5)_3NH^+][OH^-]}{[(C_2H_5)_3N]}$$

The size of the ionization constant gives us the relative strength of an acid or base. A larger ionization constant means that relatively more products are formed. An acid with a larger ionization constant is therefore a stronger proton donor and a stronger acid than one with a smaller ionization constant. Similarly, a base with a larger ionization constant is a stronger proton acceptor and a stronger base than one with a smaller ionization constant. The larger the ionization constant, the stronger the acid (or base).

EXAMPLE   Using equilibrium constants to gauge the relative strengths of acids and bases

Two acids, formic acid (HCOOH) and hydrobromous acid (HBrO), and their ionization constants are given. Write the ionization equilibrium for each, the equilibrium expression for each, and state which is the stronger acid.

$$HCOOH: K_a = 1.8 \times 10^{-4} \qquad HBrO: K_a = 2.0 \times 10^{-9}$$

SOLUTION The ionization equilibrium and equilibrium expression for each are given (we must remember that the concentration of water is not included in the ionization equilibrium expression when water is a solvent):

$$\text{HCOOH(aq)} + \text{H}_2\text{O(l)} \rightleftharpoons \text{H}_3\text{O}^+\text{(aq)} + \text{HCOO}^-\text{(aq)} \qquad K_a = \frac{[\text{H}_3\text{O}^+][\text{HCOO}^-]}{[\text{HCOOH}]}$$

$$\text{HBrO(aq)} + \text{H}_2\text{O(l)} \rightleftharpoons \text{H}_3\text{O}^+\text{(aq)} + \text{BrO}^-\text{(aq)} \qquad K_a = \frac{[\text{H}_3\text{O}^+][\text{BrO}^-]}{[\text{HBrO}]}$$

The acid with the larger equilibrium constant is the stronger of the two. Because 1.8×10^{-4} is greater than 2.0×10^{-9}, formic acid is a stronger acid than hydrobromous acid.

EXERCISE Write the ionization equilibria and equilibrium expressions for the bases pyridine (C_5H_5N, $K_b = 1.8 \times 10^{-9}$) and ammonia (NH_3, $K_b = 1.8 \times 10^{-5}$). State which is the stronger base.

[Answer:

$$\text{C}_5\text{H}_5\text{N(aq)} + \text{H}_2\text{O(l)} \rightleftharpoons \text{C}_5\text{H}_5\text{NH}^+\text{(aq)} + \text{OH}^-\text{(aq)}$$

$$K_b = \frac{[\text{C}_5\text{H}_5\text{NH}^+][\text{OH}^-]}{[\text{C}_5\text{H}_5\text{N}]}$$

$$\text{NH}_3\text{(aq)} + \text{H}_2\text{O(l)} \rightleftharpoons \text{NH}_4^+\text{(aq)} + \text{OH}^-\text{(aq)}$$

$$K_b = \frac{[\text{NH}_4^+][\text{OH}^-]}{[\text{NH}_3]}$$

Ammonia is the stronger base.]

KEY CONCEPT B pK_a, pK_b, and K_w

The so-called p function is quite common in the sciences. For any property or value,

$$\text{p(anything)} = -\log_{10}(\text{anything})$$

If we want to determine the p of some value, we first take the logarithm (to the base 10) of the value and then change the sign of the result. Thus

$$pK_w = -\log K_w$$

$$pK_a = -\log K_a$$

$$pK_b = -\log K_b$$

For most calculators, the logarithm to the base 10 of a number is found by entering the number and pushing the log key. A few examples are

$$pK_w = -\log(1.0 \times 10^{-14}) = -(-14.00) = 14.00$$

$$pK_a(\text{HCOOH}) = -\log(1.8 \times 10^{-4}) = -(-3.74) = 3.74$$

$$pK_b(\text{C}_5\text{H}_5\text{N}) = -\log(1.8 \times 10^{-9}) = -(-8.74) = 8.74$$

EXAMPLE Calculating pK_a, pK_b

K_a for lactic acid is 8.4×10^{-4}. What is pK_a for this acid?

SOLUTION The definition of pK_a is $pK_a = -\log K_a$. Thus,

$$pK_a = -\log K_a$$

$$= \log(8.4 \times 10^{-4})$$

$$= 3.08$$

EXERCISE The pK_b of hydroxylamine is 7.97. What is K_b for this base?

[Answer: 1.1×10^{-8}]

PITFALL   Using the correct logarithm on your calculator

Remember to use the base 10 logarithms on your calculator when doing calculations with the p function, like those just illustrated. The base 10 logarithms usually use a log key to obtain the logarithm of a number and the 10^x key to find the antilog. The corresponding natural logarithm functions (which we will not use until later) are typically obtained by the ln and e^x keys.

14.7   THE CONJUGATE SEESAW

KEY CONCEPT   A strong conjugate acid has a weak conjugate base

If we say that an acid HA is a strong acid, we mean that it easily donates H^+. Thus, A^- does not hold on to the H^+ very well; A^- is a weak base. Similarly, if HA is a weak acid, it does not readily donate H^+. Thus, A^- holds on to the H^+ very strongly, and A^- is a strong base. Thus, for a conjugate acid-base pair:

• A strong conjugate acid has a weak conjugate base.
• A weak conjugate acid has a strong conjugate base.

The terms *weak* and *strong* are qualitative. We can quantify this idea by noting that because $pK_a + pK_b = 14.00$ for a conjugate acid-base pair,

• If a conjugate acid has a small pK_a (is a strong acid), the conjugate base must have a large pK_b (is a weak base).
• If a conjugate acid has a large pK_a (is a weak acid), the conjugate base must have a small pK_b (is a strong base).

EXAMPLE 1   Judging the strength of conjugate acids and bases

Sulfurous acid, H_2SO_3, is a stronger acid that nitrous acid, HNO_2. Which is the stronger base, HSO_3^- or NO_2^-? Do not refer to Table 14.1 of the text.

SOLUTION Because H_2SO_3 is a stronger acid than HNO_2, NO_2^- holds on to protons more strongly than HSO_3^-; therefore, NO_2^- is a stronger proton acceptor than HSO_3^-, which means that NO_2^- is a stronger base.

EXERCISE The trimethylammonium ion, $(CH_3)_3NH^+$, is a stronger acid than the dimethylammonium ion, $(CH_3)_2NH_2^+$. Write the formula of the weaker conjugate base of these two acids. Do not refer to Table 14.1 of the text.

[*Answer:* $(CH_3)_3N$]

EXAMPLE 2   Calculating the pK_b of a conjugate base

What is the pK_a of the acid $C_6H_5NH^+$?

SOLUTION $C_6H_5NH^+$ is the conjugate acid of C_6H_5N (pyridine). The pK_b of pyridine is 8.74. Thus,

$$pK_a(C_6H_5NH^+) + pK_b(C_6H_5N) = 14.00$$
$$pK_a(C_6H_5NH^+) + 8.74 = 14.00$$
$$pK_a(C_6H_5NH^+) = 5.26$$

EXERCISE What is the pK_b of ClO^-? Use Table 14.2 of the text.

[*Answer:* 6.47]

14.8   THE SPECIAL ROLE OF WATER

KEY CONCEPT   Limiting strengths in water

Some acids are so strong that when they are placed in water, they completely donate their protons to water to form H_3O^+:

$$HA(aq) + H_2O(l) \xrightarrow{100\%} H_3O^+(aq) + A^-(aq)$$

In fact, this reaction will occur for any acid HA in which H_2O is a better proton acceptor than A^-. Thus, the proton-accepting ability of water determines whether an acid (in aqueous solution) is a strong acid or a weak acid. If H_2O is a better proton acceptor than A^-, then the acid is a strong acid; if H_2O is a weaker proton acceptor than A^-, then the acid will not donate all its protons to H_2O, and HA is a weak acid.

A similar argument can be used with bases. Some bases are so strong that when they are placed in water, they are able to completely strip protons from water to form OH^-:

$$B(aq) + H_2O(l) \xrightarrow{100\%} BH^+(aq) + OH^-(aq)$$

In fact, this will occur for any base B for which BH^+ is a weaker proton donor than H_2O. Thus, the proton-donating ability of water determines whether a base (in aqueous solution) is a strong base or a weak base. If BH^+ is a weaker proton donor than H_2O, then B is a strong base; if BH^+ is a stronger proton donor than H_2O, then B will not accept protons so strongly from H_2O, and B is a weak base.

The important point here is that the properties of the solvent play an important role in determining the strengths of acids and bases dissolved in the solvent. In water, for instance, all strong acids form H_3O^+ and all strong bases form OH^-. Water, in a sense, makes all strong acids appear the same (as H_3O^+ formers) and all strong bases appear the same (as OH^- formers), so this effect is called the **leveling effect** of water.

EXAMPLE   Determining whether an acid or base is weak or strong

When acid HA is placed in water, it undergoes 100% ionization and completely forms H_3O^+ and A^-; when acid HA' is placed in water, it also undergoes 100% ionization. Because both behave the same in water, is it correct to conclude that HA and HA' have the same acid strength?

SOLUTION No. The fact that both acids behave the same in water shows that they are both strong enough acids to donate their protons completely to H_2O. However, it is possible to imagine a solvent (Sol) that is not as good a proton acceptor as water; in such a solvent, the acids are not 100% ionized, and therefore, their proton-donating abilities are differentiated:

$$HA + Sol \xrightarrow{100\%} SolH^+ + A^-$$

$$HA + Sol \xrightarrow{5\%} SolH^+ + A^-$$

EXERCISE Suppose you have two bases, B and B', which are both strong bases in water, and you would like to see whether they have different base strengths in another solvent. Would you choose a solvent whose protonated form is a better proton donor than protonated water or one that is worse?

[*Answer:* Worse]

14.9   WHY ARE SOME ACIDS WEAK AND OTHERS STRONG?

KEY CONCEPT   Binary acid (HX) strengths

Two factors affect the strength of a binary acid (such as HF or H_2S), the polarity of the H—X bond and the strength of the H—X bond. The more polar the H—X bond, the stronger the acid; and the weaker the H—X bond, the stronger the acid. These two effects can work in opposite directions, so we must be careful when predicting acid strengths, to know which one is most important. In Groups 16 and 17, the strength of the H—X bond determines the acid strength.

EXAMPLE Predicting acid strengths

Which is likely to be the stronger acid, NH_3 or PH_3. The bond strengths are N—H, 388 kJ $\cdot mol^{-1}$, P—H, 343 kJ$\cdot mol^{-1}$. The electronegativities of the atoms are N, 3.0; P, 2.1; H, 2.1.

SOLUTION Neither of these is known as a particularly strong acid; in fact, NH_3 is a well-known base (when dissolved in water.) However, we can imagine the existence of a solvent in which these compounds act as acids. Because they are hydrides from the same group (15), we predict that the strength of the H—X bond will determine the relative acid strength. The N—H bond strength is larger than the P—H bond strength, so we predict that PH_3 will be the stronger acid.

EXERCISE Which is likely to be the stronger acid, CH_4 or H_2S? The bond strengths are C—H, 412 kJ$\cdot mol^{-1}$, S—H, 344 kJ$\cdot mol^{-1}$. The electronegativities of the atoms are C, 2.5; S, 2.5; H, 2.1.

[*Answer:* H_2S]

14.10   THE STRENGTHS OF OXOACIDS

KEY CONCEPT   Electronegative atoms make oxoacids more acidic

In oxoacids, the acidic hydrogen is always attached to an oxygen; all of the O—H bonds have the same bond strength, and it is the polarity of the O—H bond that determines the acid strength. Electron-withdrawing atoms attached to the oxygen atom make the O—H bond more polar and therefore, make the acid more acidic. It should be noted that a high oxidation number tends to make an atom electron withdrawing. These effects are summarized as follows:

• In different oxoacids that have the same central atom but a different number of oxygen atoms surrounding the central atom, the higher the oxidation number of the central atom, the stronger the acid. (It should be remembered that the more oxygen atoms around the central atom, the higher its oxidation number.)
• In different oxoacids that have the same number of oxygen atoms surrounding the central atom, the higher the electronegativity of the central atom, the stronger the acid.

EXAMPLE Predicting acid strengths of oxoacids

Which is likely to be the stronger acid, H_3PO_3 or H_3PO_4?

SOLUTION In this case, the central atom in both acids is phosphorus, so the electronegativity of the central atom is the same in both acids. The deciding factor will be the oxidation number of the central atom or, equivalently, the number of oxygen atoms attached to the central atom. Phosphoric acid (H_3PO_4), with four oxygen atoms attached to the phosphorus (phosphorus oxidation number +5) should be more acidic than phosphorus acid (H_3PO_3), with three oxygen atoms attached to the central phosphorus (phosphorus oxidation number +3).

EXERCISE $HBrO_2$ has not yet been discovered. When (and if) it is, predict whether it should be a stronger or weaker acid than $HClO_2$?

[*Answer:* Weaker]

KEY WORDS   Define or explain each term in a written sentence or two.

acid strength   basicity constant
acidity constant   proton transfer equilibrium

THE pH OF SOLUTIONS OF WEAK ACIDS AND BASES

14.11   SOLUTIONS OF WEAK ACIDS

KEY CONCEPT   Weak acid pH involves an equilibrium

The pH of a weak acid is calculated from the equilibrium H_3O^+ concentration and the definition $pH = -\log[H_3O^+]$. With a weak acid, it is necessary to first solve an equilibrium problem in order to get the equilibrium concentration of H_3O^+.

EXAMPLE   Calculating the pH of a weak acid

What is the pH of a 0.050 M solution of hypochlorous acid, HOCl? $K_a = 3.0 \times 10^{-5}$.

SOLUTION   We first recognize that when 1 mol of HClO ionizes, it forms 1 mol of H_3O^+ and 1 mol of ClO^-. Thus, if we represent the amount of HClO that ionizes by $-x$, the amount of H_3O^+ that forms is $+x$, and the amount of ClO^- that forms is $+x$. The table of concentrations and changes is

	$HClO(aq)$ + $H_2O(l)$ $\rightleftharpoons$ $H_3O^+(aq)$ + $ClO^-(aq)$		
initial concentration	0.050 M	0	0
change to reach equilibrium	$-x$	$+x$	$+x$
equilibrium concentration	0.050 M $- x$	x	x

In setting up the table, we have assumed that the initial 1.0×10^{-7} M concentration of H_3O^+ (from the autoprotolysis of water) is so small relative to the amount of H_3O^+ formed by ionization of the acid that we can consider the initial concentration of H_3O^+ to be 0. The equilibrium expression is

$$\frac{[H_3O^+][ClO^-]}{[HClO]} = K_a$$

Substituting the equilibrium concentration and the value of K_a gives

$$\frac{(x)(x)}{0.050 - x} = 3.0 \times 10^{-5}$$

We now assume that x is much smaller than 0.050 ($x \ll 0.050$), so $0.050 - x = 0.050$.

$$\frac{x^2}{0.050} = 3.0 \times 10^{-5}$$

$$x^2 = 1.5 \times 10^{-6}$$

$$x = 1.2 \times 10^{-3}$$

Before proceeding, we check to see that the assumptions made regarding the size of x are appropriate. Because x is approximately 10,000 times larger than 1.0×10^{-7}, neglecting 1.0×10^{-7} relative to x (i.e., setting the original concentration of H_3O^+ equal to 0 in the table) is justified. Second, x is 2.4% of 0.050:

$$\frac{1.2 \times 10^{-3}}{0.050} \times 100 = 2.4\%$$

As a rule of thumb, as long as x is not greater than 5% of the original concentration of 0.050, the approximation is justified (see the next pitfall).

Recognizing that $[H_3O^+] = x$, we get

$$[H_3O^+] = 1.2 \times 10^{-3} \text{ M}$$

$$pH = -\log[H_3O^+]$$

$$= -\log(1.2 \times 10^{-3})$$

$$= -(-2.92)$$

$$= 2.92$$

EXERCISE What is the pH of a 0.33 M solution of HF(aq)? $K_a(HF) = 3.5 \times 10^{-4}$.

[*Answer:* 1.98]

PITFALL When are approximations about one number being smaller than another correct?

We often make approximations such as

x is much smaller than 0.23, so that

$$0.23 - x = 0.23$$

How do we judge whether such an approximation is justified? In general, calculations involving equilibria as we do them may be as much as 5–10% off the correct values because of effects that we do not take into account. Thus, as long as any approximations do not introduce an error greater than 5%, the approximation is justified. Practically, if we say a "large value" is much greater than a "small value," so that

$$\text{Large value} - \text{small value} = \text{large value}$$

the approximation is justified if

$$\frac{\text{Small value}}{\text{Large value}} \times 100 < 5\%$$

14.12 SOLUTIONS OF WEAK BASES

KEY CONCEPT Weak base pH involves an equilibrium

The pH of a weak base is calculated from the equilibrium H_3O^+ concentration and the definition $pH = -\log[H_3O^+]$. The equilibrium concentration of H_3O^+ is obtained by solving an equilibrium problem to find the concentration of OH^- and then using the relationship $[H_3O^+][OH^-] = 1.0 \times 10^{-14}$. Alternatively, once the concentration of OH^- has been obtained, you can use the relationship $pOH = -\log[OH^-]$ to find the pOH and then calculate the pH by using $pH + pOH = 14.00$.

EXAMPLE Calculating the pH of a weak base

What is the pH of a 0.60 M $(CH_3)_3N(aq)$ solution? $K_b\{(CH_3)_3N\} = 6.5 \times 10^{-5}$.

SOLUTION The equilibrium and table of concentration values for this problem are

	$(CH_3)_3N(aq) + H_2O(l) \rightleftharpoons (CH_3)_3NH^+(aq) + OH^-(aq)$		
initial concentration	0.60 M	0	0
change to reach equilibrium	$-x$	$+x$	$+x$
equilibrium concentration	0.60 M $- x$	x	x

As we did earlier with H_3O^+, we assume that the amount of OH^- formed by the autoprotolysis of water is too small to be of any importance, and we consider the initial concentration of OH^- to be 0 M. The equilibrium expression is

$$\frac{[(CH_3)_3NH^+][OH^-]}{[(CH_3)_3N]} = K_b$$

Substituting the equilibrium values and the value of K_b gives

$$\frac{(x)(x)}{0.60 - x} = 6.5 \times 10^{-5}$$

We now assume that x is much smaller than 0.60 ($x \ll 0.60$) so that $0.60 - x = 0.60$.

$$\frac{x^2}{0.60} = 6.5 \times 10^{-5}$$

$$x^2 = 3.9 \times 10^{-5}$$

$$x = 6.2 \times 10^{-3}$$

x is large with respect to 1.0×10^{-7} and small with respect to 0.60 (x is 1.0% of 0.60), so our approximations are acceptable. Remembering that $x = [OH^-]$, we get

$$[OH^-] = 6.2 \times 10^{-3} \text{ M}$$

$$pOH = -\log[OH^-]$$

$$= -\log(6.2 \times 10^{-3})$$

$$= -(-2.21)$$

$$= +2.21$$

We finally calculate the pH:

$$pH + pOH = 14.00$$

$$pH + 2.21 = 14.00$$

$$pH = 14.00 - 2.21$$

$$= 11.79$$

EXERCISE What is the pH of a 0.080 M $NH_2NH_2(aq)$ solution? $pK_b(NH_2NH_2) = 1.7 \times 10^{-6}$.

[*Answer:* 10.57]

14.13   POLYPROTIC ACIDS AND BASES

KEY CONCEPT   Polyprotic acid equilibrium calculations

Many Brønsted acids are able to donate more than one proton. Such acids are called **polyprotic** (many proton) **acids.** A **diprotic acid** is one that can donate two protons; carbonic acid H_2CO_3, sulfuric acid H_2SO_4, and oxalic acid HOOCCOOH are examples of diprotic acids. Polyprotic acids ionize in a stepwise fashion. First one proton ionizes, then a second, then a third, and so on. Each ionization step has a smaller equilibrium constant than the previous one. For instance, for carbonic acid,

$$H_2CO_3(aq) + H_2O(l) \rightleftharpoons H_3O^+(aq) + HCO_3^-(aq) \qquad K_{a1} = 4.3 \times 10^{-7}$$

$$HCO_3^-(aq) + H_2O(l) \rightleftharpoons H_3O^+(aq) + CO_3^{2-}(aq) \qquad K_{a2} = 5.6 \times 10^{-11}$$

Because of the large difference in K_{a1} and K_{a2} for most diprotic acids, the following approximations may be used when doing equilibrium calculations for them. For a diprotic acid H_2A:

1. The $H_3O^+(aq)$ and $HA^-(aq)$ equilibrium concentrations are determined by the first ionization step only. Thus, to determine the pH, we need only consider the first ionization step.
2. $[H_3O^+] = [HA^-]$ at equilibrium. This conclusion is a direct result of approximation 1.
3. $[A^{2-}] = K_{a2}$. This conclusion is a direct result of approximation 2.

One important exception to these rules occurs for H_2SO_4, in which the first ionization step is 100% complete. The second step has a relatively large equilibrium constant and does contribute H_3O^+ ions; it must be considered in calculating the pH. Also, for any acid, these approximations will not give the correct results if the concentration of acid is extremely low; in such a case, a more rigorous approach must be used.

EXAMPLE   Calculating the pH of a sulfuric acid (H_2SO_4) solution

What is the pH of a 0.020 M H_2SO_4(aq) solution?

SOLUTION The first ionization of H_2SO_4(aq) is 100% complete. For 0.020 M H_2SO_4, this ionization leads to 0.020 M $H_3O^+(aq)$ and 0.020 M HSO_4^-(aq). In the second ionization step, the 0.020 M HSO_4^- ionizes with $K_{a2} = 0.012$. The equilibrium and table of concentration values are

	$HSO_4^-(aq)$ + $H_2O(l)$ $\rightleftharpoons$ $H_3O^+(aq)$ + $SO_4^{2-}(aq)$		
initial concentration	0.020 M	0.020 M	0
change to reach equilibrium	$-x$	$+x$	$+x$
equilibrium concentration	0.020 M $- x$	0.020 M $+ x$	x

There is a new feature in this table. One of the products (H_3O^+) starts out with a nonzero concentration. The first ionization produces 0.020 M H_3O^+, which is present when the second ionization occurs. The equilibrium expression is

$$\frac{[H_3O^+][SO_4^{2-}]}{[HSO_4^-]} = K_{a2}$$

Substituting the equilibrium concentrations and the value of K_{a2} gives

$$\frac{(0.020 + x)(x)}{(0.020 - x)} = 0.012$$

For this example, we cannot make the assumption that x is much smaller than 0.020. Because K_{a2} is relatively large, considerable ionization occurs, and x will turn out to be a significant fraction of 0.020. We must solve for x without any approximations. Multiplying both sides of the equation by $(0.020 - x)$ gives

$$(0.020 + x)(x) = 0.012(0.020 - x)$$

$$0.020x + x^2 = 2.4 \times 10^{-4} - 0.012x$$

Subtracting $(2.4 \times 10^{-4} - 0.012x)$ from both sides of the equation gives

$$0.020x + x^2 - (2.4 \times 10^{-4} - 0.012x) = 0$$

$$x^2 + 0.032x - 2.4 \times 10^{-4} = 0$$

The equation is now in the form $ax^2 + bx + c = 0$, and the quadratic formula can be used to solve it:

$$x = \frac{-b \pm \sqrt{b^2 - 4ac}}{2a}$$

For $a = 1$, $b = 0.032$, and $c = -2.4 \times 10^{-4}$, we get

$$x = \frac{-0.032 \pm \sqrt{0.032^2 - (4)(1)(-2.4 \times 10^{-4})}}{2(1)}$$

$$= \frac{-0.032 \pm \sqrt{0.00102 + 0.00096}}{2}$$

$$= \frac{-0.032 \pm \sqrt{0.00198}}{2}$$

$$= \frac{-0.032 \pm 0.044}{2}$$

The two roots are

$$x = \frac{-0.032 - 0.044}{2} \qquad x = \frac{-0.032 + 0.044}{2}$$

$$= -0.038 \qquad\qquad = 0.0060$$

The negative root would correspond to a negative molarity, which is impossible. This root has no physical meaning and is ignored. Thus, $x = 0.0060$, and from the table of concentrations,

$$[H_3O^+] = 0.020 \text{ M} + x$$

$$= 0.020 \text{ M} + 0.0060 \text{ M}$$

$$= 0.026 \text{ M}$$

$$pH = -\log[H_3O^+]$$

$$= -\log(0.026)$$

$$= 1.59$$

EXERCISE What is the pH of a 0.040 M H_2SO_4(aq) solution?

[*Answer:* 2.19]

KEY WORDS   Define or explain each term in a written sentence or two.

percentage protonated
polyprotic acid
polyprotic base

DESCRIPTIVE CHEMISTRY TO REMEMBER

- **Natural rain** has a pH of about 5.7; the low pH is due largely to dissolved CO_2.
- The pH of **stomach juices** is about 1.7.
- The relative acid strengths of the **hydrohalic acids** (in nonaqueous solution) is

    weakest acid   HF < HCl < HBr < HI   strongest acid

- **Universal indicator paper** turns different colors at different pH values.
- The United States Environmental Protection Agency (EPA) defines waste as *corrosive* if its **pH** is either lower than 3.0 or higher than 12.5.
- All **carboxylic acids** (acids containing the COOH group) are weak acids in water.
- **Ethanol**, CH_3CH_2OH, is the alcohol produced from the fermentation of grains such as wheat, barley, and rye. Its pK_a is 16, so it is not considered an acid.

CHEMICAL EQUATIONS TO KNOW

- HCl is a strong Brønsted acid that, in aqueous solution, donates a proton to water:

$$HCl(aq) + H_2O(l) \rightleftharpoons H_3O^+(aq) + Cl^-(aq)$$

- Hydrogen carbonate ion is a weak Brønsted acid that can donate a proton to water:

$$HCO_3^-(aq) + H_2O(l) \rightleftharpoons H_3O^+(aq) + CO_3^{2-}(aq)$$

- Ammonia is a weak Brønsted base that accepts a proton from water to form OH^-:

$$NH_3(g) + H_2O(l) \rightleftharpoons NH_4^+(aq) + OH^-(aq)$$

- Acetic acid is a weak Brønsted acid that can donate a proton to water:

$$CH_3COOH(aq) + H_2O(l) \rightleftharpoons H_3O^+(aq) + CH_3CO_2^-(aq)$$

- The oxide ion (O^{2-}) is a strong base and completely reacts with water to form hydroxide ion:

$$O^{2-}(aq) + H_2O(l) \longrightarrow 2OH^-(aq)$$

- Adding acid to hydrogen carbonate ion shifts the following equilibrium to the left:

$$H_2CO_3(aq) + H_2O(l) \rightleftharpoons H_3O^+(aq) + HCO_3^-(aq)$$

The carbonic acid thus formed can then decompose to water and carbon dioxide:

$$H_2CO_3(aq) \rightleftharpoons H_2O(l) + CO_2(g)$$

MATHEMATICAL EQUATIONS TO KNOW AND UNDERSTAND

$pK_a = -\log K_a$ $pK_a + pK_b = 14.00$

$pK_b = -\log K_b$ $pOH = -\log[OH^-]$

$pH = -\log[H_3O^+]$ $pH + pOH = 14.00$

$$\text{Percentage ionized} = \frac{\text{molarity of ionized HA}}{\text{initial molarity of HA}} \times 100$$

$$\text{Percentage protonated} = \frac{\text{molarity of protonated base}}{\text{initial molar concentration of base}} \times 100$$

SELF-TEST EXERCISES

What are acids and bases?

1. What is the Brønsted acid in the reaction $NH_3(g) + H_2O(l) \rightleftharpoons NH_4^+(aq) + OH^-(aq)$?
(a) NH_4^+ (b) NH_3 (c) OH^- (d) H_2O

2. What is the Brønsted base in the reaction, $O^{2-}(aq) + H_2O(l) \rightleftharpoons 2OH^-(aq)$?
(a) O^{2-} (b) OH^- (c) H_2O

3. What is the Brønsted base in the reaction, $CH_3NH_2(l) + H_2O(l) \rightleftharpoons CH_3NH_3^+(aq) + OH^-(aq)$?
(a) CH_3NH_2 (b) $CH_3NH_3^+$ (c) OH^- (d) H_2O

4. What is the Brønsted acid in the reaction $H_2S(g) + 2 NH_3(g) \rightleftharpoons (NH_4)_2S(s)$?
(a) H_2S　　　　(b) NH_3　　　　(c) NH_4^+　　　　(d) S^{2-}

5. Which of the following can act both as a Brønsted acid and as a Brønsted base?
(a) CO_2^{2-}　　　(b) CH_4　　　(c) H_2O　　　(d) H_2　　　(e) NH_4^+

6. What are the products of the Brønsted acid-base reaction: $HCN(aq) + H_2O(l) \rightleftharpoons$?
(a) H_2CN^+, OH^-　　　　(b) $HCNO, H_2$　　　　(c) CN^-, H_3O^+
(d) H_2, O^{2-}　　　　(e) CO_2, NO_2, H_3O^+

7. What the products of the Brønsted acid-base reaction $HS^-(aq) + OH^-(aq) \rightleftharpoons$?
(a) H_2S, O^{2-}　　　(b) S^{2-}, H_2O　　　(c) H_2O_2, S_8
(d) SO_2, H_2O　　　(e) H_2SO_4

8. Which of the following is a conjugate acid-base pair in the following equilibrium?

$$F^-(aq) + H_2O(l) \rightleftharpoons HF(aq) + OH^-(aq)$$

(a) F^-, HF　　　(b) HF, H_2O　　　(c) F^-, H_2O
(d) OH^-, HF　　　(e) OH^-, F^-

9. Which of the following is the conjugate acid of NH_3?
(a) NH_2^-　　　(b) NH_4^+　　　(c) HCl　　　(d) H_3O^+　　　(e) H_2O

10. What is the molarity of OH^- in an aqueous solution in which the molarity of H_3O^+ is 4.7×10^{-2} M?
(a) 1.0×10^{-7} M　　　　(b) 4.7×10^{-16} M　　　　(c) 3.3×10^{-5} M
(d) 2.1×10^{-13} M　　　　(e) 6.8×10^{-12} M

11. What is the molarity of H_3O^+ in an aqueous solution in which the molarity of OH^- is 3.8×10^{-6} M?
(a) 6.1×10^{-5} M　　　　(b) 1.0×10^{-7} M　　　　(c) 8.1×10^{-11} M
(d) 3.8×10^{-8} M　　　　(e) 2.6×10^{-9} M

12. The autoprotolysis constant, K_w, for pure water at 40°C is 2.9×10^{-14}? What is $[H_3O^+]$ in pure water at 40°C?
(a) 1.0×10^{-7} M　　　　(b) 3.5×10^{-7} M　　　　(c) 1.7×10^{-7} M
(d) 2.9×10^{-7} M　　　　(e) 6.1×10^{-7} M

13. What is the hydronium ion concentration of a 0.0052 M nitric acid (HNO_3) solution?
(a) 1.9×10^{-12} M　　　　(b) 1.0×10^{-7} M　　　　(c) 0.0104 M
(d) 0.0052 M　　　　(e) 0.0026 M

14. What is the hydronium ion concentration in a 0.025 M sodium hydroxide (NaOH) solution?
(a) 0.025 M　　　　(b) 4.0×10^{-13} M　　　　(c) 0.10 M
(d) 1.9×10^{-12} M　　　　(e) 1.0×10^{-7} M

15. What is the hydronium ion concentration in a 0.002 88 M solution of $Ba(OH)_2$?
(a) 0.002 88 M　　　(b) 174 M　　　(c) 3.5×10^{-12} M
(d) 0.005 76 M　　　(e) 1.2×10^{-9} M

16. What is the pH of lemon juice with a hydronium ion concentration of 2.40×10^{-3} M?
(a) 11.38　　　(b) 3.46　　　(c) 2.62　　　(d) 7.00　　　(e) 8.61

17. What is the pH of seawater that has a hydroxide ion concentration of 1.4×10^{-5} M?
(a) 5.22   (b) 4.85   (c) 8.78   (d) 9.15   (e) 7.00

18. Household ammonia typically has a pH of about 12.0, which classifies it as
(a) neutral.   (b) acidic.   (c) basic.

19. What is the hydronium ion concentration of a solution with pH = 4.40?
(a) 4.0×10^{-5} M   (b) 1.0×10^{-7} M   (c) 2.6×10^{-10} M
(d) 6.1×10^{-6} M   (e) 2.6×10^{4} M

20. What is the hydroxide ion concentration of a solution with pH = 5.80?
(a) 6.3×10^{-9} M   (b) 4.0×10^{-5} M   (c) 1.0×10^{-7} M
(d) 1.6×10^{-6} M   (e) 6.6×10^{-7} M

21. Which pH corresponds to the solution with the highest hydroxide ion concentration?
(a) 6.24   (b) 10.16   (c) 2.68   (d) 7.00

22. What is the pH of a 2.5×10^{-4} M KOH solution?
(a) 9.35   (b) 4.65   (c) 3.60   (d) 7.00   (e) 10.40

23. What is the pH of a 0.001 83 M $HClO_4$ solution?
(a) 11.26   (b) 7.00   (c) −2.74   (d) 2.74   (e) −11.26

24. What is the pOH of a 0.0074 M solution of HCl?
(a) 11.87   (b) 0.0074   (c) 2.13   (d) 7.00   (e) 10.47

25. What is the pOH of a 6.6×10^{-4} M solution of $Ba(OH)_2$?
(a) 10.82   (b) 11.12   (c) 14.00   (d) 2.88   (e) 3.18

26. What is the concentration of hydroxide ions in a solution with a pOH of 4.43?
(a) 1.0×10^{-4} M   (b) 3.5×10^{-12} M   (c) 3.7×10^{-5} M
(d) 2.7×10^{-10} M   (e) 1.0×10^{-4} M

27. What is the concentration of hydronium ions in a solution with a pOH of 9.38?
(a) 4.2×10^{-10} M   (b) 7.0×10^{-14} M   (c) 2.4×10^{-5} M
(d) 3.8×10^{-9} M   (e) 1.0×10^{-9} M

Weak acids and bases

28. The acidity constant for formic acid is 3.5×10^{-4}. What is the pK_a for this acid? Do not refer to your text.
(a) 14.00   (b) 3.75   (c) 7.00   (d) 10.25   (e) 4.00

29. The pK_a of hydrocyanic acid is 9.31. What is K_a for this acid? Do not refer to your text.
(a) 1.0×10^{-7} (b) 7.0×10^{-14} (c) 2.0×10^{-5}
(d) 4.9×10^{-10} (e) 3.1×10^{-9}

30. What is the ionization equilibrium expression, K_a for the following equilibrium?

$$HNO_2(aq) + H_2O(l) \rightleftharpoons H_2O^+(aq) + NO_2^-(aq)$$

(a) $[H_3O^+][NO_2^-]/[HNO_2][H_2O]$ (b) $[NO_2^-]/[HNO_2]$ (c) $[NO_2^-]/[HNO_2][H_2O]$
(d) $[H_3O^+][NO_2^-]/[HNO_2]$ (e) $[H_3O^+]/[H_2O]$

31. Which of the following acids is the strongest acid? Each is listed with its K_a.

(a) HF, 3.5×10^{-4} (b) CH_3COOH, 1.8×10^{-5} (c) HBrO, 2.0×10^{-9}

(d) HClO, 3.0×10^{-5} (e) H_2SO_3, 1.6×10^{-2}

32. A 1.0 M solution of which base has the lowest $[H_3O^+]$? Each base is listed with its K_b. It is not necessary to calculate $[H_3O^+]$ for each base.

(a) $CO(NH_2)_2$, 1.3×10^{-14} (b) NH_3, 1.8×10^{-5} (c) nicotine, 1.0×10^{-6}

(d) $C_2H_5NH_2$, 6.5×10^{-4} (e) NH_2OH, 1.1×10^{-8}

33. The pK_a's of five acids are shown. Which corresponds to the strongest acid?

(a) 10.22 (b) 8.61 (c) 6.52 (d) 12.85 (e) 9.33

34. What is the pK_b of a base with $K_b = 8.2 \times 10^{-8}$?

(a) 8.91 (b) −7.09 (c) 6.31 (d) 7.09 (e) −8.91

35. The pK_b's of five bases are shown. Which has a conjugate acid that is a weaker acid than an acid with $pK_a = 6.23$?

(a) 3.12 (b) 10.13 (c) 7.89 (d) both a and c (e) both b and c

36. The pK_a of hypobromous acid (HBrO) is 8.69. What is the pK_b of the hypobromite ion (BrO^-)?

(a) 14.00 (b) 1.31 (c) 7.00 (d) 8.69 (e) 5.31

37. The pK_b of hydrazine (NH_2NH_2) is 5.77. What is K_a for the hydrazinium ion ($NH_2NH_3^+$)?

(a) 5.9×10^{-9} (b) 1.7×10^{-6} (c) 1.0×10^{-5}

(d) 3.6×10^{-10} (e) 8.1×10^{-11}

38. The K_a of benzoic acid (C_6H_5COOH) is 6.5×10^{-5}. What is K_b for the benzoate ion ($C_6H_5COO^-$)?

(a) 1.5×10^{-10} (b) 1.5×10^{-4} (c) 6.5×10^{-5}

(d) 1.0×10^{-14} (e) 6.5×10^{-19}

39. Which solution contains the highest concentration of OH^- ions?

(a) 0.20 M NH_2NH_2 (b) 0.20 M LiOH (c) 0.20 M $Ba(OH)_2$

(d) 0.20 M NH_3 (e) 0.20 M $(C_2H_5)_3N$

40. NH_2^- is a stronger base than OH^-. Which does the following equilibrium favor?

$$NH_2^-(aq) + H_2O(l) \rightleftharpoons NH_3(aq) + OH^-(aq)$$

(a) products (b) reactants (c) neither

41. The equilibrium $HBrO(aq) + H_2O(l) \rightleftharpoons H_3O^+(aq) + BrO^-(aq)$ forms some products but strongly favors reactants. What is the strongest acid in the equilibrium?

(a) H_2O (b) HBrO (c) BrO^- (d) H_3O^+

42. What is the strongest acid that can exist in water?

(a) HCl (b) H_3O^+ (c) NH_4^+ (d) HF (e) H_2O

43. What is the strongest Brønsted base that can exist in water?

(a) NH_2^- (b) OH^- (c) NH_2NH_2 (d) NH_3 (e) CH_3^-

44. Ethylamine, $C_2H_5NH_2$, is a base. What is the equilibrium that occurs when it is placed in water?

(a) $C_2H_5NH_2(aq) + H_2O(l) \rightleftharpoons H_3O^+(aq) + C_2H_5NH^-(aq)$

(b) $C_2H_5NH_2(aq) \rightleftharpoons NH_2^-(aq) + C_2H_5^+(aq)$

(c) $C_2H_5NH_2(aq) + H_2O(l) \rightleftharpoons OH^-(aq) + C_2H_5NH_3^+(aq)$

(d) $C_2H_5NH_2(aq) + H_2O(l) \rightleftharpoons OH^-(aq) + C_2H_5NH^+(aq) + H_2(g)$

45. Which is predicted to be the strongest acid?

(a) $HMnO_4$ (b) H_3PO_4 (c) $HTcO_4$ (d) H_2TiO_4 (e) HIO_4

46. Which of the following is the strongest acid?

(a) $HClO$ (b) $HClO_2$ (c) $HClO_3$ (d) $HClO_4$

47. Without referring to any references, predict which is the strongest carboxylic acid.

(a) $CHCl_2COOH$ (b) $CH_2ClCOOH$ (c) CH_3COOH (d) CCl_3COOH

The pH of solutions

48. What is the hydronium ion concentration of a 0.30 M HBrO solution?

(a) 4.5×10^{-5} M   (b) 2.4×10^{-5} M   (c) 1.0×10^{-7} M

(d) 6.0×10^{-10} M   (e) 8.1×10^{-4} M

49. What is the pH of a 0.50 M HCN solution?

(a) 7.00   (b) 9.68   (c) 5.55   (d) 4.80   (e) 2.43

50. What is the pH of a 2.0 M hydrazine solution?

(a) 10.04   (b) 7.00   (c) 2.74   (d) 11.27   (e) 8.26

51. What is the pH of a 0.25 M ammonia solution?

(a) 11.32   (b) 9.55   (c) 7.00   (d) 1.64   (e) 8.47

52. What is the hydronium ion concentration of a 0.40 M iodic acid solution?

(a) 0.35 M   (b) 0.19 M   (c) 0.26 M   (d) 0.40 M   (e) 0.068 M

53. What is the pH of a 0.0010 M $H_2SO_4(aq)$ solution? $K_{a2} = 0.012$.

(a) 3.00   (b) 2.73   (c) 1.54   (d) 1.74   (e) 2.22

54. What is the pH of a 0.0025 M H_2CO_3 solution?

(a) 4.49   (b) 9.51   (c) 7.00   (d) 6.37   (e) 7.63

55. What is the concentration of A^{2-} in a 0.20 M solution of H_2A, where H_2A is an unknown acid with $pK_{a1} = 4.22$, $pK_{a2} = 8.15$.

(a) 3.5×10^{-3} M   (b) 6.0×10^{-5} M   (c) 8.1×10^{-10} M

(d) 7.1×10^{-9} M   (e) 3.8×10^{-5} M

Descriptive chemistry

56. Which of the following contributes to the natural acidity of streams?

(a) HCl (b) H_2SO_3 (c) $HClO_3$ (d) H_2CO_3 (e) HF

57. Which group is characteristic of carboxylic acids?

(a) $-NH_2$ (b) $-OH$ (c) $-COOH$ (d) $-PO_3$ (e) $-CH_3$

58. What species is formed when O_2^- is placed in water?

(a) O_2 (b) H_2O (c) OH^- (d) H_2O_2

(e) None of these; O_2^- simply dissolves.

CHAPTER 15

SALTS IN WATER

One of the most unexpected phenomena a chemistry student is likely to encounter is the existence of salts that act as acids and bases. We usually think of acids as compounds containing hydrogen, and compounds that contain hydroxide ions and amines as bases. In this chapter we explore both the acid-base character of salt ions and salt solubilities.

IONS AS ACIDS AND BASES

15.1   IONS AS ACIDS

KEY CONCEPT   Cations as acids

One way for a salt to change the pH of a solution is through the reaction of its cation. There are two common types of cation reactions that change the pH. In one, a cation such as a protonated amine reacts with water to form hydronium ion, for example,

$$NH_2NH_3^+(aq) + H_2O(l) \rightleftharpoons NH_2NH_2(aq) + H_3O^+(aq)$$

$$(C_2H_5)_3NH^+(aq) + H_2O(l) \rightleftharpoons (C_2H_5)_3N(aq) + H_3O^+(aq)$$

These are acid ionizations; they result in formation of H_3O^+. The pK_a of each can be calculated from the pK_b of the unprotonated amine, as listed in Table 14.2 of the text. In a second type of reaction, a hydrated metal cation loses a proton in an ionization:

$$[Al(H_2O)_6]^{3+}(aq) + H_2O(l) \rightleftharpoons [Al(H_2O)_5(OH)]^{2+}(aq) + H_3O^+(aq)$$

$$[Fe(H_2O)_6]^{2+}(aq) + H_2O(l) \rightleftharpoons [Fe(H_2O)_5(OH)]^+(aq) + H_3O^+(aq)$$

All metal cations except those from Group 1, those from Group 2, and those with a +1 charge undergo such reactions. Table 15.1 of the text lists the pK_a values of some of these cations. Both types of equilibria increase the acidity of a solution by increasing the concentration of $H_3O^+(aq)$.

EXAMPLE 1   Predicting whether a salt solution is acidic, basic, or neutral

Predict whether an aqueous solution of hydrazinium bromide, NH_2NH_3Br, is acidic, basic, or neutral.

SOLUTION We first must decide what ions are formed when the salt is dissolved in water. In this case the salt is made of $NH_2NH_3^+$ and Br^- ions. Br^- is the conjugate base of the strong acid HBr and is, therefore, a very weak base; it will not affect the pH of the solution. However, the hydrazinium ion, $NH_2NH_3^+$, possesses a proton that will ionize; that is, because hydrazine is a weak base, its conjugate acid is relatively strong.

$$NH_2NH_3^+(aq) + H_2O(l) \rightleftharpoons NH_2NH_2(aq) + H_3O^+(aq)$$

A solution of NH_2NH_3Br, is, therefore, acidic.

EXERCISE Predict whether an aqueous solution of methylammonium nitrate, $CH_3NH_3NO_3$, is acidic, basic, or neutral. Write any equilibria involved.

[*Answer:* Acidic; $CH_3NH_3^+(aq) + H_2O(l) \rightleftharpoons CH_3NH_2(aq) + H_3O^+(aq)$]

EXAMPLE 2   Predicting whether a salt solution is acidic, basic, or neutral

Is a salt solution of sodium bicarbonate ($NaHCO_3$) acidic, basic, or neutral? $pK_a(HCO_3^-) = 10.25$, $pK_a(H_2CO_3) = 3.69$.

SOLUTION Na^+ has no effect on the pH of the solution, but we must consider two possible equilibria for the hydrogen carbonate ion. In the first, it acts as an Brønsted acid; in the second, as a Brønsted base.

$$HCO_3^-(aq) + H_2O(l) \rightleftharpoons H_3O^+(aq) + CO_3^{2-}(aq) \qquad pK_a = 10.25$$

$$HCO_3^-(aq) + H_2O(l) \rightleftharpoons H_2CO_3(aq) + OH^-(aq) \qquad pK_b = 10.32$$

If the first equilibrium predominates, the solution is acidic because H_3O^+ is formed in this equilibrium. If the second predominates, the solution is basic because OH^- is formed. In this case, the equilibrium constants for both equilibria are almost the same, so neither equilibrium prevails, and the solution is neutral. (The pK_b of the second equilibrium is calculated using $pK_a + pK_b = 14.00$, and the pK_a of carbonic acid, 3.69.)

EXERCISE Predict whether a solution of $NaHSO_4$ is acidic, basic, or neutral. $pK_a(HSO_4^-) = 1.92$. Write any equilibria involved.

[*Answer:* Acidic; $HSO_4^-(aq) + H_2O(l) \rightleftharpoons H_3O^+(aq) + SO_4^{2-}(aq)$]

15.2   IONS AS BASES

KEY CONCEPT   Anions as bases

An anion that is part of a salt may react with water to form OH^- ion and, therefore, increase the pH of a solution. If the anion is the conjugate base of a weak acid, the equilibrium may result in surprisingly large amounts of OH^-.

EXAMPLE   Predicting whether a salt solution is acidic, basic, or neutral

Predict whether a solution of sodium cyanide ($NaCN$) is acidic, basic, or neutral.

SOLUTION Na^+ cannot act as either a Brønsted base or Brønsted acid; it does not affect the acidity of the solution. However, CN^- is the conjugate base of the weak acid $HCN(pK_a = 9.31)$, so CN^- reacts with H_2O to form HCN and OH^-. That is, the following equilibrium results in the formation of OH^- ion; a solution of NaCN is basic.

$$CN^-(aq) + H_2O(l) \rightleftharpoons HCN(aq) + OH^-(aq)$$

EXERCISE Predict whether a solution of $Ca(NO_3)_2$ is acidic, basic, or neutral. Write any equilibria involved.

[*Answer:* Neutral; no equilibrium occurs.]

15.3   THE pH OF SALT SOLUTIONS

KEY CONCEPT   Salt solutions can be acidic or basic

Because a salt contains an ion that may be an acid or a base, the pH of a salt can be quite different from 7. To calculate the pH of a salt solution, we must consider both the cation and the anion. We must decide whether either the cation or the anion has acid or base character, then write the equilibrium involved and solve the equilibrium problem, using the techniques learned in the previous chapter.

EXAMPLE 1   Calculating the pH of a salt solution

What is the predicted pH of a 0.20 M $CH_3NH_3^+Cl^-(aq)$ solution? $pK_b(CH_3NH_2) = 3.44$.

SOLUTION Cl^- is a very weak Brønsted base and has no effect on the pH of the solution, so we focus on the methylammonium ion, $CH_3NH_3^+$. It participates in the equilibrium

$$CH_3NH_3^+(aq) + H_2O(l) \rightleftharpoons H_3O^+(aq) + CH_3NH_2(aq)$$

We calculate K_a for the equilibrium, using the given value of $pK_b(CH_3NH_2)$:

$$pK_a(CH_3NH_3^+) + pK_b(CH_3NH_2) = 14.00$$

$$pK_a(CH_3NH_3^+) + 3.44 = 14.00$$

$$pK_a(CH_3NH_3^+) = 14.00 - 3.44 = 10.56$$

$$K_a = 10^{-10.56} = 2.8 \times 10^{-11}$$

The table of concentrations is

	$CH_3NH_3^+(aq)$ + $H_2O(l)$ $\rightleftharpoons$ $H_3O^+(aq)$ + $CH_3NH_2(aq)$		
initial concentration	0.20 M	0	0
change to reach equilibrium	$-x$	$+x$	$+x$
equilibrium concentration	0.20 M $- x$	x	x

The initial concentration of H_3O^+ is 0 because we assume the 1×10^{-7} M concentration of H_3O^+ formed by the autoprotonation of water is small enough to be neglected relative to the large amount of H_3O^+ formed in the equilibrium. The initial concentration of $CH_3NH_3^+$ (0.20 M) originates with the complete ionization of the 0.20 M CH_3NH_3Cl. Substituting the equilibrium values from the table into the equilibrium expression gives

$$\frac{[H_3O^+][CH_3NH_2]}{[CH_3NH_3^+]} = 2.8 \times 10^{-11}$$

$$\frac{(x)(x)}{0.20 - x} = 2.8 \times 10^{-11}$$

We now assume that x is so much smaller than 0.20, so that $0.20 - x = 0.20$. Thus

$$\frac{x^2}{0.20} = 2.8 \times 10^{-11}$$

$$x^2 = 0.20 \times 2.8 \times 10^{-11} = 5.6 \times 10^{-12}$$

$$x = 2.4 \times 10^{-6}$$

We note that x is very small relative to 0.20. It is also fairly large relative to 1×10^{-7} (24 times as large), so the approximations used are acceptable. x equals the equilibrium hydronium concentration, so we can determine the pH:

$$[H_3O^+] = 2.4 \times 10^{-6}$$

$$pH = -\log[H_3O^+]$$

$$= -\log(2.4 \times 10^{-6})$$

$$= 5.63$$

EXERCISE What is the pH of a 0.085 M solution of $NH_2NH_3^+Cl^-$? $pK_b(NH_2NH_2) = 5.77$.

[*Answer:* 4.65]

EXAMPLE 2   Calculating the pH of a salt solution

What is the pH of a 0.088 M potassium hypochlorite (KClO) solution? $pK_a = 7.53$ for hypochlorous acid, HClO.

SOLUTION As always in this type of problem the first task is to write the equilibrium. Potassium ion (K^+) does not act as a Brønsted acid, so it does not affect the pH of the solution. We therefore focus on the hypochlorite ion. It participates in the equilibrium shown; the table of concentration values for this problem is also included.

	$ClO^-(aq)$	+	$H_2O(l) \rightleftharpoons$	$HClO(aq)$ +	$OH^-(aq)$
initial concentration	0.088 M			0	0
change to reach equilibrium	$-x$			$+x$	$+x$
equilibrium concentration	0.088 M $- x$			x	x

We calculate K_b for the equilibrium:

$$pK_b(ClO^-) + pK_a(HClO) = 14.00$$

$$pK_b(ClO^-) + 7.53 = 14.00$$

$$pK_b(ClO^-) = 6.47$$

$$K_b = 10^{-6.47}$$

$$= 3.4 \times 10^{-7}$$

The equilibrium expression is

$$\frac{[HClO][OH^-]}{[ClO^-]} = 3.4 \times 10^{-7}$$

Substituting in the equilibrium concentrations from the table of values gives

$$\frac{(x)(x)}{0.088 - x} = 3.4 \times 10^{-7}$$

We assume that 0.088 is much greater than x, so that $0.088 - x = 0.088$.

$$\frac{x^2}{0.088} = 3.4 \times 10^{-7}$$

$$x^2 = 3.0 \times 10^{-8}$$

$$x = 1.7 \times 10^{-4}$$

The parameter x corresponds to the OH^- ion concentration. To calculate the pH in the most convenient way, we use

$$pH + pOH = 14.00$$

$$pOH = -\log[OH^-]$$

$$= -\log(x)$$

$$= -\log(1.7 \times 10^{-4})$$

$$= 3.76$$

$$pH + 3.76 = 14.00$$

$$pH = 10.24$$

The pH corresponds to a basic solution. This result is expected because the equilibrium forms OH^- as one of the products.

EXERCISE What is the pH of a 0.088 M potassium hypobromite (KBrO) solution? $pK_a(HBrO) = 8.69$.

[*Answer:* 10.82]

EXAMPLE 3 Calculating the pH of a salt solution

What is the predicted pH of a 0.23 M $Co(NO_3)_2$ solution? $pK_a(Co^{2+}) = 8.89$. Assume only one proton is lost from the hydrated ion.

SOLUTION The NO_3^- ion is a very weak Brønsted base and it does not affect the pH of the solution, so we focus on the cobalt(II) ion (Co^{2+}). We do not know how many waters of hydration solvate the Co^{2+} ion, but this information is not required because we are told that only one proton ionizes from the hydrated ion. If there are m waters of hydration, the equilibrium and table of values for this problem are as follows:

	$Co(H_2O)_m^{2+}(aq)$ + $H_2O(l)$ $\rightleftharpoons$ $H_3O^+(aq)$ + $Co(H_2O)_{m-1}(OH)^+(aq)$		
initial concentration	0.23 M	0	0
change to reach equilibrium	$-x$	$+x$	$+x$
equilibrium concentration	0.23 M $- x$	x	x

Thus,

$$\frac{[H_3O^+][Co(H_2O)_{m-1}(OH)^+]}{[Co(H_2O)_m^{2+}]} = K_a = 10^{-8.89}$$

$$= 1.3 \times 10^{-9}$$

With the approximation that x is much smaller than 0.23, and that $0.23 - x = 0.23$, we get

$$\frac{x^2}{0.023} = 1.3 \times 10^{-9}$$

$$x^2 = 3.0 \times 10^{-10}$$

$$x = 1.7 \times 10^{-5}$$

$$[H_3O^+] = 1.7 \times 10^{-5}$$

$$pH = -\log(1.7 \times 10^{-5})$$

$$= 4.76$$

EXERCISE What is the pH of a 0.30 M $Ni(NO_3)_2$ solution. $pK_a(Ni^{2+}) = 10.60$. Assume that only one proton is lost from the hydrated ion.

[*Answer: 5.56*]

15.4 THE pH OF MIXED SOLUTIONS

KEY CONCEPT Adding the salt of an acid to a solution of the acid affects the pH

For our purposes, a mixed solution is a solution of a weak acid or a weak base that also contains a salt of the acid or base; for instance, a solution of HF that also contains NaF would be a mixed solution. To calculate the pH of such a solution, we must do a standard equilibrium calculation, using a table of concentrations. Because the salt is present, one of the ions formed from the equilibrium will have a nonzero concentration before the equilibrium occurs.

EXAMPLE Calculating the pH of a mixed solution

What is the pH of a solution that is 0.25 M in C_6H_5COOH and 0.15 M in NaC_6H_5COO? $pK_a(C_6H_5COOH) = 4.19$.

SOLUTION Sodium benzoate, like any other sodium salt, ionizes completely in aqueous solution, so a solution that is 0.15 M in NaC_6H_5COO actually contains 0.15 M $Na^+(aq)$ and 0.15 M $C_6H_5COO^-(aq)$. Our solution is a mixed solution of a weak acid and its conjugate base as a common ion. The table of concentration follows.

	$C_6H_5COOH(aq) + H_2O(l) \rightleftharpoons H_3O^+(aq) + C_6H_5COO^-(aq)$		
before equilibrium	0.25 M	0 M	0.15 M
change to reach equilibrium	$-x$	$+x$	$+x$
after equilibrium	0.25 M $- x$	x	0.15 M $+ x$

Because $pK_a = 4.19$, $K_a = 10^{-4.19} = 6.5 \times 10^{-5}$. The equilibrium expression is

$$\frac{[H_3O^+][C_6H_5COO^-]}{[C_6H_5COOH]} = K_a$$

Substituting the equilibrium values from the table gives

$$\frac{(x)(0.15 + x)}{(0.25 - x)} = 6.5 \times 10^{-5}$$

We assume that $x \ll 0.15$, so $0.15 + x = 0.15$, and that $x \ll 0.25$, so $0.25 - x = 0.25$. Thus,

$$\frac{x(0.15)}{(0.25)} = 6.5 \times 10^{-5}$$

$$x = \frac{(0.25)(6.5 \times 10^{-5})}{0.15}$$

$$= 1.1 \times 10^{-4}$$

We note that because $0.25 > 1.1 \times 10^{-4}$ and $0.15 > 1.1 \times 10^{-4}$, our assumptions regarding the small size of x are justified. Also, because $1.1 \times 10^{-4} > 1.0 \times 10^{-7}$, the relatively small amount of hydronium formed by the autoprotolysis of water can be ignored. Because x equals the hydronium ion concentration at equilibrium,

$$[H_3O^+] = 1.1 \times 10^{-4} \text{ M}$$

$$pH = -\log[H_3O^+]$$

$$= -\log(1.1 \times 10^{-4})$$

$$= -(-3.96)$$

$$= 3.96$$

EXERCISE What is the pH of a solution that is 0.55 M in HF and 0.75 M in NaF? $pK_a(HF) = 3.75$.

[*Answer:* 3.88]

TITRATIONS

15.5 STRONG ACID-STRONG BASE TITRATIONS

KEY CONCEPT   Calculating the pH at various points in a titration

A strong base is titrated with a strong acid by adding to the base a carefully measured volume of acid of known concentration. An important parameter in a titration is the pH at various points, because knowledge of the pH helps in choosing the correct indicator for the titration. We shall explore how the pH changes when the concentration of the base is known in advance. For the titration of NaOH(aq) with HCl(aq), the net reaction is

$$OH^-(aq) + H_3O^+(aq) \longrightarrow 2H_2O(l)$$

We should note that the stoichiometry of the reaction tells us that for every 1 mmol of acid added (giving 1 mmol of H_3O^+), 1 mmol of base reacts. Typically, we want to know the pH at four different stages of the titration:

1. Before the addition of any acid.
2. After addition of some acid but before the stoichiometric point is reached.
3. At the stoichiometric point (the point at which just enough acid has been added to react with all the base present at the start).
4. After the stoichiometric point (i.e., when more acid is added after the stoichiometric point has been reached).

At all points of the titration, the total volume equals the sum of the initial volume of base plus the volume of acid added. The total volume of solution increases as the titration is done; the volume of added acid must be taken into account.

EXAMPLE   Calculating the pH at various points in a titration

10.00 mL of a 0.210 M NaOH(aq) solution is titrated with 0.0911 M HCl(aq). What is the pH of the titration mixture (a) before addition of any acid; (b) after addition of 12.00 mL HCl; (c) at the stoichiometric point (after addition of 23.05 mL HCl); and (d) after addition of a total of 30.00 mL HCl?

SOLUTION Throughout the solution to this problem, we shall use V_{total} to represent the total volume of titration mixture, in milliliters.

$$V_{total} = 10.00 + V_{acid}$$

where V_{acid} = the total volume of acid added and 10.00 mL is the original volume of base given in the problem.

(a) Before addition of any acid, the OH^- ion concentration is the same as that of the NaOH in solution:

$$[OH^-] = 0.210$$

$$pOH = -\log[OH^-] = -\log(0.210)$$

$$= +0.68$$

$$pH + pOH = 14.00$$

$$pH = 14.00 - 0.68$$

$$= 13.32$$

(b) After addition of 12.00 mL HCl, V_{total} = 10.00 mL + 12.00 mL = 22.00 mL. The pH at this point is determined by the millimoles of *unreacted* OH^- left in solution. This, in turn, is equal to the initial millimoles of OH^- minus the millimoles of OH^- that have reacted as a result of addition of acid:

$$\text{Millimoles unreacted } OH^- = \text{initial mmol } OH^- - \text{mmol } OH^- \text{ reacted with added acid}$$

$$\text{Initial millimoles } OH^- = 10.00 \text{ mL} \times 0.210 \frac{\text{mmol}}{\text{mL}} = 2.10 \text{ mmol}$$

$$\text{Millimoles } OH^- \text{ reacted} = 12.00 \text{ mL} \times 0.0911 \frac{\text{mmol}}{\text{mL}} = 1.09 \text{ mmol}$$

$$\text{Millimoles unreacted } OH^- = 2.10 \text{ mmol} - 1.09 \text{ mmol} = 1.01 \text{ mmol}$$

The molarity of OH^- equals the millimoles OH^- divided by the total volume of solution:

$$\text{Molarity of } OH^- = \frac{1.01 \text{ mmol}}{22.00 \text{ mL}} = 0.0459 \text{ mol} \cdot L^{-1}$$

From this, we calculate the pOH and the pH:

$$pOH = -\log[OH^-] = -\log(0.0459)$$

$$= 1.34$$

$$pH + pOH = 14.00$$

$$pH + 1.34 = 14.00$$

$$pH = 14.00 - 1.34$$

$$= 12.66$$

(c) At the stoichiometric point, the amount of acid added (23.05 mL, 2.10 mmol) is exactly that required to react with the original amount of base (10.00 mL, 2.10 mmol) present. In addition, neither Na^+ (from the NaOH) nor Cl^- (from HCl) affects the pH of the solution. Thus, there is no excess OH^- or H_3O^+ present in the titration mixture, only the amount that forms from the autoionization of water. The mixture (an NaCl solution) is neutral, with pH = 7.00.

(d) The total volume of the solution after addition of 30.00 mL HCl is 40.00 mL. (V_{total} = 10.00 mL + 30.00 mL.) At this point, a total of 2.73 mmol of acid have been added. The original 2.10 mmol of base have reacted with 2.10 mmol of this acid; and there is an extra 0.63 mmol ($2.73 - 2.10 = 0.63$) of acid. This extra acid now determines the pH of the solution:

$$\text{Molarity of } H_3O^+ = \frac{0.63 \text{ mmol}}{40.00 \text{ mL}}$$

$$= 0.016 \text{ mol} \cdot L^{-1}$$

$$pH = -\log[H_3O^+]$$

$$= 1.80$$

EXERCISE 20.00 mL of a 0.0650 M NaOH(aq) solution is titrated with 0.1044 M HCl(aq). What is the pH of the reaction mixture (a) before addition of any acid, (b) after addition of 6.00 mL HCl, (c) at the stoichiometric point (after addition of 12.45 mL HCl), and (d) after addition of a total of 20.00 mL HCl?

[*Answer:* (a) 12.81; (b) 12.38; (c) 7.00; (d) 1.17]

15.6 STRONG ACID-WEAK BASE AND WEAK ACID-STRONG BASE TITRATIONS

KEY CONCEPT   Calculating the pH at various points in a titration

The anion of a weak acid is a strong Brønsted base. As a weak acid is titrated, the anion is liberated from the acid and may affect the pH of the titration mixture. For instance, F^- ion is liberated as HF is titrated; the effect of the F^- on the pH of the solution must be taken into account. (For a strong acid, such as HCl, the anion liberated is not a strong base; it has no effect on the pH.) We are, again, interested in the pH at four stages of the titration:

1. Before the addition of any base; the calculation is the same as that for the pH of a weak acid.
2. After addition of some base but before the stoichiometric point is reached; this calculation is the same as that for the pH of a mixed solution.
3. At the stoichiometric point; this calculation is the same as that of the pH of a salt.
4. After the stoichiometric point; this calculation is the same as that for the dilution of a base.

(The calculations for the titration of a weak base with a strong acid are similar to those for the titration of a weak acid with a strong base. An extra exercise is included for practice.)

EXAMPLE   Calculating the pH during the titration of a weak acid with a strong base

Calculate the pH at the following points for the titration of 25.00 mL of 0.100 M HOCl with 0.150 M NaOH (a) before addition of any base; (b) after addition of 10.00 mL of base; (c) at the stoichiometric point (16.33 mL of base total); and (d) after addition of 5.00 mL of base beyond the stoichiometric point. K_a(HOCl) = 3.0×10^{-8} M, pK_a = 7.53.

SOLUTION First, we prepare a table showing the various amounts (in millimoles) of each of the chemical species involved at each stage of the titration after the added base reacts with acid but *before equilibrium*. To do this, we recognize from the equation for the titration,

$$HOCl(aq) + OH^-(aq) \longrightarrow H_2O(l) + OCl^-(aq)$$

that 1 mmol NaOH reacts with 1 mmol HOCl to produce 1 mmol OCL⁻. Also, the millimoles of OH⁻ are the same as the millimoles of NaOH, because NaOH is 100% ionized. (The millimoles of NaOH are calculated by using the relation millimole solute = milliliter solution × molarity.) There are 2.50 mmol HOCl at the start of the titration (25.00 mL × 0.100 M = 2.50 mmol).

NaOH Added, mL	OH⁻ Added, mmol	HOCl Left After Reaction with OH⁻, mmol	OCl⁻ Formed After Reaction, mmol	OH⁻ Left After Reaction, mmol	V_{total}
0.00	0.00	2.50	0.00	0.00	25.00
10.00	1.50	1.00	1.50	0.00	35.00
16.33	2.50	0.00	2.50	0.00	41.33
21.33	3.20	—	—	0.70	46.33

After the equivalence point (16.33 mL), there is no HOCl left to react, so two entries are left blank. The first row of entries applies to the situation before any NaOH is added, so no reaction is possible at this point, and the words *after reaction* in the column headings should be ignored. Before proceeding, you should confirm all of the entries in the table!

We also need K_b for this problem:

$$pK_b + pK_a = 14.00$$

$$pK_b = 14.00 - 7.53$$

$$= 6.47$$

$$K_b = 10^{-6.47}$$

$$= 3.4 \times 10^{-7}$$

The preceding table shows amounts before equilibrium. The calculation that follow take into account any equilibrium process that occurs at each stage of the titration.

(a) Before addition of any base, the problem is the same as that of determining the pH of a 0.100 M solution of HOCl. The equilibrium and table of values are as follows.

	HOCl(aq)	+	H₂O(l) ⇌	H₃O⁺(aq)	+	OCl⁻(aq)
initial concentration	0.100 M			0		0
change to reach equilibrium	$-x$			$+x$		$+x$
equilibrium concentration	0.100 M $- x$			x		x

Substituting the equilibrium values into the equilibrium expression gives

$$\frac{x^2}{0.100 - x} = 3.0 \times 10^{-8}$$

If we assume that x is small and that $0.100 - x = 0.100$, then

$$x^2 = 3.0 \times 10^{-9}$$

$$[H_3O^+] = x = 5.5 \times 10^{-5}$$

$$pH = 4.26$$

(b) The table at the very start of the problem indicates that, after addition of 10.00 mL of NaOH, the solution consists of a mixture of a weak acid (HOCl) and its conjugate base (OCl$^-$); so it is a mixed solution. We calculate the pH by using a table of concentrations. However, our first task is to calculate the initial concentrations of the two species involved in the equilibrium, HOCl and OCl$^-$.

$$\text{Molarity of HOCl} = \frac{1.00 \text{ mmol}}{35.00 \text{ mL}} = 0.0286 \text{ mol} \cdot L^{-1}$$

$$\text{Molarity of OCl}^- = \frac{1.50 \text{ mmol}}{35.00 \text{ mL}} = 0.0429 \text{ mol} \cdot L^{-1}$$

Now we can set up the table of concentrations.

	HOCl(aq)	+	H$_2$O(l) $\rightleftharpoons$ H$_3$O$^+$(aq)	+	OCl$^-$(aq)
before equilibrium	0.0286 M		0 M		0.0429 M
change to reach equilibrium	$-x$		$+x$		$+x$
after equilibrium	0.0286 M $- x$		x		0.0429 M $+ x$

Substituting the equilibrium values into the equilibrium expressions gives

$$\frac{(x)(0.0429 + x)}{(0.0286 - x)} = 3.0 \times 10^{-8}$$

We now assume that $x \ll 0.0286$ and that $0.0286 - x = 0.0286$; we also assume that $x \ll 0.0429$ and that $0.0429 + x = 0.0429$. Thus,

$$\frac{(x)(0.0429)}{0.0286} = 3.0 \times 10^{-8}$$

$$x = \frac{(0.0286 \text{ M})(3.0 \times 10^{-8})}{0.0429}$$

$$= 2.0 \times 10^{-8}$$

Because $x = [H_3O^+]$,

$$pH = -\log[H_3O^+]$$

$$= -\log(2.0 \times 10^{-8})$$

$$= -(-7.70)$$

$$= 7.70$$

(c) At the stoichiometric point, exactly enough base has been added to react with the original amount of acid present (2.50 mmol). A solution of Na$^+$ ions and ClO$^-$ ions remains. The pH is calculated in the same way as for a solution of the salt NaClO. Na$^+$ does not affect the pH, so we focus on the OCl$^-$ ion.

$$\text{Molarity of OCl}^- = \frac{\text{millimoles OCl}^-}{\text{milliliters solution}}$$

$$= \frac{2.50 \text{ mmol}}{41.67 \text{ mL}}$$

$$= 0.0600 \text{ mol} \cdot \text{L}^{-1}$$

The equilibrium and table of concentration values are

	OCl⁻(aq)	+	H₂O(l) ⇌	HOCl(aq) +	OH⁻(aq)
initial concentration	0.0600 M			0	0
change to reach equilibrium	$-x$			$+x$	$+x$
equilibrium concentrations	0.0600 M $- x$			x	x

The equilibrium expression is

$$\frac{[\text{HOCl}][\text{OH}^-]}{[\text{OCl}^-]} = K_b$$

Substituting in the equilibrium concentrations gives

$$\frac{(x)(x)}{0.0600 - x} = 3.4 \times 10^{-7}$$

We assume that $x \ll 0.0600$ and that $0.0600 - x = 0.0600$.

$$\frac{x^2}{0.0600} = 3.4 \times 10^{-7}$$

$$x^2 = 2.0 \times 10^{-8}$$

$$[\text{OH}^-] = x = 1.4 \times 10^{-4}$$

$$\text{pOH} = -\log[\text{OH}^-] = -\log(1.4 \times 10^{-4})$$

$$= 3.85$$

$$\text{pH} = 14.00 - \text{pOH}$$

$$= 10.15$$

(d) After the stoichiometric point, the additional amount of strong base OH⁻ added completely overwhelms the small amount of OH⁻ formed by the ionization of OCl⁻ in water. The pH of the solution is determined by the additional OH⁻ added. From the table constructed at the start of the whole problem, we have

$$\text{Molarity of OH}^- = \frac{0.70 \text{ mmol}}{46.33 \text{ mL}}$$

$$= 0.015 \text{ mol} \cdot \text{L}^{-1}$$

$$\text{pOH} = 1.82$$

$$\text{pH} = 12.18$$

EXERCISE 1 Calculate the pH at the following points for the titration of 25.00 mL of 0.200 M benzoic acid (C_6H_5COOH) with 0.150 M NaOH: (a) before the addition of any NaOH; (b) after addition of 10.00 mL of NaOH; (c) at the stoichiometric point (33.33 mL of NaOH); and (d) after addition of 5.00 mL of NaOH beyond the stoichiometric point. $K_a(C_6H_5COOH) = 6.5 \times 10^{-5}$ M, $pK_a = 4.19$

[*Answer:* (a) 2.44; (b) 3.81; (c) 8.56; (d) 12.20]

EXERCISE 2 Calculate the pH at the following points for the titration of 50.00 mL of 0.100 M NH_2NH_2 with 0.200 M HCl: (a) before the addition of any acid; (b) after addition of 15.00 mL of HCl; (c) at the stoichiometric point (25.00 mL of HCl added); (d) after addition of 5.00 mL of acid past the stoichiometric point. $K_b(NH_2NH_2) = 1.7 \times 10^{-6}$ M, $pK_b = 5.77$.

[*Answer:* (a) 10.61; (b) 8.05; (c) 4.70; (d) 1.90]

15.7 INDICATORS AS WEAK ACIDS

KEY CONCEPT   Acid-base indicators

An **acid-base indicator** is a compound that exhibits different colors in different pH ranges. A familiar indicator is phenolphthalein, which is colorless in an acid solution but pink above pH 8.2. Indicators are used to tell when to stop a titration. The choice of a specific indicator depends on the pH at the stoichiometric point and how fast the pH changes around the stoichiometric point. The *ideal* choice is one for which the pH at the stoichiometric point of the titration equals the pK of the indicator. However, if the pH $\pm$ 1 equals the pK of the indicator, the indicator will usually be suitable. Practically, we choose an indicator for which the pH at the stoichiometric point is in the middle of the range of pH values over which the indicator changes color (Table 15.3).

pH at stoichiometric point $\pm$ 1 = pK_{ind}

pH at stoichiometric point = pH at midpoint of range over which indicator changes color

For instance, in a titration in which the pH at the stoichiometric point is 5.2, an excellent choice for an indicator is methyl red, which changes color in the pH range 4.8–6.0. The midpoint of this range is 5.4, which is close to 5.2.

EXAMPLE   Choosing an indicator

Which indicator from Table 15.3 of the text would be best for a titration with pH 9.2 at the stoichiometric point?

SOLUTION We try to choose an indicator in which the midpoint of the pH range for the color change of the indicator is the same as the pH at the equivalence point of the titration. From the table, we find that thymol blue and phenolphthalein both look promising.

| Thymol blue | pH range = 8.0–9.6 | midpoint of range = 8.8 |
| Phenolphthalein | pH range = 8.2–10.0 | midpoint of range = 9.1 |

The midpoints of both are quite close to the pH of 9.2 at the stoichiometric point. Either indicator would be suitable.

EXERCISE What indicator from Table 15.3 of the text would be best for a titration with pH 5.5 at the stoichiometric point?

[*Answer:* Methyl red is the only choice from the table.]

KEY WORDS   Define or explain each term in a written sentence or two.

end point   pH curve
indicator   stoichiometric point

BUFFER SOLUTIONS

15.8   THE ACTION OF BUFFERS

KEY CONCEPT   Buffers contain both an acid and a base

A **buffer** is a solution that does not change pH very dramatically when either acid or base is added in moderate amounts. A buffer can be prepared by mixing a weak acid with a salt of the weak acid or by mixing a weak base with a salt of the weak base. In the first case (weak acid + salt of weak acid), the buffer contains a weak acid and its conjugate base. In the second case (weak base + salt of weak base), the buffer contains a weak base and its conjugate acid. For example,

Weak acid +	salt	gives	weak acid +	conjugate base
CH_3COOH	CH_3COONa		CH_3COOH	CH_3COO^-
Weak base +	salt	gives	weak base +	conjugate base
NH_3	NH_4Cl		NH_3	NH_4^+

We can see how a buffer prevents drastic changes in pH by considering the equilibrium involved. For instance, for an acetic acid–sodium acetate buffer, the equilibrium is

$$CH_3COOH(aq) + H_2O(l) \rightleftharpoons CH_3COO^-(aq) + H_3O^+(aq)$$

This is a mixed solution, with relatively large amounts of both CH_3COOH and CH_3COO^- present. If a moderate amount of an acid is added, protons are mostly transferred to the base CH_3COO^- instead of to H_2O; very little H_3O^+ is formed as a result of the addition and the pH does not change very much. That is, after addition of an acid such as HCl, we have

$$HCl(aq) + CH_3COO^-(aq) \longrightarrow CH_3COOH(aq) + Cl^-(aq) \qquad \text{predominant reaction}$$

$$HCl(aq) + H_2O(l) \longrightarrow H_3O^+(aq) + Cl^-(aq) \qquad \text{unimportant reaction}$$

On addition of a moderate amount of base, the base reacts with CH_3COOH instead of forming free OH^- ions, so again the pH does not change very much. For example, if NaOH is added, we have

$$OH^-(aq) + CH_3COOH(aq) \longrightarrow CH_3COO^-(aq) + H_2O(l) \qquad \text{predominant reaction}$$

$$\text{Unreacted } OH^-(aq) \qquad \text{unimportant amount}$$

EXAMPLE   Understanding buffers

Will a solution of hydrofluoric acid (HF) and sodium fluoride (NaF) act as a buffer? If it does, what reaction occurs when a small amount of strong acid (such as HCl) is added?

SOLUTION A solution of a weak acid and a salt of the acid will act as a buffer. Because HF is a weak acid and NaF is a salt of HF, the mixture will act as a buffer. When HCl is added, it will transfer protons mostly to F^- instead of to H_2O, so the H_3O^+ concentration will not change very much.

$$HCl(aq) + F^-(aq) \longrightarrow Cl^-(aq) + HF(aq) \qquad \text{predominant reaction}$$

$$HCl(aq) + H_2O(aq) \longrightarrow H_3O^+(aq) + Cl^-(aq) \qquad \text{unimportant reaction}$$

EXERCISE What reaction occurs in the HF-NaF buffer when a strong base is added?

[*Answer:* $OH^-(aq) + HF(aq) \rightarrow H_2O(aq) + F^-(aq)$]

15.9   SELECTING A BUFFER

KEY CONCEPT   Buffers contain both an acid and a base

The pH of a buffer solution can be estimated with the Henderson-Hasselbalch equation:

$$pH = pK_a + \log\left(\frac{[\text{conjugate base}]}{[\text{conjugate acid}]}\right)$$

The pK_a is that of the conjugate acid. The concentrations of conjugate acid and conjugate base used usually correspond to those *before* equilibrium occurs. Using the concentration before equilibrium is tantamount to assuming that x (the change in concentrations to reach equilibrium) is small with regard to the initial concentrations of both acid and base. In this regard, we should emphasize that using the Henderson-Hasselbalch equation is completely equivalent to setting up a table of concentrations and solving an equilibrium problem.

EXAMPLE Calculating the pH of a buffer

What is the pH of the buffer prepared by making up a solution that is 0.50 M acetic acid (CH_3COOH) and 0.40 M sodium acetate (CH_3COONa)? $pK_a(CH_3COOH) = 4.75$.

SOLUTION The pH of a buffer can be estimated with the Henderson-Hasselbalch equation. In this problem, CH_3COOH is the acid and CH_3COO^- is the base. We recognize that sodium acetate is 100% dissociated in solution, so the concentration of acetate is the same as that of the sodium acetate. Thus,

$$[CH_3COOH] = 0.50 \qquad [CH_3COO^-] = 0.40$$

$$pH = 4.75 + \log\frac{[CH_3COO^-]}{[CH_3COOH]}$$

$$= 4.75 + \log\left(\frac{0.40}{0.50}\right)$$

$$= 4.75 - 0.10$$

$$= 4.65$$

Thus, the pH is approximately 4.7. (You should now solve this problem as an equilibrium problem, using a table of concentrations, and assure that the answer comes out the same.)

EXERCISE What is the pH of the buffer prepared by making up a solution that is 0.30 M ammonia (NH_3) and 0.20 M ammonium chloride (NH_4Cl)? $pK_b(NH_3) = 4.75$.

[*Answer:* 4.82 or, about 4.8]

15.10   BUFFER CAPACITY

KEY CONCEPT   Buffers have a limit to their action

A buffer operates by providing a weak acid to react with added base and a weak base to react with added acid. If enough base is added to a buffer to react with all of the weak acid, the buffer is no longer able to protect against the addition of more base. If enough acid is added to a buffer to react with all of the weak base, the buffer is no longer able to protect against the addition of more acid. We conclude that any buffer has a limit to its buffering ability. As a working rule of thumb, when the ratio of conjugate base to conjugate acid in the buffer is less than 0.1 or more than 10, we assume that the buffer's capacity has been exceeded.

$$\frac{[\text{conjugate base}]}{[\text{conjugate acid}]} \geq 10 \qquad \text{buffer no longer protects against base; not enough acid is present}$$

$$10 > \frac{[\text{conjugate base}]}{[\text{conjugate acid}]} > 0.1 \qquad \text{buffer protects against both acid and base}$$

$$\frac{[\text{conjugate base}]}{[\text{conjugate acid}]} \leq 0.1 \qquad \text{buffer no longer protects against acid; not enough base is present}$$

These limits define what is called the **buffer capacity** of the buffer. If we substitute these ratios into the Henderson-Haselbalch equation (as upper and lower bounds, so that the inequalities become equal signs), we arrive at the following:

$$\frac{[\text{conjugate base}]}{[\text{conjugate acid}]} = 10 \qquad pH = pK_a + \log 10 = pK_a + 1$$

$$\frac{[\text{conjugate base}]}{[\text{conjugate acid}]} = 0.1 \qquad pH = pK_a + \log 0.1 = pK_a - 1$$

These results indicate that a buffer works best in the pH range from $pK_a - 1$ to $pK_a + 1$. So, for instance, a benzoic acid–sodium benzoate buffer, for which pK_a (benzoic acid) = 4.19, buffers in the pH range 3.19 (4.19 − 1) to 5.19 (4.19 + 1).

EXAMPLE 1   Preparing a buffer

Describe how to prepare 1.0 L of a buffer that works at pH = 2.22. Table 15.4 of the text will be helpful.

SOLUTION Buffers work best when the pH of the buffer is in the range $pK_a \pm 1$, so we first must find the appropriate buffer from Table 15.4. The $HClO_2$-ClO_2^- buffer system has $pK_a = 2.00$, so this should work at pH = 2.22, that is, the pK_a within ±1 of the desired pH. The Henderson-Hasselbalch equation will allow us to find out how much $HClO_2$ and how much ClO_2^- (probably as $NaClO_2$) to add to water to prepare the buffer. For this system,

$$pH = pK_a + \log\left(\frac{[ClO_2^-]}{[HClO_2]}\right)$$

Because pH = 2.22 and $pK_a = 2.00$,

$$2.22 = 2.00 + \log\left(\frac{[ClO_2^-]}{[HClO_2]}\right)$$

$$\log\left(\frac{[ClO_2^-]}{[HClO_2]}\right) = +0.22$$

$$\frac{[ClO_2^-]}{[HClO_2]} = 10^{0.22}$$

$$\frac{[ClO_2^-]}{[HClO_2]} = 1.66$$

Because we are asked to prepare one liter of buffer, we can add 1.00 mol of $HClO_2$ and 1.66 mol of $NaClO_2$ to enough water to make 1.0 L of solution; this gives the 1.66/1.00 mol ratio of base to acid required. (We could also use 0.100 mol of $HClO_2$ and 0.166 mol of $NaClO_2$.) The molar mass of $HClO_2$ is 68.5 g·mol^{-1} and that of $NaClO_2$ is 90.4 g·mol^{-1}, so we would add 68.5 g of $HClO_2$ and 150 g of $NaClO_2$ (1.66 × 90.4 = 150) to enough water to make 1.0 L of solution.

EXERCISE What mole ratio of NH_3 to NH_4^+ must be used to prepare a buffer that has a pH of 9.02?

[*Answer:* 0.58]

The next example problem is similar to Exercises 15.108 and 15.109 of the text. Check with your instructor to see whether it is required material in your course.

EXAMPLE 2   Estimating the buffer capacity of a buffer

An acetic acid–sodium acetate buffer is prepared by making up a solution that is 0.050 M in acetic acid and 0.050 M in sodium acetate. How many milliliters of 1.0 M NaOH can be added to 250 mL of this buffer before its buffer capacity is exceeded?

SOLUTION The reaction that occurs when NaOH is added to this buffer is

$$CH_3COOH(aq) + OH^-(aq) \rightleftharpoons CH_3COO^-(aq) + H_2O(l)$$

As NaOH is added, the CH_3COOH reacts (so its concentration decreases) and CH_3COO^- is formed (so its concentration increases). Thus, the ratio [conjugate base]/[conjugate acid] gets larger. The buffer capacity will be exceeded when the ratio gets large enough that

$$\frac{[\text{conjugate base}]}{[\text{conjugate acid}]} = 10$$

To determine when this ratio is reached, we first calculate the millimoles of acetic acid (CH_3COOH) and acetate ion (CH_3COO^-) in the 250 mL of buffer:

$$\text{Millimoles } CH_3COOH = 250 \text{ mL} \times 0.050 \text{ M} = 12.5 \text{ mmol}$$

$$\text{Millimoles } CH_3COO^- = 250 \text{ mL} \times 0.050 \text{ M} = 12.5 \text{ mmol}$$

When x mmol NaOH are added, x mmol CH_3COOH reacts and x mmol CH_3COO^- are formed. So after addition of x mmol NaOH.

$$\text{Millimoles } CH_3COOH = 12.5 \text{ mmol} - x$$

$$\text{Millimoles } CH_3COO^- = 12.5 \text{ mmol} + x$$

Because both CH_3COOH and CH_3COO^- are in the same solution, the volumes cancel in the ratio of concentrations; this is,

$$\frac{[\text{conjugate base}]}{[\text{conjugate acid}]} = \frac{\text{millimoles conjugate base}}{\text{millimoles conjugate acid}} = \frac{\text{millimoles } CH_3COO^-}{\text{millimoles } CH_3COOH}$$

For our problem, the buffer capacity is exceeded when

$$\frac{12.5 \text{ mmol} + x}{12.5 \text{ mmol} - x} = 10$$

$$12.5 \text{ mmol} + x = 10(12.5 \text{ mmol} - x)$$

$$12.5 \text{ mmol} + x = 125 \text{ mmol} - 10x$$

$$10x + x = 125 \text{ mmol} - 12.5 \text{ mmol}$$

$$11x = 112.5 \text{ mmol}$$

$$x = \frac{112.5 \text{ mmol}}{11}$$

$$= 10.2 \text{ mmol}$$

This is the number of millimoles of NaOH that just exceeds the buffer capacity. Because the concentration of NaOH being added is 1.0 M, we have

$$\text{Milliliters NaOH} = \frac{10.2 \text{ mmol}}{1.0 \text{ mmol} \cdot \text{mL}^{-1}}$$

$$= 10.2 \text{ mL} = 10 \text{ mL}$$

EXERCISE An ammonia–ammonium chloride buffer is prepared by making up a solution that is 0.40 M in ammonia (NH_3) and 0.35 M in ammonium chloride (NH_4Cl). How many milliliters of 2.0 M HCl can be added to 100 mL of this buffer before its buffer capacity is exceeded?

[*Answer:* 17 mL]

KEY WORDS   Define or explain each term in a written sentence or two.

acid buffer   buffer capacity
base buffer   Henderson-Hasselbalch equation
buffer

SOLUBILITY EQUILIBRIA

15.11   THE SOLUBILITY PRODUCT

KEY CONCEPT K_{sp} describes the solubilities of sparingly soluble salts

We learned earlier that a sparingly soluble salt can participate in a dynamic equilibrium in which the solid salt is in equilibrium with dissolved salt in saturated solution. For instance, for $CaSO_4$,

$$CaSO_4(s) \rightleftharpoons Ca^{2+}(aq) + SO_4^{2-}(aq)$$

We can write an equilibrium expression for this type of equilibrium in the standard way. Because the concentration of a pure solid does not appear in the equilibrium expression, we get

$$K_{sp} = [Ca^{2+}][SO_4^{2-}]$$

The subscript sp stands for solubility product, which is the name given to the equilibrium constants that describe this type of equilibrium. The solubility of a sparingly soluble salt can be estimated from the solubility product, and the solubility product estimated from the solubility, as illustrated in the examples that follow.

EXAMPLE 1   Estimating solubility

Estimate the solubility of $BaCO_3$ at 25°C. $K_{sp} = 8.1 \times 10^{-9}$.

SOLUTION We first construct a table of concentration values just as we have done for other equilibria. An uppercase P denotes the portion or piece of solid in the mixture; it is not actually a concentration. An S denotes solubility.

	$BaCO_3(s)$	$\rightleftharpoons$ $Ba^{2+}(aq)$	+ $CO_3^{2-}(aq)$
initial concentration	P	0	0
change to reach equilibrium	$-S$	$+S$	$+S$
equilibrium concentration	$P - S$	S	S

We place a very significant interpretation on S. As can be seen from the table of concentration values, it is the amount of $BaCO_3$ that dissolves, expressed in $mol \cdot L^{-1}$ (M). The solubility constant expression is

$$[Ba^{2+}][CO_3^{2-}] = K_{sp}$$

Substituting values from the table and the value of K_{sp} gives

$$(S)(S) = 8.1 \times 10^{-9}$$

$$S = 9.0 \times 10^{-5}$$

The solubility of $BaCO_3$ is 9.0×10^{-5} $mol \cdot L^{-1}$, or 9.0×10^{-5} M.

EXERCISE What is the solubility of CuCl? $K_{sp} = 1.0 \times 10^{-6}$.

[*Answer:* 1.0×10^{-3} M]

EXAMPLE 2   Estimating solubility

According to the value of K_{sp}, what is the solubility of Cu_2S? $K_{sp} = 2.0 \times 10^{-47}$.

SOLUTION In the table of concentration values for this equilibrium, which follows, we are careful to remember that 2 mol Cu^+ ion are formed for every 1 mol of Cu_2S that dissolves.

	$Cu_2S(s) \rightleftharpoons 2Cu^+(aq) + S^{2-}(aq)$		
initial concentration	P	0	0
change to reach equilibrium	$-S$	$+2S$	$+S$
equilibrium concentration	$P - S$	$2S$	S

S has the same interpretation as in the previous problem. It represents the amount of Cu_2S that dissolves and is its solubility in moles per liter. The solubility expression is

$$[Cu^+]^2[S^{2-}] = K_{sp}$$

Substituting values from the concentration table and the solubility constant gives

$$(2S)^2(S) = 2.0 \times 10^{-47}$$

$$4S^3 = 2.0 \times 10^{-47}$$

$$S^3 = 5.0 \times 10^{-48}$$

$$S = (5.0 \times 10^{-48})^{1/3}$$

The superscript $\frac{1}{3}$ indicates that the cube root of a number should be taken. To get the cube root of 5.0×10^{-48} on calculators with standard arithmetic notation, do the following:

Enter 5.0×10^{-48}

Depress the Y^x key

Enter 0.33333333

Depress the = key

The result is the solubility of Cu_2S:

$$Solubility = 1.7 \times 10^{-16}\ mol \cdot L^{-1}$$

EXERCISE Estimate the solubility of $Fe(OH)_2$? $K_{sp} = 1.6 \times 10^{-14}$.

[*Answer:* $1.6 \times 10^{-5}\ mol \cdot L^{-1}$]

EXAMPLE 3 Estimating the solubility product

The solubility of lead(II) bromide ($PbBr_2$) at 20°C is 0.84 g per 100 mL water. Estimate K_{sp} for $PbBr_2$.

SOLUTION The equilibrium and table of concentration values are

	$PbBr_2(s) \rightleftharpoons Pb^{2+}(aq) + 2Br^-(aq)$		
initial concentration	P	0	0
change to reach equilibrium	$-S$	$+S$	$+2S$
equilibrium concentration	$P - S$	S	$2S$

The equilibrium expression is

$$[Pb^{2+}][Br^-]^2 = K_{sp}$$

Substituting the values from the table of concentration values gives

$$(S)(2S) = K_{sp}$$

$$4S^3 = K_{sp}$$

We now need the value of S to calculate K_{sp}. The molar mass of $PbBr_2$ is 367 $g \cdot mol^{-1}$. Thus,

$$0.84\ \text{g } PbBr_2 \times \frac{1\ \text{mol } PbBr_2}{367\ \text{g}} = 2.3 \times 10^{-3}\ \text{mol } PbBr_2$$

We assume that when 0.84 g $PbBr_2$ dissolves in 100 mL water, 100 mL of solution results. That is, the amount of $PbBr_2$ is so small that the volume of solution is the same as the volume of solvent. 100 mL is the same as 0.100 L of solution, so

$$\text{Solubility of } PbBr_2 = \frac{2.3 \times 10^{-3} \text{ mol}}{0.10 \text{ L}}$$

$$S = 2.3 \times 10^{-2} \text{ mol} \cdot L^{-1}$$

Substituting $S = 2.3 \times 10^{-2}$ into the equilibrium expression gives

$$K_{sp} = 4S^3$$

$$= 4(2.3 \times 10^{-2})^3$$

$$= 4.9 \times 10^{-5}$$

This value is a little different from that in Table 15.5 of the text, because the value in the text table is determined by a more accurate method.

EXERCISE The solubility of silver carbonate (Ag_2CO_3) is 3.0×10^{-3} g per 100 mL water. (a) Estimate K_{sp} for silver carbonate and (b) compare your result with the value of 6.2×10^{-12} given in Table 15.5 of the text.

[*Answer:* (a) 5×10^{-12}; (b) close to value in Table 15.5]

15.12   THE COMMON-ION EFFECT

KEY CONCEPT   A common ion can affect solubilities

According to Le Chatelier's principle, addition of any product (that is not a pure solid or pure liquid) to a dynamic equilibrium forces the equilibrium to shift toward reactants. For instance, for the equilibrium

$$CuI(s) \rightleftharpoons Cu^+(aq) + I^-(aq)$$

addition of Cu^+ ion (by the addition of $CuNO_3$, for example) would cause precipitation of some CuI. Addition of I^- ion (by the addition of NaI, for example) would also cause precipitation of some CuI. Thus, less CuI is dissolved in the presence of an external source of Cu^+ ion or I^- ion than in pure water. The lowering of the solubility of a sparingly soluble salt by the presence of an ion in the salt is an example of the **common-ion effect.** That is, addition of an ion common to one already in solution affects the equilibrium. We can quantitatively estimate solubilities in the presence of a common ion, as shown in the next two examples.

EXAMPLE 1   Calculating solubility

Calculate the solubility of $CaSO_4$ in 0.10 M Na_2SO_4(aq). $K_{sp}(CaSO_4) = 2.4 \times 10^{-5}$.

SOLUTION We first note that 0.10 M Na_2SO_4 is fully dissociated and the concentration of SO_4^{2-} is, therefore, 0.10 M. The equilibrium and table of concentration values then become:

	$CaSO_4(s) \rightleftharpoons$	$Ca^{2+}(aq)$ +	$SO_4^{2-}(aq)$
initial concentration	P	0	0.10 M
change to reach equilibrium	$-S$	$+S$	$+S$
equilibrium concentration	$P - S$	S	0.10 M $+ S$

The equilibrium expression is

$$[Ca^{2+}][SO_4^{2-}] = K_{sp}$$

Substituting the equilibrium concentration and value of K_{sp} gives

$$(S)(0.10 + S) = 2.4 \times 10^{-5}$$

We now assume that S is very small relative to 0.10, so $0.10 + S = 0.1$.

$$(S)(0.10) = 2.4 \times 10^{-5}$$

$$S = 2.4 \times 10^{-4} \, \text{mol} \cdot \text{L}^{-1}$$

This value of S is very much smaller than 0.10, so the approximation made is justified. Because S is the solubility (it is the amount of solid that dissolves), the solubility of $CaSO_4$ in 0.10 M Na_2SO_4 is $2.4 \times 10^{-4} \, \text{mol} \cdot \text{L}^{-1}$ (M). (In pure water, the solubility is 5.0×10^{-3} M, about five times larger.)

EXERCISE What is the solubility of CuI in 0.30 M NaI? $K_{sp}(CuI) = 1.0 \times 10^{-6}$.

[*Answer:* $3.3 \times 10^{-6} \, \text{mol} \cdot \text{L}^{-1}$]

EXAMPLE 2 Calculating solubility

What is the solubility of $BaF_2(s)$ in 0.20 M NaF? $K_{sp}(BaF_2) = 1.7 \times 10^{-6}$.

SOLUTION In 0.20 M NaF, $[F^-] = 0.20$ M. The equilibrium and table of values follows:

	$BaF_2(s) \rightleftharpoons Ba^{2+}(aq)$	$+$	$2F^-(aq)$
initial concentration	P	0	0.20 M
change to reach equilibrium	$-S$	$+S$	$+2S$
equilibrium concentration	$P - S$	S	0.20 M $+ 2S$

In setting up this table, we should note that the initial F^- concentration originates with NaF and does not come from the dissolving of BaF_2. Thus, we do *not* multiply 0.20 M by 2; there is only one mole of F^- ions per mole of KF. The equilibrium expression is

$$[Ba^{2+}][F^-]^2 = K_{sp}$$

Substituting the equilibrium values and value of K_{sp} gives

$$(S)(0.20 + 2S)^2 = 1.7 \times 10^{-6}$$

We assume that $2S$ is much smaller than 0.20, so $0.20 + 2S = 0.20$.

$$(S)(0.20)^2 = 1.7 \times 10^{-6}$$

$$S = 4.3 \times 10^{-5} \, \text{mol} \cdot \text{L}^{-1}$$

The solubility of $BaF_2(s)$ in 0.20 M NaF is S, $4.3 \times 10^{-5} \, \text{mol} \cdot \text{L}^{-1}$.

EXERCISE What is the solubility of PbI_2 in 0.50 M NaI? $K_{sp}(PbI_2) = 1.4 \times 10^{-8}$.

[*Answer:* $5.6 \times 10^{-8} \, \text{mol} \cdot \text{L}^{-1}$]

15.13   PREDICTING PRECIPITATION

KEY CONCEPT   A salt precipitates if $Q_{sp} \geq K_{sp}$

Qualitative analysis involves identifying the cations and/or anions in a unknown solution, often through precipitation reactions that give precipitates with characteristic and well-known colors. One important aspect of this process is being able to predict whether a precipitate will form. To do so, we use Q_{sp}, the expression that has the same form as K_{sp} but uses the actual concentrations of ions in a given solution at some moment; these do not have to be the equilibrium concentrations. We then compare Q_{sp} with K_{sp} for the compound we want to precipitate.

- When $Q_{sp} > K_{sp}$, the compound precipitates.
- When $Q_{sp} = K_{sp}$, a saturated solution has been made and, most likely, no visible precipitate forms.
- When $Q_{sp} < K_{sp}$, no precipitate forms (and part of any precipitate already present dissolves).

EXAMPLE 1 Predicting whether or not a precipitate forms

Does $CaCO_3$ precipitate if a solution that is initially 2.0×10^{-4} M in Ca^{2+} is made 4.0×10^{-4} M in Na_2CO_3? $K_{sp}(CaCO_3) = 8.7 \times 10^{-9}$.

SOLUTION The size of Q_{sp} relative to K_{sp} determines if a precipitate forms. For $CaCO_3$,

$$CaCO_3(s) \rightleftharpoons Ca^{2+}(aq) + CO_3^{2-}(aq)$$

$$Q_{sp} = [Ca^{2+}][CO_3^{2-}]$$

$$= (2.0 \times 10^{-4})(4.0 \times 10^{-4})$$

$$= 8.0 \times 10^{-8}$$

Because 8.0×10^{-8} $M^2 > 8.7 \times 10^{-9}$ M^2, $Q_{sp} > K_{sp}$ and $CaCO_3$ precipitates.

EXERCISE Does CuS precipitate if a solution that is 5×10^{-8} M in Cu^{2+} has its pH adjusted so that the S^{2-} ion concentration is 6×10^{-20} M? $K_{sp}(CuS) = 8.5 \times 10^{-45}$.

[*Answer:* Yes]

EXAMPLE 2 Predicting whether or not a precipitate forms

Assume we add 0.050 mL (about 1 drop) of 8×10^{-2} M NaF to 1.00 mL of a solution that is 0.010 M in Ba^{2+}, Pb^{2+}, and Ca^{2+}. Which ions precipitate as fluorides? $K_{sp}(BaF_2) = 1.7 \times 10^{-6}$, $K_{sp}(PbF_2) = 3.7 \times 10^{-8}$, and $K_{sp}(CaF_2) = 4.0 \times 10^{-11}$.

SOLUTION The value of Q_{sp} for each salt determines whether or not it precipitates. To calculate Q_{sp}, we must determine the concentrations of each ion after mixing of the two solutions. We do this by using the relationship that describes the molarity change for dilution of a solution, $M_1V_1 = M_2V_2$. In this case, the original molarity and volume for each metal are $M_1 = 0.010$ M and $V_1 = 1.00$ mL. The final volume V_2 is 1.05 mL (1.00 mL + 0.050 mL = 1.05 mL). Thus,

$$M_1V_1 = M_2V_2$$

$$M_2 = \frac{M_1V_1}{V_2}$$

$$= \frac{(0.010 \text{ M})(1.00 \text{ mL})}{1.05 \text{ mL}}$$

$$= 0.0095 \text{ M}$$

For the F^- ion, $M_1 = 8 \times 10^{-2}$, $V_1 = 0.050$ mL, and $V_2 = 1.05$ mL. We calculate

$$M_2 = \frac{(8 \times 10^{-2} \text{ M})(0.050 \text{ mL})}{1.05 \text{ mL}}$$

$$= 4 \times 10^{-3} \text{ M}$$

It is now possible to calculate Q_{sp} for all of the salts. Because the concentration of each metal is the same, we get

$$Q_{sp} = [M^{2+}][F^-]^2$$

where the symbol M^{2+} represents a metal cation. Substituting the values previously calculated gives

$$Q_{sp} = (0.0095)(4 \times 10^{-3})^2$$
$$= 2 \times 10^{-7}$$

By comparing Q_{sp} with K_{sp} for each salt, we finally determine which ions precipitate:

BaF_2: $2 \times 10^{-7} < 1.7 \times 10^{-6}$, so $Q_{sp} < K_{sp}$ and BaF_2 does not precipitate.

PbF_2: $2 \times 10^{-7} > 3.7 \times 10^{-8}$, so $Q_{sp} > K_{sp}$ and PbF_2 precipitates.

CaF_2: $2 \times 10^{-7} > 4.0 \times 10^{-11}$, so $Q_{sp} > K_{sp}$ and CaF_2 precipitates.

EXERCISE Does adding another 0.45 mL NaF (for a total of 0.50 mL) in the preceding problem succeed in precipitating BaF_2?

[*Answer:* Yes]

15.14 DISSOLVING PRECIPITATES

KEY CONCEPT   Dissolving a salt by removal of an ion

It is possible to dissolve a sparingly soluble solid MX, by removing the anion X^- from solution:

$$MX(s) \rightleftharpoons M^+(aq) + X^-(aq)$$

According to Le Chatelier's principle, if $X^-(aq)$ is removed from solution, the equilibrium will shift toward products. This shift means that MX(s) will dissolve. Three situations in which this technique is used are the following:

1. When the solid is a hydroxide such as $Fe(OH)_3$, addition of an acid removes OH^- from solution (because H_3O^+ reacts with OH^-), so the hydroxide dissolves.

$H_3O^+(aq) + OH^-(aq) \rightleftharpoons H_2O(l)$
       Acid removes OH^- from solution

$Fe(OH)_3(s) \rightleftharpoons Fe^{3+}(aq) + 3OH^-(aq)$
       Equilibrium shifts to right and $Fe(OH)_3$ dissolves

2. When the solid has a basic anion, such as F^-, CO_3^{2-}, SO_3^{2-}, or S^{2-}, addition of an acid removes the anion from solution (because H_3O^+ reacts with any basic anion), so the salt dissolves. For example, with zinc carbonate:

$2H_3O^+(aq) + CO_3^{2-}(aq) \rightleftharpoons 3H_2O(l) + CO_2(aq)$
       Acid removes CO_3^{2-} from solution

$ZnCO_3(aq) \rightleftharpoons Zn^{2+}(aq) + CO_3^{2-}(aq)$
       Equilibrium shifts to right and $ZnCO_3$ dissolves

3. It is sometimes possible to change the oxidation number of an ion in a salt in order to remove the ion from solution. This change shifts the solubility equilibrium toward products and, therefore, dissolves the salt. For example, with insoluble sulfides (such as CuS), we can oxidize the sulfide ion to solid sulfur (which precipitates out of solution) and thereby dissolves the sulfide.

$3S^{2-}(aq) + 8H^+(aq) + 2NO_3^-(aq) \longrightarrow 3S(s) + 2NO(g) + 4H_2O(l)$
       Oxidation removes S^{2-}

$CuS(s) \rightleftharpoons Cu^{2+}(aq) + S^{2-}(aq)$
       Equilibrium shifts to right and CuS dissolves

EXAMPLE   Dissolving a precipitate by removal of an ion from solution

Develop a method to dissolve the insoluble salt cobalt(II) cyanide, $Co(CN)_2$. Write the reactions involved.

SOLUTION   The cyanide ion, which is the anion of a weak acid, is a base. If we add an acid (such as HCl) to the solution, the cyanide ion will react with the acid and be removed from solution. Thus, $Co(CN)_2$ will dissolve.

$$H_3O^+(aq) + CN^-(aq) \rightleftharpoons H_2O(l) + HCN(aq) \qquad \text{Acid removes } CN^- \text{ from solution}$$

$$Co(CN)_2(aq) \rightleftharpoons Co^{2+}(aq) + 2CN^-(aq) \qquad \begin{array}{l}\text{Equilibrium shifts to right and} \\ Co(CN)_2 \text{ dissolves}\end{array}$$

EXERCISE   Develop a method to dissolve copper(I) sulfite (Cu_2SO_3). Write the equations involved.

Answer: When acid is reacted with sulfite ion, a gas (SO_2) that bubbles out of solution forms. Thus, addition of acid will remove sulfite from solution and the Cu_2SO_3 will dissolve.

$$SO_3^{2-}(aq) + H_3O^+(aq) \rightleftharpoons SO_2(g) + H_2O(l)$$
$$\text{Sulfite removed from solution}$$

$$Cu_2SO_3(s) \rightleftharpoons Cu^+(aq) + SO_3^{2-}(aq)$$
$$Cu_2SO_3 \text{ dissolves}$$

15.15   COMPLEX IONS AND SOLUBILITIES

KEY CONCEPT A   Dissolving a salt by formation of a complex

It is often possible to produce a complex between a metal ion and an anion. Table 15.6 illustrates the formation of a variety of complexes. When it is possible to form a complex ion of one of the ions in a salt, the ion will be removed from solution and the salt will dissolve. For instance, with AgCl, ammonia can be used to remove Ag^+ from solution so that the AgCl dissolves.

$$Ag^+(aq) + 2NH_3(aq) \rightleftharpoons Ag(NH_3)_2^+(aq) \qquad Ag^+ \text{ is removed from solution}$$

$$AgCl(s) \rightleftharpoons Ag^+(aq) + Cl^-(aq) \qquad \text{Equilibrium shifts to right and AgCl dissolves}$$

EXAMPLE   Dissolving salts through complex formation

What reagent might be used to dissolve solid nickel(II) sulfide through complex formation? Explain your answer.

SOLUTION   We first consult Table 15.6 and discover that Ni^{2+} can form a complex with ammonia (NH_3). Thus, addition of NH_3 will result in formation of $Ni(NH_3)_6^{2+}$ and removal of Ni^{2+} from solution. This means that the solubility equilibrium shifts to the right and NiS dissolves. These reactions are shown below.

$$Ni^{2+}(aq) + 6NH_3(aq) \rightleftharpoons Ni(NH_3)_6^{2+}(aq) \qquad \text{Removal of } Ni^{2+} \text{ from solution}$$

$$NiS(s) \rightleftharpoons Ni^{2+}(aq) + S^{2-}(aq) \qquad \text{Shifts to right, NiS dissolves}$$

EXERCISE   Develop a method to dissolve the insoluble salt gold(III) bromide, $AuBr_3$. Write the reactions involved.

Answer: Remove Au^{3+} from solution by adding CN^- (typically as KCN) to form the complex $Au(CN)_4^-$.

$$Au^{3+}(aq) + CN^-(aq) \rightleftharpoons Au(CN)_4^-(aq)$$
$$Au^{3+} \text{ is removed from solution}$$

$$AuBr_3(s) \rightleftharpoons Au^{3+}(aq) + 3Br^-(aq)$$
$$\text{Equilibrium shifts to right and AuBr dissolves}$$

KEY CONCEPT B   Formation constants

We now consider how to quantitatively account for the solubility of a compound that contains an ion that forms a complex. Let us assume that we want to know the solubility of AgCl in the presence of ammonia. The two equilibria involved are

$$AgCl(s) \rightleftharpoons Ag^+(aq) + Cl^-(aq) \qquad K_{sp} = [Ag^+][Cl^-]$$

$$Ag^+(aq) + 2NH_3(aq) \rightleftharpoons Ag(NH_3)_2^+(aq) \qquad K_f = \frac{[Ag(NH_3)_2^+]}{[Ag^+][NH_3]^2}$$

K_f is the **formation constant** of the complex $Ag(NH_3)_2^+$. Adding these two chemical equations allows us to write the equilibrium that shows how the AgCl dissolves in ammonia. In addition, when two chemical equilibria are added, K for the new equilibrium equals the product of the two original equilibrium constants. Thus,

$$AgCl(s) + 2NH_3(aq) \rightleftharpoons Ag(NH_3)_2^+(aq) + Cl^-(aq) \qquad K = K_s \times K_f = \frac{[Ag(NH_3)_2^+][Cl^-]}{[NH_3]^2}$$

It is now possible to solve an equilibrium problem by constructing a table of concentrations and by using the new equilibrium constant.

EXAMPLE   Calculating the solubility of AgCl in the presence of ammonia

What is the solubility of AgCl in 0.0050 M NH_3?

SOLUTION We set up a table of equilibrium concentrations, using the symbol P to represent a solid precipitate. We note that the initial concentration of ammonia is 0.0050 M, and we use the symbol S to represent the solubility of AgCl.

	AgCl(s)	+	2NH₃(aq)	⇌	Ag(NH₃)₂⁺(aq)	+ Cl⁻(aq)
before equilibrium	P		0.0050 M		0 M	0 M
change to reach equilibrium	$-S$		$-2S$		$+S$	$+S$
after equilibrium	$P - S$		0.0050 M $- 2S$		S	S

We substitute the equilibrium values into the equilibrium expression and solve for S. Because $K_{sp} = 1.6 \times 10^{-10}$ and $K_f = 1.6 \times 10^7$, we get

$$\frac{[Ag(NH_3)_2^+][Cl^-]}{[NH_3]^2} = K_s \times K_f$$

$$\frac{(S)(S)}{(0.0050 - 2S)^2} = (1.6 \times 10^{-10})(1.6 \times 10^7)$$

$$\frac{S^2}{0.0050 - 2S)^2} = 2.6 \times 10^3$$

We now take the square root of both sides of the equation

$$\frac{S}{0.0050 - 2S} = 5.1 \times 10^{-2}$$

To solve for S, we multiply both sides of the equation by $0.0050 - 2S$ and then collect all terms with S on the left side of the equation:

$$S = (5.1 \times 10^{-2})(0.0050 - 2S)$$
$$S = 2.6 \times 10^{-4} - 0.10S$$
$$1.10S = 2.6 \times 10^{-4}$$
$$S = \frac{2.6 \times 10^{-4}}{1.10}$$
$$= 2.4 \times 10^{-4}$$

Because S stands for the solubility of AgCl, the answer is 2.4×10^{-4} mol·L^{-1}.

EXERCISE What is the solubility of AgBr in 0.0025 M KCN?

[**Answer:** 5.0×10^{-5} mol·L^{-1}]

KEY WORDS   Define or explain each term in a written sentence or two.

common-ion effect   reaction quotient
formation constant   solubility product

DESCRIPTIVE CHEMISTRY TO REMEMBER

- All **metal cations,** except those of Group 1, Group 2, and those with +1 charge, produce acid solutions.
- **Formic acid** (HCOOH) is the acid present in ant venom.
- An **indicator** is a weak Brønsted acid that has one color in acid form (HIn) and another color in basic form (In$^-$).
- Blood is maintained at pH = 7.4 by **buffer action.**

CHEMICAL EQUATIONS TO KNOW

- An example of a metal ion acting as an acid is illustrated by the behavior of chromium(III) in aqueous solution:

$$Cr(H_2O)_6^{3+}(aq) \rightleftharpoons Cr(H_2O)_5OH^{2+}(aq) + H_3O^+(aq)$$

- The ammonium ion is acidic:

$$H_2O(l) + NH_4^+(aq) \rightleftharpoons NH_3(aq) + H_3O^+(aq)$$

- Nitric acid can oxidize sulfide ion to elemental sulfur:

$$3S^{2-}(aq) + 8H^+(aq) + 2NO_3^-(aq) \longrightarrow 3S(s) + 2NO(g) + 4H_2O(l)$$

- Formation of the complex ion $Ag(S_2O_3)_2^{3-}$ is used in photographic processing to remove silver halide from exposed film after it has been developed:

$$Ag^+(aq) + 2S_2O_3^{2-}(aq) \rightleftharpoons Ag(S_2O_3)_2^{3-}(aq)$$

- The hydrogen sulfate ion (HSO_4^-) is a weak acid that can donate a proton to water:

$$HSO_4^-(aq) + H_2O(l) \rightleftharpoons H_3O^+(aq) + SO_4^{2-}(aq)$$

- The carbonate ion is a weak base that can accept a proton from water:

$$CO_3^{2-}(aq) + H_2O(l) \rightleftharpoons OH^-(aq) + HCO_3^-(aq)$$

- The carbonate ion reacts with acid to form CO_2:

$$CO_3^{2-}(aq) + 2HCl(l) \rightleftharpoons CO_2(g) + H_2O(l) + 2Cl^-(aq)$$

- Silver forms a well-known complex with ammonia.

$$Ag^+(aq) + 2NH_3(aq) \rightleftharpoons Ag(NH_3)_2^+(aq)$$

MATHEMATICAL EQUATIONS TO KNOW AND UNDERSTAND

$$pH = pK_a + \log\frac{[A^-]}{[HA]}$$ Henderson-Hasselbalch equation

SELF-TEST EXERCISES

Ions as acids and bases

1. What is the pH of a 0.25 M solution of NH_4Br?

(a) 7.00   (b) 0.62   (c) 13.41   (d) 9.06   (e) 4.93

2. What is the pH of a 0.14 M solution of morphine hydrochloride? Morphine hydrochloride can be represented as $RNH_3^+Cl^-$, where R is a large organic part of the molecule. pK_b (morphine) = 5.79.

(a) 0.90   (b) 4.53   (c) 9.46   (d) 13.21   (e) 7.00

3. For equal concentrations of each cation, which one results in the most acidic solution?

(a) NH_4^+, $pK_b(NH_3)$ = 4.75   (b) $CH_3NH_3^+$, $pK_b(CH_3NH_2)$ = 3.44

(c) $(C_2H_5)_3NH^+$, $pK_b[(C_2H_5)_3N]$ = 2.99

4. What is the pH of a 0.12 M solution of $Co(NO_3)_2$? $pK_a(Co^{2+})$ = 8.89. Assume only one proton is lost from the hydrated ion.

(a) 4.91   (b) 9.09   (c) 0.94   (d) 7.00   (e) 13.06

5. What is the pH of a 0.25 M solution of potassium formate, NaHCOO? $pK_a(HCOOH)$ = 3.75.

(a) 8.56   (b) 2.29   (c) 11.71   (d) 7.00   (e) 5.44

6. What is the pH of a solution of 0.050 M calcium acetate $Ca(CH_3COO)_2$? $pK_a(CH_3COOH)$ = 4.75.

(a) 10.32   (b) 5.28   (c) 8.72   (d) 3.74   (e) 7.66

7. What is the pH of a solution that is 0.55 M in hypochlorous acid (HClO) and 0.35 M in potassium hypochlorite? $pK_a(HClO)$ = 7.53.

(a) 7.91   (b) 7.33   (c) 7.70   (d) 7.00   (e) 7.58

8. What is the pH of a solution that is 0.10 M in hydrazinium chloride (NH_2NH_3Cl) and 0.080 M in hydrazine (NH_2NH_2)? $pK_b(NH_2NH_2)$ = 5.77.

(a) 8.13   (b) 5.49   (c) 5.67   (d) 5.92   (e) 8.33

9. A solution made up with equimolar concentrations of a weak acid and its conjugate base has a pH of 6.05. What is the pK_a for the acid?

(a) 5.95   (b) 6.35   (c) 6.05   (d) need more information to tell

10. What percentage of anilinium ion ($C_6H_5NH_3^+$) is deprotonated in a 0.030 M solution of anilinium chloride ($C_6H_5NH_3Cl$)? $K_a(C_6H_5NH_3^+) = 2.3 \times 10^{-5}$.

(a) 0.077%   (b) 1.4%   (c) 0.82%   (d) 0.23%   (e) 2.7%

11. What percentage of hypobromite ion (BrO^-) is protonated in a 0.16 M solution of sodium hypobromite? $K_a(HBrO) = 3.0 \times 10^{-5}$.

(a) 0.019%   (b) 3.0%   (c) 4.6×10^{-3}%

(d) 2.1×10^{-7}%   (e) 0.033%

12. What is the pH of the solution prepared by mixing 25.0 mL of 0.024 M HF with 15.0 mL of 0.15 M NaF? $K_a(HF) = 3.5 \times 10^{-4}$.

(a) 6.41   (b) 7.00   (c) 7.59   (d) 4.03   (e) 9.97

13. What is the pH of the solution prepared by mixing 25.0 mL of 0.0040 M HCl with 60.0 mL of 0.025 M NaCl?

(a) 7.00   (b) 2.93   (c) 8.44   (d) 11.12   (e) 6.14

Titrations

14. What is the pH of the solution that results when 10.0 mL of 0.0835 M HCl and 10.0 mL of 0.0850 M NaOH are mixed?

(a) 10.81   (b) 1.07   (c) 7.00   (d) 12.93   (e) 3.12

15. What is the pH of the solution that results when 12.5 mL of 0.0962 M HNO_3 and 5.75 mL of 0.102 M $Ba(OH)_2$ are mixed?

(a) 11.20   (b) 7.00   (c) 2.80   (d) 6.11   (e) 7.89

16. How many milliliters of 0.0250 M NaOH are required to neutralize 10.0 mL of 0.108 M HBr?

(a) 12.6   (b) 43.2   (c) 8.44   (d) 21.1   (e) 2.31

Exercises 17–20 concern a titration in which 20.00 mL of 0.145 M NaOH are titrated with 0.100 M HCl.

17. What is the pH before any HCl has been added?

(a) 13.04   (b) 7.00   (c) 13.16   (d) 1.06   (e) 0.84

18. What is the pH after the addition of 15.00 mL HCl?

(a) 1.20   (b) 12.60   (c) 7.00   (d) 1.44   (e) 12.85

19. What is the pH at the equivalence point (after addition at a total of 29.00 mL HCl)?

(a) 1.24   (b) 5.83   (c) 8.26   (d) 7.00   (e) 12.81

20. What is the pH after the addition of a total of 40.00 mL HCl?

(a) 1.74   (b) 12.46   (c) 7.00   (d) 1.06   (e) 13.09

Exercises 21–24 concern a titration in which 20.00 mL of 0.15 M benzoic acid (C_6H_5COOH) are titrated with 0.10 M NaOH. $pK_a(C_6H_5COOH) = 4.19$.

21. What is the pH before any base is added?

(a) 1.46   (b) 2.51   (c) 4.22   (d) 7.00   (e) 11.50

22. What is the pH after addition of 20.00 mL NaOH?

(a) 7.00   (b) 4.26   (c) 4.49   (d) 3.96   (e) 7.30

23. What is the pH at the stoichiometric point (30.00 mL NaOH)?
(a) 8.36   (b) 8.43   (c) 5.52   (d) 7.00   (e) 6.29

24. What is the pH after addition of 10.00 mL NaOH past the stoichiometric point?
(a) 12.22   (b) 1.84   (c) 7.00   (d) 10.67   (e) 3.45

Exercises 25–28 concern a titration in which 15.00 mL of 0.120 M morphine are titrated with 0.100 M HCl. pK_b(morphine) = 5.79.

25. What is the pH before the addition of any acid?
(a) 3.46   (b) 0.94   (c) 13.18   (d) 10.65   (e) 7.00

26. What is the pH after addition of 5.00 mL HCl?
(a) 7.79   (b) 7.00   (c) 6.20   (d) 8.24   (e) 8.62

27. What is the pH at the stoichiometric point (18.00 mL HCl added)?
(a) 3.00   (b) 0.35   (c) 4.74   (d) 9.76   (e) 11.04

28. What is the pH after addition of 5.00 mL HCl beyond the stoichiometric point?
(a) 1.88   (b) 7.00   (c) 5.15   (d) 3.12   (e) 12.11

29. Which indicator of the following five would be the best to use for a titration with pH 7.4 at the stoichiometric point? Each indicator is shown with the pH range in which it changes color.
(a) methyl red (4.8–6.0)   (b) bromthymol blue (6.0–7.6)
(c) thymol blue (9.0–9.6)   (d) phenol red (6.8–8.0)
(e) phenolphthalein (8.2–10.0)

Buffer solutions

30. What is the pH of a formic acid–sodium formate buffer that is 0.80 M in formic acid (HCOOH) and 0.50 M in sodium formate (HCOONa)?
(a) 4.14   (b) 4.00   (c) 4.39   (d) 3.55   (e) 3.81

31. What is the pH of a trimethylammonium chloride–trimethylamine buffer that is 0.20 M in trimethylammonium chloride [$(CH_3)_3NHCl$] and 0.25 M in trimethylamine [$(CH_3)_3N$]? $pK_b[(CH_3)_3N] = 4.19$.
(a) 9.91   (b) 9.88   (c) 9.65   (d) 10.00   (e) 9.71

32. Which two species, when mixed together in aqueous solution, will act as a buffer?
(a) HCl, Cl^- (b) $NaOH$, OH^- (c) NH_3, NH_4^+
(d) $NaCl$, Na^+ (e) OH^-, O^{2-}

33. What is the role of CH_3COO^- in an acetic acid–sodium acetate (CH_3COOH-$NaCH_3COO$) buffer?
(a) donates H^+ (b) provides negative charge   (c) reacts with OH^-
(d) accepts H^+ (e) acts as indicator

34. Which buffer system would be best to use to prepare a buffer with pH = 6.82?
(a) CH_3COOH-CH_3COO^-, $pK_a = 4.74$ (b) HNO_2-NO_2^-, $pK_a = 3.37$
(c) $HClO_2$-ClO_2^-, $pK_a = 2.00$ (d) NH_3-NH_4^+, $pK_a = 9.26$
(e) $H_2PO_4^-$-HPO_4^{2-}, $pK_a = 7.21$

35. A $HClO_2$-ClO_2^- buffer is prepared with pH = 2.35. What ratio of base to acid concentrations, that is $[ClO_2^-]/[HClO_2]$, was used? pK_a = 2.00.

(a) 2.2   (b) 0.45   (c) 0.85   (d) 1.2   (e) 1.0

36. A NH_3-NH_4^+ buffer is prepared with $[NH_3]/[NH_4^+]$ = 1.54. What is the pH of the buffer? pK_a = 9.26.

(a) 9.45   (b) 9.26   (c) 9.07   (d) 10.80   (e) 7.72

37. How many milliliters of 2.0 M HCl can be added to 100 mL of a buffer that is 0.20 M in ammonia and 0.20 M in ammonium chloride before the buffer capacity is exceeded?

(a) 8.2   (b) 6.0   (c) 9.3   (d) 5.3   (e) 7.5

38. How many milliliters of 1.5 M NaOH can be added to 250 mL of a buffer that is 0.050 M in acetic acid and 0.040 M in sodium acetate before the buffer capacity is exceeded?

(a) 12   (b) 7.0   (c) 16   (d) 5.0   (e) 10 .

Solubility equilibria

39. What is the solubility of iron(II) sulfide (FeS) in $mol \cdot L^{-1}$? K_{sp} = 6.3×10^{-18}.

(a) $4.0 \times 10^{-35} \, mol \cdot L^{-1}$ (b) $6.3 \times 10^{-18} \, mol \cdot L^{-1}$ (c) $1.2 \times 10^{-17} \, mol \cdot L^{-1}$

(d) $2.5 \times 10^{-9} \, mol \cdot L^{-1}$ (e) $6.3 \times 10^{-10} \, mol \cdot L^{-1}$

40. What is the solubility of copper(II) iodate $[Cu(IO_3)_2]$ in $mol \cdot L^{-1}$? K_{sp} = 1.4×10^{-7}.

(a) $3.8 \times 10^{-4} \, mol \cdot L^{-1}$ (b) $1.9 \times 10^{-4} \, mol \cdot L^{-1}$ (c) $1.4 \times 10^{-7} \, mol \cdot L^{-1}$

(d) $5.2 \times 10^{-3} \, mol \cdot L^{-1}$ (e) $3.3 \times 10^{-3} \, mol \cdot L^{-1}$

41. What is the solubility of bismuth(III) sulfide (Bi_2S_3) in $mol \cdot L^{-1}$? K_{sp} = 1.0×10^{-97}.

(a) $2.9 \times 10^{-33} \, mol \cdot L^{-1}$ (b) $1.6 \times 10^{-20} \, mol \cdot L^{-1}$ (c) $3.2 \times 10^{-49} \, mol \cdot L^{-1}$

(d) $4.0 \times 10^{-20} \, mol \cdot L^{-1}$ (e) $2.0 \times 10^{-98} \, mol \cdot L^{-1}$

42. The solubility of magnesium carbonate ($MgCO_3$) at 20°C is 1×10^{-2} g per 100 g water. Estimate K_{sp} for magnesium carbonate.

(a) 4×10^{-4} (b) 1×10^{-1} (c) 5×10^{-7}

(d) 1×10^{-6} (e) 7×10^{-9}

43. The solubility of calcium fluoride (CaF_2) at 20°C is 2×10^{-3} g per 100 g water. Estimate K_{sp} from this information.

(a) 7×10^{-11} (b) 7×10^{-8} (c) 4×10^{-4}

(d) 2×10^{-3} (e) 7×10^{-14}

44. What is the solubility of AgCl in 0.20 M NaCl? $K_{sp}(AgCl)$ = 1.6×10^{-10}.

(a) $1.3 \times 10^{-5} \, mol \cdot L^{-1}$ (b) $8.0 \times 10^{-10} \, mol \cdot L^{-1}$ (c) $3.5 \times 10^{-10} \, mol \cdot L^{-1}$

(d) $1.6 \times 10^{-10} \, mol \cdot L^{-1}$ (e) $3.2 \times 10^{-11} \, mol \cdot L^{-1}$

45. What is the solubility of BaF_2 in 0.25 M NaF? $K_{sp}(BaF_2)$ = 1.7×10^{-6}.

(a) $1.7 \times 10^{-6} \, mol \cdot L^{-1}$ (b) $2.7 \times 10^{-5} \, mol \cdot L^{-1}$ (c) $6.8 \times 10^{-6} \, mol \cdot L^{-1}$

(d) $1.3 \times 10^{-3} \, mol \cdot L^{-1}$ (e) $4.4 \times 10^{-7} \, mol \cdot L^{-1}$

46. What is the solubility of Sb_2S_3 in a solution in which $[S^{2-}]$ = 5.0×10^{-3} M? K_{sp} = 1.7×10^{-93}.

(a) $5.6 \times 10^{-20} \, mol \cdot L^{-1}$ (b) $2.2 \times 10^{-21} \, mol \cdot L^{-1}$ (c) $5.8 \times 10^{-44} \, mol \cdot L^{-1}$

(d) $3.4 \times 10^{-94} \, mol \cdot L^{-1}$ (e) $2.8 \times 10^{-19} \, mol \cdot L^{-1}$

47. What is the solubility of $Al(OH)_3(s)$ when the pH = 5.12? $K_{sp} = 1.0 \times 10^{-33}$.
(a) $2.5 \times 10^{-9} \, mol \cdot L^{-1}$ (b) $1.2 \times 10^{-7} \, mol \cdot L^{-1}$ (c) $8.1 \times 10^{-5} \, mol \cdot L^{-1}$
(d) $5.2 \times 10^{-8} \, mol \cdot L^{-1}$ (e) $4.4 \times 10^{-7} \, mol \cdot L^{-1}$

48. What is the solubility of $Al(OH)_3$ when the solution pH equals 10.55?
(a) $1.4 \times 10^{-2} \, mol \cdot L^{-1}$ (b) $2.0 \times 10^{-9} \, mol \cdot L^{-1}$ (c) $2.8 \times 10^{-12} \, mol \cdot L^{-1}$
(d) $3.4 \times 10^{-8} \, mol \cdot L^{-1}$ (e) $2.2 \times 10^{-23} \, mol \cdot L^{-1}$

49. What happens to the solubility of PbS when the pH is increased?
(a) solubility increases   (b) solubility decreases   (c) solubility remains the same

50. What happens to the solubility of $PbCl_2$ when the pH is decreased?
(a) solubility increases   (b) solubility decreases   (c) solubility remains the same

51. A solution is prepared that is 2×10^{-4} M in Ag^+ and 5×10^{-9} M in Br^-. Does AgBr precipitate? $K_{sp}(AgBr) = 7.7 \times 10^{-13}$.
(a) Yes   (b) No

52. 0.20 mL of 6×10^{-3} M Na_2CO_3 is added to 1.0 mL of a solution that is 5×10^{-5} M in Ag^+, Mg^{2+}, and Ba^{2+}. Which ions precipitate? $K_{sp}(Ag_2CO_3) = 6.2 \times 10^{-12}$, $K_{sp}(MgCO_3) = 1.0 \times 10^{-5}$, $K_{sp}(BaCO_3) = 8.1 \times 10^{-9}$.
(a) Ag^+ only   (b) Mg^{2+} only   (c) Ba^{2+} only
(d) Mg^{2+} and Ba^{2+} (e) all precipitate

53. What is the solubility of AgCl(s) in 0.25 M NH_3? $K_f\{[Ag(NH_3)_2]^+\} = 1.6 \times 10^7$, $K_{sp}(AgCl) = 1.6 \times 10^{-10}$.
(a) $0.25 \, mol \cdot L^{-1}$ (b) $1.6 \times 10^7 \, mol \cdot L^{-1}$ (c) $4.0 \times 10^{-3} \, mol \cdot L^{-1}$
(d) $0.051 \, mol \cdot L^{-1}$ (e) $0.013 \, mol \cdot L^{-1}$

54. Which reagent could be used to dissolve copper(II) hydroxide $[Cu(OH)_2(s)]$?
(a) CO_2 (b) HCl   (c) NaCl   (d) NH_3 (e) KNO_3

55. Which reagent could be used to dissolve radium carbonate $(RaCO_3)$?
(a) CaO   (b) $CaCl_2$ (c) NH_3 (d) KCl   (e) HNO_3

56. Based on the information in Table 15.6, predict which solid will dissolve in KCN?
(a) WO_3 (b) $CaCO_3$ (c) HgO   (d) $Fe(OH)_2$ (e) Ir_2O_3

57. Given the equilibrium constants for equilibriums I and II, what is the equilibrium constant for equilibrium III?

$$\text{I:} \qquad AgI(s) \rightleftharpoons Ag^+(aq) + I^-(aq) \qquad K_s = 1.5 \times 10^{-16}$$
$$\text{II:} \ Ag^+(aq) + 2CN^-(aq) \rightleftharpoons Ag(CN)_2^-(aq) \qquad K_f = 5.6 \times 10^8$$
$$\text{III:} \quad AgI(s) + 2CN^-(aq) \rightleftharpoons Ag(CN)_2^-(aq) + I^-(aq) \qquad K = ?$$

(a) $K = 7.1 \times 10^{-10}$ (b) $K = 1.4 \times 10^9$ (c) $K = 3.7 \times 10^{24}$
(d) $K = 8.4 \times 10^{-8}$ (e) $K = 2.7 \times 10^{-25}$

Descriptive chemistry

58. Which of the following cations would you expect to be acidic when dissolved in water?
(a) Fe^{3+} (b) Mg^{2+} (c) Na^+ (d) Ag^+ (e) Ba^{2+}

59. Which ion is basic?

(a) NH_4^+ (b) O^{2-} (c) Cl^- (d) K^+ (e) NO_3^-

60. What is the approximate pH of blood and much of the cellular fluids in humans?

(a) 7.4   (b) 8.6   (c) 2.0   (d) 4.5   (e) 6.9

61. What is the formula (with the correct charge) of the common complex formed from Ag^+ and NH_3?

(a) $Ag(NH_3)_2$ (b) $Ag(NH_3)^+$ (c) $Ag(NH_3)$

(d) $Ag(NH_3)_2^+$ (e) $Ag(NH_3)_3^-$

CHAPTER 16

ENERGY IN TRANSITION: THERMODYNAMICS

Thermodynamics is the science of the transformation of energy. It is a critical aspect of the study of chemistry because the occurrence of all chemical reactions and many physical processes can be explained by analyzing the energy changes that occur. It is probably not an exaggeration to say that the field of thermodynamics is one of the foundations of all science.

THE FIRST LAW OF THERMODYNAMICS

16.1 SYSTEMS AND SURROUNDINGS

KEY CONCEPT There are three types of systems

The definitions of the **system** and the **surroundings** are somewhat arbitrary. Typically, we define them in order to make analysis of energy movement and energy transformations as simple as possible. The system is usually a very small piece of apparatus containing the chemicals of interest; for instance, a Styrofoam cup with a solution and thermometer may be "the system." The surroundings are everything else in the universe; practically, only objects or substances in direct contact with the system undergo measurable changes and are thus considered to be the surroundings. Systems are often classified as isolated, open, or closed on the basis of whether they can exchange energy and/ or matter with the surroundings:

Type of System	Matter Exchange with Surroundings Possible?	Energy Exchange with Surroundings Possible?
isolated	no	no
closed	no	yes
open	yes	yes

EXAMPLE Understanding the system

Assume we run a chemical reaction in a thermos bottle with a snug top and a thermometer in it. We assume the thermos works perfectly well in insulating the chemical reaction from the atmosphere and that we take care not to add or remove any chemicals during the experiment. What type of system does this apparatus approximate?

SOLUTION Because the thermos acts as a perfect insulator, exchange of energy (generally as heat) with the surroundings cannot occur. In addition, care is taken to assure that no matter is added or removed during the reaction, so there is no exchange of matter between the system and the surroundings. Thus, the thermos is an isolated system.

EXERCISE It is important to understand the thermodynamics of automobile engines in order to maximize their efficiency. What type of system is an automobile engine?

[*Answer:* Open]

16.2 HEAT AND WORK

KEY CONCEPT   Changing the energy of the system

There are three common ways to change the energy of a system. One way is to add to the system matter, such as gasoline, that provides the energy. This can occur only for an open system, because an open system is the only one that can exchange matter with the surroundings. A second way is to heat or cool the system; in other words, we add or remove energy directly as heat. This change can occur for both an open system and a closed system because both are allowed to exchange energy with the surroundings. The third way to change the energy of a system is by doing work on the system (which adds energy) or by having the system do work on the surroundings (which removes energy from the system). This change can occur for both an open system or a closed system. An isolated system, by definition, has a constant energy.

Heat is the natural transfer of internal energy from hot to cold objects; it is not an energy, it is a process. There are many kinds of **work,** which is done when an object, such as a piston, pushes against an opposing force. A common kind of work we often encounter is that done by pushing against the pressure of a gas.

EXAMPLE   Changing the energy of the system

Assume we have a tightly closed pressure cooker on a stove and the stove is turned on to raise the temperature of the contents. What type of system is this and in which of the three ways mentioned above is the energy of the system being changed?

SOLUTION Because the pressure cooker is tightly closed, it cannot exchange matter with the surroundings, but energy (as heat) can be directly added; this is a closed system. We are changing its energy by the direct addition of energy from a hot object (the stove) to a cooler one (the pressure cooker).

EXERCISE A common toy is a paper or balsa wood airplane that has a propeller attached to a rubber band. The propeller is wound up by turning it until the rubber band becomes taut; when the propeller is released, it spins rapidly as the rubber band unwinds and the plane can fly, powered by the spinning propeller. If the propeller is the system, how are we adding energy to it as we wind it?

[*Answer:* By doing work on the system]

16.3   INTERNAL ENERGY

KEY CONCEPT A   Heat or work can change the internal energy of a system

In chemical systems we are concerned with the energy of a collection of particles (atoms, molecules, or formula units). The total energy of all the particles in a sample is called the **internal energy** (U) of the sample; it is a state property. If heat is added to a system or work is done on the system, the internal energy increases; if heat is removed from the system or the system does work on the surroundings, the internal energy decreases:

$$\text{Change in internal energy} = \text{heat} + \text{work}$$
$$\Delta U \qquad\qquad q \qquad w$$

The sign conventions we use in this equation are shown in the table.

Heat Change	Sign of Heat and/or Work	Internal Energy
heat leaves system	−	decreases
heat added to system	+	increases
work done on system	+	increases
work done by system	−	decreases

EXAMPLE Calculating a change in internal energy

A chemical reaction is run in such a way that 844 J of heat is evolved and, at the same time, 118 J of work is done on the surroundings by the chemical system. What is the change in the internal energy of the system?

SOLUTION When heat leaves the system, it is assigned a negative sign, and when work is done by the system, it is assigned a negative sign. Thus, for our problem,

$$q = -844 \text{ J} \qquad w = -118 \text{ J}$$

The change in internal energy is given by

$$\Delta U = q + w$$

We now substitute the given values of q and w:

$$\Delta U = -844 \text{ J} + (-118 \text{ J})$$
$$= -962 \text{ J}$$

EXERCISE A chemical reaction is run in such a way that 1.55 kJ of heat is evolved and, at the same time, 328 J of work is done on the chemical system. What is the change in the internal energy of the system?

[*Answer:* −1.23 kJ]

KEY CONCEPT B   The internal energy of an isolated system is constant

Earlier we defined an isolated system as one that has no contact with the surroundings. It cannot exchange heat with the surroundings and cannot do work or have work done on it. Thus, for an isolated system, $q = 0$ and $w = 0$. If we remember that for any system, $\Delta U = q + w$, then it must be true that for an isolated system,

$$\Delta U = 0 \qquad \text{(isolated system)}$$

This is a mathematical statement of the **first law of thermodynamics:** The internal energy of an isolated system is constant. That is, because $\Delta U = 0$, the internal energy does not change.

EXAMPLE Understanding the first law of thermodynamics

Is the Earth an isolated system? Explain.

SOLUTION For an isolated system, $\Delta U = 0$; this is because $q = 0$ and $w = 0$. The Earth, however, receives a good deal of energy from the Sun; so, for the Earth, $q \neq 0$. Thus, the Earth is not an isolated system.

EXERCISE Is a tightly corked, high-quality thermos bottle a reasonable model of an isolated system (if we consider only a short period of time)?

[*Answer:* Yes, because ΔU is close to 0 for such a system.]

16.4   HEAT TRANSFERS AT CONSTANT VOLUME

KEY CONCEPT   At constant volume, $\Delta U = q$

One of the most important forms of work is that accomplished by the expansion of a gas against a resisting pressure; one way this can occur is when a gas, trapped in a cylinder by a piston, expands. As long as the piston offers some resistance, some work is done by the system. In a system at constant volume, expansions and contractions cannot occur, so $w = 0$ for this type of work. Because, for all systems, $\Delta U = q + w$, it follows for a constant volume system that $\Delta U = q$. For a constant volume system, **changes in the internal energy** can occur only by the gain or loss of heat.

EXAMPLE   Calculating changes in internal energy at constant volume

A chemical reaction in a bomb calorimeter (Section 6.3 of the text) releases 732 J of heat through its walls. What is ΔU for the calorimeter for this process?

SOLUTION In general, $\Delta U = q + w$. A bomb calorimeter is a constant volume system, so $w = 0$ and, therefore, $\Delta U = q$. Because 732 J of heat were released, $q = -732$ J. Thus, $\Delta U = -732$ J.

EXERCISE A tightly capped tea kettle has 8.92 kJ of heat added to it. What is ΔU for this process, if we consider the tea kettle the system?

[*Answer:* $\Delta U = +8.92$ kJ]

16.5   ENTHALPY

KEY CONCEPT   Enthalpy and constant pressure processes

The change in enthalpy is the heat transfer that occurs when a reaction is run at constant pressure (as in an open container). If we use the symbol q_P to indicate the heat transfer at constant pressure, then $\Delta H = q_P$. Because the volume is allowed to change in such a situation, the heat transfer that occurs comes from two possible sources: (a) any change in internal energy that occurs and (b) any work that occurs. Thus, the enthalpy change is determined by both changes in internal energy and work done.

Constant pressure: $\Delta H = q_P$ Enthalpy change is determined by transfer of heat; the heat accounts for changes in internal energy and any work that must be accomplished.

EXAMPLE   Using the definition of enthalpy

A chemical reaction run in an open beaker releases 1.23 kJ of heat. The same reaction run in a closed stainless steel cylinder releases 1.25 kJ. What is ΔH for the reaction?

SOLUTION The change in enthalpy for a chemical reaction is the heat released or absorbed when the reaction is run at constant pressure. When a reaction is run in an open container, the pressure is constant at atmospheric pressure, and the heat released or absorbed is ΔH. Because in this example heat is released, a negative sign is associated with the change. Thus, $\Delta H = -1.23$ kJ. (The difference of 0.02 kJ is the work used to make room in the atmosphere for the products to form.)

EXERCISE A reaction run at constant volume absorbs 3.55 kJ of heat from the surroundings. A separate experiment indicates that the work that accompanies the reaction when it is run at constant pressure is -0.08 kJ. What is ΔH for the reaction?

[*Answer:* 3.47 kJ]

KEY WORDS   Define or explain each term in a written sentence or two.

closed system   internal energy   surroundings
enthalpy change   isolated system   work
first law of thermodynamics open system
heat   system

THE DIRECTION OF SPONTANEOUS CHANGE

16.6 SPONTANEOUS CHANGE

KEY CONCEPT   Many processes are like a "one-way" street

Most processes always occur in one direction. For example,

1. When a hot piece of metal is placed in cool water, the metal cools and the water warms until both are at the same temperature. When a piece of metal at some temperature is placed in water at that same temperature, there is never a change in which the metal warms and the water cools.
2. When a bottle of gas is opened in a room, the gas molecules spontaneously spread throughout the room. Never will the spread-out molecules suddenly rush back and gather inside the bottle.
3. When an ice cube is placed in a beaker of water at 25°C, the ice melts and the water cools. A beaker of water at the cooler temperature never suddenly warms a little as an ice cube spontaneously forms in the water.

These are examples of **spontaneous changes**, which are changes that have a tendency to occur without outside intervention.

EXAMPLE   Understanding spontaneous change

Describe a situation in which the direction of spontaneous change reverses because of a change in temperature. (*Hint:* Use water as the substance involved.)

SOLUTION If we take a sample of water at 10°C, it would spontaneously change from solid (ice) to liquid. However, if we lower the temperature to −10°C, it would spontaneously change from liquid to solid. The direction of spontaneous change depends on the temperature.

EXERCISE Does the function of a spray can of paint depend on any spontaneous change?

[*Answer:* Yes, expansion of the propellant gas when you push the button]

16.7   ENTROPY AND DISORDER

KEY CONCEPT A   Spontaneous change increases disorder

Let's consider the three examples of the previous section to clarify the relationship between spontaneous change and disorder. In each example, as in all spontaneous processes, the direction of change is one that leads from order to disorder. By disorder, we mean that there are more choices for the location of molecules or energy. As an analogy, imagine that you take a handful of marbles out of a bag and throw them on the floor; the marbles have more freedom to be in different locations as they come to rest on the floor than when they are confined in the bag. It is this "choice" of location that defines the notion of disorder. In molecular systems, it is the change from order to disorder that is actually the driving force that makes spontaneous processes occur.

In the first example of the preceding section, the thermal energy originally confined to the atoms of the hot metal disperses throughout the water as chaotic thermal motion of the water molecules; because the energy has spread over more quantum levels, an increase in disorder has occurred. In the second example, the molecules originally contained in a small volume disperse and occupy a greater volume; because the molecules in the greater volume have more opportunity to move from place to place and be in different locations, an increase in disorder has occurred. In the third example, both effects occur; thermal energy becomes dispersed over more quantum levels and molecules spread from a smaller to larger volume. Thus, when energy spreads out among more molecules (and hence more quantum levels) as chaotic thermal motion (first and second examples), or when molecules disperse from a smaller to a larger volume (second and third examples), disorder increases.

EXAMPLE   Understanding spontaneous processes

Explain how the melting ice in the third example results in an increase in the disorder of matter.

SOLUTION When the ice melts, the water molecules originally trapped at one site in the solid have access to the entire volume of the liquid. This change results in an increase in the disorder of the water molecules.

EXERCISE When a solid dissolves in a solvent, is there an increase or a decrease in molecular disorder?

[*Answer:* Increase]

KEY CONCEPT B   Entropy as a measure of disorder

The entropy (S) of a sample is a measure of the disorder (both energy disorder and molecular disorder) of the sample. The greater the entropy, the greater the disorder.

EXAMPLE   Qualitatively predicting the change in entropy

Is there an increase, decrease, or no change in entropy (as measured by molecular disorder, neglecting energy disorder) in the chemical reaction

$$2SO_2(g) + O_2(g) \longrightarrow 2SO_3(g)$$

SOLUTION There are 3 mol of gaseous reactants and 2 mol of gaseous products. When the component atoms are contained in 3 mol of gas, they are more spread out than when they are contained in 2 mol of gas. The reactants are more disordered than the products, so there is a decrease in disorder in the chemical reaction. There is a decrease in entropy for the reaction ($\Delta S < 0$), if energy disorder is ignored.

EXERCISE Is there an increase, decrease, or no change in entropy (as measured by molecular disorder) in the chemical reaction

$$2Na(s) + 2H_2O(l) \longrightarrow 2Na^+(aq) + 2OH^-(aq) + H_2(g)$$

[*Answer:* Increase; $\Delta S > 0$]

16.8   STANDARD ENTROPIES

KEY CONCEPT A   The entropy of a substance depends on many factors

The standard molar entropy of a substance is the entropy per mole of the pure substance at 1 atm pressure. Some features of standard molar entropies are apparent from an examination of Appendix 2A of the text.

1. As the temperature of a substance increases, its entropy increases.
2. The entropy of a substance depends on the physical state of the substance:

$$S(\text{gas}) > S(\text{liquid}) > S(\text{solid})$$

3. Larger, more complex substances tend to have a greater entropy than smaller, simpler substances (for the same physical state at the same temperature).
4. All entropies at 298 K are positive. (At 0 K, the entropy of all substances is 0 or near 0, and as the temperature increases, the entropy increases.)

EXAMPLE   Predicting entropies of substances

Which has a higher entropy, a sugar cube or the same sugar dissolved in a cup of tea?

SOLUTION When a sample occupies a larger volume, the molecules of the sample have more positions in space that they can occupy than when they are in a smaller volume. (For example, several horses have more choices of where to be in a larger corral than in a smaller corral.) The larger volume, therefore, corresponds to more disorder. The sugar, when dissolved, occupies a larger volume then when it is a solid sugar cube; its entropy is higher when it is dissolved.

EXERCISE Which has a higher entropy, a sample of gas in a balloon or the same gas after it has leaked out of the balloon?

[*Answer:* The gas after it has leaked out]

KEY CONCEPT B   Standard reaction entropies

The entropies in Table 16.1 (and Appendix 2A) of the text can be used to calculate the change in entropy for a reaction at standard conditions:

$$\text{Reactants at standard conditions} \longrightarrow \text{products at standard conditions}$$

Pure, 1 atm pressure   Pure, 1 atm pressure

$$\Delta S_r^\circ = S^\circ(\text{products}) - S^\circ(\text{reactants})$$

This calculation is done in precisely the same manner as calculations of reaction enthalpy from ΔH_f° values. As before, we must remember that each entropy in the table is a per-mole quantity, so the number of moles of each reactant and product in the chemical equation must be taken into account.

EXAMPLE   Calculating a standard reaction entropy

Calculate the standard reaction entropy for the following reaction and comment on the sign of the result.

$$2H_2(g) + CO(g) \longrightarrow CH_3OH(l) \qquad \Delta S_r^\circ = ?$$

SOLUTION $\Delta S_r^\circ = S^\circ(\text{products}) - S^\circ(\text{reactants})$. We first retrieve the value of S° for each reactant and product from Appendix 2A or Table 16.1 of the text.

$$\begin{array}{ccc} 130.7 \text{ J(K·mol)}^{-1} & 197.7 \text{ J(K·mol)}^{-1} & 126.8 \text{ J(K·mol)}^{-1} \\ 2H_2(g) \quad + & CO(g) \quad \longrightarrow & CH_3OH(l) \end{array}$$

Substituting into the equation for calculating the standard reaction entropy gives

$\Delta S_r^\circ = S^\circ(\text{products}) - S^\circ(\text{reactants})$

$= [1 \text{ mol } CH_3OH \times S^\circ\{CH_3OH(l)\}] - [1 \text{ mol } CO \times S^\circ\{CO(g)\} + 2 \text{ mol } H_2 \times S^\circ\{H_2(g)\}]$

$= \left[1 \text{ mol } CH_3OH \times 126.8 \dfrac{J}{mol \cdot K} \right] - \left[\left(1 \text{ mol } CO \times 197.7 \dfrac{J}{mol \cdot K} \right) + \left(2 \text{ mol } H_2 \times 130.7 \dfrac{J}{mol \cdot K} \right) \right]$

$= [126.8 \text{ J·K}^{-1}] - [459.1 \text{ J·K}^{-1}]$

$= -332.3 \text{ J·K}^{-1}$

In this reaction, 3 mol of a gas become 1 mol of a liquid; so a decrease in disorder and a decrease in entropy occur. This is reflected in the negative sign of ΔS_r°.

EXERCISE First predict the sign of the entropy change and then calculate the standard reaction entropy for the reaction

$$2SO_2(g) + O_2(g) \longrightarrow 2SO_3(g)$$

[*Answer:* Negative; -188.1 J·K^{-1}]

PITFALL   Elements have finite values of $S°$

Unlike the values of the standard enthalpies of formation, the standard molar entropies of the elements are not 0. Do not forget to include them in calculations of standard reaction entropies.

16.9   THE SURROUNDINGS

KEY CONCEPT   The entropy change of the surroundings is important

To determine whether a chemical reaction or physical process is spontaneous or not, we must take into consideration the entropy change of the surroundings. The entropy change of the surroundings is calculated with the following relationship:

$$\text{Change in entropy of surroundings} = \frac{\text{heat transfer to or from surroundings}}{T \text{ (in kelvins) of surroundings}}$$

$$\Delta S_{surr} = \frac{\text{heat transfer to or from surroundings}}{T_{surr}}$$

For both endothermic and exothermic reactions, the heat flow (from the surroundings' point of view) has a sign opposite to that of the reaction enthalpy. This leads to

$$\Delta S_{surr} = \frac{-\Delta H}{T_{surr}}$$

Heat flow into the surroundings increases the entropy of the surroundings because the chaotic thermal motion of the molecules in the surroundings increases, whereas heat flow out of the surroundings decreases its entropy.

$$\text{Exothermic reaction} \rightarrow \Delta H < 0 \rightarrow \begin{array}{c}\text{heat flow into} \\ \text{surroundings}\end{array} \rightarrow \begin{array}{c}\text{disorder in surroundings} \\ \text{increases}\end{array} \rightarrow \Delta S_{surr} > 0$$

$$\text{Endothermic reaction} \rightarrow \Delta H > 0 \rightarrow \begin{array}{c}\text{heat flow out of} \\ \text{surroundings}\end{array} \rightarrow \begin{array}{c}\text{disorder in surroundings} \\ \text{decreases}\end{array} \rightarrow \Delta S_{surr} < 0$$

EXAMPLE   Calculating the entropy change of the surroundings

When a single ice cube freezes in your freezer, approximately 6.7 kJ of energy, as heat, is released to the surroundings. First, predict the sign of the entropy change for the surroundings and then calculate the change.

SOLUTION Because heat is moving into the surroundings, we expect the temperature of the surroundings and the chaotic thermal motion of the air molecules in the surroundings to increase. This leads to an increase in entropy for the surroundings and a positive sign for the entropy change. To calculate the change in entropy, we use the following equation with $\Delta H = -6.7$ kJ and assume the temperature is the freezing point of water, 273 K:

$$\Delta S_{surr} = -\frac{\Delta H}{T}$$

$$\Delta S_{surr} = -\left[\frac{-6.7 \times 10^3 \text{ J}}{273 \text{ K}}\right]$$

$$\Delta S_{surr} = +24 \text{ J} \cdot \text{K}^{-1}$$

EXERCISE A reaction that is endothermic by 83.0 kJ is run at 62°C. What is the entropy change of the surroundings for this reaction?

[*Answer:* −248 J·K^{-1}]

16.10   THE OVERALL CHANGE IN ENTROPY

KEY CONCEPT   The second law of thermodynamics

The **second law of thermodynamics** provides an explanation for the occurrence of any process in terms of the total entropy change that accompanies the process. We first recall that the total entropy change is the sum of the entropy change of the system and the entropy change of the surroundings:

$$\Delta S_{tot} = \Delta S_{sys} + \Delta S_{surr}$$

The second law states that a spontaneous process is accompanied by an increase in the total entropy of the system and surroundings combined, that is,

$$\text{Spontaneous process:} \quad \Delta S_{tot} > 0$$

It is important to realize that whether a process is exothermic or endothermic does not necessarily determine the spontaneity of the process. For instance, an exothermic reaction increases the entropy of the surroundings by increasing the chaotic thermal motion of the surroundings. Thus, even if the entropy of the system decreases as the reactants become products, this decrease may be offset by an increase in the entropy of the surroundings, and the change may still be a spontaneous change. An endothermic reaction decreases the entropy of the surroundings by decreasing the chaotic thermal motion of the surroundings. This decrease may be offset by an increase in the entropy change of the system as the reactants become products, and the change may still be a spontaneous change. For chemical reactions, the entropy change of the system consists of the entropy change, $\Delta S° = S°(\text{products}) - S°(\text{reactants})$. Very often the entropy change of the system is relatively small, so the entropy change of the surroundings determines whether the change is a spontaneous change. That is, it is often the exothermicity or endothermicity of a reaction that determines whether or not the change is a spontaneous change.

EXAMPLE 1   Using the second law qualitatively

The following reaction is exothermic. Without detailed calculations, comment on whether it is a spontaneous reaction.

$$2C_4H_{10}(l) + 13O_2(g) \longrightarrow 8CO_2(g) + 10H_2O(g)$$

SOLUTION Because the reaction is exothermic, the entropy change in the surroundings is positive and contributes to a spontaneous change. In addition, the reaction involves forming 18 mol of gas from 10 mol of gas and 2 mol of liquid, so the entropy change of the system is almost certainly positive. This also contributes to making the reaction a spontaneous change. The entropy effects in the surroundings and in the system are positive, so the reaction is spontaneous. In summary,

$$\Delta S_{surr} > 0 \quad \text{and} \quad \Delta S_{system} > 0 \quad \text{so the process is spontaneous}$$

EXERCISE The reaction below is endothermic. Without detailed calculations, comment on whether it is a spontaneous reaction.

$$3H_2O(g) + 2CO_2(g) \longrightarrow 3O_2(g) + C_2H_5OH(l)$$

[*Answer:* $\Delta S_{surr} < 0$ and $\Delta S_{system} < 0$; not a spontaneous reaction]

EXAMPLE 2   Using the second law qualitatively

The following reaction is endothermic. Without detailed calculations, comment on whether it is a spontaneous reaction.

$$SiO_2(s) + 2C(s) \longrightarrow Si(s) + 2CO(g)$$

SOLUTION An endothermic reaction decreases the entropy of the surroundings, which contributes to a nonspontaneous change. In the reaction, 2 mol of gas and 1 mol of solid are formed from 3 mol of solid, so the entropy change of the system is most likely positive. The two factors have opposite effects, so whether the reaction is spontaneous depends on which effect is larger.

$\Delta S_{surr} < 0$ and $\Delta S_{system} > 0$ Opposite effects; cannot tell whether reaction is spontaneous without detailed calculation

EXERCISE The following reaction is exothermic. Without detailed calculations, comment on whether it is a spontaneous reaction.

$$C_2H_4(g) + Cl_2(g) \longrightarrow C_2H_4Cl_2(l)$$

[*Answer:* $\Delta S_{surr} > 0$ and $\Delta S_{system} < 0$; cannot tell whether reaction is a spontaneous process without detailed calculations]

EXAMPLE 3 Calculating the total entropy change of a reaction

Calculate the standard entropy change at 298 K of the system and of the surroundings for the reaction

$$3Fe(s) + 2O_2(g) \longrightarrow Fe_3O_4(s)$$

Is this reaction (which is a major one that occurs when your car rusts) a spontaneous process?

SOLUTION We anticipate the need to calculate $\Delta S°$ and $\Delta H°$ for the reaction by assembling a table of enthalpies of formations and standard molar entropies.

	3Fe(s)	**+**	**2O$_2$(g)**	$\longrightarrow$	**Fe$_3$O$_4$(s)**
$\Delta H_f°$ per mole	0		0		−1118 kJ
$S°$ per mole	27.28 J·K^{-1}		205.14 J·K^{-1}		146.4 J·K^{-1}

The entropy change for the system (the standard reaction entropy) is

$$\Delta S° = S°(\text{products}) - S°(\text{reactants})$$

$$= \left[(1 \text{ mol Fe}_3O_4) \times \left(146.4 \frac{J}{mol \cdot K} \right) \right]$$

$$- \left[\left(2 \text{ mol O}_2 \times 205.14 \frac{J}{mol \cdot K} \right) + (3 \text{ mol Fe}) \times \left(27.28 \frac{J}{mol \cdot K} \right) \right]$$

$$= [146.4 \text{ J·K}^{-1}] - [492.12 \text{ J·K}^{-1}]$$

$$= -345.7 \text{ J/K}$$

To calculate the entropy change in the surroundings, we need the standard enthalpy of reaction:

$$\Delta H° = \left[(1 \text{ mol Fe}_3O_4) \times \left(-1118 \frac{kJ}{mol} \right) \right] - \left[(2 \text{ mol O}_2) \times \left(0 \frac{kJ}{mol} \right) + (2 \text{ mol Fe}) \times \left(0 \frac{kJ}{mol} \right) \right]$$

$$= -1118 \text{ kJ} = -1.118 \times 10^6 \text{ J}$$

The entropy change in the surroundings is

$$\Delta S_{surr} = \frac{-\Delta H°}{T_{surr}}$$

$$= \frac{-(-1.118 \times 10^6 \text{ J})}{298 \text{ K}}$$

$$= +3.75 \times 10^3 \text{ J·K}^{-1}$$

The total entropy change for the reaction is

$$\Delta S°(\text{system + surroundings}) = -345.7 \text{ J·K}^{-1} + 3.75 \times 10^3 \text{ J·K}^{-1}$$

$$= +3.40 \times 10^3 \text{ J·K}^{-1}$$

The large positive change in entropy indicates that the reaction is spontaneous.

EXERCISE Calculate the total standard entropy change at 298 K for the reaction shown and state whether it is a spontaneous reaction.

$$CuSO_4 \cdot 5H_2O(s) \longrightarrow CuSO_4(s) + 5H_2O(g)$$

[*Answer:* $-251 \, J \cdot K^{-1}$; not a spontaneous reaction]

KEY WORDS   Define or explain each term in a written sentence or two.

entropy

second law of thermodynamics

spontaneous change

standard molar entropy

standard reaction entropy

FREE ENERGY

16.11   FOCUSING ON THE SYSTEM

KEY CONCEPT A   The free energy change of the system

Whether or not a process is spontaneous depends on the total entropy change of the process. The **Gibbs free energy** change (ΔG) of the system is defined in such a way that it depends directly on the total entropy change:

$$\Delta G = -T \, \Delta S_{total}$$

Here, ΔS_{total} is the entropy change of the universe:

$$\Delta S_{total} = \Delta S_{surr} + \Delta S_{system}$$

Because any spontaneous change has a positive ΔS_{total}, it must also have a negative ΔG.

$$\text{Spontaneous change:} \quad \Delta S_{total} > 0$$
$$\Delta G < 0$$

Notice that ΔG does not have the subscript "total" because the free energy change is for the system only. The free energy change of the system "keeps track of" the entropy change of the universe. Any spontaneous change occurs so as to maximize the entropy of the universe and minimize the free energy of the system. From the definition of the Gibbs free energy, $G = H - TS$, its change during a process is

$$\Delta G_{system} = \Delta H_{system} - T \, \Delta S_{system}$$

We write this simply as

$$\Delta G = \Delta H - T \, \Delta S$$

It must be remembered that ΔS now refers to the change in the *system* only. In earlier sections, we used this symbol to indicate the entropy change of the universe.

EXAMPLE   Calculating the free energy change

A reaction has $\Delta H = +36.2 \, kJ$ and $\Delta S = 115 \, J \cdot K^{-1}$ at 25°C. Calculate ΔG and state whether the reaction is spontaneous.

SOLUTION The free energy change is given by

$$\Delta G = \Delta H - T\Delta S$$

Substituting the given values

$$\Delta H = 36.2 \text{ kJ} = 3.62 \times 10^4 \text{ J}$$

$$T = 25 + 273 = 298 \text{ K}$$

$$\Delta S = 115 \text{ J} \cdot \text{K}^{-1}$$

gives

$$\Delta G = 3.62 \times 10^4 \text{ J} - (298 \text{ K})\left(115\frac{\text{J}}{\text{K}}\right)$$

$$= 3.62 \times 10^4 \text{ J} - 3.43 \times 10^4 \text{ J}$$

$$= 0.19 \times 10^4 \text{ J}$$

$$= 1.9 \times 10^3 \text{ J}$$

The reaction is not spontaneous because ΔG is positive.

EXERCISE What is ΔG for the reaction in the example at 100°C. (Assume that ΔH and ΔS do not change in going from 25 to 100°C.) Is the reaction spontaneous at 100°C?

[*Answer:* −6.70 kJ; yes]

PITFALL   The proper units for $\Delta G = \Delta H - T\Delta S$

Because values of ΔH are typically expressed in kilojoules and values of ΔS are usually expressed in joules per kelvin (not kilojoules per kelvin), the units of either ΔH or ΔS must be adjusted so that the two are consistent. In the preceding example, for instance, the units of ΔH are changed from kilojoules to joules. The temperature must be expressed in kelvins when using the relation $\Delta G = \Delta H - T\Delta S$.

KEY CONCEPT B   The free energy change for an equilibrium

There is no net change for a system at equilibrium, so the free energy change is 0; that is,

$$\text{System at equilibrium:} \quad \Delta G = 0$$

Two examples of equilibrium are melting at the melting point (T_m) and boiling at the boiling point (T_b). Because $\Delta G = 0$ for melting and boiling, we get

$$T_m = \frac{\Delta H_{fus}}{\Delta S_{fus}}$$

$$T_b = \frac{\Delta H_{vap}}{\Delta S_{vap}}$$

We can use these equations to calculate any one of the quantities in the equations if the other two are known.

EXAMPLE   Calculating the molar entropy of a phase change

What is the molar entropy of vaporization for $CCl_4(g)$, which has a molar enthalpy of vaporization of 30.00 kJ·mol^{-1} and boils at 349.9 K?

SOLUTION The molar entropy of vaporization can be calculated with the formula

$$T_b = \frac{\Delta H_{vap}}{\Delta S_{vap}}$$

This expression rearranges to

$$\Delta S_{vap} = \frac{\Delta H_{vap}}{T_b}$$

Substituting the given data gives

$$\Delta S_{vap} = \frac{30.00 \times 10^3 \text{ J} \cdot \text{mol}^{-1}}{349.9 \text{ K}}$$

$$= 85.7 \text{ J}(\text{K} \cdot \text{mol})^{-1}$$

EXERCISE The molar heat of melting of O_2 is 444 J·mol^{-1} and its melting temperature is $-218.8°C$. What is ΔS for the melting of O_2?

[*Answer:* 8.17 J(K·mol)$^{-1}$]

16.12   STANDARD REACTION FREE ENERGIES

KEY CONCEPT   Standard free energies of formation

If a reaction is run under standard conditions (i.e., with pure reactants and products at 1 atm pressure), the free energy change associated with the reaction is called the **standard free energy**:

Reactants (pure, 1 atm) $\longrightarrow$ products (pure, 1 atm)

$\Delta G = $ standard free energy ($\Delta G°$)

The **standard free energy of formation** of a compound is the standard reaction free energy for synthesis of the compound from its elements:

Elements (pure, 1 atm) $\longrightarrow$ compound (pure, 1 atm)

$\Delta G = $ standard free energy of formation of compound ($\Delta G_f°$)

A table of standard free energies of formation is given in Appendix 2A of the text. A compound is classified as **thermodynamically unstable** if its standard free energy of formation is positive. This means that there is a *tendency* for the compound to decompose into its elements. A **thermodynamically stable** compound is one with a negative $\Delta G_f°$, because the standard reaction free energy for decomposition of the compound into its elements is then positive, and the decomposition is not a spontaneous process.

EXAMPLE   Calculating the standard free energy of formation

Calculate the standard free energy of formation of $CO_2(g)$ at 25°C from its standard enthalpy of formation and standard molar entropy. Is $CO_2(g)$ thermodynamically stable or thermodynamically unstable?

SOLUTION The standard free energy of formation is the free energy change associated with the following reaction, where it is understood that all products and reactants are pure and at 1 atm pressure:

$$C(\text{graphite}) + O_2(g) \longrightarrow CO_2(g) \qquad \Delta G_f° = ?$$

To calculate $\Delta G_f°$, we use the following steps:

1. $\Delta G_f° = \Delta H_f° - T\Delta S°$
2. $\Delta H_f° = \Delta H_f°(\text{products}) - \Delta H_f°(\text{reactants})$

$$= \left\{1 \text{ mol } CO_2 \times \left(-393.51\frac{\text{kJ}}{\text{mol}}\right)\right\} - \left\{1 \text{ mol } C \times \left(0\frac{\text{kJ}}{\text{mol}}\right) + 1 \text{ mol } O_2 \times \left(0\frac{\text{kJ}}{\text{mol}}\right)\right\}$$

$$= -393.51 \text{ kJ}$$

$$= -3.9351 \times 10^5 \text{ J}$$

3. $\Delta S° = S°(\text{products}) - S°(\text{reactants})$

$$= \left\{ 1 \text{ mol } CO_2 \times \left(213.74 \frac{J}{K \cdot mol} \right) \right\}$$

$$- \left\{ 1 \text{ mol } C \times \left(5.740 \frac{J}{K \cdot mol} \right) + 1 \text{ mol } O_2 \times \left(205.1 \frac{J}{K \cdot mol} \right) \right\}$$

$$= +2.9 \text{ J/K}$$

Substituting into the required equation gives

$$\Delta G° = \Delta H° - T \, \Delta S°$$

$$= -3.9351 \times 10^5 \text{ J} - (298 \text{ K})(2.9 \text{ J} \cdot \text{K}^{-1})$$

$$= -3.9351 \times 10^5 \text{ J} - 8.6 \times 10^2 \text{ J}$$

$$= -3.9437 \times 10^5 \text{ J}$$

$$= -394.37 \text{ kJ}$$

The answer -394.37 kJ is the standard free energy of formation for 1 mol of CO_2, so $\Delta G_f°(CO_2, g) = -394.37 \text{ kJ} \cdot \text{mol}^{-1}$. Because $CO_2(g)$ has a negative standard free energy of formation, it is thermodynamically stable.

EXERCISE Calculate the standard free energy of formation of $CCl_4(l)$ at 25°C from its standard enthalpy of formation and standard molar entropy. Is $CCl_4(l)$ a thermodynamically stable compound?

[*Answer:* $-65.2 \text{ kJ} \cdot \text{mol}^{-1}$; yes]

16.13   USING FREE ENERGIES OF FORMATION

KEY CONCEPT   Calculating free energies of reaction

Standard free energies of formation can be used to calculate a standard reaction free energy, in a manner analogous to using standard enthalpies of formation to calculate standard reaction enthalpies and standard molar entropies to calculate standard reaction entropies:

$$\Delta G° = \Delta G_f°(\text{products}) - \Delta G_f°(\text{reactants})$$

In using standard free energies of formation, we must remember to take into account the number of moles of reactants and products expressed in the balanced equation.

EXAMPLE   Calculating the free energy of formation

In the production of butane gas $\{C_4H_{10}(g)\}$ at 298 K from the reaction of methane $\{CH_4(g)\}$ with ethyne $\{C_2H_2(g)\}$ a spontaneous reaction?

SOLUTION To answer this question, we must calculate ΔG for the reaction

$$2CH_4(g) + C_2H_2(g) \longrightarrow C_4H_{10}(g)$$

We assume that all the gases are at a partial pressure of 1 atm so that we can use $\Delta G°$ to answer the question.

$$\Delta G° = \Delta G_f°(\text{products}) - \Delta G_f°(\text{reactants})$$

$$= \left\{ 1 \text{ mol } C_4H_{10} \times \left(-17.03 \frac{kJ}{mol} \right) \right\}$$

$$- \left\{ 2 \text{ mol } CH_4 \times \left(-50.72 \frac{kJ}{mol} \right) + 1 \text{ mol } C_2H_2 \times \left(209.20 \frac{kJ}{mol} \right) \right\}$$

$$= \{-17.03 \text{ kJ}\} - \{107.76 \text{ kJ}\}$$

$$= -124.79 \text{ kJ}$$

Because $\Delta G° < 0$, the production of butane from the reaction of methane and ethyne at the specified conditions (1 atm pressure) is a spontaneous process. This does not mean it is easy to do so! We may still have to devise a way to make the reaction occur at a reasonable rate. The negative $\Delta G°$ tells us the reaction is a spontaneous process and tends to occur, but does not guarantee it will occur at a finite rate.

EXERCISE An entrepreneur asks you to invest your life savings in a brand new energy-producing device that depends on using sand (SiO_2) as a fuel in the following reaction:

$$2SiO_2(s) + N_2(g) \longrightarrow 2Si(s) + N_2O_4(g)$$

With the high availability of sand and atmospheric nitrogen, not to mention the lucrative sales of silicon to the semiconductor industry, the entrepreneur guarantees you will get rich. How much money should you invest?

[*Answer:* None. The reaction has a positive $\Delta G(+1811$ kJ$)$, so it does not occur as a spontaneous process. The scheme will not work.]

16.14   FREE ENERGY AND COMPOSITION

KEY CONCEPT   Calculating free energies of reaction

The free energies we have considered up to now have all been calculated for standard conditions, with very specific concentrations assumed for all reactants and products. But as you might expect, the free energy of a system depends on the concentration of the substances present. For any reaction or physical process, the actual free energy of the system (ΔG_r) is related to the standard free energy ($\Delta G_r°$) and is given by

$$\Delta G_r = \Delta G_r° + RT \ln Q$$

Q is the reaction quotient first introduced in Section 13.6 of the text; remember that the reaction quotient has the same form as the equilibrium constant but uses the actual concentrations of substances present, not the equilibrium concentrations. Also remember the following rules for calculating Q:

1. For any equilibrium involving gases only, Q_p should be used.
2. For any equilibrium involving a nondissolved gas, the concentration of the gas must be expressed as a partial pressure and not as a molar concentration.
3. Gas partial pressures must be expressed in atmospheres and not in Torrs.

EXAMPLE   Calculating ΔG_r for a reaction

The standard reaction free energy for the reaction that follows is -141.74 kJ. What is the reaction free energy for this reaction at 25.0°C when the partial pressure of the gases are SO_2, 333 Torr; O_2, 187 Torr; SO_3, 467 Torr. In what direction will the reaction proceed under these conditions?

$$2SO_2(g) + O_2(g) \rightleftharpoons 2SO_3(g)$$

SOLUTION We use the equation $\Delta G_r = \Delta G_r° + RT \ln Q$ to solve the problem, with $\Delta G_r° = -141.74$ kJ and ΔG_r the unknown. Q in this situation is Q_p, with all partial pressures expressed in atmospheres.

$$Q_p = \frac{P_{SO_3}^2}{P_{SO_2}^2 P_{O_2}}$$

Using the conversion 760 Torr = 1 atm, we get for each of the partial pressures:

$$SO_2: \ 333 \ \text{Torr} \times \frac{1 \ \text{atm}}{760 \ \text{Torr}} = 0.438 \ \text{atm}$$

$$O_2: \ 187 \ \text{Torr} \times \frac{1 \ \text{atm}}{760 \ \text{Torr}} = 0.246 \ \text{atm}$$

$$SO_3: \ 467 \ \text{Torr} \times \frac{1 \ \text{atm}}{760 \ \text{Torr}} = 0.614 \ \text{atm}$$

This gives

$$Q_p = \frac{0.614^2}{0.438^2 \times 0.246}$$

$$= 7.99$$

Substituting $Q_p = 7.99$ and $\Delta G_r^\circ = -141.74$ kJ into the original equation gives

$$\Delta G_r = \Delta G_r^\circ + RT \ln Q$$

$$= -141.74 \ \text{kJ} + (8.315 \ \text{J} \cdot \text{K}^{-1} \cdot \text{mol}^{-1}) \times (298.2 \ \text{K}) \times \ln(7.99)$$

$$= -141.74 \ \text{kJ} + (8.315 \ \text{J} \cdot \text{K}^{-1} \cdot \text{mol}^{-1}) \times (298.2 \ \text{K}) \times 2.078$$

$$= -141.74 \ \text{kJ} + 5152 \ \text{J}$$

$$= -141.74 \ \text{kJ} + 5.152 \ \text{kJ}$$

$$= -136.59 \ \text{kJ}$$

Because ΔG_r is negative, the formation of products is spontaneous; we expect that SO_2 and O_2 will react to form SO_3.

EXERCISE The standard reaction free energy for the reaction that follows is -75.21 kJ. What is the reaction free energy for this reaction at 25.0°C when the partial pressure of the gases are NO, 333 Torr; O_2, 187 Torr; N_2O_4, 467 Torr. In what direction will the reaction proceed under these conditions?

$$2NO(g) + O_2(g) \rightleftharpoons N_2O_4(g)$$

[*Answer:* -134.64 kJ; reaction proceeds to products]

16.15   FREE ENERGY AND EQUILIBRIUM

KEY CONCEPT   The equilibrium constant can be calculated from ΔG_r°

At equilibrium, $Q = K$ and $\Delta G_r = 0$. When these values are substituted into the equation $\Delta G_r = \Delta G_r^\circ + RT \ln Q$, we get the following

$$0 = \Delta G_r^\circ + RT \ln K$$

which rearranges to

$$\Delta G_r^\circ = -RT \ln K$$

This expression applies to any of the equilibria we have encountered up to now. However, there are some rules regarding how gases are handled:

1. For any equilibrium involving gases only, K_P should be used.
2. For any equilibrium involving a nondissolved gas, the concentration of the gas must be expressed as a partial pressure and not as a molar concentration.
3. Gas partial pressures must be expressed in atmospheres and not in Torrs.

EXAMPLE 1 Calculating an equilibrium constant from $\Delta G°$

Use the data in Appendix 2A of the text to calculate the equilibrium constant K_P at 25°C for the equilibrium

$$2SO_2(g) + O_2(g) \rightleftharpoons 2SO_3(g)$$

SOLUTION For any equilibrium involving gases only,

$$\Delta G° = -RT \ln K_P$$

We use the standard free energies of formation of SO_3 (-371.06 kJ·mol^{-1}) and SO_2 (-300.19 kJ·mol^{-1}) to calculate $\Delta G°$:

$$\Delta G° = \Delta G_f°(\text{products}) - \Delta G_f°(\text{reactants})$$

$$= \left\{ 2 \text{ mol } SO_3 \times \left(-371.06 \frac{\text{kJ}}{\text{mol}} \right) \right\} - \left\{ 2 \text{ mol } SO_2 \times \left(-300.19 \frac{\text{kJ}}{\text{mol}} \right) \right\}$$

$$= \{-742.12 \text{ kJ}\} - \{-600.38 \text{ kJ}\}$$

$$= -742.12 \text{ kJ} + 600.38 \text{ kJ}$$

$$= -141.74 \text{ kJ}$$

Thus, at 25°C (298 K),

$$-141.74 \text{ kJ} = -RT \ln K_P$$

$$= -\left(8.314 \frac{\text{J}}{\text{K}} \right)(298 \text{ K}) \ln K_P$$

$$= -2.48 \text{ kJ} \ln K_P$$

Dividing both sides of the equation by 2.48 kJ gives

$$\ln K_P = \frac{141.74 \text{ kJ}}{2.48 \text{ kJ}}$$

$$= 57.2$$

This gives

$$K_P = e^{57.2}$$

$$= 7 \times 10^{24}$$

(The value for this answer is very sensitive to how and where you round off in the calculation.)

EXERCISE Calculate K_P at 25°C for the equilibrium shown from the free energies of formation of the substances involved.

$$H_2(g) + I_2(g) \rightleftharpoons 2HI(g)$$

[*Answer:* 0.254]

EXAMPLE 2 Calculating a vapor pressure from $\Delta G°$

Calculate the vapor pressure of chloroform ($CHCl_3$) at 25°C from the standard free energies of formation of liquid chloroform (-71.84 kJ·mol^{-1}) and gaseous chloroform (-70.12 kJ·mol^{-1}).

SOLUTION A vapor pressure is measured when the vapor and liquid are in equilibrium with each other:

$$CHCl_3(l) \rightleftharpoons CHCl_3(g)$$

The equilibrium constant for this equilibrium is

$$K_P = P_{CHCl_3}$$

$$P_{CHCl_3} = \text{partial pressure of } CHCl_3 = \text{vapor pressure of } CHCl_3$$

Thus, if we determine the value of the equilibrium constant, we shall have the vapor pressure of chloroform. We use the relationship between K_P and $\Delta G°$:

$$\Delta G° = -RT \ln K_P$$

First, we calculate $\Delta G°$ from the standard free energies of formation:

$$\Delta G° = \Delta G_f^{\circ}(\text{products}) - \Delta G_f^{\circ}(\text{reactants})$$

$$= \left\{ 1 \text{ mol } CHCl_3(g) \times \left(-70.12 \frac{kJ}{mol} \right) \right\} - \left\{ 1 \text{ mol } CHCl_3(l) \times \left(-71.84 \frac{kJ}{mol} \right) \right\}$$

$$= \{-70.12 \text{ kJ}\} - \{-71.84 \text{ kJ}\}$$

$$= -70.12 \text{ kJ} + 71.84 \text{ kJ}$$

$$= +1.72 \text{ kJ}$$

We now substitute this value into the required equation, in which we use the fact that $RT = 2.48$ kJ at 298 K:

$$+1.72 \text{ kJ} = -2.48 \text{ kJ} \ln K_P$$

$$\ln K_P = -\frac{1.72}{2.48} = -0.694$$

$$K_P = e^{-0.694}$$

$$= 0.500 \text{ atm}$$

Because of the fact that the standard states refer to 1 atm pressure, K_P comes out in atmospheres. Therefore, the vapor pressure of chloroform at 25°C is 0.500 atm, or 380 Torr.

EXERCISE The vapor pressure of methanol at 25°C is 124 Torr. The standard free energy of formation of liquid methanol at 25°C is -166.22 kJ·mol^{-1}. What is the standard free energy of formation of gaseous methanol at 25°C? (Remember that pressure must be expressed in atmospheres.)

[*Answer:* -162 kJ·mol^{-1}]

16.16   THE EFFECT OF TEMPERATURE

KEY CONCEPT   The spontaneity of a reaction depends on the temperature

The factors that determine the sign of ΔG are (a) the sign of ΔH, (b) the sign of ΔS, and (c) the temperature. An exothermic process has $\Delta H < 0$. Because $\Delta G = \Delta H - T\Delta S$, a negative ΔH contributes to making ΔG negative. An endothermic process, however, has $\Delta H > 0$, and the positive ΔH contributes to making ΔG positive. Because of the minus sign in $\Delta G = \Delta H - T\Delta S$, a positive ΔS tends to make ΔG negative, whereas a negative ΔS tends to make ΔG positive. The effect of temperature is illustrated in the previous example and the accompanying exercise. At 25°C, the reaction described is not spontaneous, but at 100°C it is. The temperature is a weighting factor for ΔS; the higher the temperature, the more important is ΔS in determining the sign of ΔG. These ideas are summarized in the following table.

Sign of ΔH	Sign of ΔS	Temperature	Sign of ΔG	Nature of Change
−	+	all	−	spontaneous process
+	−	all	+	not a spontaneous process
−	−	low	−	spontaneous process
		high	+	not a spontaneous process
+	+	low	+	not a spontaneous process
		high	−	spontaneous process

It is possible for ΔG to be 0. In such a circumstance, the system is at equilibrium. When a system is at equilibrium, the change represented by the chemical equation is not a spontaneous process, nor is the reverse process a spontaneous process.

EXAMPLE   Predicting the spontaneity of a process

Formation of product (the melting of ice) for the following process has $\Delta H > 0$ and $\Delta S > 0$. Is the formation of product spontaneous or not spontaneous at a relatively high temperature? Is your answer what you expect from you common experience?

$$H_2O(s) \rightleftharpoons H_2O(l)$$

SOLUTION   In this particular situation, there are competing effects. The positive ΔH contributes to making the process nonspontaneous and , in competition, the positive ΔS contributes to making it spontaneous. At a high enough temperature, the ΔS term overwhelms the ΔH term and the process would become spontaneous; so at a "relatively high temperature" the process is spontaneous. This does agree with common experience, because ice melts when the temperature is raised above 0°C. In this case, a "relatively high temperature" is any temperature above 0°C!

EXERCISE   Most batteries (including flashlight and car batteries) do not work as well when it is cold as when it is warm. What would the signs of ΔH and ΔS be for the overall chemical reactions that take place in a battery if the only factor that determined the way the battery responds to the temperature were the signs of ΔH and ΔS?

[Answer: $\Delta H > 0$ and $\Delta S > 0$]

KEY WORDS   Define or explain each term in a written sentence or two.

spontaneous process

standard free energy of reaction

standard free energy of formation

thermodynamically unstable compound

CHEMICAL EQUATIONS TO KNOW

- Magnesium reacts with oxygen to form magnesium oxide:

$$2Mg(s) + O_2(g) \longrightarrow 2MgO(s)$$

- Hydrogen reacts with fluorine to form hydrogen fluoride:

$$H_2(g) + F_2(g) \longrightarrow 2HF(g)$$

- Carbon monoxide reacts with oxygen to form carbon dioxide:

$$2CO(g) + O_2(g) \longrightarrow 2CO_2(g)$$

- Carbon can reduce iron(III) oxide to iron:

$$2Fe_2O_3(s) + 3C(s) \longrightarrow 4Fe(s) + 3CO_2(g)$$

MATHEMATICAL EQUATIONS TO KNOW AND UNDERSTAND

$\Delta U = q + w$ \hspace{2cm} change in internal energy

$\Delta U = q_v$ \hspace{2cm} heat at constant volume

$\Delta U = q_p$ \hspace{2cm} heat at constant pressure

$\Delta U = \Delta H - P\,\Delta V$ \hspace{2cm} change in internal energy

$\Delta H_r^\circ = \Sigma n\,\Delta H_f^\circ(\text{products}) - \Sigma n\,\Delta H_f^\circ(\text{reactants})$ \hspace{1cm} reaction enthalpy

$\Delta S_{surr} = \dfrac{-\Delta H}{T}$ \hspace{2cm} entropy of surroundings

$\Delta S_{tot} = \Delta S + \Delta S_{surr}$ \hspace{2cm} total entropy

$\Delta S_r^\circ = \Sigma n\,\Delta S_m^\circ(\text{products}) - \Sigma n\,\Delta S_m^\circ(\text{reactants})$ \hspace{1cm} reaction entropy

$\Delta G = \Delta H - T\,\Delta S$ \hspace{2cm} change in free energy

$\Delta G^\circ = \Delta H^\circ - T\,\Delta S^\circ$ \hspace{2cm} change in standard free energy

$\Delta G_r^\circ = \Sigma n\,\Delta G_f^\circ(\text{products}) - \Sigma n\,\Delta G_f^\circ(\text{reactants})$ \hspace{1cm} reaction free energy

$\Delta G = \Delta G^\circ + RT \ln Q$ \hspace{2cm} dependence of free energy on concentration

$\Delta G^\circ = RT \ln K$ \hspace{2cm} free energy relation to any equilibrium constant

SELF-TEST EXERCISES

The first law of thermodynamics

1. Which of the following is an open system?
(a) the Earth \hspace{2cm} (b) a tightly stoppered vacuum bottle of apple cider
(c) the whole universe \hspace{1.5cm} (d) a capped Styrofoam cup of tea
(e) a capped bottle of soda pop

2. What type of system can exchange neither energy nor mass with the surroundings?
(a) open \hspace{1cm} (b) closed \hspace{1cm} (c) isolated

3. What type of system is a flashlight battery?
(a) open \hspace{1cm} (b) closed \hspace{1cm} (c) isolated

4. Which of the following ways to transfer energy requires a difference in temperature?
(a) heating \hspace{1cm} (b) work \hspace{1cm} (c) exchange of matter

5. Which of the following is one of the correct units for work?
(a) $m \cdot s^{-1}$ \hspace{1cm} (b) atm \hspace{1cm} (c) $g \cdot m \cdot s^{-1}$ \hspace{1cm} (d) J \hspace{1cm} (e) $L \cdot kg$

6. A reaction is run during which 1.23 kJ of heat is released and the reaction system does 235 J of work on the surroundings. What is ΔU for this process?

(a) 10×10^3 J
(b) -10×10^3 J
(c) 1.57 kJ

(d) -1.47 kJ
(e) 1.23 kJ

7. Under what conditions is it possible to create or destroy energy;

(a) constant pressure
(b) constant volume

(c) constant temperature
(d) in isolated systems

(e) energy cannot be created or destroyed

8. A reaction run at constant volume liberates 6.24 kJ; when run at constant pressure, 6.44 kJ is liberated. What is ΔH for the reaction?

(a) -0.20 kJ
(b) -6.44 kJ
(c) $+0.20$ kJ

(d) -12.68 kJ
(e) 6.24 kJ

9. When a reaction is run at constant volume, 8.69 kJ of heat is absorbed; when run at constant pressure, 8.15 kJ is absorbed. What is ΔU for the reaction?

(a) -16.84 kJ
(b) -0.54 kJ
(c) -4.35 kJ

(d) 8.69 kJ
(e) 8.15 kJ

10. During a reaction, 2.87 kJ of heat is absorbed and the reaction system does 445 J of work on the surroundings. What is ΔU for the reaction?

(a) 448 kJ
(b) 0 kJ
(c) 2.87 kJ
(d) 3.32 kJ
(e) 2.42 kJ

11. What is ΔU for an isolated system?

(a) 100 kJ
(b) -25 kJ
(c) 0 kJ
(d) 15 kJ

(e) need more information to tell

12. A reaction run at constant volume releases 46.9 kJ; when the same reaction is run at constant pressure, it releases 45.8 kJ. What is the work done by the system for the reaction when it is run at constant pressure?

(a) $+1.1$ kJ
(b) -1.1 kJ
(c) 92.7 kJ
(d) -92.7 kJ
(e) 0 kJ

The direction of spontaneous change

13. Which process results in an increase in energy disorder?

(a) A deck of cards is shuffled.

(b) A beaker of hot water cools in a cold room.

(c) A piece of copper at 25°C is dropped into water at 25°C.

(d) Ideal gas molecules spread throughout a room.

14. When ideal gas molecules escape from a small volume and disperse throughout a room,

(a) the disorder of energy increases.
(b) the disorder of energy decreases.

(c) the disorder of matter decreases.
(d) the disorder of matter increases.

(e) there is no change in disorder.

15. Which of the following changes reflects the second law of thermodynamics?

(a) energy increases
(b) energy decreases
(c) entropy increases

(d) entropy decreases
(e) none of these

16. Which of the following reactions has a positive entropy change? (Do not calculate the standard reaction entropy for each.)

(a) $CaO(s) + H_2O(g) \longrightarrow Ca(OH)_2(s)$

(b) $2NO_2(g) \longrightarrow N_2O_4(g)$

(c) $CuSO_4(s) + 5H_2O(l) \longrightarrow CuSO_4 \cdot 5H_2O(s)$

(d) $PCl_5(s) \longrightarrow PCL_3(l) + Cl_2(g)$

(e) $H_2O(g) \longrightarrow H_2O(l)$

17. Without referring to any tables, predict which has the greatest standard molar entropy at 298 K?

(a) $H_2O(l)$ (b) $Ca(s)$ (c) $Ca(OH)_2(s)$ (d) $CH_4(g)$ (e) $C_2H_4(g)$

18. Predict which has the greatest standard molar entropy at the given temperature.

(a) $H_2O(l, 300\ K)$ (b) $H_2O(s, 260\ K)$ (c) $H_2O(g, 450\ K)$

(d) $H_2O(l, 400\ K)$ (e) $H_2O(g, 385\ K)$

19. Calculate the standard reaction entropy for the reaction

$$2H_2O(l) + O_2(g) \longrightarrow 2H_2O_2(l)$$

(a) $-125.8\ J \cdot K^{-1}$ (b) $-165.6\ J \cdot K^{-1}$ (c) $564.2\ J \cdot K^{-1}$

(d) $-30.2\ J \cdot K^{-1}$ (e) $+39.7\ J \cdot K^{-1}$

20. Calculate the standard reaction entropy for the reaction

$$2HCl(g) + F_2(g) \longrightarrow 2HF(g) + Cl_2(g)$$

(a) $-26.26\ J \cdot K^{-1}$ (b) $-35.11\ J \cdot K^{-1}$ (c) $-5.97\ J \cdot K^{-1}$

(d) $7.16\ J \cdot K^{-1}$ (e) $+3.19\ J \cdot K^{-1}$

21. What is the entropy change in the surroundings at 298 K for a reaction for which $\Delta H = -32.2$ kJ?

(a) $-108\ J \cdot K^{-1}$ (b) $-32.2\ J \cdot K^{-1}$ (c) $9.60\ J \cdot K^{-1}$

(d) $+32.2\ J \cdot K^{-1}$ (e) $108\ J \cdot K^{-1}$

22. Calculate the standard reaction entropy for the reaction

$$CuSO_4 \cdot 5H_2O(s) \longrightarrow CuSO_4(s) + 5H_2O(g)$$

(a) $-122\ J \cdot K^{-1}$ (b) $244\ J \cdot K^{-1}$ (c) $-3\ J \cdot K^{-1}$

(d) $158\ J \cdot K^{-1}$ (e) $753\ J \cdot K^{-1}$

23. An endothermic reaction _____ the entropy of the surroundings by _____ the chaotic thermal motion in the surroundings.

(a) increases, increasing   (b) increases, decreasing

(c) decreases, decreasing   (d) decreases, increasing

24. For a system at equilibrium, the total entropy change in the system plus surroundings

(a) is zero.   (b) is positive.

(c) is negative.   (d) depends on the temperature.

25. What is the sign of the entropy change for the following process?

$$H_2O(l, 25°C) \longrightarrow H_2O(g, 1\ atm, 25°C)$$

(a) zero   (b) positive   (c) negative

26. What is the molar entropy of melting of CCl_4? Its enthalpy of melting is 2.47 kJ/mol and its melting point is 250 K.

(a) 618 kJ·K^{-1} (b) 2.47 kJ·K^{-1} (c) 15.3 J·K^{-1}

(d) 9.88 J·K^{-1} (e) 21.2 J·K^{-1}

27. What is the boiling point of methanol? Its enthalpy of vaporization is 35.27 kJ·mol^{-1} and its entropy of vaporization is 104.6 J·K^{-1}.

(a) 297.6 K   (b) 337.2 K   (c) 481.2 K   (d) 368.9 K   (e) 412.6 K

28. A hydrogen-bonded liquid is typically _____ disorderly than a non−hydrogen-bonded liquid and therefore has a _____ entropy of vaporization than a non−hydrogen-bonded liquid.

(a) less, higher   (b) less, lower   (c) more, higher   (d) more, lower

29. The reaction shown is exothermic:

$$H_2(g) + O_2(g) \longrightarrow H_2O_2(l)$$

Which statement is correct? (Answer without making detailed calculations.)

(a) The reaction is a spontaneous process.

(b) The reaction is not a spontaneous process.

(c) It is impossible to judge whether it is a spontaneous process without calculations.

30. The reaction shown is exothermic:

$$2CO(g) \longrightarrow 2C(s) + O_2(g)$$

Which statement is correct? (Answer without making detailed calculations.)

(a) The reaction is a spontaneous process.

(b) The reaction is not a spontaneous process.

(c) It is impossible to judge whether it is a spontaneous process without calculations.

Free energy

31. Calculate ΔG at 100°C for a reaction for which $\Delta H = -35.1$ kJ and $\Delta S = -225$ J·K^{-1}. Also state whether the reaction is spontaneous or not.

(a) +57.6 kJ, spontaneous   (b) +48.8 kJ, spontaneous

(c) +48.8 kJ, not spontaneous   (d) −119.0 kJ, spontaneous

(e) +57.6 kJ, not spontaneous

32. A change with a positive change in total entropy has a _____ change in free energy and _____ a spontaneous change.

(a) negative, is   (b) positive, is   (c) negative, is not

(d) positive, is not   (e) zero, is

33. Which of the following is correct for any system at equilibrium?

(a) $\Delta G < 0$ (b) $\Delta G = 0$ (c) $\Delta G > 0$

(d) ΔG varies   (e) none of these

34. A reaction that is spontaneous at all temperatures must have a _____ reaction enthalpy and _____ reaction entropy.

(a) positive, positive   (b) negative, negative

(c) positive, negative   (d) negative, positive

35. A reaction with $\Delta H = 361$ kJ and $\Delta S = +121$ J·K^{-1} is spontaneous at

(a) low temperatures only.   (b) high temperatures only.   (c) all temperatures.

36. What is the minimum temperature at which a reaction with $\Delta H = 182$ kJ and $\Delta S = 203$ $J \cdot K^{-1}$ is a spontaneous process?

(a) 1115 K   (b) 897 K   (c) 0.897 K   (d) 1.12 K   (c) 36.9 K

37. Which of the compounds shown is the most thermodynamically stable? Each is shown with its standard free energy of formation.

(a) $CaCl_2(s)$, -748.1 kJ $\cdot$ mol^{-1} (b) HBr(g), -53.45 kJ $\cdot$ mol^{-1}

(c) $SO_2(g)$, -300.19 kJ $\cdot$ mol^{-1} (d) $H_2O(l)$, -237.13 kJ $\cdot$ mol^{-1}

(e) $NH_3(g)$, $+328.1$ kJ $\cdot$ mol^{-1}

38. Does the following reaction tend to occur at 25°C? $\Delta G_f^\circ(PbO) = -188.89$ kJ $\cdot$ mol^{-1}.

$$Pb(s) + H_2O(g) \longrightarrow PbO(s) + H_2(g)$$

(a) yes   (b) no

39. What is $\Delta G°$ for the following reaction? $\Delta G_f^\circ(NaHCO_3) = -851.9$ kJ $\cdot$ mol^{-1}, $\Delta G_f^\circ(Na_2CO_3) = -1048.2$ kJ $\cdot$ mol^{-1}.

$$2NaHCO_3(s) \longrightarrow Na_2CO_3(s) + CO_2(g) + H_2O(g)$$

(a) -1256.2 kJ   (b) 32.7 kJ   (c) 884.6 kJ

(d) -3374.9 kJ   (e) -2523.0 kJ

40. Calculate the reaction free energy at 25°C for the following reaction when the partial pressure of the gases are CO, 0.880 atm; O_2, 0.670 atm; CO_2, 1.11 atm.

$$2CO(g) + O_2(g) \rightleftharpoons 2CO_2(g)$$

(a) -512.2 kJ   (b) 516.5 kJ   (c) -514.4 kJ

(d) 515.3 kJ   (e) -513.5 kJ

41. Calculate the reaction free energy at 25°C for the following reaction when the partial pressure of the gases are H_2, 1.23 atm; Cl_2, 1.08 atm; HCl, 0.956 atm.

$$H_2(g) + Cl_2(g) \rightleftharpoons 2HCl(g)$$

(a) -189.5 kJ   (b) -191.0 kJ   (c) -191.6 kJ

(d) -190.2 kJ   (e) -190.6 kJ

42. What is the calculated equilibrium constant, at 25°C, for the equilibrium shown? $\Delta G_f^\circ(SO_2) = -300.19$ kJ $\cdot$ mol^{-1}, $\Delta G_f^\circ(SO_3) = -371.06$ kJ $\cdot$ mol^{-1}.

$$2SO_2(g) + O_2(g) \rightleftharpoons 2SO_3(g)$$

(a) 2.5×10^{-25} atm^{-1} (b) 2.6×10^{12} atm^{-1} (c) 3.8×10^{-13} atm^{-1}

(d) 7.0×10^{24} atm^{-1} (e) 1.03 atm^{-1}

43. What is the equilibrium constant, at 25°C, for

$$PCl_5(g) \rightleftharpoons PCl_3(g) + Cl_2(g)$$

(a) 1.1×10^{13} atm   (b) 9.4×10^{-14} atm   (c) 3.1×10^{-7} atm

(d) 3.3×10^6 atm   (e) 4.2×10^3 atm

44. What is the vapor pressure of $POCl_3(l)$ at 25°C? The standard free energy of formation of the liquid is -520.9 kJ $\cdot$ mol^{-1} and that of the gas is -513.0 kJ $\cdot$ mol^{-1}.

(a) 31.4 Torr   (b) 24.2 Torr   (c) 15.8 Torr

(d) 81.6 Torr   (e) 2.2 Torr

CHAPTER 17

ELECTRONS IN TRANSITION: ELECTROCHEMISTRY

The field of electrochemistry is concerned with using electricity to accomplish chemical change or, conversely, using chemical reactions to generate an electric current. When you start your automobile, an electron transfer reaction in your car battery occurs and provides the electric current needed to power the starter motor. Similarly, flashlight batteries use electron transfer reactions to provide an electric current; the battery is "dead" when the reactants have been used up. In a similar vein, the charging of your car battery (or rechargeable batteries of any sort) involves using an electric current to drive a chemical reaction. Electrochemical reactions are extremely important in the production of many common metals; it is no exaggeration to say that our modern technology relies, in part, on a thorough understanding of the material in this chapter.

TRANSFERRING ELECTRONS

17.1   HALF-REACTIONS

KEY CONCEPT   A redox reaction can be broken down into two half-reactions

In a typical redox reaction (Section 3.12), one atom loses one or more electrons and another atom gains these electrons. For instance, in a well-known reaction, copper transfers electrons to silver ions:

$$Cu(s) + 2Ag^+(aq) \longrightarrow Cu^{2+}(aq) + 2Ag(s)$$

It is sometimes useful to think of the electron transfer as occurring in steps. So, for instance, we can imagine that copper loses two electrons in one step and the silver ions accept the electrons in a separate step:

$$Cu(s) \longrightarrow Cu^{2+}(aq) + 2e^- \quad \text{(oxidation half-reaction)}$$
$$2Ag^+(aq) + 2e^- \longrightarrow 2Ag(s) \quad \text{(reduction half-reaction)}$$

Each of these imaginary reactions, in which electrons are written as reactants or products, is called a **half-reaction.** One of the two half-reactions shows the loss of electrons that occurs (oxidation), the other the gain of electrons (reduction). The two species in a half-reaction are often called a **redox couple** and are written as

Oxidized species (species without electrons)/reduced species (species with electrons)

$$Cu^{2+}/Cu$$

$$Ag^+/Ag$$

It should be remembered that half-reactions are imaginary constructs; redox reactions do not (usually) occur in steps. Nonetheless, the half-reaction is a useful concept, as we shall see.

EXAMPLE Writing half-reactions and redox couples

Write the two half-reactions associated with the equation

$$Mg(s) + Zn^{2+}(aq) \longrightarrow Mg^{2+}(aq) + Zn(s)$$

SOLUTION In this equation, Mg loses two electrons to form Mg^{2+} (it is oxidized) and Zn^{2+} gains two electrons to form Zn. The electron loss and electron gain can be written as

$$Mg(s) \longrightarrow Mg^{2+}(aq) + 2e^- \quad \text{(oxidation half-reaction)}$$

$$Zn^{2+}(aq) + 2e^- \longrightarrow Zn(s) \quad \text{(reduction half-reaction)}$$

The two redox couples, written as oxidized species/reduced species are Mg^{2+}/Mg and Zn^{2+}/Zn.

EXERCISE Write the two half-reactions and the redox couples associated with the equation $Zn(s) + 2H^+(aq) \rightarrow Zn^{2+}(aq) + H_2(g)$.

[*Answer:* $Zn(s) \rightarrow Zn^{2+}(aq) + 2e^-$, oxidation half-reaction, Zn^{2+}/Zn; $2H^+(aq) + 2e^- \rightarrow H_2(g)$, reduction half-reaction, H^+/H_2]

17.2 BALANCING REDOX EQUATIONS

KEY CONCEPT Balancing redox equations in acid solution

Some redox equations require a highly systematic approach for balancing. Because many of these reactions are run in aqueous solution with an acid or base present, we are allowed to add H_2O and H^+ (acid) or OH^- (base) in the balancing process. We outline the method for balancing with an example. Usually you will be provided with an unbalanced equation showing only the redox process that occurs:

$$As_2O_3 + HNO_3 \longrightarrow H_3AsO_4 + NO_2 \quad \triangle$$

Step 1. Write the net reaction for the unbalanced equation:

$$As_2O_3 + NO_3^- \longrightarrow AsO_4^{3-} + NO_2 \quad \triangle$$

Step 2. Determine the oxidation numbers of all of the elements in the equation, and indicate what has been oxidized and what reduced:

Atom	Left	Right	Process
As	+3	+5	oxidation
O	−2	−2	no change
N	+5	+4	reduction

Step 3. Write unbalanced half-reactions, omitting electrons (the skeletal equations):

Oxidation half-reaction: $As_2O_3 \longrightarrow AsO_4^{3-}$

Reduction half-reaction: $NO_3^- \longrightarrow NO_2$

Step 4. Mass balance the atom oxidized in the oxidation half-reaction and the atom reduced in the reduction half-reaction:

Oxidation half-reaction: $As_2O_3 \longrightarrow 2AsO_4^{3-}$

Reduction half-reaction: $NO_3^- \longrightarrow NO_2$

Step 5. Balance O by adding the proper amount of H_2O to the right or left:

Oxidation half-reaction: $5H_2O + As_2O_3 \longrightarrow 2AsO_4^{3-}$

Reduction half-reaction: $NO_3^- \longrightarrow NO_2 + H_2O$

Step 6. Balance H by adding H^+ to the right or left:

Oxidation half-reaction: $5H_2O + As_2O_3 \longrightarrow 2AsO_4^{3-} + 10H^+$

Reduction half-reaction: $2H^+ + NO_3^- \longrightarrow NO_2 + H_2O$

Step 7. Balance the actual charge by adding electrons to the right or left. Charge is balanced when the total charge on the left equals the total charge on the right. The total charge may be positive, negative, or zero.

Oxidation half-reaction: $5H_2O + As_2O_3 \longrightarrow 2AsO_4^{3-} + 10H^+ + 4e^-$
                Total charge = 0   Total charge = 0

Reduction half-reaction: $e^- + 2H^+ + NO_3^- \longrightarrow NO_2 + H_2O$
                Total charge = 0   Total charge = 0

Notice that the electrons added are consistent with our notion that electron transfer can be used to describe redox processes. Four electrons are lost in the oxidation half-reaction as two arsenic atoms change from +2 to +4 oxidation number, and one electron is gained in the reduction half-reaction as a nitrogen atom changes from +5 to +4.

Step 8. Add the two half-reactions so that the electrons lost in the oxidation equal the electrons gained in the reduction. In this example, the reduction half-reaction must be multiplied by 4:

$$5H_2O + As_2O_3 \longrightarrow 2AsO_4^{3-} + 10H^+ + \cancel{4e^-}$$
$$\underline{\cancel{4e^-} + 8H^+ + 4NO_3^- \longrightarrow 4NO_2 + 4H_2O}$$
$$5H_2O + As_2O_3 + 8H^+ + 4NO_3^- \longrightarrow 2AsO_4^{3-} + 10H^+ + 4NO_2 + 4H_2O$$

Step 9. Cancel any species that appear on both sides of the equation. In our example, the four H_2O on the right cancel four of the five H_2O on the left and the eight H^+ on the left cancel eight of the ten on the right:

$$H_2O(l) + As_2O_3(s) + 4NO_3^-(aq) \longrightarrow 2AsO_4^{3-}(aq) + 2H^+(aq) + 4NO_2(g)$$

Step 10. Check mass balance and charge balance to make certain you have made no mistakes:

Atom	Left	Right
H	2	2
O	16	16
As	2	2
N	4	4
charge	−4	−4

EXAMPLE   Balancing a complex redox reaction in acid solution

Balance the following reaction in acid solution:

$$KIO_3(aq) + KHSO_3(aq) \longrightarrow I_2(s) + K_2SO_4(aq) \quad \triangle$$

SOLUTION

Step 1. Write the net reaction for the unbalanced equation:

$$IO_3^- + HSO_3^- \longrightarrow I_2 + SO_4^{2-}$$

Step 2. Determine the oxidation numbers of all the elements in the equation, and find what has been oxidized and what reduced:

Atom	Left	Right	Process
I	+5	0	reduction
O	−2	−2	no change
S	+4	+6	oxidation
H	+1	+1	no change*

*We anticipate that H^+ will appear on the right after the equation is balanced.

Step 3. Write the half-reactions:

$$\text{Oxidation half-reaction:} \quad HSO_3^- \longrightarrow SO_4^{2-}$$

$$\text{Reduction half-reaction:} \quad IO_3^- \longrightarrow I_2$$

Step 4. Mass balance the atom oxidized in the oxidation half-reaction and the atom reduced in the reduction half-reaction:

$$\text{Oxidation half-reaction:} \quad HSO_3^- \longrightarrow SO_4^{2-}$$

$$\text{Reduction half-reaction:} \quad 2IO_3^- \longrightarrow I_2$$

Step 5. Balance O by adding the proper amount of H_2O to the right or left:

$$\text{Oxidation half-reaction:} \quad H_2O + HSO_3^- \longrightarrow SO_4^{2-}$$

$$\text{Reduction half-reaction:} \quad 2IO_3^- \longrightarrow I_2 + 6H_2O$$

Step 6. Balance H by adding H^+ to the right or left:

$$\text{Oxidation half-reaction:} \quad H_2O + HSO_3^- \longrightarrow SO_4^{2-} + 3H^+$$

$$\text{Reduction half-reaction:} \quad 12H^+ + 2IO_3^- \longrightarrow I_2 + 6H_2O$$

Step 7. Balance the actual charge by adding electrons to the right or left. Charge is balanced when the total charge on the left equals the total charge on the right.

$$\text{Oxidation half-reaction:} \quad H_2O + HSO_3^- \longrightarrow SO_4^{2-} + 3H^+ + 2e^-$$
$$\text{Total charge} = -1 \qquad\qquad \text{Total charge} = -1$$

$$\text{Reduction half-reaction:} \quad 10e^- + 12H^+ + 2IO_3^- \longrightarrow I_2 + 6H_2O$$
$$\text{Total charge} = 0 \qquad\qquad \text{Total charge} = 0$$

Step 8. Add the two half-reactions so that the electrons lost in the oxidation equal the electrons gained in the reduction:

$$5H_2O + 5HSO_3^- \longrightarrow 5SO_4^{2-} + 15H^+ + \cancel{10e^-}$$
$$\underline{\cancel{10e^-} + 12H^+ + 2IO_3^- \longrightarrow I_2 + 6H_2O}$$
$$5H_2O + 5HSO_3^- + 12H^+ + 2IO_3^- \longrightarrow 5SO_4^{2-} + 15H^+ + I_2 + 6H_2O$$

Step 9. Cancel any species that appear on both sides of the equation. In this example, the five H_2O on the left cancel five of the six on the right, and the twelve H^+ on the left cancel twelve of the fifteen on the right:

$$5HSO_3^-(aq) + 2IO_3^-(aq) \longrightarrow 5SO_4^{2-}(aq) + 3H^+(aq) + I_2(s) + H_2O(l)$$

Step 10. Check mass balance and charge balance to make certain you have made no mistakes:

Atom	Left	Right
H	5	5
O	21	21
S	5	5
I	2	2
charge	−7	−7

EXERCISE Balance the equation $MnO_2(s) + 4HCl(aq) \rightarrow MnCl_2(aq) + Cl_2(g)$, in acid solution.

[*Answer:* $MnO_2(s) + 4H^+(aq) + 2Cl^-(aq) \rightarrow$
$Mn^{2+}(aq) + Cl_2(g) + 2H_2O(l)$]

PITFALL   Charge balancing a chemical equation

To be balanced, an equation must have the same total charge on the left as on the right. Balance does not mean that the charge on the right cancels the charge on the left but that charge on the left equals that on the right. For example, the following equation is mass balanced because the number of atoms of each element is the same on both sides of the equation. However, it is not charge balanced because the total charge on the left is +1 and the total charge on the right is +2:

$$Ag^+(aq) + Cu(s) \longrightarrow Cu^{2+}(aq) + Ag(s) \quad \triangle$$
Total charge +1   ≠   Total charge +2

To balance this equation, we multiply $Ag^+(aq)$ by 2 and $Ag(s)$ by 2:

$$2Ag^+(aq) + Cu(s) \longrightarrow Cu^{2+}(aq) + 2Ag(s)$$
Total charge +2   =   Total charge +2

Atom	Left	Right
Ag	2	2
Cu	1	1
charge	+2	+2

KEY WORDS   Define or explain each term in a written sentence or two.

half-reaction   redox couple
oxidation half-reaction   reduction half-reaction

GALVANIC CELLS

17.3   EXAMPLES OF GALVANIC CELLS

KEY CONCEPT   Galvanic cells produce an electric current

A galvanic cell uses a spontaneous electron transfer chemical reaction ($\Delta G < 0$) to produce an electric current. For example, if a piece of zinc is dropped into a solution containing Cu^{2+} ions, a spontaneous reaction occurs:

$$Zn(s) + Cu^{2+}(aq) \longrightarrow Zn^{2+}(aq) + Cu(s)$$

In this reaction, electrons move from Zn to Cu^{2+}. When the zinc is in the Cu^{2+} solution, the electrons move just a short distance from the zinc atoms to the copper(II) ions (see Figure 17.2 of the text). Now, if we can devise a way, by physically separating the zinc atoms from the copper(II) solution, to force the electrons to move through a wire to get from the Zn to Cu^{2+}, we will have created a **galvanic cell**. (Remember that an electric current is a stream of electrons moving through a wire.) One way to accomplish this is shown in the accompanying figure. In this cell, the zinc atoms in a zinc electrode give up electrons and form Zn^{2+} ions (which dissolve in the electrolyte solution in which the zinc electrode is placed). These electrons then travel through an external wire to a second electrode that dips into a Cu^{2+} solution; there Cu^{2+} ions collide with the electrode and pick up the electrons to form copper atoms. Because a solution must remain electrically neutral, a means must be provided to balance the charge of the Zn^{2+} ions entering solution and the Cu^{2+} ions removed from solution. For instance, a glass tube containing a salt in a gel (called a "salt bridge") can accomplish this purpose; as shown in the figure, two Cl^- ions move to the beaker at the left for each Zn^{2+} ion that enters solution. Similarly, two K^+ ions move to the beaker at the right for every Cu^{2+} ion lost from solution.

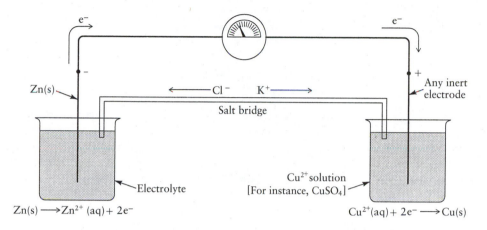

EXAMPLE Writing half-reactions and redox couples

The following reaction is spontaneous. Could a galvanic cell use this reaction to produce an electric current? If so, show how such a cell could be constructed and the half-reactions that would occur.

$$Mg(s) + Zn^{2+}(aq) \longrightarrow Mg^{2+}(aq) + Zn(s)$$

SOLUTION Because the reaction is spontaneous, it could be used in a galvanic cell to produce an electric current. To construct the cell, a magnesium electrode would be placed in a beaker of an electrolyte solution; the electrolyte chosen (for example, NaCl) must not react with the magnesium. An inert electrode (for example, one made of platinum) would be placed in a solution containing Zn^{2+} ions (for example, $ZnCl_2$) and a wire connected between the two electrodes. A salt bridge would have to be used to assure electrical neutrality in each beaker. The cell and half-reactions are pictured here.

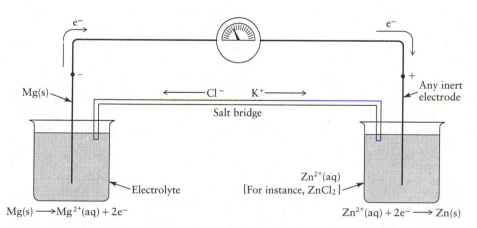

EXERCISE The following reaction is not spontaneous. Could a galvanic cell use this reaction to produce an electric current? If so, show how such a cell could be constructed and the half-reactions that would occur.

$$Cu(s) + 2H_3O^+(aq) \longrightarrow Cu^{2+}(aq) + H_2(g) + 2H_2O(l)$$

[*Answer:* Because the reaction is not spontaneous, it could not be used to construct a galvanic cell.]

17.4   THE NOTATION FOR CELLS

KEY CONCEPT   A cell diagram symbolizes the components of a cell

A **cell diagram** is a shorthand way of writing the essential chemistry and components of an electro-chemical cell. It is written as

Anode compartment ‖ cathode compartment

The double vertical bar represents the salt bridge. Each compartment is written to show the electrodes and any chemicals present. The electrode is written first for the anode compartment and last for the cathode compartment, so the electrodes are always the first and last symbols in the cell diagram. A single vertical line indicates an area of contact between cell components in different phases. Two chemicals present in a single solution are separated by a comma.

EXAMPLE   Writing cell diagrams

A cell is constructed by dipping a tin electrode into a tin(II) chloride ($SnCl_2$) solution in one beaker and a platinum electrode into a silver nitrate ($AgNO_3$) solution in another beaker. A salt bridge is used. What is the cell diagram for this cell? The reaction is

$$Sn(s) + 2Ag^+(aq) \longrightarrow 2Ag(s) + Sn^{2+}(aq)$$

SOLUTION The two half-reactions are

Anode: $\qquad\qquad\qquad Sn(s) \longrightarrow Sn^{2+}(aq) + 2e^-$

Cathode: $2Ag^+(aq) + 2e^- \longrightarrow 2Ag(s)$

The tin anode is a reactant. The cell diagram is

$$Sn(s) \,|\, Sn^{2+}(aq) \,\|\, Ag^+(aq) \,|\, Ag(s) \,|\, Pt(s)$$

The solid silver metal that forms in the reaction plates the cathode; thus, there is an area of contact between Ag(s) and Pt(s), as indicated in the cell diagram. In such a diagram, the anode is the first substance listed [a piece of tin, Sn(s), in our example], and the cathode is the last substance listed [a piece of platinum, Pt(s), in our example].

EXERCISE Write the cell diagram for a cell that uses the reaction

$$Mg(s) + 2H^+(aq) \longrightarrow H_2(g) + Mg^{2+}(aq)$$

A platinum electrode is used with the H^+/H_2 couple, and a salt bridge also is used.

[*Answer:* $Mg(s) \,|\, Mg^{2+}(aq) \,\|\, H^+(aq) \,|\, H_2(g) \,|\, Pt(s)$]

PITFALL   What is (and isn't) in a cell diagram?

A cell diagram gives the essential components and chemicals in a cell but does not tell us how the cell is actually constructed. Building a working chemical cell is a subtle art. Each cell has its own idiosyncracies and problems that must be considered, and many different cell designs exist.

17.5   CELL POTENTIAL

KEY CONCEPT   The cell potential is measured in volts

The **cell potential,** which is measured in units of volts (V), is a measure of how "hard" the electrons in the external circuit are pushed from the anode to the cathode. The potential of a cell depends on the reaction that is used in the cell as well as on the concentration of reactants and products. Because, inside the cell, oxidation occurs at the anode, the anode (denoted by −) is a source of electrons in the external circuit. The electrons in the external circuit travel to the cathode (denoted by +), where they enter the cell again and are used for reduction. When measuring the cell potential with a voltmeter, the negative terminal of the voltmeter must be connected to the anode and the positive terminal to the cathode. If the voltmeter gives a negative potential, it means the terminals are connected backward; there is no such thing as a negative cell potential.

EXAMPLE   Accounting for the cell potential

A galvanic cell is pictured in Figure 17.7 of the text. What cell potential would be measured if the salt bridge were removed?

SOLUTION For a cell to work, a salt bridge (or some equivalent cell component) must be provided to maintain electrical neutrality in the different compartments of the cell. When the salt bridge is removed, the oxidation and reduction processes in the cell immediately stop and the cell stops working. The measured potential would be 0 V.

EXERCISE A fresh flashlight D cell has a potential of about 1.5 V. What causes this voltage to fall as the battery is used?

[*Answer:* The concentration of reactants in the redox reaction decreases as the reactants react to form products.]

17.6   CELL POTENTIAL AND REACTION FREE ENERGY

KEY CONCEPT A   Cell potential and free energy

For a cell operating at any conditions of concentration, there exists a relationship between the cell potential E(cell) and ΔG for the reaction:

$$\Delta G = -nFE(\text{cell}) \qquad \text{(all conditions)}$$

where n is the number of electrons transferred for the reaction as written and F is the **Faraday constant,** which gives the charge in coulombs associated with a mole of electrons {96,485 C · (mol e⁻)⁻¹}. For a cell in which the reactants and products are all at standard conditions (1 atm pressure, pure), E(cell) becomes $E°$(cell) and ΔG becomes $\Delta G°$.

$$\Delta G° = -nFE°(\text{cell})$$

$E°$(cell) is called the **standard cell potential.** These relationships give the expected correlations between ΔG and E(cell).

1. A cell works only for a spontaneous process ($\Delta G < 0$), and the cell voltage is always positive (E(cell) > 0). The minus sign in the equation $\Delta G = -nFE$(cell) assures the difference in the signs of ΔG and E(cell).
2. At equilibrium $\Delta G = 0$. No net reaction occurs, so E(cell) also must equal 0. The equation predicts this.
3. A large negative ΔG means the reaction has a very strong tendency to occur. This implies that E(cell) should be large and positive, as predicted by $\Delta G = -nFE$(cell).

When using $\Delta G = -nFE(\text{cell})$, recall that $1 \text{ J} = 1 \text{ C} \cdot \text{V}$.

EXAMPLE 1 Calculating ΔG from E(cell)

$E(\text{cell}) = 0.25 \text{ V}$ for the following reaction. What is ΔG for this reaction?

$$\text{Cd}(s) + \text{Pb}^{2+}(\text{aq}, 0.020 \text{ M}) \longrightarrow \text{Cd}^{2+}(\text{aq}, 0.10 \text{ M}) + \text{Pb}(s)$$

SOLUTION The relationship between ΔG and $E(\text{cell})$ is

$$\Delta G = -nFE(\text{cell})$$

Cd(s) loses two electrons when it changes to Cd^{2+}, so $n = 2$. [This fact is confirmed by the fact that Pb^{2+} gains two electrons as it changes to Pb(s).] Thus,

$$n = 2 \text{ mol e}^-$$

$$E(\text{cell}) = +0.25 \text{ V}$$

$$\Delta G = -(2 \text{ mol e}^-)\left(9.65 \times 10^4 \frac{\text{C}}{\text{mol e}^-}\right)(0.25 \text{ V})$$

$$= -4.8 \times 10^4 \text{ C} \cdot \text{V}$$

$$= -4.8 \times 10^4 \text{ J}$$

$$= -48 \text{ kJ}$$

EXERCISE What is ΔG for the reaction in the preceding example when $[\text{Pb}^{2+}] = 1.0 \text{ M}$ and $[\text{Cd}^{2+}] = 1.0 \text{ M}$? For these conditions (standard conditions), $\Delta G = \Delta G°$, $E(\text{cell}) = E°(\text{cell})$ and the measured cell potential is 0.27 V.

[*Answer:* −52 kJ]

EXAMPLE 2 Calculating E(cell) from ΔG

$\Delta G = -718 \text{ kJ}$ at 25°C for the following reaction. What is $E(\text{cell})$ for this reaction?

$$2\text{Al}(s) + 3\text{Fe}^{2+}(\text{aq}, 2.0 \text{ M}) \rightleftharpoons 2\text{Al}^{3+}(\text{aq}, 0.030 \text{ M}) + 3\text{Fe}(s)$$

SOLUTION The equation that relates ΔG to $E(\text{cell})$ is

$$\Delta G = -nFE(\text{cell})$$

$$E(\text{cell}) = \frac{\Delta G}{-nF}$$

In the chemical equation, two Al(s) atoms lose six electrons to become two Al^{3+} ions, so $n = 6$. [This fact is confirmed by noting that three Fe^{2+} ions must gain six electrons to become three Fe(s) atoms.] Because $1 \text{ V} = 1 \text{ J} \cdot \text{C}^{-1}$, it is convenient to express ΔG in joules rather than kilojoules. Thus, we have

$$\Delta G = -7.18 \times 10^5 \text{ J}$$

$$n = 6$$

$$E(\text{cell}) = \frac{-7.18 \times 10^5 \text{ J}}{-(6 \text{ mol e}^-)[9.65 \times 10^4 \text{ C} \cdot (\text{mol e}^-)^{-1}]}$$

$$= +1.24 \text{ J} \cdot \text{C}^{-1}$$

$$= 1.24 \text{ V}$$

EXERCISE At standard conditions, $\Delta G° = -706 \text{ kJ}$ for the reaction in the preceding example. What is $E°(\text{cell})$?

[*Answer:* +1.22 V]

KEY CONCEPT B   Electrode potentials

Just as a cell reaction can be thought of as the result of two half-reactions, a cell potential can be thought of as the sum of the two half-reaction potentials. The half-reaction potentials are called **electrode potentials,** or, if the cell is at standard conditions, **standard electrode potentials.** A table of standard electrode potentials can be constructed by assigning a value of 0 V (exactly) to the electrode potential of the **standard hydrogen electrode (SHE)** and comparing other electrode potentials to the SHE.

Standard hydrogen electrode half-reaction: $2H^+(aq, 1\ \text{M}) + 2e^- \longrightarrow H_2(g, 1\ \text{atm})$
$$E° = 0\ \text{V (exactly)}$$

Any half-reaction can be written as either a reduction process or an oxidation process. The electrode potential for an oxidation half-reaction is called an **oxidation potential** and that for a reduction half-reaction a **reduction potential.** For instance,

$$Zn(s) \longrightarrow Zn^{2+}(aq) + 2e^-   \quad E° = \text{oxidation potential} = +0.76\ \text{V}$$

$$Zn^{2+}(aq) + 2e^- \longrightarrow Zn(s)   \quad E° = \text{reduction potential} = -0.76\ \text{V}$$

As illustrated for the Zn^{2+}/Zn couple, reversing a half-reaction changes the sign of the electrode potential. Thus, for any couple,

$$\text{Oxidation potential} = -\text{reduction potential}$$

It is quite possible for one of the electrode potentials in a working cell to be negative if the sum of the two electrode potentials for the cell is positive. If the calculated cell potential, $E(\text{cell})$, for a proposed cell is negative, the cell will not work as proposed. The reverse reaction will have a positive cell potential, so a cell using the reverse reaction will function.

EXAMPLE 1   Calculating $E°(\text{cell})$ from electrode potentials

What is $E°(\text{cell})$ for a proposed cell that uses the following reaction? Will the cell function at standard conditions?

$$Ti(s) + Fe^{2+}(aq) \longrightarrow Fe(s) + Ti^{2+}(aq)$$

SOLUTION To calculate $E°(\text{cell})$, we must first determine which half-reactions are involved and what their standard electrode potentials are. The two half-reactions are

Anode: $Ti(s) \longrightarrow Ti^{2+}(aq) + 2e^-$

Cathode: $Fe^{2+}(aq) + 2e^- \longrightarrow Fe(s)$

The relevant reduction potentials given in Appendix 2B of the text are

Ti^{2+}/Ti couple: $Ti^{2+}(aq) + 2e^- \longrightarrow Ti(s) \quad E° = -1.63\ \text{V}$

Fe^{2+}/Fe couple: $Fe^{2+}(aq) + 2e^- \longrightarrow Fe(s) \quad E° = -0.44\ \text{V}$

To get the correct voltage, we must write each half-reaction as it functions in the overall cell reaction and sum the electrode potentials, $E°(\text{cell}) = E°(\text{anode}) + E°(\text{cathode})$.

Anode:	$Ti(s) \longrightarrow Ti^{2+}(aq) + 2e^-$	$E° = +1.63\ \text{V}$
Cathode:	$Fe^{2+}(aq) + 2e^- \longrightarrow Fe(s)$	$E° = -0.44\ \text{V}$
	$Ti(s) + Fe^{2+}(aq) \longrightarrow Fe(s) + Ti^{2+}(aq)$	$E°(\text{cell}) = \quad 1.19\ \text{V}$

Note that the Ti^{2+}/Ti couple appears as an oxidation half-reaction, so its $E°$ (+1.63 V) is the negative of the standard reduction potential (−1.63 V). Because the proposed cell has a positive standard cell potential (+1.19 V), it will function at standard conditions.

EXERCISE What is $E°$(cell) for a proposed cell that uses the following reaction? Will the cell function at standard conditions?

$$Ni(s) + V^{2+}(aq) \longrightarrow Ni^{2+}(aq) + V(s)$$

[*Answer*: -0.96 V; no]

EXAMPLE 2   Predicting $E°$(cell) for a proposed cell

What is $E°$(cell) for a proposed cell that uses the following reaction? Will the cell function at standard conditions?

$$2In(s) + 6H_2O(l) \longrightarrow 2In^{3+}(aq) + 3H_2(g) + 6OH^-(aq)$$

SOLUTION The two half-reactions involved in the cell reaction are

Cathode: $2H_2O(l) + 2e^- \longrightarrow H_2(g) + 2OH^-(aq)$

Anode: $In(s) \longrightarrow In^{3+}(aq) + 3e^-$

The standard reduction potentials for these half-reactions are

H_2O/H_2 couple: $2H_2O(l) + 2e^- \longrightarrow H_2(g) + 2OH^-(aq)$ $E° = -0.83$ V

In^{3+}/In couple: $In^{3+}(aq) + 3e^- \longrightarrow In(s)$ $E° = -0.34$ V

To get the standard cell potential, we add the half-reactions and sum the standard electrode potentials for the reactions as written. Remember that multiplying the half-reactions by a number (so that electrons cancel when the half-reactions are added) does not affect the standard electrode potentials. Reversing a half-reaction changes the sign of the standard electrode potential. Thus, we use 3 times the H_2O/H_2 couple written as reduction (reaction 1) and 2 times the In^{3+}/In couple written as oxidation (reaction 2):

$6H_2O(l) + 6e^- \longrightarrow 3H_2(g) + 6OH^-(aq)$	$E° = -0.83$ V	(1)
$2In(s) \longrightarrow 2In^{3+}(aq) + 6e^-$	$E° = +0.34$ V	(2)
$6H_2O(l) + 2In(s) \longrightarrow 3H_2(g) + 6OH^-(aq) + 2In^{3+}(aq)$	$E°$(cell) $= -0.49$ V	

The standard cell potential for the proposed cell is negative, so the cell would *not* function at standard conditions. A cell constructed to take advantage of the reverse reaction would function, however, with a cell potential of 0.49 V.

EXERCISE A fuel cell is proposed that would take advantage of the following reaction. What is $E°$(cell) for the proposed cell? Would it function at standard conditions?

$$2H_2(g) + O_2(g) \longrightarrow 2H_2O(l)$$

[*Answer*: 1.23 V; yes]

17.7   THE SIGNIFICANCE OF STANDARD POTENTIALS

KEY CONCEPT   The standard potential is a measure of reducing ability

Imagine that we consider the possibility of a metal (denoted by M) reacting with H^+ to form H_2.

$$M(s) + 2H^+(aq) \longrightarrow M^{2+}(aq) + H_2(g)$$

Because the standard electrode potential for the H^+/H_2 couple is defined to be exactly 0 V, the sign of the standard reduction potential of the metal indicates whether the reaction has a thermodynamic tendency to occur or not to occur under standard conditions. For instance, if the standard reduction potential of the metal is -0.76 V [$M^{2+}(aq) + 2e^- \rightarrow M(s)$, $E° = -0.76$ V], then the standard potential of the cell reaction $M(s) + 2H^+(aq) \rightarrow M^{2+}(aq) + H_2(g)$ is $+0.76$ V and the reaction has a thermodynamic tendency to occur. On the other hand, if the standard reduction potential of the metal is +0.34 V [M^{2+}(aq) + $2e^-$ → M(s), $E° = +0.34$ V], then the standard potential of the cell reaction M(s) + $2H^+$(aq) → M^{2+}(aq) + H_2(g) is −0.34 V and the reaction is thermodynamically unfavorable. Thus, the sign of the standard reduction potential of a metal indicates its thermodynamic tendency to reduce H^+ to H_2 at standard conditions (1 M solution concentrations, gases at 1 atm partial pressure).

EXAMPLE   Understanding standard reduction potentials

The standard reduction potential for the Fe^{2+}/Fe couple is −0.44 V. Without doing a detailed calculation, comment on both the thermodynamic tendency for the following reaction to occur and whether it will actually occur if a piece of iron is placed into a 1 M H^+ solution.

$$Fe(s) + 2H^+(aq) \longrightarrow Fe^{2+}(aq) + H_2(g)$$

SOLUTION Because the standard reduction potential of the Fe^{2+}/Fe couple is −0.44 V, the standard reaction potential of the reaction is +0.44 V and it has a thermodynamic tendency to occur under standard conditions. However, we cannot tell whether the reaction will actually occur. The metal may be passivated or other factors may make the reaction fail to occur.

EXERCISE When a sample of an unknown metal is dropped into a 1 M H^+ solution under standard conditions, bubbles are observed. Could the unknown metal be silver? (Consult Table 17.1 of the text.)

[*Answer:* No]

17.8   THE ELECTROCHEMICAL SERIES

KEY CONCEPT   The electrochemical series ranks reducing agents

The half-reactions in Table 17.1 of text are all written in the form

Oxidized form of substance + electron(s) $\longrightarrow$ reduced form of substance

The oxidized form may act as an oxidizing agent in a reaction (it takes electrons from another substance); in which case, it is convenient to think of the half-reaction as

Oxidizing agent + electron(s) $\longrightarrow$ reduced form

The substances at the left of Table 17.1 of the text act as oxidizing agents. Because the half-reactions high in the table have very positive reduction potentials, these half-reactions tend to go very strongly. The oxidizing agents high in the table (such as F_2 and Ce^{4+}) are therefore very strong oxidizing agents. Any oxidizing agent from a couple high in the table can oxidize the reduced form of any couple lower in the table. For example, F_2 can oxidize Au. In a similar fashion, the reduced form may act as a reducing agent in a reaction (it gives electrons to another substance); in which case, it is convenient to think of the reverse of the half-reaction as

Reducing agent $\longrightarrow$ oxidized form + electron(s)

Reducing agents low in the table (such as Li and K) have very large oxidation potentials for this reverse reaction, so they reduce other substances very strongly. They are good reducing agents. The reverse of a half-reaction low in the table, when combined with a half-reaction higher in the table, has a positive $E°$(cell). For example, Li can reduce Na^+.

The table allows us (a) to predict which proposed oxidations and reductions from the half-reactions in the table are possible and (b) to rank the relative oxidizing power of the oxidizing agents and the relative reducing power of the reducing agents. Table 17.1 of the text is an **electrochemical series,** which is a list of reduction half-reactions, with the best oxidizing agents (on the left of the half-reactions) at the top of the list and the best reducing agents (on the right of the half-reactions) at the bottom. Table 17.1 is required for the examples and exercises that follow.

EXAMPLE 1   Relative oxidizing strengths

Is it possible to use Br_2 to oxidize Ag?

SOLUTION Oxidizing agents higher in the electrochemical series will oxidize the *reduced form* of couples lower in the series. Br_2 is above Ag, so Br_2 should oxidize Ag. We show the two reduction half-reactions involved to stress the point that the oxidizing agent is on the left and the substance oxidized on the right, as they appear in the table.

$$(Br_2, \text{higher}) \qquad Br_2 + 2e^- \longrightarrow 2Br^-$$

$$Ag^+ + e^- \longrightarrow Ag \qquad (Ag, \text{lower})$$

We can check the result by determining $E°(\text{cell})$ for the proposed reaction in the usual way. The positive value obtained for the cell potential confirms that the reaction should proceed spontaneously.

$$
\begin{array}{ll}
Br_2 + 2e^- \longrightarrow 2Br^- & E° = +1.09 \text{ V} \\
2Ag \longrightarrow 2Ag^+ + 2e^- & E° = -0.80 \text{ V} \\
\hline
Br_2 + 2Ag \longrightarrow 2Ag^+ + 2Br^- & E°(\text{cell}) = +0.29 \text{ V}
\end{array}
$$

EXERCISE Is it possible to oxidize Fe to Fe^{3+} with Pb?

[*Answer:* No]

EXAMPLE 2   Relative reducing strengths

A cell is proposed that would use Ag to reduce I_2 for the cell reaction. Predict whether the cell will work.

SOLUTION Ag (the reducing agent in the Ag^+/Ag couple) is higher than I_2 (the oxidized form in the I_2/I^- couple) in the electrochemical series. A reducing agent higher in the table will not reduce the oxidized form of a substance lower in the table, so we predict the cell will not work. The reduction half-reactions, as they appear in Table 17.1, are shown here to stress that, in the table, the reducing agent is on the right of a half-reaction and the species to be reduced on the left of a half-reaction.

$$Ag^+ + e^- \longrightarrow Ag \qquad (Ag, \text{higher})$$

$$(I_2, \text{lower}) \qquad I_2 + 2e^- \longrightarrow 2I^-$$

To confirm the prediction, we can calculate $E°(\text{cell})$ in the normal way. The negative standard cell potential confirms that the reaction will not work at standard conditions.

$$
\begin{array}{ll}
2Ag \longrightarrow 2Ag^+ + 2e^- & E° = -0.80 \text{ V} \\
I_2 + 2e^- \longrightarrow 2I^- & E° = +0.54 \text{ V} \\
\hline
I_2 + 2Ag \longrightarrow 2Ag^+ + 2I^- & E°(\text{cell}) = -0.26 \text{ V}
\end{array}
$$

EXERCISE Can Zn be used to reduce Fe^{3+} to Fe?

[*Answer:* Yes]

17.9   STANDARD POTENTIALS AND EQUILIBRIUM CONSTANTS

KEY CONCEPT   Equilibrium constants can be calculated from $E°$

The equilibrium constant of a reaction is related to the standard cell potential of the reaction through the relation

$$\ln K = \frac{nFE°(\text{cell})}{RT}$$

At 25°C, this becomes

$$\ln K = \frac{nE°(\text{cell})}{0.02569 \text{ V}}$$

This relation gives us a method for calculating the equilibrium constant for any reaction that can be written as the sum of two half-reactions and for which $E°(\text{cell})$ is known.

EXAMPLE   Calculating an equilibrium constant

What is the equilibrium constant at 25°C for the reaction

$$Fe^{3+}(aq) + Cu^{+}(aq) \longrightarrow Fe^{2+}(aq) + Cu^{2+}(aq)$$

SOLUTION We first determine $E°(\text{cell})$ for the reaction from the standard reduction potentials in Table 17.1 of the text.

$$\begin{array}{ll}
Fe^{3+}(aq) + e^{-} \longrightarrow Fe^{2+}(aq) & E° = +0.77 \text{ V} \\
\underline{Cu^{+}(aq) \longrightarrow Cu^{2+}(aq) + e^{-}} & \underline{E° = -0.15 \text{ V}} \\
Fe^{3+}(aq) + Cu^{+}(aq) \longrightarrow Fe^{2+}(aq) + Cu^{2+}(aq) & E°(\text{cell}) = +0.62 \text{ V}
\end{array}$$

From the half-reactions, $n = 1$. Substituting into the equation for K gives

$$\ln K = \frac{(1)(+0.62 \cancel{V})}{0.02569 \cancel{V}}$$

$$= 24$$

$$K = e^{24}$$

$$= 3 \times 10^{10} \quad \text{(no units)}$$

EXERCISE What is the equilibrium constant for the reaction

$$Ni^{2+}(aq) + 2In(s) \longrightarrow Ni(s) + 2In^{+}(aq)$$

[*Answer:* 9×10^{-4}]

17.10   THE NERNST EQUATION

KEY CONCEPT   The cell potential depends on concentrations

The **Nernst equation** relates the concentrations of the substances in a cell to the measured cell potential. If $E(\text{cell})$ is the cell potential and $E°(\text{cell})$ the standard cell potential,

$$E(\text{cell}) = E°(\text{cell}) - \frac{RT}{nF} \ln Q$$

or, at 25°C,

$$E(\text{cell}) = E°(\text{cell}) - \frac{0.02569 \text{ V}}{n} \ln Q$$

Q, the mass action quotient we encountered earlier, is an expression that resembles the equilibrium expression but uses the actual concentrations present at some time (not necessarily the equilibrium concentrations); n is the number of electrons transferred for the reaction as written. By using an electrode for which Q depends on the hydronium ion concentration, we can directly relate the cell potential to the pH. For instance, as shown in the text (Box 17.1), when $n = 2$ and the hydronium ion is a product of the cell reaction used:

$$E(\text{cell}) = E' + (0.0592)(\text{pH})$$

The value of E' depends on the details of the cell.

EXAMPLE   Calculating E(cell) for a cell at nonstandard conditions

Use the Nernst equation to calculate the cell potential at 25°C for the reaction

$$Fe(s) + 2Ag^+(aq, 0.050 \text{ M}) \longrightarrow Fe^{2+}(aq, 2.0 \text{ M}) + 2Ag(s)$$

SOLUTION $E°$(cell) is calculated with the standard electrode potentials of the two half-reactions involved:

$$
\begin{array}{ll}
Fe(s) \longrightarrow Fe^{2+}(aq) + 2e^- & E° = +0.44 \text{ V} \\
\underline{2Ag^+(aq) + 2e^- \longrightarrow 2Ag(s)} & \underline{E° = +0.80 \text{ V}} \\
Fe(s) + 2Ag^+(aq) \longrightarrow Fe^{2+}(aq) + 2Ag(s) & E°(cell) = \quad 1.24 \text{ V}
\end{array}
$$

Q has the same form as the equilibrium expression:

$$Q = \frac{[Fe^{2+}]}{[Ag^+]^2}$$

Because $[Fe^{2+}] = 2.0$ and $[Ag^+] = 0.050$,

$$Q = \frac{(2.0)}{(0.050)^2}$$

$$= 8.0 \times 10^2$$

Substituting $n = 2$ and the values calculated above into the Nernst equation gives

$$E(cell) = 1.24 \text{ V} - \frac{0.02569 \text{ V}}{2} \ln(8.0 \times 10^2)$$

$$= 1.24 \text{ V} - \frac{0.02569 \text{ V}}{2}(6.68)$$

$$= 1.24 \text{ V} - 0.0859 \text{ V}$$

$$= 1.15 \text{ V}$$

EXERCISE Calculate E(cell) at 25°C for the reaction

$$Sn^{4+}(aq, 2.0 \text{ M}) + 2In^{2+}(aq, 2.0 \text{ M}) \longrightarrow Sn^{2+}(aq, 0.10 \text{ M}) + 2In^{3+}(aq, 0.10 \text{ M})$$

[*Answer:* 0.76 V]

17.11   PRACTICAL CELLS

See Descriptive Chemistry to Remember in this chapter.

17.12   CORROSION

KEY CONCEPT   Corrosion of iron requires oxygen

Corrosion is the unwanted oxidation of a metal; for example, the rusting of iron is a form of corrosion. When oxidation occurs, electrons are lost and some substance must be present to accept the electrons. In most corrosive processes, the electron acceptor is oxygen (O_2) itself. To decide whether oxidation is thermodynamically favored, we calculate E(cell) under the actual conditions that occur, which are not necessarily standard conditions. For the rusting of iron in the presence of water, we can use the Nernst equation to determine E(cell) for the reduction process (for which $E° = +1.23$ V).

$$O_2(g) + 4H^+(aq) + 4e^- \longrightarrow 2H_2O(l) \qquad E(cell) = +0.81 \text{ V}$$

The first step in the oxidation process is the oxidation of the iron (Fe) to Fe^{2+}, for which the oxidation potential is the negative of the Fe^{2+}/Fe standard electrode potential (-0.44 V).

$$Fe(s) \longrightarrow Fe^{2+}(aq) + 2e^- \qquad E(cell) = +0.44 \text{ V}$$

Thus, in the presence of oxygen and water, iron corrodes to Fe^{2+}. As described in the text, later steps in the rusting process result in the eventual formation of the familiar reddish-brown solid $Fe_2O_3 \cdot H_2O$ ("rust"). One method of preventing the corrosion of materials is to use a wire to connect a sample of a more easily oxidizable metal (a **sacrificial anode**) to the structure you are trying to protect so that the sacrificial anode loses electrons in preference to the structural metal. Magnesium, with a standard electrode reduction potential of -2.36 V, is often used as a sacrificial anode; it loses electrons far more easily than iron, for which the standard electrode potential is -0.44 V.

EXAMPLE   Analyzing corrosive processes

Acid raid has sometimes been blamed for accelerating rust formation on automobiles. Based on the chemical reactions involved, is this a justifiable complaint?

SOLUTION Yes. H^+ is a reactant in the oxidation process in which oxygen accepts electrons (see the preceding Key Concept section). The presence of acids is likely to accelerate this reaction.

EXERCISE Could a piece of tin be used as a sacrificial anode?

[*Answer:* No]

KEY WORDS   Define or explain each term in a written sentence or two.

anode	electrochemical series	sacrificial anode
cathode	galvanic cell	standard cell potential
cell diagram	Nernst equation	standard electrode potential
cell potential	oxidation potential	standard reduction potential
corrosion	reduction potential	

ELECTROLYSIS

17.13   ELECTROLYTIC CELLS

KEY CONCEPT   In an electrolytic cell, an electric current drives a nonspontaneous process

Unlike a galvanic cell, the chemical reaction that occurs in an **electrolytic cell** is not spontaneous. Electrical energy, provided by a battery or power supply, is used to force the reaction to occur. In an electrolytic cell, as in galvanic cells, oxidation occurs at the anode ($+$) and reduction at the cathode ($-$); thus, the cathode must be connected to a source of electrons and the anode to a sink for electrons. A schematic of an electrolytic cell that might be used to purify silver follows. A piece of impure silver is used as the anode; as the anode is oxidized, solid impurities in the anode fall to the bottom of the cell and silver enters the solution as Ag^+. The Ag^+ then migrates to the cathode, is reduced, and is plated out as pure silver.

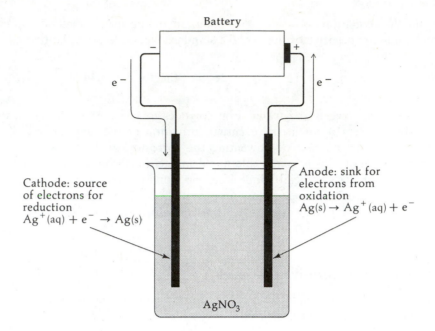

Battery

e^-

e^-

Cathode: source
of electrons for
reduction
$Ag^+(aq) + e^- \rightarrow Ag(s)$

Anode: sink for
electrons from
oxidation
$Ag(s) \rightarrow Ag^+(aq) + e^-$

$AgNO_3$

Thus, in this electrolysis cell, oxidation of Ag(s) occurs at a silver anode, and aqueous Ag^+ ion is reduced (and plated out) at the cathode. The cathode is a source of electrons and has a negative sign, whereas the anode attracts electrons and has a positive sign. Because no spontaneous chemical reaction occurs, both electrodes can be placed in the same solution.

EXAMPLE   Predicting the products from electrolysis

Suppose we wish to produce calcium metal from molten calcium chloride by using an electrolytic cell. Are there likely to be any other products produced in the cell? If so, what product(s) would you expect?

SOLUTION To produce Ca(s) from $Ca^{2+}(l)$, two electrons must be supplied for each calcium atom. The electrons must come from another chemical entity. (Remember, the external battery cannot be a net source of electrons. The battery can push and pull electrons, but must gain an electron for each one it gives up.) The likely source of electrons is the chloride ion (Cl^-) in the molten $CaCl_2$. When Cl^- gives up electrons, Cl_2 is formed, thus chlorine (Cl_2) is the other likely electrolytic product.

EXERCISE In the electrolysis of molten $CaCl_2$ described in the preceding example, how many moles of Cl_2 will be produced when one mole of calcium metal is produced?

[*Answer:* 1 mol]

17.14   THE POTENTIAL NEEDED FOR ELECTROLYSIS

KEY CONCEPT   The voltage required for an electrolytic cell can be estimated

Electrolysis forces a reaction to run in a direction opposite to its spontaneous direction. For instance, let us assume concentrations are such that $E(cell)$ for the following reaction is $+0.15$ V:

$$Ag(s) + Fe^{3+}(aq) \longrightarrow Fe^{2+}(aq) + Ag^+(aq) \qquad E(cell) = -0.15 \text{ V}$$

If we want to force this reaction to run in the reverse direction in an electrolysis cell, we must apply a minimum voltage of 0.15 V to the cell:

$$Fe^{2+}(aq) + Ag^+(aq) \xrightarrow{\;E(applied)\,=\,0.15\text{ V}\;} Ag(s) + Fe^{3+}(aq)$$

In practice, the voltage calculated in this manner is not enough to cause an electrolysis to occur. For complex reasons, a higher applied voltage must often be used, especially when a gas is produced.

This excess voltage, above and beyond the theoretical minimum voltage, is called an **overpotential.** Reported overpotentials usually refer to the reaction at a single electrode half-reaction. Overpotentials depend on a number of factors; two important ones are the electrode material and the substance produced at the electrode.

EXAMPLE 1   Calculating the minimum voltage required for an electrolysis

What is the minimum voltage needed to run the following reaction in an electrolytic cell at standard conditions at 25°C?

$$2Cu^{2+}(aq) + 2H_2O(l) \longrightarrow 2Cu(s) + 4H^+(aq) + O_2(g)$$

SOLUTION An electrolysis forces a reaction to run in the direction opposite to that of the spontaneous process. The minimum voltage needed for the electrolysis equals the cell potential generated by the reverse reaction. Thus, we calculate E(cell) at standard conditions at 25°C for the reverse of the electrolysis reaction:

$$
\begin{array}{ll}
2Cu(s) \longrightarrow 2Cu^{2+}(aq) + 4e^- & E° = -0.34 \text{ V} \\
\underline{4H^+(aq) + O_2(g) + 4e^- \longrightarrow 2H_2O(l)} & \underline{E° = +1.23 \text{ V}} \\
4H^+(aq) + O_2(g) + 2Cu(s) \longrightarrow 2Cu^{2+}(aq) + 2H_2O(l) & E°(\text{cell}) = \quad 0.89 \text{ V}
\end{array}
$$

As required, this reaction is the reverse of the electrolysis reaction. Because it has a cell potential of 0.89 V, a minimum of 0.89 V is required to force the electrolysis reaction to occur:

$$2Cu^{2+}(aq) + 2H_2O(l) \xrightarrow{E(\text{applied}) = 0.89 \text{ V}} 2Cu(s) + 4H^+(aq) + O_2(g)$$

EXERCISE What is the minimum voltage required to run the following reaction in an electrolysis cell at standard conditions at 25°C?

$$4Ag^+(aq) + 2H_2O(l) \longrightarrow 4Ag(s) + 4H^+(aq) + O_2(g)$$

[*Answer:* 0.43 V]

EXAMPLE 2   Calculating the voltage needed for an electrolysis

What is the minimum voltage needed to run the electrolysis in the preceding example at 25°C when $[Cu^{2+}] = 0.10$ M and $[H^+] = 1.0 \times 10^{-7}$ M? Assume the pressure of O_2 remains at (exactly) 1 atm.

SOLUTION To determine the minimum voltage for an electrolysis, we must calculate E(cell) for the reverse spontaneous process:

$$2Cu(s) + 4H^+(aq, 1.0 \times 10^{-7} \text{ M}) + O_2(g, 1 \text{ atm}) \longrightarrow 2Cu^{2+}(aq, 0.10 \text{ M}) + 2H_2O(l)$$

For this reaction, $n = 4$ and $E°(\text{cell}) = 0.89$ V (see preceding example), and

$$Q = \frac{[Cu^{2+}]^2}{[H^+]^4 P_{O_2}}$$

$$= \frac{(0.10)^2}{(1.0 \times 10^{-7})^4(1)}$$

$$= 1.0 \times 10^{26}$$

The Nernst equation now gives

$$E(\text{cell}) = E°(\text{cell}) - \frac{0.02569 \text{ V}}{n} \ln Q$$

$$= 0.89 \text{ V} - \frac{0.02569 \text{ V}}{4} \ln(1.0 \times 10^{26})$$

$$= 0.89 \text{ V} - \frac{0.02569 \text{ V}}{4}(59.87)$$

$$= 0.51 \text{ V}$$

Because $E(cell) = 0.51$ V for the spontaneous reverse process, 0.51 V is the minimum voltage that must be applied to get the original electrolysis reaction to occur:

$$2Cu^{2+}(aq, 0.10\ M) + 2H_2O(l) \xrightarrow{\ E(applied)\ =\ 0.51\ V\ } 2Cu(s) + 4H^+(aq, 1.0 \times 10^{-7}\ M) + O_2(g, 1\ atm)$$

EXERCISE What is the minimum voltage needed to run the electrolysis in the preceding exercise at 25°C when $[Ag^+] = 0.20$ M and $[H^+] = 1.0 \times 10^{-7}$ M. Assume the pressure of O_2 is exactly 1 atm.

[*Answer:* 0.06 V]

EXAMPLE 3   Calculating the voltage needed for electrolysis with overpotential

Assume the overpotential for formation of O_2 in the preceding example is 0.66 V and that the overpotential for formation of Ag is negligible. What is the actual voltage needed to get the electrolysis to work?

SOLUTION The overpotential is the voltage above and beyond the minimum theoretical voltage that must be applied for an electrolysis cell to function. Thus, 0.66 V must be added to the calculated 0.51 V. A total of 1.17 V (0.66 V + 0.51 V = 1.17 V) is required.

EXERCISE Assume the overpotential for formation of O_2 in the cell in the previous exercise is 0.53 V and that for formation of Ag is negligible. (The overpotential for O_2 is different from that in the example because we assume a different electrode material is used.) What is the actual voltage needed to get the electrolysis to work?

[*Answer:* 0.59 V]

KEY CONCEPT B   Competing oxidations and reductions

When more than one reducible species is present, there is a choice regarding which is reduced during the electrolysis. The species with the most positive reduction potential is most easily reduced and is reduced first. If more than one oxidizable species is present, the one with the most positive oxidation potential is oxidized first.

Water abounds in aqueous solutions, so the possibility exists that water will be oxidized and/or reduced during an electrolysis. The reduction and oxidation potentials of neutral water (pH = 7) are

Reduction:	$2H_2O(l) + 2e^- \longrightarrow H_2(g) + 2OH^-(aq)$	$E = -0.42$ V
Oxidation:	$2H_2O(l) \longrightarrow 4e^- + 4H^+(aq) + O_2(g)$	$E = -0.81$ V

If we ignore overvoltage effects, any reducible species with a reduction potential more positive than -0.42 V (e.g., $+0.61$ V, -0.31 V) is reduced before water, but water is reduced before a species with a reduction potential less positive than -0.42 V (e.g., -0.96 V, -1.11 V). A species with an oxidation potential more positive than -0.81 V (e.g., $+1.16$ V, -0.62 V) is oxidized before water, whereas water is oxidized before a species with an oxidation potential less positive than -0.81 V (e.g., -1.03 V, -1.86 V).

EXAMPLE   Predicting the product of an electrolysis

What products form when an aqueous solution of $MgBr_2$ is electrolyzed? Assume standard state conditions for $MgBr_2$ and no overpotentials.

SOLUTION The first task is to recognize all the possibilities for oxidation and reduction. Mg^{2+} is not easily oxidized, but it can be reduced. Water can also be reduced. The relevant half-reactions and reduction potentials are

$Mg^{2+}(aq) + 2e^- \longrightarrow Mg(s)$	$E° = -2.36$ V
$2H_2O(l) + 2e^- \longrightarrow H_2(g) + 2OH^-(aq)$	$E° = -0.42$ V

Because water has a more positive reduction potential than Mg^{2+} ion (-0.42 V > -2.36 V), the water is reduced first. Br^- is not easily reduced, but it can be oxidized. Water can also be oxidized. The relevant half-reactions and oxidation potentials are

$$2Br^-(aq) \longrightarrow Br_2(l) + 2e^- \qquad\qquad E° = -1.09 \text{ V}$$

$$2H_2O(l) \longrightarrow 4e^- + 4H^+(aq) + O_2(g) \qquad E° = -0.81 \text{ V}$$

Because H_2O has a more positive oxidation potential than Br^- (-0.81 V > -1.09 V), H_2O is oxidized first. If we neglect overpotential, the overall electrolysis reaction and required voltage are

$$2H_2O(l) \xrightarrow{E(\text{applied}) = 1.23 \text{ V}} 2H_2(g) + O_2(g)$$

The $E(\text{applied})$ is calculated with the techniques of the previous section (0.42 V + 0.81 V = 1.23 V).

EXERCISE What products form when an aqueous solution of CuF_2 is electrolyzed? Write the electrolysis reaction.

[**Answer:** $Cu(s)$, $O_2(g)$, H^+; $2Cu^{2+}(aq) + 2H_2O(l) \rightarrow$
$\qquad\qquad 2Cu(s) + 4H^+(aq) + O_2(g)$]

PITFALL Use of $E°$ to predict what gets oxidized or reduced

Often the $E°$ values used in predicting products are based on the assumption that standard state concentrations are present. If concentrations are different from standard state concentrations, $E°$ values may not be reliable predictors of what gets oxidized or reduced. Also, we must consider overpotentials in order to determine the actual reactions that occur in an electrolysis.

17.15   THE PRODUCTS OF ELECTROLYSIS

KEY CONCEPT   How much product formed depends on a variety of factors

The amount of product formed during an electrolysis depends on three factors: (a) the current used, (b) how long the electrolysis runs, and (c) the moles of electrons required per mole of product in the relevant half-reaction. Because 1 ampere = 1 coulomb·second^{-1}, the current multiplied by time gives the total charge that flows through an electrolytic cell:

Current (in amperes, A) $\times$ time (in seconds, s) = charge (in coulombs, C)

The charge is related by the Faraday constant to the moles of electrons that flow through the cell,

$$1 \text{ mol } e^- = 96,485 \text{ C}$$

The moles of electrons can be treated like a reactant or product in a half-reaction and, as such, can be related to the moles of other substances in the half-reaction. For instance, in the reduction of Al^{3+},

$$Al^{3+} + 3e^- \longrightarrow Al$$

the production of 3 mol of electrons by an electric current results in the production of 1 mol of Al. The connections and relations among the various quantities are shown in the accompanying figure:

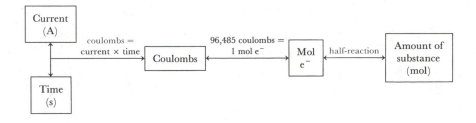

EXAMPLE 1   Predicting the amount of product formed in an electrolysis

How many grams of Al are produced by a 25-A current running for 2.0 h? The cathode reaction is $Al^{3+}(l) + 3e^- \rightarrow Al(s)$.

SOLUTION The amount of Al produced depends on the number of moles of electrons that flow through the electrolysis cell, which, in turn, depends on the total charge that flows. Using the fact that $1 \text{ A} = 1 \text{ C} \cdot \text{s}^{-1}$, we get the total charge:

$$25\frac{C}{s} \times 2.0 \text{ h} \times \frac{3600 \text{ s}}{1 \text{ h}} = 1.8 \times 10^5 \text{ C}$$

The fact that 1 mol of electrons carries 96,485 C of charge allows us to calculate the moles of electrons that carry 1.8×10^5 C of charge:

$$1.8 \times 10^5 \text{ C} \times \frac{1 \text{ mol e}^-}{96,485 \text{ C}} = 1.9 \text{ mol e}^-$$

The grams of aluminum produced can now be calculated. The half-reaction indicates that 3 mol of electrons are required to produce 1 mol of aluminum:

$$1.9 \text{ mol e}^- \times \frac{1 \text{ mol Al}}{3 \text{ mol e}^-} \times \frac{26.98 \text{ g Al}}{1 \text{ mol Al}} = 17 \text{ g Al}$$

EXERCISE How many grams of Cu(s) are produced from $Cu^{2+}(aq)$ by a 15.0-A current running for 24.0 h?

[*Answer:* 427 g]

EXAMPLE 2   Calculating the time required for an electrolysis

How much time (in seconds) is required to produce 150 g Na from a NaCl melt by a 25-A current?

SOLUTION The reduction involved is $Na^+(l) + e^- \rightarrow Na(s)$. The moles of electrons needed to produce 150 grams is

$$150 \text{ g Na} \times \frac{1 \text{ mol Na}}{23.0 \text{ g Na}} \times \frac{1 \text{ mol e}^-}{1 \text{ mol Na}} = 6.52 \text{ mol e}^-$$

With this result, we can determine the total charge that must flow:

$$6.52 \text{ mol e}^- \times \frac{96,485 \text{ C}}{1 \text{ mol e}^-} = 6.29 \times 10^5 \text{C}$$

Because the current is 25 A, the charge is delivered at the rate of $25 \text{ C} \cdot \text{s}^{-1}$:

$$6.29 \times 10^5 \text{ C} \times \frac{1s}{25 \text{ C}} = 2.5 \times 10^4 \text{ s}$$

EXERCISE How much time (in seconds) is required to plate 1.20 kg of Cr from Cr^{3+} with a 125-A current?

[*Answer:* 5.34×10^4 s]

17.16   APPLICATIONS OF ELECTROLYSIS

See Descriptive Chemistry to Remember in this chapter.

KEY WORDS   Define or explain each term in a written sentence or two.

electrolysis	Faraday's law of electrolysis
electrolytic cell	overpotential
electroplating	

DESCRIPTIVE CHEMISTRY TO REMEMBER

- A **primary cell** is a sealed cell that is not designed to be recharged. A **secondary cell** is a cell that is designed to be rechargeable. A **fuel cell** is a primary cell that operates on a continually renewed supply of fuel.

Some practical cells

- **Leclanché** cell (standard dry cell), a primary cell: Zn anode, MnO_2 cathode, electrolyte contains NH_4Cl, cell potential = 1.5 V.

Anode reaction: $4NH_4^+(aq) + Zn(s) \longrightarrow [Zn(NH_3)_4]^{2+}(aq) + 4H^+(aq)$

Cathode reaction: $MnO_2(s) + H_2O(l) + e^- \longrightarrow MnO(OH)(s) + OH^-(aq)$

- **Alkaline** cell, a primary cell: Zn anode, MnO_2 cathode, electrolyte contains KOH, cell potential = 1.5 V.

Anode reaction: $Zn(s) + 2OH^-(aq) \longrightarrow Zn(OH)_2(s) + 2e^-$

Cathode reaction: $MnO_2(s) + H_2O(l) + e^- \longrightarrow MnO(OH)(s) + OH^-(aq)$

- **Mercury** cell, a primary cell: Zn anode, HgO cathode, electrolyte contains KOH, cell potential = 1.3 V.

Anode reaction: $Zn(s) + 2OH^-(aq) \longrightarrow Zn(OH)_2(s) + 2e^-$

Cathode reaction: $HgO(s) + H_2O(l) + 2e^- \longrightarrow Hg(l) + 2OH^-(aq)$

- **Lead-acid** cell, a secondary cell: Pb anode, PbO_2 cathode, electrolyte contains $H_2SO_4(aq)$, cell potential = 2 V.

Anode reaction: $Pb(s) + SO_4^{2-}(aq) \longrightarrow PbSO_4(s) + 2e^-$

Cathode reaction: $PbO_2(s) + 4H^+(aq) + SO_4^{2-}(aq) + 2e^- \longrightarrow PbSO_4(s) + 2H_2O(l)$

- **Nickel-cadmium** cell (nicad cell), a secondary cell: Cd anode, $Ni(OH)_3$ cathode, electrolyte contains $OH^-(aq)$, cell potential = 1.4 V.

Anode reaction: $2OH^-(aq) + Cd(s) \longrightarrow Cd(OH)_2(aq) + 2e^-$

Cathode reaction: $Ni(OH)_3(s) + e^- \longrightarrow Ni(OH)_2(s) + OH^-(aq)$

- **Passivation** is the formation on a metal of a surface film that protects the metal from further reaction. As an example, aluminum and zinc exposed to air both form oxide coats that prevent further oxidation.
- **Corrosion** is the oxidation of metals by water (H_2O/H_2, OH^- couple) or oxygen (O_2, H^+/H_2O couple).
- In the **Downs process**, molten NaCl is electrolyzed to produce sodium metal.
- **Electroplating** is a process in which the object to be plated is made the cathode in an electrolytic cell, and a surface coating of a metal is deposited on the object as metal ions in solution are reduced. Chromium plating for automobile parts is done by electrolysis.
- Attempts to electroplate metallic **chromium** from $Cr^{3+}(aq)$ solutions have been unsuccessful because the hydrated ion $[Cr(H_2O)_6]^{3+}$ is so stable that it cannot be reduced.

CHEMICAL EQUATIONS TO KNOW

- In the **Daniell cell,** $Cu^{2+}(aq)$ is reduced to $Cu(s)$, and $Zn(s)$ is oxidized to $Zn^{2+}(aq)$:

$$Zn(s) + Cu^{2+}(aq) \longrightarrow Cu(s) + Zn^{2+}(aq)$$

- The overall reaction in the familiar **commercial dry cell** is

$$Zn(s) + 2MnO_2(s) + 4NH_4^+(aq) + 2OH^-(aq) \longrightarrow$$
$$[Zn(NH_3)_4]^{2+}(aq) + 2H_2O(l) + 2MnO(OH)(s)$$

- The overall reaction in a **mercury cell** is

$$Zn(s) + HgO(s) + H_2O(l) \longrightarrow Zn(OH)_2(s) + Hg(l)$$

- The overall reaction in a **lead-acid cell** (during discharge) is

$$Pb(s) + PbO_2(s) + 2H_2SO_4(aq) \longrightarrow 2PbSO_4(s) + 2H_2O(l)$$

- In the **electrolysis** of aqueous NaCl, Cl_2 is produced at the anode and OH^- at the cathode. If these products are allowed to mix, hypochlorite ion is produced:

$$Cl_2(g) + 2OH^-(aq) \longrightarrow 2ClO^-(aq) + H_2(g)$$

- The **electroplating** of chromium—a process that uses a lot of electricity because six electrons are required for every atom of chromium—uses the following reduction half-reaction:

$$CrO_3(aq) + 6H^+(aq) + 6e^- \longrightarrow Cr(s) + 3H_2O(l)$$

MATHEMATICAL EQUATIONS TO KNOW AND UNDERSTAND

$\Delta G = -nFE(cell)$	free energy of electrochemical cell
$\Delta G° = -nFE°(cell)$	standard free energy of electrochemical cell
$E°(cell) = E°(anode) + E°(cathode)$	standard potential of cell
$nFE°(cell) = RT \ln K$	relation of equilibrium constant to standard cell potential
$E(cell) = E°(cell) + \dfrac{RT}{nF} \ln Q$	cell potential of cell at any conditions

coulombs delivered
   = current in amperes $\times$ time in seconds

SELF-TEST EXERCISES

Transferring electrons

1. What is the oxidation half-reaction in the following reaction?

$$Zn(s) + 2H^+(aq) \longrightarrow Zn^{2+}(aq) + H_2(g)$$

(a) $Zn(s) \rightarrow Zn^{2+}(aq) + 2e^-$ (b) $H_2(g) \rightarrow 2H^+(aq) + 2e^-$
(c) $Zn^{2+}(aq) + 2e^- \rightarrow Zn(s)$ (d) $2H^+(aq) + 2e^- \rightarrow H_2(g)$

2. What is the reduction half-reaction in the following reaction?

$$Ce^{4+}(aq) + Fe^{2+}(aq) \longrightarrow Ce^{3+}(aq) + Fe^{3+}(aq)$$

(a) $Ce^{4+}(aq) + e^- \rightarrow Ce^{3+}(aq)$ (b) $Fe^{2+}(aq) \rightarrow Fe^{3+}(aq) + 2e^-$

(c) $Ce^{4+}(aq) \rightarrow Fe^{3+}(aq) + 2e^-$ (d) $Fe^{2+}(aq) \rightarrow Ce^{3+}(aq) + 2e^-$

3. What coefficient appears in front of H^+ when the equation shown is balanced in acid?

$$Cl^-(aq) + CrO_4^{2-}(aq) \longrightarrow HClO_2(aq) + Cr^{3+}(aq) \quad \triangle$$

(a) 5   (b) 1   (c) 8   (d) 23   (e) 14

4. What coefficient appears in front of OH^- when the equation shown is balanced in base?

$$IO_3^-(aq) + Cl_2(g) \longrightarrow IO_4^-(aq) + Cl^-(aq) \quad \triangle$$

(a) 3   (b) 4   (c) 2   (d) 17   (e) 5

5. What species is reduced in the following reaction?

$$2Cr(s) + 3MnO_2(s) + 12H^+(aq) \longrightarrow 2Cr^{3+}(aq) + 3Mn^{2+}(aq) + 6H_2O(l)$$

(a) Cr   (b) MnO_2 (c) Cr^{3+} (d) H^+ (e) H_2O

Galvanic cells

6. For the following cell reaction, what is produced at the cathode?

$$Zn(s) + Pb^{2+}(aq) \longrightarrow Zn^{2+}(aq) + Pb(s)$$

(a) Zn   (b) Pb^{2+} (c) Zn^{2+} (d) Pb

Use the following schematic of a galvanic cell for Exercises 7–9. The reaction that occurs is

$$Mg(s) + 2Ag^+(aq) \longrightarrow Mg^{2+}(aq) + 2Ag(s)$$

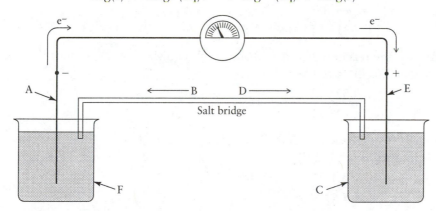

7. What is a suitable material for the electrode on the left (A)?

(a) Ag   (b) C   (c) Pt   (d) Mg   (e) Au

8. What is a possible correct identity for the species (B) moving to the left in the salt bridge?

(a) KBr   (b) Cl^- (c) Ca^{2+} (d) NH_4^+ (e) NH_3

9. What is a suitable solute (C) for the beaker on the right?

(a) $MgCl_2$ (b) $Mg(NO_3)_2$ (c) $AgNO_3$ (d) CO_2 (e) $CaCO_3$

10. What is the cell diagram for the cell that uses the following reaction; the Fe^{3+}/Fe^{2+} couple uses a platinum electrode?

$$2Fe^{3+}(aq) + Cu(s) \longrightarrow Cu^{2+}(aq) + 2Fe^{2+}(aq)$$

(a) $Cu(s) | Cu^{2+}(aq) \| Fe^{3+}(aq) | Pt(s)$

(b) $Pt(s) | Fe^{3+}(aq), Fe^{2+}(aq) \| Cu^{2+}(aq) | Cu(s)$

(c) $Pt(s) | Fe^{3+}(aq) \| Cu^{2+}(aq) | Pt(s)$

(d) $Pt(s) | Cu(s) | Cu^{2+}(aq) \| Fe^{3+}(aq), Fe^{2+}(aq) | Cu(s)$

(e) $Cu(s) | Cu^{2+}(aq) \| Fe^{3+}(aq), Fe^{2+}(aq) | Pt(s)$

11. What is the anode reaction for the cell represented by the cell diagram that follows?

$$Pt(s) | H_2(g) | HCl(aq) \| Hg_2Cl_2(s) | Hg(l)$$

(a) $H_2(g) \rightarrow 2H^+(aq) + 2e^-$

(b) $Pt(s) \rightarrow Pt^{2+}(aq) + 2e^-$

(c) $2Hg(l) + 2Cl^-(aq) \rightarrow Hg_2Cl_2(s) + 2e^-$

(d) $Hg_2Cl_2(s) + 2e^- \rightarrow 2Hg(l) + 2Cl^-(aq)$

(e) $2HCl(aq) + 2e^- \rightarrow H_2(g) + 2Cl^-(aq)$

12. Which of the following is a reasonable cathode reaction for a cell using the given reaction?

$$CuS(s) \longrightarrow Cu^{2+}(aq) + S^{2-}(aq)$$

(a) $CuS(s) \rightarrow Cu^{2+}(aq) + S^{2-}(aq) + 2e^-$

(b) $Cu^{2+}(aq) + 2e^- \rightarrow Cu(s)$

(c) $CuS(s) + 2e^- \rightarrow Cu(s) + S^{2-}(aq)$

(d) $Cu(s) + S^{2-}(aq) \rightarrow CuS(s) + 2e^-$

13. 1 joule = which of the following?

(a) $1 \text{ volt} \cdot \text{coulomb}^{-1}$

(b) $1 \text{ coulomb} \cdot \text{volt}^{-1}$

(c) $1 \text{ volt} \cdot \text{coulomb}$

(d) $1 \times 10^{-5} \text{ volt} \cdot \text{coulomb}$

14. A cell uses an oxidation half-reaction that is a strongly spontaneous process and a reduction half-reaction that is not spontaneous. What is a reasonable estimate for the cell potential?

(a) $+2.5$ V   (b) -1.3 V   (c) 0.0 V   (d) $+0.15$ V   (e) -0.18 V

15. What is ΔG for the reaction $Mg(s) + 2H^+(aq) \rightarrow Mg^{2+}(aq) + H_2(g)$ if it is run under conditions such that $E(cell) = +1.65$ V?

(a) $+159$ kJ   (b) $+318$ kJ   (c) 0 kJ   (d) -159 kJ   (e) -318 kJ

16. What is ΔG for the reaction $2Ce(s) + 3Mg^{2+}(aq) \rightarrow 3Mg(s) + 2Ce^{3+}(aq)$ if it is run under conditions such that $E(cell) = +0.562$ V?

(a) -325 kJ   (b) $+325$ kJ   (c) -163 kJ

(d) $+163$ kJ   (e) -108 kJ

17. What is $E(cell)$ for the reaction $Zn(s) + Cu^{2+}(aq) \rightarrow Zn^{2+}(aq) + Cu(s)$ if it is run under conditions such that $\Delta G = -532$ kJ.

(a) $+1.10$ V   (b) -3.16 V   (c) -0.03 V

(d) $+2.76$ V   (e) $+5.51$ V

18. Assume $E(cell) = 1.20$ V for $A \rightarrow B$. What is $E(cell)$ for the reaction $2A \rightarrow 2B$?

(a) $+0.30$ V   (b) $+4.80$ V   (c) $+0.60$ V

(d) $+2.40$ V   (e) $+1.20$ V

19. What is $E°(cell)$ for $Cd(s) + Sn^{2+}(aq) \rightarrow Sn(s) + Cd^{2+}(aq)$?

(a) $+0.0$ V   (b) $+0.26$ V   (c) -0.40 V

(d) -0.55 V   (e) $+0.14$ V

20. What is $E°(cell)$ for $2Na(s) + 2H_2O(l) \rightarrow 2NaOH(aq) + H_2(g)$?

(a) $+2.71$ V   (b) -0.83 V   (c) $+1.88$ V

(d) -3.54 V   (e) $+1.62$ V

21. What is $E°$(cell) for $Au(s) + 3Ag^+(aq) \rightarrow Au^{3+}(aq) + 3Ag(s)$?

(a) $+0.20$ V   (b) -1.40 V   (c) -0.60 V

(d) $+1.00$ V   (e) $+0.80$ V

22. What is $E°$(cell) for the cell $Pt(s) \,|\, Fe^{3+}(aq), Fe^{2+}(aq) \,\|\, Cu^+(aq) \,|\, Cu(s)$?

(a) $+0.27$ V   (b) $+0.83$ V   (c) -0.25 V

(d) $+1.29$ V   (e) $+1.01$ V

23. Which of the following is the strongest oxidizing agent? (Consult Table 17.1 of the text.)

(a) Cl_2 (b) H_2O (c) Pb   (d) Li   (e) Ag^+

24. Which of the following is the strongest reducing agent? (Consult Table 17.1.)

(a) Ca   (b) Al^{3+} (c) H_2 (d) Cu   (e) Br_2

25. Which of the following will oxidize H_2O? (Consult Table 17.1.)

(a) K   (b) Au^+ (c) Ag^+ (d) Fe^{3+} (e) Pb

26. Which of the following will reduce Pb^{2+}? (Consult Appendix 2B of the text.)

(a) Cl_2 (b) Zn   (c) Ca^{2+} (d) Ag   (e) Fe^{3+}

27. Which metal reacts with an acid to form H_2? (Consult Table 17.1.)

(a) Mg   (b) Cu   (c) Ag   (d) Au

28. What is K at 25°C for the reaction $Cl_2(g) + 2F^-(aq) \rightarrow F_2(g) + 2Cl^-(aq)$?

(a) 7.8×10^{-52} (b) 8.7×10^{-18} (c) 1.2×10^{17}

(d) 7.6×10^{-35} (e) 2.6×10^{15}

29. What is E(cell) for the following reaction?

$$2Fe^{3+}(aq, 0.20 \text{ M}) + Cu(s) \longrightarrow Cu^{2+}(aq, 0.10 \text{ M}) + 2Fe^{2+}(aq, 0.10 \text{ M})$$

(a) $+0.37$ V   (b) $+0.43$ V   (c) $+0.48$ V

(d) $+0.55$ V   (e) $+0.31$ V

30. A lead-acid cell is an example of a

(a) secondary cell   (b) primary cell   (c) fuel cell

31. The corrosion of iron requires which two substances?

(a) N_2, O_2 (b) H_2O, N_2 (c) HCl, O_2

(d) H_2O, O_2 (e) H_2O, CO_2

32. When calculating the potential required for rusting on an automobile, is it appropriate to use standard conditions?

(a) yes   (b) no

33. Only one of the following metals could (in principle) be used for a sacrificial anode for an iron structure. Which one? (Consult Table 17.1 of the text.)

(a) Ag   (b) Au   (c) Cu   (d) Fe   (e) Zn

Electrolysis

34. Neglecting overpotentials, what voltage must we apply to run the electrolysis reaction $2Na^+(aq) + 2I^-(aq) \rightarrow 2Na(s) + I_2(s)$ at standard conditions at 25°C?

(a) 0.54 V   (b) 2.71 V   (c) 1.67 V   (d) 3.25 V   (e) 2.17 V

35. Neglecting overpotentials, what voltage must we apply to run the following electrolysis reaction at 25°C?

$$2I^-(aq, 0.20 \text{ M}) + 2H_2O(l) \longrightarrow I_2(s) + H_2(g, 1 \text{ atm}) + 2OH^-(aq, 5.0 \times 10^{-3} \text{ M})$$

(a) 1.17 V   (b) 1.37 V   (c) 1.46 V   (d) 1.28 V   (e) 1.15 V

36. What voltage must be applied to run the following electrolysis at 25°C? Assume that the overpotential for production of $O_2(g)$ is 0.48 V and that for production of $Zn(s)$ is negligible.

$$2Zn^{2+}(aq, 0.50 \text{ M}) + 2H_2O(l) \longrightarrow 2Zn(s) + O_2(g, 1 \text{ atm}) + 4H^+(aq, 0.20 \text{ M})$$

(a) 0.76 V   (b) 1.93 V   (c) 1.96 V   (d) 1.23 V   (e) 2.44 V

37. What products should be formed during the electrolysis of aqueous $NiBr_2$? Assume that no overpotentials are present and standard concentrations of solute are used.

(a) Br_2, H_2, OH^- (b) Ni, O_2, H^+ (c) Br_2, O_2, H^+
(d) Ni, H_2, OH^- (e) O_2, H_2

38. An aqueous solution containing Ag^+, Cu^{2+}, and Zn^{2+} is electrolyzed. What product should form at the cathode?

(a) Ag   (b) Cu   (c) Zn   (d) H_2, OH^- (e) O_2, H^+

39. How many grams of aluminum can be plated out of a Al^{3+} solution with a current of 100 A running for 12.0 h?

(a) 0.112   (b) 154   (c) 402   (d) 1.21×10^3 (e) 0.335

40. Using a 75.0-A current, how long should it take to plate out 5.00 g of nickel from a $NiCl_2$ solution?

(a) 124 s   (b) 62.0 s   (c) 31.0 s   (d) 496 s   (e) 219 s

Descriptive chemistry

41. What is the anode reaction in a dry cell?
(a) $MnO_2(s) + H_2O(l) + e^- \longrightarrow MnO(OH)(s) + OH^-(aq)$
(b) $4NH_4^+(aq) + Zn(s) \longrightarrow [Zn(NH_3)_4]^{2+}(aq) + 4H^+(aq) + 2e^-$
(c) $Pb(s) + SO_4^{2-}(aq) \longrightarrow PbSO_4(s) + 2e^-$
(d) $2H^+(aq) + 2e^- \longrightarrow H_2(g)$
(e) $Cd(s) + 2OH^-(aq) \longrightarrow Cd(OH)_2(s) + 2e^-$

42. The anode and cathode reactions in the lead-acid battery both yield the same product under discharge. What is it?
(a) Pb   (b) PbO_2 (c) H_2O (d) $PbSO_4$ (e) H^+

43. Which of the following is the cell that uses a constantly renewed supply of fuel to generate its current?
(a) primary cell   (b) nicad cell   (c) secondary cell
(d) fuel cell   (e) lead-acid cell

CHAPTER 18

KINETICS: THE RATES OF REACTIONS

Some reactions, such as the explosion of a stick of dynamite, are relatively fast; others, such as the rusting of an automobile body, are much slower. Biochemical reactions in our bodies must occur at rates fast enough to allow us to think, see, smell, move—indeed, to stay alive. In this chapter we consider the speed with which reactions occur, the main concern of chemical kinetics.

CONCENTRATION AND RATE

18.1   THE DEFINITION OF REACTION RATE

KEY CONCEPT   A change in concentration is used to measure reaction rate

When we talk about a rate, we usually are thinking about how long it takes for something to happen. The same is true of reaction rates. A reaction in which a lot of product is formed and a lot of reactant consumed in a short amount of time is a fast reaction. Similarly, a slow reaction is one in which very little product is formed and very little reactant consumed over a long time. The quantitative definition of rate follows; the rate has units of concentration divided by time.

$$\text{Rate of reaction} = \frac{\text{change in concentration of reactant or product}}{\text{time it takes for the change to occur}}$$

For instance, for the following reaction, the rate could be defined in any of three ways:

$$2NOBr(g) \longrightarrow 2NO(g) + Br_2(g)$$

$$\text{Rate of reaction} = \frac{\text{increase in concentration of } Br_2}{\text{time for the increase to occur}} = \frac{\Delta[Br_2]}{\Delta t}$$

$$\text{Rate of reaction} = \frac{\text{increase in concentration of NO}}{\text{time for the increase to occur}} = \frac{\Delta[NO]}{\Delta t}$$

$$\text{Rate of reaction} = \frac{\text{decrease in concentration of NOBr}}{\text{time for decrease to occur}} = -\frac{\Delta[NOBr]}{\Delta t}$$

EXAMPLE   Calculating the rate of a reaction from changes in concentration

When the reaction shown is run in a 1.0-L flask, 8.0×10^{-4} mol of Br_2 is formed during the first 5.0 s of reaction. What is the rate of reaction, as defined by the change in concentration of Br_2?

$$2NOBr(g) \longrightarrow 2NO(g) + Br_2(g)$$

SOLUTION The concentration of Br_2 at the start of the reaction is 0, that is, $[Br_2] = 0$ M. After formation of 8.0×10^{-4} mol Br_2 in a 1.0-L flask, $[Br_2] = 8.0 \times 10^{-4}$ mol·L^{-1} or 8.0×10^{-4} M. Thus,

$$\text{Change in concentration of Br}_2 = 8.0 \times 10^{-4} \, \text{M} - 0 \, \text{M}$$
$$= 8.0 \times 10^{-4} \, \text{M}$$

This change occurs in a 5.0-s time interval, so $\Delta t = 5.0$ s. We substitute these values into the equation for rate that involves Br_2:

$$\text{Rate of reaction} = \frac{\text{increase in concentration of Br}_2(g)}{\text{time for the increase to occur}}$$
$$= \frac{8.0 \times 10^{-4} \, \text{M}}{5.0 \, \text{s}}$$
$$= 1.6 \times 10^{-4} \, \text{mol} \cdot \text{L}^{-1} \cdot \text{s}^{-1}$$

EXERCISE In the same reaction, what is the rate of reaction, as measured by the loss of NOBr?

[*Answer:* $3.2 \times 10^{-4} \, \text{mol} \cdot \text{L}^{-1} \cdot \text{s}^{-1}$]

PITFALL   The consistency of reaction rates

How fast a reaction proceeds does not change as we change the method of measurement. It is the same no matter which product or reactant concentration we measure. The number we obtain for the rate of reaction, however, may depend on what product or reactant concentration is measured.

PITFALL   The sign of the rate of reaction

The rate of reaction is always a positive quantity. If we remember that the rate is determined by the *decrease* in concentration of a reactant over time, or the *increase* in concentration of a product over time, we always end up with a positive rate.

18.2   THE INSTANTANEOUS RATE OF REACTION

KEY CONCEPT   The instantaneous rate is the rate at an instant in time

The instantaneous rate of reaction is the rate at some instant in time. At the beginning of a reaction, the reaction rate is relatively fast. As the reaction proceeds, the speed of the reaction slows down, that is, the rate decreases. Finally, the reaction stops. At every instant in time, the reaction is different from that at the previous instant in time. To determine the instantaneous rate, we use the equation

$$\text{Rate} = \frac{\Delta[\text{reactant or product}]}{\Delta t}$$

and make both the change in concentration and the time interval as small as possible. One way to accomplish this graphically is to plot the concentration of reactant (or product) as a function of time and draw a tangent line on the curve at the point where you want to know the instantaneous rate. The slope of the tangent is the instantaneous rate. The plot must have concentration on the y-axis and time on the x-axis in order to get the correct answer for the instantaneous rate.

EXAMPLE   Determining the instantaneous rate of reaction

When the reaction $2N_2O_5 \rightarrow 4NO_2 + O_2$ is run in liquid bromine as a solvent at 55°C, the following results are obtained for the concentration of N_2O_5 in $\text{mol} \cdot \text{L}^{-1}$ as a function of time.

Time, s	0	100	200	300	400	500	600	700	800
[N_2O_5]	0.100	0.081	0.066	0.054	0.044	0.035	0.029	0.023	0.019

Plot the data, with time as the x-axis and the molarity of N_2O_5 as the y-axis. From the plot, determine the instantaneous rate of reaction at $t = 300$ s by drawing a tangent at $t = 300$ s and determining the slope of the tangent.

SOLUTION The figure shows the required plot, with a tangent line drawn at $t = 300$ s.

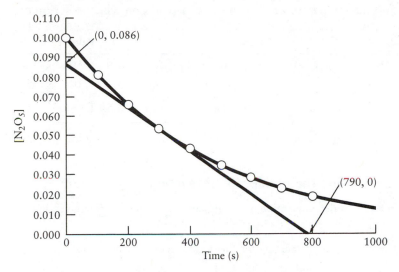

The point at which the tangent crosses the y-axis is (0 s, 0.086 M); it crosses the x-axis at (790 s, 0 M). We can use these values to determine the slope of the line:

$$\text{Slope} = \frac{\text{rise}}{\text{run}} = \frac{\Delta y}{\Delta x}$$

$$= \frac{0.086 \text{ M} - 0 \text{ M}}{0 \text{ s} - 790 \text{ s}}$$

$$= -1.1 \times 10^{-4} \text{ mol} \cdot \text{L}^{-1} \cdot \text{s}^{-1}$$

The negative sign reflects the fact that we are dealing with the loss of reactant. The rate of loss of N_2O_5, which must be positive, is the magnitude of the slope:

$$\text{Rate of loss of } N_2O_5 = 1.1 \times 10^{-4} \text{ mol} \cdot \text{L}^{-1} \cdot \text{s}^{-1}$$

EXERCISE Assume the reaction A $\rightarrow$ (products) gives the following results for the molarity of A as a function of time.

Time, s	0	10	20	30	40	50	60	70
[A]	0.20	0.15	0.11	0.078	0.057	0.041	0.030	0.022

Determine the instantaneous rate of reaction at $t = 30$ s by plotting the values, drawing a tangent at $t = 30$ s, and using the slope of the tangent line to get the instantaneous rate. Your answer may be a *little* different from the one given because drawing a tangent line is not an exact procedure.

[*Answer:* 2.6×10^{-3} mol $\cdot$ L$^{-1} \cdot$ s^{-1}]

18.3   RATE LAWS

KEY CONCEPT A   Initial rates and the concentration dependence of initial rates

The initial rate of a reaction is the instantaneous rate of reaction right at the start of the reaction, before any significant amount of product has formed. It has been found that the initial rate of reaction almost always depends on the concentration of reactant at the start of the experiment. The quantitative dependence is

$$\text{Initial rate of reaction} = \text{constant} \times (\text{concentration of reactant})^n$$

where n is usually a small integer. The value of the constant (the rate constant, k) depends on which reaction occurs and the temperature; the value of n depends on which reaction occurs. For example,

Reaction	Rate	Value of k	Temperature, °C
$2N_2O_5(g) \longrightarrow 4NO_2(g) + O_2(g)$	initial rate $= k[N_2O_5]_0$	$4.9 \times 10^{-3}\,s^{-1}$	65
		$3.1 \times 10^{-5}\,s^{-1}$	25
$2NO_2(g) \longrightarrow 2NO(g) + O_2(g)$	initial rate $= k[NO_2]_0^2$	$0.54\,L\cdot mol^{-1}\cdot s^{-1}$	300

The equation for the initial rate can be used to calculate the initial rate for any starting concentration of reactant.

EXAMPLE Calculating the initial rate of reaction

Calculate the initial rate of decomposition of $NO_2(g)$ at 300°C with the data given in the previous table when the initial concentration of NO_2 is 0.012 M.

SOLUTION The initial rate of reaction for the decomposition of $NO_2(g)$ is

$$\text{Initial rate} = k[NO_2]_0^2$$

At 300°C, $k = 0.54\,L\cdot mol^{-1}\cdot s^{-1}$, and we are given $[NO_2]_0 = 0.012$ M; therefore,

$$\text{Initial rate} = (0.54\,L\cdot mol^{-1}\cdot s^{-1})(0.012\,mol\cdot L^{-1})^2$$
$$= (0.54\,\cancel{L}\cdot\cancel{mol^{-1}}\cdot s^{-1})(0.000144\,mol\cancel{^2}\cdot L\cancel{^{-2}})$$
$$= 7.8 \times 10^{-5}\,mol\cdot L^{-1}\cdot s^{-1}$$

EXERCISE Calculate the initial rate of reaction for the decomposition of N_2O_5 at 25°C when the initial concentration of N_2O_5 is 0.045 M.

[*Answer:* $1.4 \times 10^{-6}\,mol\cdot L^{-1}\cdot s^{-1}$]

KEY CONCEPT B   Rate laws and reaction order

If the products formed in a reaction do not affect the rate of the reaction, the equation that describes the initial rate applies for the rate of reaction throughout the course of the reaction. As an example, for the decomposition of NO_2,

$$\text{Instantaneous rate at any time} = k[NO_2]^2$$

Here the symbol $[NO_2]$ represents the concentration at the time that the rate is calculated. This equation, called the **rate law** for the reaction, is usually shortened to the form

$$\text{Rate} = k[NO_2]^2$$

The constant k is called the **rate constant** and depends on both the temperature and the identity of the reactants, but not on their concentrations. The power to which the concentration is raised in the rate law defines the **order** of the reaction. Some examples of reaction orders and rate laws are given in the following table.

Reaction	Rate Law	Order
$SO_2Cl_2(g) \longrightarrow SO_2(g) + Cl_2(g)$	rate $= k[SO_2Cl_2]$	first
$2NH_3(g) \longrightarrow N_2(g) + 3H_2(g)$	rate $= k$	zero
$2N_2O(g) \longrightarrow 2N_2(g) + O_2(g)$	rate $= k[N_2O]$	first
$2NO_2(g) \longrightarrow NO(g) + NO_3(g)$	rate $= k[NO_2]^2$	second

Many reactions involve more than one reactant in the rate law. For example, consider the following reaction and rate law:

$$2NO(g) + H_2(g) \longrightarrow N_2O(g) + H_2O(g) \qquad \text{Rate} = k[NO]^2[H_2]$$

The rate law is said to be "second order in NO" because the concentration of NO is raised to the power 2, and "first order in H_2" because the concentration of H_2 is raised to the power 1. The **overall order** of the reaction is the sum of all the powers of all the reactant concentrations in the rate law. For our example, $2 + 1 = 3$, so the overall order of the reaction is 3; it is third order overall.

EXAMPLE Recognizing the order of reaction

A reaction and the rate equation associated with the reaction are

$$2NO_2(g) + F_2(g) \longrightarrow 2NO_2F(g) \qquad \text{Rate} = k[NO_2][F_2]$$

What is the order of reaction for each reactant and the overall rate of reaction?

SOLUTION In the rate equation given, the concentration of NO_2 is raised to the power 1 and that of F_2 to the power 1. This means that the order in NO_2 is 1 and the order in F_2 is 1. The sum of the orders is 2 $(1 + 1 = 2)$, so the overall order is 2. The reaction is first order in NO_2, first order in F_2, and second order overall.

EXERCISE A reaction and the rate law for the reaction are

$$2NO(g) + Br_2(g) \longrightarrow 2NOBr(g) \qquad \text{Rate} = k[NO]^2[Br_2]$$

What is the order for each reactant and the overall order?

[*Answer:* Second order in NO; first order in Br_2; third order overall]

PITFALL   Do the coefficients in a chemical equation tell us the orders of the rate?

In general the coefficients in the balanced chemical equation *do not* determine the orders expressed in the rate equation. For most chemical reactions, the *only* way to determine the order for each reactant is through an experiment. There is frequently a coincidental correspondence between the orders and the coefficients in the balanced chemical equation, but because there is no way of knowing in advance when such a coincidence will occur, an experiment must still be done to determine reaction orders. An important exception to this general idea will be encountered later in the chapter.

18.4   MORE COMPLICATED RATE LAWS

KEY CONCEPT   How order of reaction affects rate of reaction

One of the interesting aspects of the order of a reactant is that it tells us how the rate changes as the concentration of the reactant changes. For example, consider the rate law

$$\text{Rate} = k[N_2O]$$

Let us assume we have an experimental situation in which $[N_2O] = 0.20$ M. The rate is then

$$\text{Rate} = 0.20k$$

Now, if we double the concentration of N_2O, so that $[N_2O] = 0.40$ M, we get for the new rate

$$\text{Rate} = 0.40k$$

which is twice the original rate. When the concentration of reactant is doubled, the rate doubles. Now consider the rate equation

$$\text{Rate} = k[NOCl]^2$$

Let us assume that we have an experiment in which the concentration of reactant is the same as in the previous example, that is, [NOCl] = 0.20 M. The rate is then

$$\text{Rate} = (0.20)^2 k = 0.040k$$

If we now double the concentration of NOCl, so that [NOCl] = 0.40 M, we get

$$\text{Rate} = (0.40)^2 k = 0.16k$$

The rate has now quadrupled; that is, the new rate is four times the old one when the concentration is doubled ($0.16k/0.04k = 4$). The point of these two calculations is that the change in rate when the concentration is changed depends on the order of reaction. We can use this fact to determine the order of a reaction. By changing the concentration of the reactant, or of one reactant at a time for a more complex reaction, we can often determine the order of reaction.

EXAMPLE 1 Determining the order of a reaction from rate data

The data in the table show the rate of reaction for different concentrations of reactant for the following reaction. Write the rate law for the reaction and state the order with respect to each reactant and the overall order.

$$2NO(g) + Cl_2(g) \longrightarrow 2NOCl(g)$$

Experiment	Initial concentration, $mol \cdot L^{-1}$		Initial rate, $mol \cdot L^{-1} \cdot h^{-1}$
	NO_2	Cl_2	
1	0.30	0.30	0.85
2	0.30	0.60	1.70
3	0.60	0.30	3.40

SOLUTION The change in the rate when the concentration of reactant changes tells us the order of reaction for each reactant. In experiments 1 and 2, the concentration of NO doesn't change, so all of the change in rate is attributable to the change in concentration of Cl_2. For these experiments, the Cl_2 concentration doubles; at the same time the rate doubles, so the reaction must be first order in Cl_2. In experiments 1 and 3, the Cl_2 concentration is held constant as the NO concentration is doubled. In these two experiments, the rate quadruples ($4 \times 0.85 = 3.40$). Because the rate quadruples as the NO concentration alone doubles, the reaction must be second order in NO. The rate law is

$$\text{Rate} = k[NO]^2[Cl_2]$$

The reaction is third order overall ($2 + 1 = 3$).

EXERCISE The data shown give the rate of reaction for different concentrations of reactant for the following reaction. Write the rate equation and state the order with respect to each reactant. Also state the overall order of the reaction.

$$2NO(g) + H_2(g) \longrightarrow N_2O(g) + H_2O(g)$$

Experiment	Initial concentration, $mol \cdot L^{-1}$		Initial rate, $mol \cdot L^{-1} \cdot s^{-1}$
	NO	H_2	
1	0.10	0.10	1.6
2	0.10	0.20	3.2
3	0.20	0.10	6.4

[*Answer:* Rate = $k[NO]^2[H_2]$; second order in NO; first order in H_2; third order overall]

EXAMPLE 2 Determining the order of a reaction from rate data

The data shown give the rate of reaction for different concentrations of reactant for the following reaction. What is the rate equation for the reaction? What is the overall order of the reaction?

$$A + B \longrightarrow C + D$$

Experiment	Initial concentration, mol·L^{-1}		Initial rate, mol·L^{-1}·s^{-1}
	A	B	
1	0.20	0.20	2.60
2	0.20	0.25	4.06
3	0.25	0.20	3.25

SOLUTION In this case, the concentration of reactants are not doubled or tripled, so the relationship of rate to the order for each reactant is not apparent. It is necessary to substitute the data into the rate equation, rate $= k[A]^m[B]^n$, with the two orders m and n as unknowns. We start with the first two experiments:

$$2.60 \text{ mol·L}^{-1}\text{·s}^{-1} = k(0.20)^m(0.20)^n$$

$$4.06 \text{ mol·L}^{-1}\text{·s}^{-1} = k(0.20)^m(0.25)^n$$

We now divide the second equation by the first.

$$\frac{4.06 \text{ mol·L}^{-1}\text{·s}^{-1}}{2.60 \text{ mol·L}^{-1}\text{·s}^{-1}} = \frac{k(0.20)^m(0.25)^n}{k(0.20)^m(0.20)^n}$$

$$1.56 = \frac{(0.25)^n}{(0.20)^n}$$

$$= \left(\frac{0.25}{0.20}\right)^n$$

$$= 1.25^n$$

To solve for n, we take the log of both sides and use the fact that $\log x^b = b \log x$.

$$\log 1.56 = \log 1.25^n$$

$$= n \log 1.25$$

$$0.193 = n(0.0969)$$

$$n = \frac{0.193}{0.0969}$$

$$= 1.99$$

$$= 2$$

To solve for m, we do the same manipulations using experiments 1 and 3. This gives us $m = 1$. Thus, the rate equation is

$$\text{Rate} = k[A]^2[B]$$

Because $2 + 1 = 3$, the overall order is 3.

EXERCISE The data show the rate of reaction for different concentrations of reactant for the reaction

$$A + B \longrightarrow C + D$$

Experiment	Initial concentration, mol·L^{-1}		Initial rate, mol·L^{-1}·s^{-1}
	A	B	
1	3.0	3.0	0.55
2	2.2	3.0	0.40
3	3.0	2.3	0.32

What is the rate equation for the reaction? What is the overall order of the reaction?

[**Answer:** Rate = $k[A][B]^2$; overall order = 3]

18.5 FIRST-ORDER INTEGRATED RATE LAWS

KEY CONCEPT   The integrated rate law gives the concentrations as a function of time

For a first-order reaction, that is, one with the rate law

$$\text{Rate} = k[A]$$

where A is the reactant, it can be shown with the methods of calculus that

$$\ln\left(\frac{[A]_0}{[A]}\right) = kt$$

In this equation, t is the time at which the concentration of reactant A is measured, $[A]$ is the concentration of A at that time, $[A]_0$ is the concentration at the start of the reaction, when $t = 0$, and k is the rate constant; ln symbolizes the natural log function. This equation, called the **integrated rate law**, tells us how the concentration of reactant $[A]$ varies with time t for a first-order reaction. The form of the equation lends itself to a useful graphic technique for determining whether a reaction of unknown order is first order or not. To make this determination, the concentration of A is measured experimentally as a function of time and a plot is made with $\ln[A]$ on the y-axis and time on the x-axis. If the plot is a straight line, the reaction is first order and the slope of the line equals $-k$. If the resulting plot is not a straight line, the reaction is not first order.

EXAMPLE 1   Using the integrated rate equation

The reaction given below is first order in sucrose, with a rate constant equal to 3.65×10^{-3} min^{-1} at 23°C. Water does not appear in the rate equation because it is the solvent. Assume an experiment is run with a starting concentration of sucrose equal to 0.310 M. What is the concentration of sucrose after 120 min?

$$\text{Sucrose} + \text{water} \longrightarrow \text{glucose} + \text{fructose}$$

SOLUTION We are given $k = 3.65 \times 10^{-3}$ min^{-1} and $t = 120$ min, so the integrated rate law becomes

$$\ln\left(\frac{[A]_0}{[A]}\right) = kt = (3.65 \times 10^{-3} \text{ min}^{-1})(120 \text{ min})$$

$$= 0.438$$

To get rid of the ln function, we must take the antilog of both sides of the equation. (The antilog of a number x corresponds to e^x on most calculators.) Antilog(ln x) = x for any x, and antilog(0.438) = $e^{0.438}$ = 1.550.

$$\text{Antilog}\left\{\ln\left(\frac{[A]_0}{[A]}\right)\right\} = \text{antilog } 0.438$$

$$\frac{[A]_0}{[A]} = e^{0.438}$$

$$= 1.550$$

The problem gives $[A]_0 = 0.310$ M; multiplying both sides of the equation by $[A]$ and dividing both sides by 1.550 gives

$$[A] = \frac{0.310 \text{ M}}{1.550}$$

$$= 0.200 \text{ M}$$

EXERCISE The decomposition of H_2O_2 in aqueous solution at 20°C is first order, with $k = 1.06 \times 10^{-3}$ min^{-1}. Assume that a starting concentration of 0.0600 M is used in the reaction. What is the concentration after 400 min?

[*Answer:* 0.0393 M]

EXAMPLE 2 Using the integrated rate equation

Assume the reaction

$$X \longrightarrow \text{products}$$

is first order, with rate constant 0.0444 s^{-1}, and a reaction starts with [X] = 0.227 M. How many seconds does it take for 75.0% of X to react?

SOLUTION The dependence of concentration on time for a first-order reaction is given by the integrated rate equation for a first-order reaction:

$$\ln\left(\frac{[A]_0}{[A]}\right) = kt$$

We do not actually need the initial concentration of X because the problem states that 75.0% of X reacts. This means that 25.0% of X remains:

$$\frac{[A]}{[A]_0} = 0.250$$

$$\frac{[A]_0}{[A]} = 4.00$$

Substituting this value and $k = 0.0444$ s^{-1} into the integrated rate equation gives

$$\ln(4.00) = 0.0444 \text{ s}^{-1}(t)$$

$$1.39 = 0.0444 \text{ s}^{-1}(t)$$

Dividing both sides of the equation by 0.0444 s^{-1} gives

$$t = \frac{1.39}{0.0444 \text{ s}^{-1}}$$

$$= 31.2 \text{ s}$$

EXERCISE For the reaction given in the example, how long does it take for 50.0% of X to react?

[*Answer:* 15.6 s]

EXAMPLE 3 Using the integrated rate equation to determine k

The decomposition of di-*t*-butyl peroxide, $(CH_3)_3COOC(CH_3)_3$, gives the following time versus concentration data at 160°C. The symbol X is used for the peroxide.

$$X \longrightarrow \text{products}$$

Time, min	0.00	1.00	2.00	3.00	4.00	5.00
[X]	0.244	0.239	0.234	0.229	0.224	0.219

Plot a graph of ln[A] versus time and determine (a) whether the reaction is first order and (b) the value of the rate constant if it is first order.

SOLUTION Using the data, we take the natural log of each concentration to construct a plot of ln[A] versus time.

Time, min	0.00	1.00	2.00	3.00	4.00	5.00
[X]	0.244	0.239	0.234	0.229	0.224	0.219
ln[X]	−1.411	−1.431	−1.452	−1.474	−1.496	−1.519

The resulting plot is

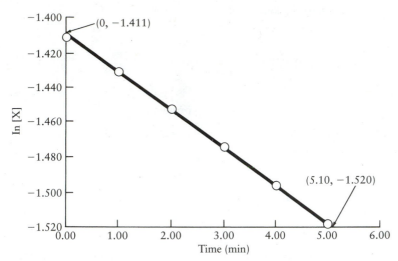

Because the points lie on a straight line, the reaction is first order. The value of k is equal to the negative of the slope of the line. The slope can be calculated with any two points on the line; we choose, for convenience, the points on the x- and y-axis (0 min, -1.411) and (5.10 min, -1.520).

$$\text{Slope} = \frac{\Delta x}{\Delta y}$$

$$= \frac{-1.411 - (-1.520)}{0 \text{ min} - 5.10 \text{ min}}$$

$$= \frac{0.109}{-5.10 \text{ min}}$$

$$= -0.0214 \text{ min}^{-1}$$

Because k is equal to the negative of the slope, we obtain

$$k = -\text{slope} = 0.0214 \text{ min}^{-1}$$

EXERCISE The time versus concentration $(\text{mol} \cdot \text{L}^{-1})$ data for the given reaction are shown in the table.

$$Y \longrightarrow \text{products}$$

Time, s	0	10	20	30	40	50
[Y]	0.155	0.137	0.125	0.109	0.0969	0.0859

Use a plot to determine (a) whether the reaction is first order and (b) the value of k if it is first order.

[*Answer:* (a) Reaction is first order; (b) $k = 0.012 \text{ s}^{-1}$]

18.6   HALF-LIVES FOR FIRST-ORDER REACTIONS

KEY CONCEPT   Half of the reactant reacts in one half-life

The **half-life** $(t_{1/2})$ of a reaction is the time it takes for half of the reactants to react. For instance, suppose that in a first-order reaction, we start with an initial concentration of of 0.80 $\text{mol} \cdot \text{L}^{-1}$ of reactant and the half-life is 5.0 s. In the first 5.0 s of reaction (one half-life), 0.40 $\text{mol} \cdot \text{L}^{-1}$ (one-half of 0.80 $\text{mol} \cdot \text{L}^{-1}$) of the reactant reacts. In the next 5.0 s (the second half-life), 0.20 $\text{mol} \cdot \text{L}^{-1}$ (one half of 0.40 $\text{mol} \cdot \text{L}^{-1}$) reacts. In the next 5.0 s, 0.10 $\text{mol} \cdot \text{L}^{-1}$ reacts and so on. The half-life for a first-order reaction is given by the following equation, in which k is the first-order rate constant.

$$t_{1/2} = \frac{\ln 2}{k}$$

EXAMPLE   Calculating and using the half-life of a reaction

At 1000 K, the decomposition of N_2O is first order, with rate constant 0.76 s^{-1}. If the reaction starts with a concentration of $0.20 \text{ mol} \cdot L^{-1}$, how long will it take for the concentration of N_2O to fall to $0.050 \text{ mol} \cdot L^{-1}$?

SOLUTION We wish to determine the time it takes for the concentration to fall to one-fourth of its original value ($0.050 \text{ mol} \cdot L^{-1}/0.20 \text{ mol} \cdot L^{-1} = \frac{1}{4}$). Because the concentration drops by a factor of $\frac{1}{2}$ for each half-life, it will take two half-lives to fall to one-fourth of the original concentration. The half-life is given by

$$t_{1/2} = \frac{\ln 2}{k}$$

Substituting $k = 0.76 \text{ s}^{-1}$ and $\ln 2 = 0.693$ into this equation gives

$$t_{1/2} = \frac{0.693}{0.76 \text{ s}^{-1}}$$

$$t_{1/2} = 0.91 \text{ s}$$

The desired time, two half-lives, is twice this number, or 1.8 s.

EXERCISE The decomposition of N_2O_5 is a first-order reaction with $k = 5.0 \times 10^{-4} \text{ s}^{-1}$ at 318 K. Assume that we start a reaction with $[N_2O_5] = 0.24$ M. How long will it take for the concentration of N_2O_5 to reach 0.030 M?

[*Answer:* 4.2×10^3 s]

18.7   SECOND-ORDER REACTIONS

KEY CONCEPT   Second-order reactions have their own integrated rate law

The integrated rate law for a second-order reaction with one reactant (rate $= k[A]^2$) relates the concentration of A to time. The rate law follows; $[A]_0$ is the concentration of A at the start of the reaction (when $t = 0$), and $[A]_t$ is the concentration of A at any time t:

$$[A]_t = \frac{[A]_0}{1 + [A]_0 kt}$$

EXAMPLE   Using the second-order integrated rate law

The decomposition of HI(g) is second order, with a rate constant of $0.060 \text{ L} \cdot \text{mol}^{-1} \cdot \text{s}^{-1}$ at 630 K. What is the concentration of HI after 250 s if the initial concentration is $0.20 \text{ mol} \cdot L^{-1}$?

SOLUTION We are given $[A]_0 = 0.20 \text{ mol} \cdot L^{-1}$ and $k = 0.060 \text{ L} \cdot \text{mol}^{-1} \cdot \text{s}^{-1}$, and asked for $[A]_t$ when $t = 250$ s. Substituting these values into the integrated rate equation for second-order reactions gives

$$[A]_t = \frac{0.20 \text{ mol} \cdot L^{-1}}{1 + (0.20 \text{ mol} \cdot L^{-1})(0.060 \text{ L} \cdot \text{mol}^{-1} \cdot \text{s}^{-1})(250 \text{ s})}$$

$$= \frac{0.20 \text{ mol} \cdot L^{-1}}{1 + 3.0}$$

$$= \frac{0.20 \text{ mol} \cdot L^{-1}}{4.0}$$

$$= 0.050 \text{ mol} \cdot L^{-1}$$

EXERCISE The decomposition of $NO_2(g)$ is second order, with a rate constant of $0.54 \text{ L} \cdot \text{mol}^{-1} \cdot \text{s}^{-1}$ at 573 K. What is the concentration of NO_2 after 75 s if the initial concentration is $1.05 \text{ mol} \cdot \text{L}^{-1}$?

[*Answer:* $0.024 \text{ mol} \cdot \text{L}^{-1}$]

KEY WORDS   Define or explain each term in a written sentence or two.

chemical kinetics	integrated rate law	reaction order
first-order reaction	pseudo–first-order reaction	second-order reaction
half-life	rate constant	
instantaneous rate of reaction	rate law	

CONTROLLING REACTION RATES

18.8   THE EFFECT OF TEMPERATURE

KEY CONCEPT   Reactions speed up as the temperature increases

We shall explore the quantitative relation between reaction rates and temperature in the next few sections. However, there are two qualitative facts that should be remembered. First, with few exceptions, the rate of reaction increases as the temperature increases. Second, as a rule of thumb for many reactions, the rate of reaction approximately doubles when the temperature increases by 10°C.

EXAMPLE   Exploring the qualitative effect of temperature on rate

People who live in cold climates in the United States learn that it is very important for their automobile's longevity to give it a thorough washing in the spring. Suggest why this might be so.

SOLUTION Salt used on the roads in winter accelerates the rusting of car bodies. However, because it is cold in the winter, the reactions that cause the rusting process are fairly slow. However, as the warm weather arrives and the rate of these reactions increase, it is important to remove all the residual salt from your car.

EXERCISE Why do cold-blooded animals need to bask in the sun before they can undertake their normal daily routine?

[*Answer:* The biochemical reactions needed to provide normal life functions do not occur at a fast enough rate until the animal warms up.]

18.9   COLLISION THEORY

KEY CONCEPT A   In order to react, molecules must collide

If two molecules are to react with each other, they must get close enough for bond making and bond breaking to occur. That is, they must collide in order for a reaction to occur. However, it takes much energy to rearrange bonds, and this energy comes from the collision itself. If two molecules collide gently, they simply rebound, like billiard balls. However, if they collide with a lot of energy, the energy of the collision can be used to make (or break) bonds. The minimum energy required for a particular reaction to occur is called the **activation energy** (E_a); E_a is different for different reactions. In addition, for a reaction to occur, molecules must also collide with the correct orientation. For instance, let's consider two NO_2 molecules reacting to produce N_2O_4 by forming a new N—N bond. The two molecules must collide "head-on" with a nitrogen on one molecule smashing into the nitrogen on the other in order for the bond to form. If they collide with an oxygen smashing into a nitrogen, no bond can form and no reaction occurs. Returning to the idea of a minimum activation energy, the rate constant can be calculated from E_a by using the following equation; A, called the **pre-exponential factor,** takes into account factors other than the activation energy (such as collision frequency and orientation factors).

$$k = Ae^{-E_a/RT}$$

It should be noted that a reaction with a small activation energy is a fast reaction, whereas one with a large activation energy is a slow reaction.

EXAMPLE   Using collision theory

Assume the pre-exponential factor is $2 \times 10^8 \ s^{-1}$ and the activation energy is 25 kJ·mol^{-1} for a particular reaction. What is k for the reaction at 25°C?

SOLUTION We must substitute the appropriate values of A, E_a, and T into the equation for the rate constant. We must remember that the temperature should be expressed in kelvins. Thus, $A = 2 \times 10^8 \ s^{-1}$, $E_a = 25$ kJ·mol^{-1}, and $T = 273.15 + 25.0 = 298.2$ K and $R = 8.314$ J·K^{-1}·mol^{-1}. The equation is

$$k = Ae^{-E_a/RT}$$

Let us first determine the term $-E_a/RT$. We will express the activation energy in joules (rather than in kilojoules) to be consistent with the unit of joules in R:

$$\frac{-E_a}{RT} = \frac{-25.0 \times 10^3 \ \cancel{J \cdot mol^{-1}}}{(8.314 \ \cancel{J \cdot K^{-1} \cdot mol^{-1}})(298 \ \cancel{K})} = -1.01$$

We now substitute this and the value of A into the original equation:

$$k = (2 \times 10^8 \ s^{-1})e^{-1.01}$$

$$= (2 \times 10^8 \ s^{-1})(0.364)$$

$$k = 7 \times 10^7 \ s^{-1}$$

EXERCISE For which reaction is it easier for the molecules to collide with the correct orientation for the reaction to occur?

$$2NO_2(g) \longrightarrow N_2O_4(g)$$

$$O(g) + O_2(g) \longrightarrow O_3(g)$$

[*Answer:* $O(g) + O_2(g) \rightarrow O_3(g)$]

KEY CONCEPT B   Activation barriers

According to **activated complex theory**, when two molecules collide with energy greater than the activation energy, they "stick together" momentarily to form an **activated complex**. Bond making and bond breaking may occur in the activated complex to form products, which then separate and move away from each other. If bond making and breaking do not occur, the activated complex breaks apart into the original reactant molecules. The activated complex has a relatively high potential energy, gathered from the kinetic energy of the colliding molecules. The following graph shows the change of the potential energy of the reaction system as the reaction progresses. The activated complex exists at the highest point of the curve.

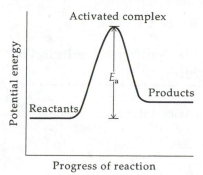

The energy hill between reactants and products is called the **activation barrier.** Molecules that collide with less than the required energy "roll back down" the activation energy hill to reactants. Thus, the energy difference between the potential energy of the reactants and that of the activated complex is E_a, the activation energy for the reaction; the higher the activation barrier, the slower is the reaction. A graph such as the preceding one, which shows the potential energy changes for the complete reaction, is called a **reaction profile.**

EXAMPLE   Interpreting a reaction profile

Three reaction profiles for three single-step reactions follow. Which reaction is fastest?

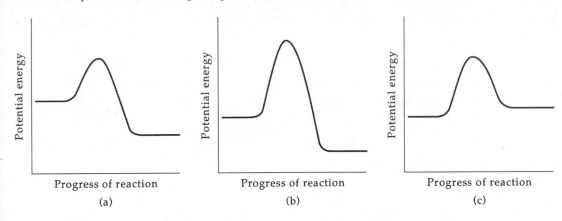

(a)   (b)   (c)

SOLUTION Profile (a) corresponds to the fastest reaction because it has the lowest activation energy.

EXERCISE Sketch the reaction profile for a two-step reaction with a fast first step and slow second step. Be certain to label the x- and y-axes.

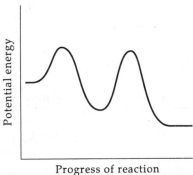

[*Answer:* This profile is one of many possible correct answers. Any profile with two activation energy hills, with the second barrier higher than the first (measured from the reactants for that step to the top of the activation energy hill), is a correct answer.]

18.10   ARRHENIUS BEHAVIOR

KEY CONCEPT   The temperature behavior of k allows us to determine E_a

The relationship between the rate constant k and temperature is often given by the Arrhenius equation:

$$\ln k = \ln A - \frac{E_a}{RT}$$

In the equation, A is the **frequency factor**, E_a is the **activation energy**, and R is the gas constant, $8.314 \, \text{J} \cdot \text{K}^{-1} \cdot \text{mol}^{-1}$. A and E_a are called the **Arrhenius parameters**. Reactions for which this relationship is obeyed are said to display **Arrhenius behavior**. There are three important ways to use this relationship.

1. A plot of $\ln k$ versus $1/T$ gives a straight line with slope equal to $-E_a/R$. Thus, by measuring the value of k at a series of different temperatures, E_a can be determined.
2. If the rate constant at one temperature, k, and the activation energy are known, the rate constant at a second temperature, k', can be calculated by a form of the relationship

$$\ln \left(\frac{k'}{k} \right) = \frac{E_a}{R} \left[\frac{1}{T} - \frac{1}{T'} \right]$$

3. If E_a and A are known, the rate constant can be calculated at any temperature.

EXAMPLE 1 Measuring an activation energy

The rate constant for the following second-order reaction in methanol solution depends on the temperature as shown. What is the activation energy for the reaction?

$$CH_3I + Br^- \longrightarrow CH_3Br + I^-$$

t, °C	3	13	23	33
k, $\text{L} \cdot \text{mol}^{-1} \cdot \text{s}^{-1}$	5.0×10^{-6}	1.7×10^{-5}	6.1×10^{-5}	2.0×10^{-4}

SOLUTION We assume that the reaction displays Arrhenius behavior; therefore, a plot of $\ln k$ versus $1/T$ should give a straight line with slope $= -E_a/R$. We first set up a table showing the values of $\ln k$ and $1/T$.

t, °C	T, K	$1/T$, K^{-1}	k, $\text{L}(\text{mol} \cdot \text{s})^{-1}$	$\ln k$
3	276	3.62×10^{-3}	5.0×10^{-6}	-12.21
13	286	3.50×10^{-3}	1.7×10^{-5}	-10.98
23	296	3.38×10^{-3}	6.1×10^{-5}	-9.70
33	306	3.27×10^{-3}	2.0×10^{-4}	-8.52

The required plot is shown in the following figure.

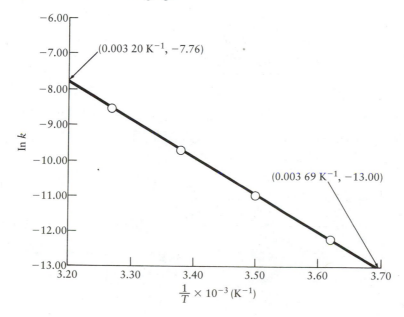

The slope of the line can be calculated by using any two points on the line; we choose the two intercepts $(0.003\,20\text{ K}^{-1}, -7.76)$ and $(0.003\,69\text{ K}^{-1}, -13.00)$ for convenience.

$$\text{Slope} = \frac{\Delta y}{\Delta x}$$

$$= \frac{-7.76 - (-13.00)}{(0.003\,20\text{ K}^{-1} - 0.003\,69\text{ K}^{-1})}$$

$$= \frac{5.24}{-4.9 \times 10^{-4}\text{ K}^{-1}}$$

$$= -1.07 \times 10^{4}\text{ K}$$

$$= -1.1 \times 10^{4}\text{ K}$$

Because the slope $= -E_a/R$, we obtain

$$E_a = -\text{slope} \times R$$

$$= -(-1.1 \times 10^{4}\text{ K}) \times 8.314\text{ J}\cdot\text{mol}^{-1}\cdot\text{K}^{-1}$$

$$= 91\text{ kJ}\cdot\text{mol}^{-1}$$

EXERCISE The temperature dependence of k for a first-order reaction is given in the table

t, °C	39.0	52.0	81.0	98.0
k, s^{-1}	3.4×10^{-8}	2.4×10^{-7}	1.0×10^{-5}	7.3×10^{-5}

What is the activation energy for this reaction?

[*Answer:* 125 kJ·mol^{-1}]

EXAMPLE 2 Calculating k at one temperature when k at a second temperature and E_a are known

For the second-order gas-phase reaction between H_2 and I_2, $k = 1.32 \times 10^{-4}\text{ L}\cdot\text{mol}^{-1}\cdot\text{s}^{-1}$ at 302°C. What is k at 348°C? $E_a = 163\text{ kJ}\cdot\text{mol}^{-1}$.

SOLUTION The equation that relates two values of k at different temperatures is

$$\ln\left(\frac{k'}{k}\right) = \frac{E_a}{R}\left[\frac{1}{T} - \frac{1}{T'}\right]$$

The parameters for the equation are (using k' for the unknown k)

$$k = 1.32 \times 10^{-4}\text{ L}\cdot\text{mol}^{-1}\cdot\text{s}^{-1} \qquad T = 302 + 273 = 575\text{ K}$$

$$k' = ? \qquad T' = 348 + 273 = 621\text{ K}$$

We first calculate the temperature factor in brackets, using the lowest common denominator TT' for the subtraction:

$$\frac{1}{T} - \frac{1}{T'} = \frac{T' - T}{TT'}$$

$$= \frac{(621 - 575)\text{ K}}{621\text{ K} \times 575\text{ K}}$$

$$= \frac{46\text{ K}}{3.57 \times 10^{5}\text{ K}^2}$$

$$= 1.3 \times 10^{-4}\text{ K}^{-1}$$

Substituting this result, $E_a = 163\ \text{kJ} \cdot \text{mol}^{-1} = 1.63 \times 10^5\ \text{J} \cdot \text{mol}^{-1}$, and $R = 8.314\ \text{J} \cdot \text{mol}^{-1} \cdot \text{K}^{-1}$ into the original equation gives

$$\ln\left(\frac{k'}{k}\right) = \frac{1.63 \times 10^5\ \text{J} \cdot \text{mol}^{-1} \times 1.3 \times 10^{-4}\ \text{K}^{-1}}{8.314\ \text{J} \cdot \text{mol}^{-1} \cdot \text{K}^{-1}}$$

$$= 2.55$$

$$\frac{k'}{k} = e^{2.55}$$

$$= 12.8$$

$$k' = 12.8\ k$$

Finally, the value of k, $1.32 \times 10^{-4}\ \text{L} \cdot \text{mol}^{-1} \cdot \text{s}^{-1}$ is substituted into this last equation to get k'

$$k' = 12.8 \times 1.32 \times 10^{-4}\ \text{L} \cdot \text{mol}^{-1} \cdot \text{s}^{-1}$$

$$= 1.70 \times 10^{-3}\ \text{L} \cdot \text{mol}^{-1} \cdot \text{s}^{-1}$$

The value of the rate constant at the higher temperature is larger, as expected.

EXERCISE The rate constant for a first-order reaction at 12°C is $5.32 \times 10^{-5}\ \text{min}^{-1}$. What is the rate constant at 41°C? The activation energy for the reaction is $96.3\ \text{kJ} \cdot \text{mol}^{-1}$.

[*Answer:* $2.2 \times 10^{-3}\ \text{min}^{-1}$]

18.11   ACTIVATED COMPLEXES

KEY CONCEPT   The activated complex exists at the top of the activation barrier

As we indicated earlier, bond breaking and/or bond making occur when the reactant(s) reach the top of the activation barrier. The short-lived **activated complex** is a species that resembles the reactants but has been distorted in a fashion that elongates the bonds being broken and brings the atoms engaging in bond making close to each other. The activated complex can roughly be thought of as the species that exists half-way between reactants and products; it is neither reactants nor products, but the entity that exists as reactants transforms to products (think of it as the half-way point of the "morphing" of reactants to products). At times, the solvent in a solution reaction is part of the activated complex.

EXAMPLE   Drawing the activated complex

Draw a Lewis structure of the N_2O_4 molecule and the NO_2 molecule. Then draw the Lewis structure of the likely activated complex for the reaction of N_2O_4 to NO_2. Remember that NO_2 is an odd-electron molecule.

$$N_2O_4(g) \longrightarrow 2NO_2(g)$$

SOLUTION The required structures are

It is clear that the N—N bond must elongate and eventually break to form the two NO_2 molecules. The activated complex is drawn with a dashed line representing the bond that is about to break. This bond is elongated in the activated complex; the continued elongation eventually results in bond rupture.

$$\begin{array}{c}
\ddot{\text{:}} \ddot{\text{O}} \text{:} \qquad \ddot{\text{:}} \ddot{\text{O}} \text{:} \\
\text{N} \text{-----} \text{N} \\
\ddot{\text{:}} \ddot{\text{O}} \text{:} \qquad \ddot{\text{:}} \ddot{\text{O}} \text{:}
\end{array}$$

EXERCISE At high temperature, the reaction of hydrogen with iodine to form hydrogen iodide occurs in one step; an H_2 collides with an I_2 and two HI molecules directly form. Draw the activated complex for this reaction, using dashed lines to show bonds being made and bonds being broken.

Answer:
$$\begin{array}{c} \text{H---I} \\ \text{|} \quad \text{|} \\ \text{H---I} \end{array}$$

18.12   CATALYSIS

KEY CONCEPT   A catalyst is used to speed up a reaction

A **catalyst** is a substance that speeds up the rate of a reaction without itself being consumed in the reaction. A **homogeneous catalyst** is one that is present in the same phase as the reactants. A **heterogeneous catalyst** is one that is present in a different phase from that of the reactants. A common type of heterogeneous catalyst is a solid that has a very large surface area that gas-phase or liquid-phase molecules can attach to in a process called **adsorption** (not absorption). When the molecules adsorb onto the surface, they may be brought into the correct orientation for reaction and/or a bond that must be broken in the reaction is weakened. A substance that irreversibly adsorbs onto a catalytic surface prevents reactant molecules from adsorbing and **poisons** the catalyst. A catalyst works by providing a new path for a reaction; the new path has a lower activation energy than the uncatalyzed path and, therefore, the reaction speeds up. It should be understood that not every reaction can be catalyzed and that, in most cases, each reaction needs its own catalyst (although there are some classes of reactions that can be catalyzed by a single type of catalyst).

EXAMPLE 1   Recognizing a catalyst

The two steps for a reaction that occurs in aqueous solution follow. What is the catalyst in the reaction? Is it homogeneous or heterogeneous? What is the overall chemical reaction?

$$V^{3+}(aq) + Cu^{2+}(aq) \longrightarrow V^{4+}(aq) + Cu^{+}(aq)$$
$$Cu^{+}(aq) + Fe^{3+}(aq) \longrightarrow Cu^{2+}(aq) + Fe^{2+}(aq)$$

SOLUTION A catalyst is a substance that speeds up a reaction without itself being consumed. In the reaction shown, Cu^{2+} is used in the first step but is regenerated in the second step; it is not consumed, so it is the catalyst. Because the Cu^{2+} catalyst is a dissolved solute and the reactants are also dissolved solutes, it is a homogeneous catalyst. The overall reaction can be derived by summing the two given reactions

$$V^{3+}(aq) + Fe^{3+}(aq) \longrightarrow V^{4+}(aq) + Fe^{2+}(aq)$$

EXERCISE The two steps for a reaction that occurs in the gas phase follow. What is the catalyst in the reaction? Is it homogeneous or heterogeneous? What is the overall chemical reaction?

$$2NO(g) + Br_2(g) \longrightarrow 2NOBr(g)$$
$$2NOBr(g) + Cl_2(g) \longrightarrow 2NOCl(g) + Br_2(g)$$

[*Answer:* Br_2 is the homogeneous catalyst; overall reaction is $2NO(g) + Cl_2(g) \longrightarrow 2NOCl(g)$.]

EXAMPLE 2   Recognizing a heterogeneous catalysis

The decomposition of NH_3 is catalyzed by the presence of a tungsten surface.

$$2NH_3(g) \longrightarrow N_2(g) + 3H_2(g)$$

Addition of H_2 to the reaction vessel slows down the reaction whereas addition of N_2 does not. Explain this observation.

SOLUTION Because addition of N_2 has no effect on the rate of reaction, we can conclude that it is not simply the addition of a product molecule that is important. The observation is explained if H_2 poisons the tungsten catalyst by adsorbing onto the tungsten more strongly than the reactant NH_3 molecules.

EXERCISE What happens to the activation energy of the preceding reaction when H_2 is added to the reaction vessel?

[*Answer:* Because the reaction slows down, the effective activation energy has likely increased.]

18.13 LIVING CATALYSTS: ENZYMES

KEY CONCEPT   Enzymes speed up biochemical reactions

An **enzyme** is a biological catalyst; it is typically an extremely large molecule with many crevices, pockets, and ridges on its surface (see Figure 18.27 of the text). In the **induced-fit** model of enzyme activity, a reactant molecule (the **substrate**) fits into a properly shaped pocket on the enzyme molecule (the **active site**). The interaction of the substrate with enzyme atoms weakens and strengthens the appropriate bonds, as does subtle shifts in the shape of the enzyme (and the active site). After reaction is complete, the newly formed product molecule leaves the active site and the enzyme is ready to accept another molecule for reaction. The effect of enzymes can be dramatic; catalase, an enzyme that catalyzes the decomposition of H_2O_2 to H_2O and O_2, speeds up the reaction by a factor of 10^{15}!

EXAMPLE   Understanding enzyme activity

The enzyme succinic dehydrogenase catalyzes the reduction of succinic acid to fumaric acid.

One of the following molecules acts as a modest poison (called a competitive inhibitor) for succinic dehydrogenase. Which one? Explain your answer.

SOLUTION One way in which an enzyme poison acts is that it occupies the active site and does not permit the proper molecule to enter the site. The most reasonable candidate for such a poison is one that closely resembles the proper substrate. Molecule A (malonic acid) is very similar in structure to succinic acid and is the compound that can poison succinic dehydrogenase.

EXERCISE Assume that catalase speeds up the decomposition of hydrogen peroxide by lowering the activation energy for the decomposition. If the activation energy of the uncatalyzed reaction is approximately $94 \text{ kJ} \cdot \text{mol}^{-1}$ at 25°C, what is the activation energy for the catalyzed reaction? Assume that the pre-exponential factor is the same for both the catalyzed and uncatalyzed reactions.

[*Answer:* $3 \text{ kJ} \cdot \text{mol}^{-1}$]

KEY WORDS   Define or explain each term in a written sentence or two.

activated complex   catalyst   pre-exponential factor
activation energy   catalyst poison   reaction profile
adsorb   collision theory   substrate
Arrhenius equation   induced-fit mechanism

REACTION MECHANISMS

18.14 ELEMENTARY REACTIONS

KEY CONCEPT   An elementary reaction shows one step in a mechanism

Most reactions occur through a series of two or more simple steps. For example, the reaction

$$2NO(g) + H_2(g) \longrightarrow N_2O(g) + H_2O(g)$$

does not occur through the collision of an H_2 molecule with NO molecules to result in products. Instead it occurs through two steps:

$$2NO \longrightarrow N_2O_2$$

$$N_2O_2 + H_2 \longrightarrow N_2O + H_2O_2$$

In the first step, two NO molecules collide to produce the intermediate N_2O_2. In the second step, the N_2O_2 collides with an H_2 molecule to produce the products N_2O and H_2O. These steps are called **elementary reactions** because they show a single simple process that occurs for the species portrayed in the step. In an elementary reaction, when two species react, the product is formed because of a direct collision between the reactants. In this respect, an elementary reaction differs from an overall chemical reaction. The *elementary reaction* shows what actually happens in a single step; an *overall reaction* shows the result of a series of steps. If one molecule only participates in an elementary reaction, the reaction is called a **unimolecular** ("one molecule") reaction. If two molecules participate, it is called **bimolecular,** and three, **termolecular.** A series of elementary steps that describes how the reactants become products for a reaction is called a **mechanism** for the reaction.

> **EXAMPLE**   Deciding whether a reaction is an elementary process
>
> In the reaction of a deuterium atom with a hydrogen molecule, a deuterium atom collides with a hydrogen molecule. At the instant of collision, an H—D bond formed as an H—H bond is broken, resulting in the direct formation of products. Is this an elementary reaction?
>
> $$D + H_2 \longrightarrow HD + H$$
>
> **SOLUTION** This is an elementary reaction. One type of elementary reaction portrays a single simple step in which the products are formed as a direct result of a collision. The reaction in the problem is this kind of elementary reaction.
>
> **EXERCISE** The reaction between nitrogen dioxide and carbon monoxide is
>
> $$NO_2(g) + CO(g) \longrightarrow CO_2(g) + NO(g)$$
>
> The reaction works in the following way. Two NO_2 molecules collide to form NO_3 and NO (NO is one of the products). The NO_3 thus formed collides with CO to form NO_2 and CO_2 (CO_2 is the other product).
>
> $$2NO_2 \longrightarrow NO_3 + NO$$
>
> $$NO_3 + CO \longrightarrow NO_2 + CO_2$$

The overall result is the formation of one NO molecule and one CO_2 molecule from one NO_2 molecule and one CO molecule, as portrayed in the chemical equation shown. Is the reaction $NO_2(g) + CO(g) \rightarrow CO_2(g) + NO(g)$ an elementary reaction?

[*Answer:* No; the products CO_2 and NO are not formed in a single step as the result of a collision between the reactants NO_2 and CO.]

PITFALL   There is more than one type of elementary reaction

The preceding example and exercise focus on one type of elementary reaction, in which two reactants collide to form product(s). We should remember that another type is considered in the text, one in which a single molecule breaks apart into fragments. There are still other types of elementary reactions, but they will not be discussed in this course.

18.15   THE RATE LAWS OF ELEMENTARY REACTIONS

KEY CONCEPT A   Unimolecular and bimolecular reactions

A **unimolecular reaction** is an *elementary* reaction in which the reactant is a single molecule. The rate law for a unimolecular reaction is always a first-order rate law:

Unimolecular reaction: A $\longrightarrow$ products   Rate = $k[A]$

A **bimolecular reaction** is an *elementary* reaction in which two reactants collide to form products. The rate of reaction is always first order in each reactant:

Bimolecular reaction: A + B $\longrightarrow$ products   Rate = $k[A][B]$

EXAMPLE   Writing the rate equation for an elementary reaction

The decomposition of acetone (C_3H_6O) starts with the reaction

$$C_3H_6O \longrightarrow CH_3 + C_2H_3O$$

In this reaction, acetone is heated to a very high temperature so that it starts vibrating violently. Eventually, the molecule falls apart into two fragments. What is the rate law for this reaction?

SOLUTION From the description of the reaction, we recognize it as a unimolecular elementary reaction. Such a reaction is always described by a first-order rate law.

$$Rate = k[C_3H_6O]$$

EXERCISE One of the steps in the decomposition of nitrous acid (HNO_2) is

$$NO_2 + NO_2 \longrightarrow N_2O_4$$

In this reaction, two NO_2 molecules collide and bond together to form a N_2O_4 molecule. What is the rate law for this reaction?

[*Answer:* Rate = $k[NO_2]^2$]

PITFALL   When is it possible to deduce a rate law from a balanced equation?

We must be careful when attempting to write a rate law from a balanced equation. This can be done reliably *only* for an elementary reaction. For reactions that are not elementary, the rate law must be determined by an experiment. It cannot be deduced from the balanced equation.

KEY CONCEPT B   The overall reaction

One of the goals of studying the rates of reaction is to deduce the mechanism for the reaction, that is, the series of elementary processes that occur when the reactants change to products. There are two basic requirements for any mechanism:

1. The sequence of elementary reactions in the mechanism must account for the overall chemical change from reactants to products. Practically, this means that the elementary steps in a mechanism must add up to the overall reaction.
2. The mechanism must be consistent with the experimental rate law that has been determined for the overall reaction.

Many mechanisms involve formation of one or more **intermediates.** An intermediate is a chemical species that is formed as a product in an elementary reaction and then becomes a reactant in another elementary reaction, so it is used up and does not appear as a product in the overall chemical equation. Intermediates are generally highly reactive.

EXAMPLE   Analyzing a reaction mechanism

A proposed mechanism for the reaction of nitrogen dioxide with fluorine is shown. Does the proposed mechanism account for the overall chemical change from reactants to products? Are there any intermediates in the mechanism?

$$\text{Overall reaction:} \qquad 2NO_2(g) + F_2(g) \longrightarrow 2NO_2F(g)$$

$$\text{Proposed mechanism:} \qquad NO_2 + F_2 \longrightarrow NO_2F + F$$

$$F + NO_2 \longrightarrow NO_2F$$

SOLUTION We check to see whether the mechanism accounts for the overall chemical changes in the chemical equation by adding the equations representing the elementary reactions in the mechanism:

$$NO_2 + F_2 \longrightarrow NO_2F + F$$
$$\underline{F + NO_2 \longrightarrow NO_2F}$$
$$NO_2 + F_2 + \cancel{F} + NO_2 \longrightarrow NO_2F + \cancel{F} + NO_2F$$

F appears on both sides of the chemical equation, so it is canceled out of the overall reaction. There are two NO_2 molecules on the left-hand side of the equation and two NO_2F on the right,

$$2NO_2 + F_2 \longrightarrow 2NO_2F$$

which is the overall chemical equation. The proposed mechanism accounts for the overall chemical change from reactants to products. We note that F is formed as a product in the first step of the mechanism but is then consumed in the second step and does not appear in the overall equation. It is an intermediate.

EXERCISE Is the following mechanism consistent with the overall reaction given? What intermediates (if any) are present?

$$\text{Overall reaction:} \qquad 2NO(g) + O_2(g) \longrightarrow 2NO_2(g)$$

$$\text{Proposed mechanism:} \qquad NO + O_2 \longrightarrow NO_2 + O$$

$$O + NO \longrightarrow NO_2$$

[*Answer:* Mechanism is consistent with overall equation; O is an intermediate.]

PITFALL   Adding the elementary reactions in a mechanism

The steps in a mechanism show what happens to different molecules and atoms in the mechanism but don't always explicitly show the number of times each step occurs as reactants become products.

For instance, the following mechanism is generally accepted as the one that accounts for the reaction of oxygen to form ozone in the upper atmosphere.

Overall reaction: $3O_2(g) \longrightarrow 2O_3(g)$

Accepted mechanism: $O_2 \longrightarrow 2O(g)$

$$O(g) + O_2(g) \longrightarrow O_3(g)$$

If the two steps portrayed in the mechanism are added, the result is not the overall equation. In actuality, the second step occurs twice each time the reaction occurs. Taking this into consideration gives the correct result when the reactions in the mechanism are added.

$$O_2(g) \longrightarrow 2O(g)$$
$$O(g) + O_2(g) \longrightarrow O_3(g) \quad \text{(once)}$$
$$O(g) + O_2(g) \longrightarrow O_3(g) \quad \text{(twice)}$$

These three reactions sum to the overall reaction. We *may not* write

$$2O(g) + 2O_2(g) \longrightarrow 2O_3(g) \quad \text{(incorrect!)}$$

to express the step that occurs twice because this would imply that two O atoms collide with two O_2 molecules in a single step to form two O_3 molecules. Such a four-particle collision is improbable.

KEY CONCEPT C   Rate-determining steps

Many reaction mechanisms contain one step that is much slower than the rest. When this occurs, the slow step acts like a bottleneck and determines how fast the whole reaction can occur. The slow step is called the **rate-determining step** in the mechanism.

EXAMPLE   Writing a rate law from a mechanism

An overall reaction and mechanism are given. What rate law is predicted for the overall reaction?

Overall reaction: $(CH_3)_3CBr(aq) + OH^-(aq) \longrightarrow (CH_3)_3COH(aq) + Br^-(aq)$

Accepted mechanism: $(CH_3)_3CBr \longrightarrow (CH_3)_3C^+ + Br^- \quad \text{(slow)}$

$$(CH_3)_3C^+ + OH^- \longrightarrow (CH_3)_3COH \quad \text{(fast)}$$

SOLUTION In this mechanism, there is a slow step followed by a fast one. The slow step determines how fast the reaction can go, that is, it is the rate-determining step. The rate of the overall reaction will be the same as the rate of the slow step. Thus, the rate law for the overall reaction is

$$\text{Rate} = k[(CH_3)_3CBr]$$

EXERCISE What is the predicted rate law for the overall reaction shown if it proceeds by the following mechanism.

Overall reaction: $2NO(g) + Cl_2(g) \longrightarrow 2NOCl(g)$

Accepted mechanism: $NO + Cl_2 \longrightarrow NOCl_2 \quad \text{(slow)}$

$$NOCl_2 + NO \longrightarrow 2NOCl \quad \text{(fast)}$$

[*Answer:* Rate $= k[NO][Cl_2]$]

KEY CONCEPT D   Elementary reactions that involve a dynamic equilibrium

In many mechanisms, a dynamic equilibrium occurs for one of the elementary reactions in the mechanism. As an example, consider the following elementary reaction (the "forward" reaction):

$$A \longrightarrow B \quad \text{Rate} = k[A]$$

Let's assume that after a certain amount of product B forms, it reacts to reform the reactant A,

$$B \longrightarrow A \qquad \text{Rate} = k'[B]$$

where the rate constant k' refers to the reverse reaction. A dynamic equilibrium exists when the rate of the reverse reaction equals the rate of the forward reaction; so, for every A molecule that becomes a B molecule, a B molecule reverts back to an A molecule. ("Whatever is done is undone.") The concentration of A and B do not change. When dynamic equilibrium is achieved, we have

$$\text{Rate of forward reaction} = \text{rate of reverse reaction}$$

$$k[A] = k'[B]$$

It is now possible, by dividing both sides of this equation by k', to express the molarity of B in terms of k, k', and the molarity of A:

$$[B] = \frac{k}{k'}[A]$$

EXAMPLE Predicting the rate law for a mechanism with a dynamic equilibrium

What is the predicted rate law for the overall reaction if it proceeds by the following mechanism. (These mechanistic steps have only recently been proposed to be important in the high temperature reaction of H_2 with I_2).

Overall reaction: $\qquad H_2(g) + I_2(g) \longrightarrow 2HI(g)$

Proposed mechanism: $\qquad\qquad I_2 \rightleftharpoons 2I \qquad$ (rapid dynamic equilibrium)

$$H_2 + 2I \longrightarrow 2HI \qquad \text{(slow)}$$

SOLUTION The rate of the overall reaction is determined by the slow step in the mechanism

$$\text{Rate} = k_2[H_2][I]^2$$

The rate constant is called k_2 to distinguish it from k_1, the rate constant for the first step in the mechanism. The rate law is correct but not acceptable as a rate law for the overall reaction because it contains the molarity of I, and I is not one of the reactants in the overall reaction. It is an intermediate and should not appear in the rate law for the overall reaction. To eliminate [I], we recognize that the rapid dynamic equilibrium in the first step gives

$$\text{Rate of forward reaction} = k_1[I_2] \qquad \text{Rate of reverse reaction} = k'_1[I]^2$$

$$k_1[I_2] = k'_1[I]^2$$

$$[I]^2 = \frac{k_1}{k'_1}[I_2]$$

This value of $[I]^2$ is now substituted in the original rate equation for the slow step to give

$$\text{Rate} = \frac{k_2 k_1}{k'_1}[H_2][I_2]$$

Recognizing that the factor $k_2 k_1 / k'_1$ equals a constant, we define an overall rate constant k with the relation $k = k_2 k_1 / k'_1$ and end up, finally, with the rate law for the overall reaction:

$$\text{Rate} = k[H_2][I_2]$$

EXERCISE What is the rate law for the following overall reaction if it proceeds by the given mechanism?

Overall reaction: $\qquad 2NO(g) + H_2(g) \longrightarrow N_2O(g) + H_2O(g)$

Proposed mechanism: $\qquad\qquad 2NO \rightleftharpoons N_2O_2 \qquad$ (rapid dynamic equilibrium)

$$N_2O_2 + H_2 \longrightarrow N_2O + H_2O \qquad \text{(slow)}$$

[*Answer:* Rate = $k[H_2][NO]^2$]

18.16 CHAIN REACTIONS

KEY CONCEPT   Chain reactions involve many steps

A chain reaction is a reaction in which an intermediate reacts to form another intermediate, which then proceeds to form some of the first intermediate. This cyclic process occurs over and over again, forming some product in each cycle. A distinctive feature of a chain reaction is related to how the reaction is sustained. In a nonchain reaction, the reaction proceeds by having the mechanism occur, from the first step to the last, over and over again. A chain reaction works somewhat differently. After the first step in the mechanism (**initiation**), two or more product-producing steps cycle through many times in a self-sustaining way (**propagation**). This occurs because an intermediate produced as a product in one of the cycling steps becomes a reactant for another cycling step. The cycling continues until a random reaction consumes one of the intermediates required for the cycling process (**termination**). In a nonchain reaction, if the first step in the mechanism occurs 1000 times, 1000 sets of product molecules are formed. In a chain reaction, 1000 initiations result in many more than 1000 sets of product molecules, because each initiation is followed by many cycles of the product-producing steps. **Retardation** occurs when an intermediate reacts to form something other than product, but also forms a reactive chain carrier. **Inhibition** occurs when an intermediate reacts with a foreign substance to produce nonreactive substances. In a **chain-branching** step, a single radical intermediate reacts to form two or more radical intermediates. Chain branching causes a rapid increase in the overall rate of reaction because the new radicals formed start new chains, more branching occurs in the new chains, more new radicals start more new chains, and so on.

> ### EXAMPLE   Identifying and understanding the steps in a chain reaction
>
> The overall reaction and mechanism for the chlorination of methane—a radical chain reaction—is
>
> Overall reaction: $CH_4(g) + Cl_2(g) \longrightarrow CH_3Cl(g) + HCl(g)$
>
> Accepted mechanism: $Cl_2 \longrightarrow 2Cl\cdot$
>
> $Cl\cdot + CH_4 \longrightarrow CH_3\cdot + HCl$
>
> $CH_3\cdot + Cl_2 \longrightarrow CH_3Cl + Cl\cdot$
>
> $Cl\cdot + CH_3\cdot \longrightarrow CH_3Cl$
>
> Identify and briefly discuss the initiation, propagation, and termination steps. Be certain to carefully explain how propagation works.
>
> **SOLUTION** The first step, in which a chlorine molecule breaks up into two chlorine atoms, is the initiation step. The chlorine atom is a radical intermediate. The next two steps are the propagation steps:
>
> $Cl\cdot + CH_4 \longrightarrow CH_3\cdot + HCl$
>
> $CH_3\cdot + Cl_2 \longrightarrow CH_3Cl + Cl\cdot$
>
> In the first of these steps, a chlorine atom formed in initiation removes a hydrogen atom from methane to form HCl and a methyl radical, $CH_3\cdot$. The methyl radical, which is an intermediate, reacts with a chlorine molecule in the next step to form a product molecule, CH_3Cl, and a chlorine atom. Now, the chlorine atom can start the propagation sequence over again by finding another methane molecule to react with. The propagation steps can cycle, over and over, as long as reactants are available and termination doesn't occur. The last step is a termination step because it involves the combination of two radicals to produce a nonradical product.
>
> **EXERCISE** Suggest a possible retardation step for the mechanism given in the example.
>
> [*Answer:* $CH_3\cdot + HCl \longrightarrow CH_4 + Cl\cdot$ and
> $Cl\cdot + CH_3Cl \longrightarrow CH_3\cdot + Cl_2$]

KEY CONCEPT   Equilibrium occurs when forward and backward rates are equal

Assume the following chemical system is at equilibrium:

$$A + B \rightleftharpoons C + D \qquad K = \frac{[C][D]}{[A][B]}$$

The reason that equilibrium exists is that the rate of forward reaction and the rate of the reverse reaction are equal. Because the two rates are equal, there is no net formation of either A, B, C, or D. Thus, we have equilibrium, by which we mean that there is no net change in the system. If the forward reaction and reverse reaction are both elementary processes with k the rate constant for the forward reaction, that is, rate = $k[A][B]$ and k' the rate constant for the reverse reaction, that is, rate = $k'[C][D]$, then the equilibrium constant K is given by

$$K = \frac{k}{k'}$$

EXAMPLE   Understanding equilibrium and rates

Assume an equilibrium occurs in which a single molecule breaks down into two identical molecules. Is it still true that the $K = k/k'$? Prove your answer.

$$A \rightleftharpoons 2B$$

SOLUTION   The equilibrium constant for the equilibrium is $K = [B]^2/[A]$. The rate of the forward reaction is rate = $k[A]$ and that for the reverse reaction is rate = $k'[B]^2$. Equilibrium occurs when the two rates are equal:

$$k[A] = k'[B]^2$$

We can rearrange this equation to get

$$\frac{[B]^2}{[A]} = \frac{k}{k'}$$

Because $K = [B]^2/[A]$, where [B] and [A] are molar concentrations expressed without molarity units, we can conclude that $K = k/k'$ and thus, the expression relating the equilibrium constant to the forward and reverse rates is still true for the chemical equilibrium A $\rightleftharpoons$ 2B.

EXERCISE   Assume that the chemical equilibrium A + B $\rightleftharpoons$ 2C occurs through a multistep mechanism. Is it true that $K = k/k'$?

[*Answer:* No]

KEY WORDS   Define or explain each term in a written sentence or two.

bimolecular reaction	initiation	reaction intermediate
branching	propagation	reaction mechanism
chain carrier	radical chain reaction	termination
elementary reaction	rate-determining step	unimolecular reaction

DESCRIPTIVE CHEMISTRY TO REMEMBER

• **Ozone** molecules (O_3) in the upper atmosphere absorb much of the intense ultraviolet radiation from the sun.

- Finely divided particles, such as iron dust (or other dusts), burn very rapidly in air because of the **large surface area** exposed to the air.
- Catalytic converters in automobiles contain solid **heterogeneous catalysts** that oxidize carbon monoxide to carbon dioxide and reduce nitrogen oxides to nitrogen (N_2).
- **Sulfur** in automobile fuels is a problem because the catalysts in catalytic converters convert SO_2 to SO_3, which can then react with water to form sulfuric acid.
- The action of many biological poisons is thought to stem from their ability to block the action of **enzymes**.

CHEMICAL EQUATIONS TO KNOW

Gas-phase reactions

- $2HI(g) \longrightarrow H_2(g) + I_2(g)$ $H_2(g) + Br_2(g) \longrightarrow 2HBr(g)$
- $2N_2O_5(g) \longrightarrow 4NO_2(g) + O_2(g)$ $C_3H_6(g) \longrightarrow CH_3{-}CH{=}CH_2(g)$
- $2NO_2(g) \longrightarrow 2NO(g) + O_2(g)$
- $2O_3(g) \longrightarrow 3O_2(g)$

Solution reactions

- $S_2O_8^{2-}(aq) + 3I^-(aq) \longrightarrow 2SO_4^{2-}(aq) + I_3^-(aq)$
- $BrO_3^-(aq) + 5Br^-(aq) + 6H^+(aq) \longrightarrow 3Br_2(aq) + 3H_2O(l)$
- $CH_3Br(aq) + OH^-(aq) \longrightarrow CH_3OH(aq) + Br^-(aq)$
- $C_2H_5Br(aq) + OH^-(aq) \longrightarrow C_2H_5OH(aq) + Br^-(aq)$

Reactions that work best when catalyzed (heterogeneous catalyst)

- $2NH_3(g) \xrightarrow{\Delta,\ Pt} N_2(g) + 3H_2(g)$
- $2KClO_3(s) \xrightarrow{\Delta,\ MnO_2} 2KCl(s) + 3O_2(g)$
- $N_2(g) + 3H_2(g) \xrightarrow{\Delta,\ Fe} 2NH_3(g)$
- $2SO_2(g) + O_2(g) \xrightarrow{V_2O_5} 2SO_3(g)$

Reactions that work best when catalyzed (homogeneous catalyst)

- $2H_2O_2(aq) \xrightarrow{Br_2} 2H_2O(l) + O_2(g)$

MATHEMATICAL EQUATIONS TO KNOW AND UNDERSTAND

$$\text{Rate} = \frac{\Delta[A]}{\Delta t}$$ definition of rate

$$\ln\left(\frac{[A]_0}{[A]}\right) = kt$$ first-order reaction integrated rate law

$$\ln\left(\frac{k'}{k}\right) = \frac{E_a}{R}\left[\frac{1}{T} - \frac{1}{T'}\right]$$ dependence of k on temperature

$$\ln k = \ln A - E_a/RT$$ Arrhenius equation for k

$$t_{1/2} = 0.693/k$$ half-life for first-order reaction

$$\frac{1}{[A]} - \frac{1}{[A]_0} = kt$$
second-order reaction integrated rate law

$$k = Ae^{-E_a/RT}$$
rate constant as function of activation energy

$$K = \frac{k}{k'}$$
equilibrium constant as function of forward and reverse reaction rate constants

SELF-TEST EXERCISES

Concentration and rate

1. The molarity of Cl_2 as a function of time is shown for the following reaction. What is the rate of reaction during the time interval $t = 5.0$ s to $t = 10.0$ s as measured by the decrease in $[Cl_2]$?

$$2NO(g) + Cl_2(g) \longrightarrow 2NOCl(g)$$

Time, s	0.0	5.0	10.0	15.0	20.0
$[Cl_2]$	0.121	0.108	0.100	0.094	0.091

(a) 1.6×10^{-3} mol·L^{-1}·s^{-1}
(b) 1.0×10^{-2} mol·L^{-1}·s^{-1}
(c) 2.2×10^{-2} mol·L^{-1}·s^{-1}
(d) 3.2×10^{-2} mol·L^{-1}·s^{-1}
(e) 2.1×10^{-3} mol·L^{-1}·s^{-1}

2. For the reaction shown, what is the rate of disappearance of H_2 when the rate of appearance of N_2 is 5.2 mmol·L^{-1}·s^{-1}?

$$2NO(g) + 2H_2(g) \longrightarrow 2H_2O(g) + N_2(g)$$

(a) 1.3 mmol·L^{-1}·s^{-1}
(b) 2.6 mmol·L^{-1}·s^{-1}
(c) 5.2 mmol·L^{-1}·s^{-1}
(d) 21 mmol·L^{-1}·s^{-1}
(e) 10 mmol·L^{-1}·s^{-1}

The following data for the reaction "A → products" are for Exercises 3–5.

Time, min	0.0	1.0	2.0	3.0	4.0	5.0
[A]	1.30	1.01	0.788	0.614	0.478	0.372

3. From a plot of [A] versus t, determine the rate of reaction at $t = 2.5$ min. (Because plotting involves a graphical technique, your answer may be somewhat different from the author's. Choose the closest answer.)
(a) 0.26 mol·L^{-1}·min^{-1}
(b) 1.2 mol·L^{-1}·min^{-1}
(c) 0.17 mol·L^{-1}·min^{-1}
(d) 5.7 mol·L^{-1}·min^{-1}
(e) 3.8 mol·L^{-1}·min^{-1}

4. From the plot used in Exercise 3, determine the initial rate. (Because this involves a graphical technique, your answer may be somewhat different from the author's. Choose the closest answer.)
(a) 5.6 mol·L^{-1}·min^{-1}
(b) 0.32 mol·L^{-1}·min^{-1}
(c) 0.44 mol·L^{-1}·min^{-1}
(d) 2.3 mol·L^{-1}·min^{-1}
(e) 1.7 mol·L^{-1}·min^{-1}

5. Use the initial rate determined in Exercise 4 and the initial concentration to determine the rate constant. The rate is given by (initial rate) $= k[A]$.
(a) 4.0 min^{-1}
(b) 0.25 min^{-1}
(c) 0.11 min^{-1}
(d) 3.2×10^{-3} min^{-1}
(e) 5.44×10^{-3} min^{-1}

6. Which of the following is a correct unit for a second-order rate constant?
(a) $mol \cdot L^{-1} \cdot min^{-1}$ (b) hr^{-1} (c) $mol \cdot s^{-1}$
(d) $L \cdot mol^{-1} \cdot s^{-1}$ (e) $s \cdot L \cdot mol^{-1}$

For Exercises 7 and 8, use the reaction $2NO(g) + O_2(g) \rightarrow 2NO_2(g)$, which follows the rate law, rate of reaction $= k[NO]^2[O_2]$.

7. What is the reaction order with respect to NO?
(a) first   (b) second   (c) third   (d) zero

8. What is the overall order of reaction?
(a) second   (b) first   (c) third   (d) zero

9. The following data show how the initial rate of reaction changes with initial concentration of reactants for the reaction $2A + 2B \rightarrow$ products. What is the rate law for the reaction?

$[A]_0$	$[B]_0$	Initial Rate, $mol \cdot L^{-1} \cdot s^{-1}$
0.22	0.22	0.816
0.44	0.22	3.26
0.22	0.44	1.63

(a) rate $= k[A][B]$ (b) rate $= k[A]^2[B]$
(c) rate $= k[B]^2[A]^2$ (d) rate $= k[A][B]^2$

10. The following data show how the initial rate of reaction changes with initial concentration of reactants for the reaction $2A + B \rightarrow$ products. What is the rate law for the reaction?

$[A]_0$	$[B]_0$	Initial Rate, $mol \cdot L^{-1} \cdot s^{-1}$
0.52	0.46	2.24
0.52	0.70	2.24
0.68	0.73	3.83

(a) rate $= k[A][B]$ (b) rate $= k[A]^2[B]$ (c) rate $= k[B]$
(d) rate $= k[A]^2[B]^2$ (e) rate $= k[A]^2$

11. At high temperatures, the decomposition of N_2O is first order with $k = 3.4 \times 10^{-5}\ s^{-1}$. What percentage of N_2O is left 1.0 h after the start of the reaction?
(a) 88.4%   (b) 71.6%   (c) 23.8%   (d) 41.1%   (e) 2.65%

12. A first-order reaction with $[A]_0 = 0.84$ M and $k = 0.112\ min^{-1}$ is run until $[A] = 0.62$ M. For how long was the reaction run?
(a) 2.7 min   (b) 12 min   (c) 1.4 min   (d) 0.15 min   (e) 6.6 min

13. What is $t_{1/2}$ for a first-order reaction with $k = 1.22\ h^{-1}$ and $[A]_0 = 0.44$ M?
(a) 0.820 h   (b) 1.22 h   (c) 0.845 h   (d) 0.568 h   (e) 1.76 h

14. A reaction with $t_{1/2} = 10.0$ min starts with $[A]_0 = 3.2$ M. What is $[A]$ after 40.0 min of reaction?
(a) 0.20 M   (b) 0.80 M   (c) 1.6 M   (d) 0.32 M   (e) 0.16 M

15. The following data were collected for the concentration of A as a function of time for the hypothetical reaction $2A \rightarrow B$. Using graphical techniques, show that the reaction is first order and determine the value of the rate constant k.

Time, s	0	20	40	60	80	100
Concentration, mol·L^{-1}	0.223	0.195	0.170	0.148	0.130	0.113

(a) $k = 0.165$ s^{-1} (b) $k = 0.002\ 24$ s^{-1} (c) $k = 1.45$ s^{-1}
(d) $k = 102$ s^{-1} (e) $k = 0.006\ 78$ s^{-1}

16. The reaction "$A + B \rightarrow$ products" is pseudo–first order, with rate $= k'[A]$ and $k' = 6.29$ s^{-1} when $[B] = 6.0$ M. What is k for the rate law, rate $= k[A][B]$?
(a) 12 L·mol^{-1}·s^{-1} (b) 6.3 L·mol^{-1}·s^{-1} (c) 38 L·mol^{-1}·s^{-1}
(d) 16 L·mol^{-1}·s^{-1} (e) 1.0 L·mol^{-1}·s^{-1}

Controlling reaction rates

17. $E_a = 132$ kJ·mol^{-1} and $A = 4.9 \times 10^8$ L·mol^{-1}·s^{-1} for the second-order reaction $NO_2(g) + CO(g) \longrightarrow NO(g) + CO_2(g)$. What is k at 300°C?
(a) 3.6×10^5 L·mol^{-1}·s^{-1} (b) 2.1×10^{-9} L·mol^{-1}·s^{-1}
(c) 5.1×10^{-15} L·mol^{-1}·s^{-1} (d) 4.8×10^8 L·mol^{-1}·s^{-1}
(e) 4.5×10^{-4} L·mol^{-1}·s^{-1}

18. The rate constant for a second-order reaction shows the following dependence on temperature. What is the activation energy for the reaction? (Choose the closest answer.)

Temperature, °C	0	5	10	15	20
k, L·mol^{-1}·s^{-1}	5.60×10^{-5}	9.50×10^{-5}	19.8×10^{-5}	37.0×10^{-5}	73.3×10^{-5}

(a) 151 kJ·mol^{-1} (b) 86.5 kJ·mol^{-1} (c) 15.2 kJ·mol^{-1}
(d) 66.3 kJ·mol^{-1} (e) 108 kJ·mol^{-1}

19. What is k at 59°C for a reaction with $k = 7.3 \times 10^{-2}$ s^{-1} at 25°C and $E_a = 82.6$ kJ·mol^{-1}?
(a) 5.5×10^{-2} s^{-1} (b) 0.15 s^{-1} (c) 3.2×10^{-2} s^{-1}
(d) 2.2 s^{-1} (e) 0.17 s^{-1}

20. Under a specific set of conditions, the rate of reaction of OH$^-$ with C_2H_5Br is 4.2×10^{-5} mol·L^{-1}·s^{-1} at 25°C. Estimate the approximate rate of reaction at 35°C. Detailed calculations are not necessary.
(a) 4.2×10^{-5} mol·L^{-1}·s^{-1} (b) 2.1×10^{-5} mol·L^{-1}·s^{-1}
(c) 8.5×10^{-5} mol·L^{-1}·s^{-1} (d) 12.6×10^{-5} mol·L^{-1}·s^{-1}
(e) 1.4×10^{-5} mol·L^{-1}·s^{-1}

21. What activation energy corresponds to the slowest reaction?
(a) 89 kJ·mol^{-1} (b) 101 kJ·mol^{-1} (c) 34 kJ·mol^{-1}

22. Which reaction profile corresponds to a reaction with a fast first step followed by a slow second step?

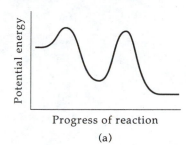

Progress of reaction

(a)

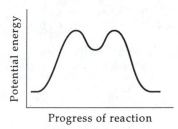

Progress of reaction

(b)

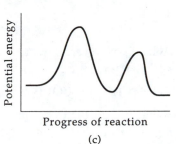

Progress of reaction

(c)

23. Which of the following best represents the activated complex for the following reaction?

$$Cl(g) + NOCl(g) \longrightarrow Cl_2(g) + NO(g)$$

(a)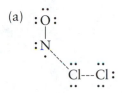

(b)

(c)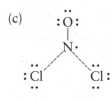

24. Addition of a catalyst to a reaction _____ the activation energy of the reaction profile.

(a) raises   (b) lowers   (c) has no effect on

25. What is the catalyst in the following mechanism?

$$Ce^{4+}(aq) + Ag^+(aq) \longrightarrow Ce^{3+}(aq) + Ag^{2+}(aq)$$
$$Ag^{2+}(aq) + Tl^+(aq) \longrightarrow Tl^{2+}(aq) + Ag^+(aq)$$
$$Tl^{2+}(aq) + Ce^{4+}(aq) \longrightarrow Tl^{3+}(aq) + Ce^{3+}(aq)$$

(a) Ce^{4+} (b) Ag^+ (c) Ag^{2+} (d) Tl^+ (e) Tl^{2+}

26. An enzyme is

(a) a reactant molecule in a biological reaction.

(b) the active site on a biological catalyst.

(c) any molecule found in a living cell.

(d) a biological catalyst.

27. In an enzymatic reaction, the molecule that combines with the enzyme in order to react is called the

(a) activated complex   (b) active site   (c) poison

(d) substrate   (e) catalyst

Reaction mechanisms

28. What is the molecularity of the following elementary reaction?

$$CH_3 + NO \longrightarrow CH_3NO$$

(a) unimolecular   (b) bimolecular   (c) termolecular

29. What is the rate law for the elementary reaction shown in Exercise 28?

(a) rate = $k[CH_3]$ (b) rate = $k[CH_3][NO]^2$ (c) rate = $k[NO]$

(d) rate = $k[CH_3][NO]$ (e) need more information to tell

The following mechanism is for Exercises 30 and 31.

$$Cl_2 \longrightarrow 2Cl$$
$$Cl + CO \longrightarrow COCl$$
$$COCl + Cl \longrightarrow COCl_2$$

30. What is the overall reaction for the mechanism?

(a) $Cl_2 + 2CO \longrightarrow 2COCl$ (b) $2Cl + CO \longrightarrow COCl_2$

(c) $CO + Cl_2 \longrightarrow COCl_2$

31. What intermediates are present?

(a) $COCl_2$, CO (b) Cl, Cl_2 (c) Cl, $COCl$

(d) CO, Cl_2 (e) CO, Cl

32. The following mechanism has been proposed for the reaction between NO and Br_2 to form NOBr.

$$Br_2 + NO \rightleftharpoons NOBr_2 \quad \text{(fast equilibrium)}$$
$$NOBr_2 + NO \longrightarrow 2NOBr \quad \text{(slow)}$$

What is the rate law for this mechanism?

(a) rate $= k[NO]^2[Br_2]$ (b) rate $= k[NOBr_2][NO]$ (c) rate $= k[NO][Br_2]$

(d) rate $= k[Br_2][NOBr_2]$ (e) rate $= k$

Use the following mechanism for Exercises 33–35:

Step 1:	$Br_2 \longrightarrow 2Br$	
Step 2:	$Br + H_2 \longrightarrow HBr + H$	
Step 3:	$H + Br_2 \longrightarrow HBr + Br$	
Step 4:	$H + HBr \longrightarrow H_2 + Br$	
Step 5:	$Br + Br \longrightarrow Br_2$	

Overall reaction: $H_2(g) + Br_2(g) \longrightarrow 2HBr(g)$

33. What steps are chain-propagating steps?

(a) 2, 3   (b) 1   (c) 1, 2   (d) 2, 4   (e) 4

34. What step is a chain-retardation step?

(a) 1   (b) 2   (c) 3   (d) 4   (e) None exists.

35. What step is a chain-termination step?

(a) 3   (b) 2   (c) 5   (d) 1   (e) 4

36. With a chain reaction, the occurrence of an explosion often is due to

(a) chain inhibition.   (b) chain termination.   (c) chain retardation.

(d) chain branching.   (e) chain initiation.

37. Suppose that an elementary process and its reverse elementary process are given by the following reactions with $k = 3.44 \times 10^{-7}\,L \cdot mol^{-1} \cdot s^{-1}$ and $k' = 6.12 \times 10^{-9}\,L \cdot mol^{-1} \cdot s^{-1}$.

$$2A \longrightarrow B + C \quad \text{Rate} = k[A]^2$$
$$B + C \longrightarrow 2A \quad \text{Rate} = k'[B][C]$$

What is K for the equilibrium $2A \rightleftharpoons B + C$?

(a) 3.50×10^{-7} (b) 56.2   (c) 0.0178

(d) 2.10×10^{-15} (e) 4.75×10^{14}

Descriptive chemistry

38. Which compound in the upper atmosphere absorbs intense ultraviolet light from the Sun?

(a) CO_2 (b) O_2 (c) O_3 (d) N_2 (e) H_2O

39. What catalyst can be used in the following reaction?

$$H_2O_2(aq) \longrightarrow 2H_2O(l) + O_2(g)$$

(a) Pt(s) (b) $MnO_2(s)$ (c) Fe(s) (d) $V_2O_5(s)$ (e) $Br_2(l)$

40. One of the conversions (which are not shown as balanced reactions) is not an important one for catalytic converters. Which one?

(a) $NO_x \longrightarrow N_2$ (b) $CO \longrightarrow CO_2$

(c) $SO_2 \longrightarrow SO_3$ (d) $H_2O \longrightarrow H_2O_2$

41. What products are formed by the decomposition of N_2O_5 in the gas phase?

(a) NO, O_2 (b) N_2, O_2 (c) NO_2, N_2O_4

(d) NO_2, O_2 (e) N_2O, NO

42. What is the principal organic (carbon-containing) product of the following reaction?

$$CH_3OH(aq) + Br^-(aq) \longrightarrow ?$$

(a) CH_4 (b) CH_2Br_2 (c) CH_3Br (d) CH_2CH_2 (e) CO_2

CHAPTER 19

THE MAIN-GROUP ELEMENTS:

I THE FIRST FOUR FAMILIES

The chemical and physical properties of any element are related to its position in the periodic table. In this chapter we explore the properties of hydrogen and the elements of Groups 1, 2, 13, and 14. We shall see that each element has its own unique properties and that there are smooth trends as we go down each group. We shall also correlate and rationalize the properties of the elements with their electron configurations.

HYDROGEN

19.1 THE ELEMENT

KEY CONCEPT   The uses and manufacture of hydrogen

About 3×10^8 kg of hydrogen per year are used in a variety of industrial processes in the United States.

Uses	Chemical Equations
rocket fuel	$2H_2(l) + O_2(l) \longrightarrow 2H_2O(g)$
conversion to ammonia	$3H_2(g) + N_2(g) \longrightarrow 2NH_3(g)$
conversion to methanol	$2H_2(g) + CO(g) \longrightarrow CH_3OH(l)$
hydrometallurgical extraction	$H_2(g) + Cu^{2+}(aq) \longrightarrow Cu(s) + 2H^+$ (example)
hydrogenation	$H_2(g) + {-}C{=}C{-} \longrightarrow {-}CH{-}CH{-}$

The principal manufacturing method for hydrogen employs two reactions, the re-forming reaction followed by the shift reaction:

$$\text{Re-forming reaction:} \quad CH_4(g) + H_2O(g) \longrightarrow CO(g) + 3H_2(g)$$

$$\text{Shift reaction:} \quad CO(g) + H_2O(g) \longrightarrow CO_2(g) + H_2(g)$$

Two typical laboratory preparations are (a) reaction of an active metal with an acid and (b) reaction of a hydride with water:

$$\text{Active metal with acid:} \quad Zn(s) + 2HCl(aq) \longrightarrow ZnCl_2(aq) + H_2(g)$$

$$\text{Hydride with water:} \quad CaH_2(s) + 2H_2O(l) \longrightarrow Ca(OH)_2(aq) + 2H_2(g)$$

EXAMPLE   Calculating the mass of the production of H₂

How many kilograms of H_2 can be produced from 10.0 kg of CH_4 and excess H_2O?

SOLUTION This synthesis involves the re-forming reaction and the shift reaction in a standard calculation of the grams of product that result from a given amount of reactant. In this case, methane (CH_4) is the limiting reagent, so we first calculate the moles of methane with its molar mass of 16.04 g·mol^{-1}:

$$10.0 \text{ kg} \times \frac{1000 \text{ g}}{1 \text{ kg}} \times \frac{1 \text{ mol}}{16.04 \text{ g}} = 623 \text{ mol } CH_4$$

We now use the balanced equation to calculate the moles CO and moles H_2 produced in the re-forming reaction:

$$623 \text{ mol } CH_4 \times \frac{1 \text{ mol CO}}{1 \text{ mol } CH_4} = 623 \text{ mol CO}$$

$$623 \text{ mol } CH_4 \times \frac{3 \text{ mol } H_2}{1 \text{ mol } CH_4} = 1.87 \times 10^3 \text{ mol } H_2$$

The CO is converted to H_2 in the shift reaction, so we calculate the moles of H_2 produced:

$$623 \text{ mol CO} \times \frac{1 \text{ mol } H_2}{1 \text{ mol CO}} = 623 \text{ mol } H_2$$

The total of 2.49×10^3 mol H_2 produced (1.87×10^3 mol + 623 mol = 2.49×10^3 mol) is converted to kilograms to get the desired answer:

$$2.49 \times 10^3 \text{ mol } H_2 \times \frac{2.02 \text{ g}}{1 \text{ mol } H_2} \times \frac{1 \text{ kg}}{1000 \text{ g}} = 5.03 \text{ kg}$$

EXERCISE Which laboratory preparation of H_2 gives the higher yield of hydrogen per gram of the more expensive starting material (Zn, CaH_2)? Try to do this exercise without detailed calculations.

[*Answer:* Reaction of CaH_2 with H_2O]

19.2   COMPOUNDS OF HYDROGEN

KEY CONCEPT   The hydrides

We consider binary hydrides only, compounds in which hydrogen combines with only one other element.

Saline hydrides.   Hydrogen forms ionic hydrides (also known as saline hydrides) when it combines with a strongly electropositive metal. They are formed by heating the element with H_2. The hydrides all contain the hydride ion, H$^-$.

$$2K(s) + H_2(g) \longrightarrow 2KH(s) \qquad (K^+ \text{ and } H^- \text{ ions combined})$$

$$Ca(s) + H_2(g) \longrightarrow CaH_2(s) \qquad (Ca^{2+} \text{ and } H^- \text{ ions combined})$$

The ionic hydrides are powerful reducing agents. They lose electrons easily.

$$2H^-(aq) \longrightarrow 2e^- + H_2(g) \qquad E° = 2.25 \text{ V}$$

The large positive electrode potential indicates an extremely favorable process. H$^-$ cannot exist in water because its oxidation potential (+2.25 V), when added to the reduction potential of neutral water (−0.42 V), gives a positive cell potential:

$$\begin{array}{ll} 2H^-(aq) \longrightarrow 2e^- + H_2(g) & E° = 2.25 \text{ V} \\ \underline{2H_2O(l) + 2e^- \longrightarrow 2OH^-(aq) + H_2(g)} & \underline{E° = -0.42 \text{ V}} \\ 2H^-(aq) + 2H_2O(l) \longrightarrow 2OH^-(aq) + 2H_2(g) & E° = +1.83 \text{ V} \end{array}$$

From the Brønsted acid-base point of view, H$^-$ is a stronger base than OH$^-$, so H$^-$ forms OH$^-$ when it combines with water.

Molecular compounds of hydrogen. The common covalent hydrides are compounds in which hydrogen is combined with a nonmetal; examples are H_2O, NH_3, H_2S, C_2H_2, and PH_3. They are synthesized with a variety of techniques, as shown by the following examples.

Direct combination of elements: $H_2(g) + Cl_2(g) \xrightarrow{\text{light}} 2HCl(g)$

Protonation of Brønsted base S^{2-}: $S^{2-}(aq) + 2HCl(aq) \longrightarrow H_2S(g) + 2Cl^-(aq)$

C_2^{2-}: $C_2^{2-}(s) + 2H_2O(l) \longrightarrow C_2H_2(g) + 2OH^-(aq)$

Br^-: $KBr(aq) + H_3PO_4(aq) \longrightarrow HBr(g) + KH_2PO_4(aq)$

Deprotonation of Brønsted acid: $PH_4^+(aq) + OH^-(aq) \longrightarrow PH_3(g) + H_2O(l)$

The polymeric hydrides aluminum hydride and beryllium hydride also are covalent hydrides.

Interstitial hydrides. Heating certain *d*-block metals in the presence of hydrogen (H_2) results in the formation of hydrides in which the hydride ions occupy holes or gaps between the metal atoms in the structure.

$$2Cu(s) + H_2(g) \longrightarrow 2CuH(s) \quad \text{(an interstitial hydride)}$$

EXAMPLE   Synthesizing a covalent hydride

Suggest a method for the preparation of ammonia (NH_3) that uses protonation of a Brønsted base.

SOLUTION One approach to this problem is to decide which Brønsted base becomes ammonia when the base is protonated. Removal of a proton from NH_3 leaves NH_2^-, so protonation of NH_2^- results in formation of ammonia:

$$NH_2^-(aq) + H_2O(l) \longrightarrow NH_3(g) + OH^-(aq)$$

EXERCISE Suggest a method for the preparation of ammonia that uses deprotonation of a Brønsted acid.

[*Answer:* $NH_4^+(aq) + OH^-(aq) \longrightarrow NH_3(g) + H_2O(l)$]

GROUP 1: THE ALKALI METALS

19.3   THE ELEMENTS

KEY CONCEPT   The alkali metals are highly reactive

The alkali metals have relatively low ionization energies and easily form cations with a +1 charge; Li^+, Na^+, and K^+ are examples. Thus, they are very easily oxidized and are therefore never found in nature in their elemental form (with oxidation state 0). The elements can be formed by electrolysis of a molten salt; for instance, sodium is produced by the electrolysis of NaCl:

$$2NaCl(l) \xrightarrow{\text{electrolysis}} 2Na(s) + Cl_2(g)$$

They also can be made by reduction with a powerful reducing agent, such as sodium, which can be used to produce potassium:

$$KCl(l) + Na(g) \xrightarrow{750\,K} NaCl(s) + K(g)$$

The Group 1 elements are soft, silvery gray metals (cesium actually has a golden cast) that can be cut with a knife. Lithium is important in the manufacture of thermonuclear weapons (Chapter 22).

EXAMPLE   Understanding the alkali metals

Potassium is called a silvery gray metal in both the text and this study guide; but, in actual laboratories, it often appears to be dirty-white, with a textured (rather than smooth) surface. Offer a possible explanation for this apparent contradiction.

SOLUTION All of the Group 1 metals are highly reactive and are easily oxidized by the oxygen in the atmosphere, even at room temperature. The textured off-white color is due to surface oxidation, resulting in the formation, in potassium, of KO_2. When the superoxide coat is scraped off, the silvery gray potassium beneath is revealed.

EXERCISE Lithium is the hardest of the alkali metals. On the basis of the expected trends in group properties, state which is the softest.

[*Answer:* Francium]

19.4   CHEMICAL PROPERTIES OF THE ALKALI METALS

KEY CONCEPT   The alkali metals show periodic trends in their reactivity

The alkali metals do not all react with a particular substance in exactly the same manner. Periodic trends in the nature of the product formed and/or the speed of reaction occur. The size of the M^+ cation formed and the strength of the bonding that holds the metal atoms together are often responsible for observed differences in reactivity.

Reaction with water. All the alkali metals react with water, with the speed of the reaction increasing down the group.

$$2Li(s) + 2H_2O(l) \longrightarrow 2LiOH(aq) + H_2(g) \qquad \text{(slow)}$$

$$2Na(s) + 2H_2O(l) \longrightarrow 2NaOH(aq) + H_2(g) \qquad \text{(fast; } H_2 \text{ may ignite; dangerous)}$$

$$2K(s) + 2H_2O(l) \longrightarrow 2KOH(aq) + H_2(g) \qquad \text{(very fast; } H_2 \text{ ignites; dangerous)}$$

$$2Rb(s) + 2H_2O(l) \longrightarrow 2RbOH(aq) + H_2(g) \qquad \text{(dangerously explosive)}$$

$$2Cs(s) + 2H_2O(l) \longrightarrow 2CsOH(aq) + H_2(g) \qquad \text{(dangerously explosive)}$$

The reaction with lithium is anomalously slow, because the small lithium atom is bound tightly in the solid, and also because lithium, relative to the other Group 1 elements, has a high ionization potential and high sublimation energy.

Reaction with nonmetals. All the alkali metals react directly with most of the nonmetals (other than the noble gases). Lithium, in anomalous fashion, is the only alkali metal that burns in nitrogen. The main reactions for burning in air are

Li: $\quad 6Li(s) + N_2(g) \longrightarrow 2Li_3N(s) \quad$ (lithium nitride formed)

$\qquad 4Li(s) + O_2(g) \longrightarrow 2Li_2O(s) \quad$ (lithium oxide formed)

Na: $\quad 2Na(s) + O_2(g) \longrightarrow Na_2O_2(s) \quad$ (sodium peroxide formed)

K: $\qquad K(s) + O_2(g) \longrightarrow KO_2(s) \quad$ (potassium superoxide formed)

Rb: $\qquad Rb(s) + O_2(g) \longrightarrow RbO_2(s) \quad$ (rubidium superoxide formed)

Cs: $\qquad Cs(s) + O_2(g) \longrightarrow CsO_2(s) \quad$ (cesium superoxide formed)

Group 1 superoxides can be used to purify air by reacting with the unwanted CO_2 and producing the desired O_2. For instance,

Potassium superoxide: $\qquad 4KO_2(s) + 2CO_2(g) \longrightarrow 2K_2CO_3(s) + 3O_2(g)$

Sodium superoxide: $\qquad 4NaO_2(s) + 2CO_2(g) \longrightarrow 2Na_2CO_3(s) + 3O_2(g)$

EXAMPLE   Understanding cohesive forces in the alkali metals

Based on the speed of reaction with water, which metal is held together by stronger cohesive forces, Na or K?

SOLUTION Lithium reacts relatively slowly with water because the lithium atoms are held very tightly together in the metal, so removing them from the metal takes a lot of energy. This strong metallic bonding results in a high activation energy (and slow speed) for the reaction. Because sodium reacts more slowly than potassium with water, we conclude that the reaction with sodium has a higher activation energy because the sodium atoms are held more tightly in the solid than potassium atoms are. In other words, sodium is held together by stronger cohesive forces.

EXERCISE What is the oxidation state of oxygen in the oxide ion, the peroxide ion, and the superoxide ion?

[*Answer:* Oxide, -2; peroxide, -1; superoxide, $-\frac{1}{2}$]

19.5   COMPOUNDS OF LITHIUM, SODIUM, AND POTASSIUM

KEY CONCEPT   Alkali metal compounds are commercially important

Lithium compounds.   The small size and high polarizing power of the Li^+ cation results in a notable tendency toward covalent bonding for lithium. It also results in a strong ion-dipole interaction, so lithium salts are often hydrated.

Sodium chloride, NaCl (table salt).   Sodium chloride is obtained by mining or evaporation of brine (seawater). NaCl is used for a vast number of industrial processes such as the production of NaOH, Na_2SO_4, and Na_2CO_3. It is also a required nutritional mineral and is used to season food. For a variety of reasons, Na^+ ions are found in seawater at a much higher concentration than K^+ ions are, even though the two ions have similar abundances on Earth.

Sodium hydroxide, NaOH (caustic soda).   Sodium hydroxide is produced in large quantities by the electrolysis of aqueous NaCl. It is used, among other things, as an inexpensive starting material for the production of other sodium salts.

$$2NaCl(aq) + 2H_2O(l) \xrightarrow{\text{electrolysis}} 2NaOH(aq) + Cl_2(g) + H_2(g)$$

Sodium sulfate, Na_2SO_4 (salt cake); $Na_2SO_4 \cdot 10H_2O$ (Glauber's salt).   Sodium sulfate is mined or produced by reaction of NaCl with sulfuric acid. Na_2SO_4 is used in papermaking and as a substitute for phosphates in detergents.

$$H_2SO_4(aq) + 2NaCl(s) \longrightarrow Na_2SO_4(aq) + 2HCl(g)$$

Sodium carbonate, Na_2CO_3 (soda ash); $Na_2CO_3 \cdot 10H_2O$ (washing soda).   Sodium carbonate is mined or produced by the Solvay process, which uses the overall reaction

$$2NaCl(aq) + CaCO_3(s) \longrightarrow Na_2CO_3(s) + CaCl_2(s)$$

It is used in large amounts in glassmaking as a source of sodium oxide, which is produced in a Lewis acid-base decomposition:

$$Na_2CO_3(s) \longrightarrow Na_2O(s) + CO_2(g)$$

$$CO_3^{2-} \longrightarrow \underset{\text{Lewis base}}{O^{2-}} + \underset{\text{Lewis acid}}{CO_2}$$

Washing soda, which is quite soluble, was formerly used to precipitate Mg^{2+} and Ca^{2+} from hard water:

$$Ca^{2+}(aq) + CO_3^{2-}(aq) \longrightarrow CaCO_3(s)$$

$$Mg^{2+}(aq) + CO_3^{2-}(aq) \longrightarrow MgCO_3(s)$$

It also provides a basic environment for aqueous solutions:

$$CO_3^{2-}(aq) + H_2O(l) \longrightarrow HCO_3^-(aq) + OH^-(aq)$$

Potassium compounds. Potassium is obtained (in the form of K^+ ions) principally through mining of KCl and $KCl \cdot MgCl_2 \cdot 6H_2O$ (a mixed salt). Potassium nitrate (KNO_3) is obtained by mixing KCl with $NaNO_3$ in an aqueous solution. Potassium nitrate is more soluble than sodium chloride; so NaCl can be crystallized out of a solution containing Na^+, K^+, Cl^-, and NO_3^- ions, and potassium nitrate then obtained by evaporation of the remaining solution. Potassium nitrate releases O_2 when heated:

$$2KNO_3(s) \longrightarrow 2KNO_2(s) + O_2(g)$$

Thus, KNO_3 is a good oxidizing agent (when heated). It is the oxidizing agent in gunpowder, which uses the two reactions:

$$2KNO_3(s) + 4C(s) \longrightarrow K_2CO_3(s) + 3CO(g) + N_2(g)$$

$$2KNO_3(s) + 2S(s) \longrightarrow K_2SO_4(s) + SO_2(g) + N_2(g)$$

The rapid formation of six moles of gas from a solid results in a rapid increase in volume, which we call an *explosion.*

EXAMPLE   Understanding the role of reagents from Group I elements

What chemical role does NaCl play (a) in the production of NaOH and (b) in the production of Na_2SO_4?

SOLUTION (a) In the manufacture of NaOH, NaCl makes the water an electrolyte and provides the Na^+ ion in the NaOH. (The hydroxide ion originates with the water and is produced by the reduction of H_2O.)

$$2NaCl(aq) + 2H_2O(l) \xrightarrow{\text{electrolysis}} 2NaOH(aq) + Cl_2(g) + H_2(g)$$

(b) In the manufacture of Na_2SO_4, the chloride ion in NaCl acts as a Brønsted base that removes protons from sulfuric acid. It also provides the Na^+ ion for the product.

$$H_2SO_4(aq) + 2NaCl(s) \longrightarrow Na_2SO_4(aq) + 2HCl(g)$$

EXERCISE In the following reaction, which occurs when black gunpowder explodes, state what is oxidized, what is reduced, what the oxidizing agent is, and what the reducing agent is.

$$2KNO_3(s) + 4C(s) \longrightarrow K_2CO_3(s) + 3CO(g) + N_2(g)$$

[*Answer:* C, oxidized; N, reduced; KNO_3, oxidizing agent; C(s), reducing agent]

GROUP 2: THE ALKALINE EARTH METALS

19.6   THE ELEMENTS

KEY CONCEPT   The production, properties, and uses of the elements

All the alkaline earths (Group 2 elements) are too reactive to be found in ores as pure metals with oxidation state 0. The elements calcium, strontium, and barium are called the *alkaline earth metals.*

Production. Beryllium occurs in ores mainly as beryl ($3BeO \cdot Al_2O_3 \cdot 6SiO_2$) and is obtained by converting beryl to the chloride and electrolyzing the chloride.

$$BeCl_2(l) \xrightarrow{\text{electrolysis}} Be(s) + Cl_2(g)$$

Magnesium occurs as the mixed carbonate dolomite ($CaCO_3 \cdot MgCO_3$) and is obtained either by electrolytic reduction or chemical reduction. The electrolytic reduction starts with seawater (which contains Mg^{2+}) and uses the following steps:

$$\text{Precipitate } Mg(OH)_2: \qquad Mg^{2+}(aq) + 2OH^-(aq) \longrightarrow Mg(OH)_2(s) \qquad (1)$$

$$\text{Synthesize } MgCl_2: \qquad Mg(OH)_2(s) + 2HCl(aq) \longrightarrow MgCl_2(s) + 2H_2O(l) \qquad (2)$$

$$\text{Electrolyze to Mg:} \qquad MgCl_2(l) \xrightarrow{\text{electrolysis}} Mg(s) + Cl_2(g) \qquad (3)$$

Calcium, strontium, and barium are found mostly in the sea as dolomite ($CaCO_3 \cdot MgCO_3$) and in limestone ($CaCO_3$), strontianite ($SrSO_4$), and barite ($BaSO_4$). All three are obtained by either electrolytic or chemical reduction. Two examples are

$$\text{Electrolytic reduction for Ca:} \qquad CaCl_2(l) \xrightarrow{\text{electrolysis}} Ca(l) + Cl_2(g)$$

$$\text{Chemical reduction for Ba:} \qquad 3BaO(s) + 2Al(s) \longrightarrow Al_2O_3(s) + 3Ba(s)$$

Properties. The ionization of two electrons from the alkaline earth ns^2 configuration (leaving an inert-gas core) requires relatively little energy, but the lattice enthalpies of the Group 2 ionic solids are rather large. Thus, the elements (with the exception of beryllium) generally exist in their compounds as ions with +2 charge. They always have +2 oxidation number in their compounds. Because of its high polarizing power, beryllium has a tendency toward covalency; it also has a diagonal relationship with aluminum. Thus, Be_2O_3 is amphoteric. Beryllium is extremely toxic, both as a metal and in its compounds.

All the Group 2 cations have strongly negative reduction potentials. The metals therefore have large positive oxidation potentials, so they are oxidized easily and are powerful reducing agents. All except beryllium reduce water:

$$Be(s) + 2H_2O(l) \longrightarrow \text{no reaction}$$

$$Mg(s) + 2H_2O(l) \longrightarrow Mg(OH)_2(aq) + H_2(g) \qquad \text{(reacts with hot water)}$$

$$Ca(s) + 2H_2O(l) \longrightarrow Ca(OH)_2(aq) + H_2(g) \qquad \text{(reacts with cold water)}$$

$$Sr(s) + 2H_2O(l) \longrightarrow Sr(OH)_2(aq) + H_2(g) \qquad \text{(reacts with cold water)}$$

$$Ba(s) + 2H_2O(l) \longrightarrow Ba(OH)_2(aq) + H_2(g) \qquad \text{(reacts with cold water)}$$

Even though the reaction of Be and Mg with water has a high positive standard potential E(cell), both metals are passivated by an oxide coat, so the reaction does not occur with beryllium at all and occurs with magnesium only if hot water is used. This difference indicates that Be(s) is more strongly passivated than Mg(s).

Uses. Some of the uses of the Group 2 elements are shown in the following table.

Element	Uses
Be	construction of missiles, satellites, x-ray tube windows, nonsparking tools
Mg	construction of airplanes; in fireworks and incendiary devices
Ca	as a getter in the production of steel (to remove unwanted oxides)
Ba	as a getter in vacuum tube construction (to remove O_2)

EXAMPLE Predicting the properties of Group 2 compounds

Suggest a reason, based on the lattice enthalpies of their respective oxides, why Be(s) is more strongly passivated than Mg(s).

SOLUTION The lattice enthalpy of BeO is 4293 kJ·mol^{-1}; that of MgO is 3889 kJ·mol^{-1}. This difference is largely due to the smaller size of the Be^{2+} ion relative to the Mg^{2+} ion. The higher lattice enthalpy means that it is more difficult for water to chemically break through the passivating BeO coat on Be than through the MgO coat on Mg.

19.7 COMPOUNDS OF BERYLLIUM, MAGNESIUM, AND CALCIUM

KEY CONCEPT   Group 2 compounds have many uses

Beryllium compounds.   The properties of beryllium compounds are mostly the result of the small size and high polarizing power of the Be^{2+} cation. The small size limits the coordination number of the cation to 4, so tetrahedral BeX$_4$ units occur in many beryllium compounds. BeH$_2$ and BeCl$_2$ are polymeric, with neighboring Be atoms held together by bonding molecular orbitals in a Be—H—Be and Be—Cl—Be group, respectively. The chloride is produced by reaction 1, which follows; the hydrated Be^{2+} ion is acidic (reaction 2).

$$C(s) + BeO(s) + Cl_2(g) \longrightarrow BeCl_2(g) + CO(g) \tag{1}$$

$$[Be(H_2O)_4]^{2+}(aq) + H_2O(l) \longrightarrow [Be(H_2O)_3OH]^+(aq) + H_3O^+(aq) \tag{2}$$

Magnesium oxide, MgO.   Burning magnesium in air yields both the nitride and the oxide:

$$2Mg(s) + O_2(g) \longrightarrow 2MgO(s)$$

$$3Mg(s) + N_2(g) \longrightarrow Mg_3N_2(s)$$

Pure oxide is formed by heating and decomposition of the carbonate. The oxide, with its small Mg^{2+} and O^{2-} ions, has a very high lattice enthalpy. It is extremely stable and melts at the very high temperature of 2800°C.

$$MgCO_3(s) \longrightarrow MgO(s) + CO_2(g)$$

Magnesium hydroxide, Mg(OH)$_2$ (milk of magnesia).   Magnesium hydroxide is not very soluble and is a mild base. It is used as an antacid to neutralize stomach acid (HCl). The MgCl$_2$ formed acts as a purgative.

$$Mg(OH)_2(s) + 2HCl(aq) \longrightarrow MgCl_2(aq) + 2H_2O(l)$$

Chlorophyll.   Chlorophyll contains magnesium in the form of +2 ions. In chlorophyll, four nitrogen atoms on a single large organic part of the molecule act as Lewis bases and bond to the Mg^{2+}, which acts as a Lewis acid. Chlorophyll is one of the key compounds in the photosynthetic process in green plants. It captures light from the Sun and funnels it (as free energy) into the photosynthesis reaction:

$$6CO_2(g) + 6H_2O(l) \longrightarrow C_6H_{12}O_6(aq) + 6O_2(g)$$

Calcium carbonate, CaCO$_3$ (chalk, limestone, marble, aragonite, and calcite are all forms of CaCO$_3$).   Like most carbonates, CaCO$_3$ decomposes to carbon dioxide and a metal oxide when heated. The oxide formed in this case is CaO, or quicklime.

$$CaCO_3 \xrightarrow{\Delta} CO_2(g) + CaO(s)$$

Quicklime is used in metallurgy, in the manufacture of iron, to remove silica (SiO$_2$) impurities from the ore.

$$CaO(s) + SiO_2(s) \longrightarrow CaSiO_3(l)$$

Structural calcium.   Calcium compounds are rigid because of the large electrostatic attraction between the small Ca^{2+} ion and anions. This rigidity makes calcium compounds useful as structural components. For example, common mortar contains $Ca(OH)_2$, which sets to a hard mass through the reaction with atmospheric CO_2:

$$Ca(OH)_2(s) + CO_2(g) \longrightarrow CaCO_3(s) + H_2O(l)$$

Tooth enamel is hydroxyapatite, a calcium-containing mineral $\{Ca_5(PO_4)_3OH\}$. Tooth decay occurs when certain bacteria that thrive in the mouth produce acids that dissolve the enamel:

$$Ca_5(PO_4)_3OH(s) + 4H_3O^+(aq) \longrightarrow 5Ca^{2+}(aq) + 3HPO_4^{2-}(aq) + 5H_2O(l)$$

Fluorapatite $\{Ca_5(PO_4)_3F\}$, in which a fluoride ion replaces the hydroxide ion in hydroxyapatite, is more resistant to this type of attack:

$$Ca_5(PO_4)_3OH(s) + F^-(aq) \longrightarrow Ca_5(PO_4)_3F(s) + OH^-(aq)$$

EXAMPLE Accounting for the properties of Group 2 compounds

Use a thermodynamic argument to suggest a reason why $CaCO_3$ must be heated to make it decompose to CaO and CO_2.

SOLUTION The reaction is $CaCO_3(s) \longrightarrow CaO(s) + CO_2(g)$. We shall consider the factors that contribute to ΔG ($= \Delta H - T\Delta S$) for the reaction. Because the net process in the decomposition is the breaking of a C—O bond, ΔH must be positive. The formation of a gas from a solid results in a large positive ΔS. Thus, we have

$$\Delta H > 0 \quad \text{and} \quad \Delta S > 0$$

A reaction with positive ΔH and positive ΔS is spontaneous only at relatively high temperatures, so $CaCO_3$ must be heated for it to decompose.

EXERCISE Suggest two reasons why fluorapatite is more resistant to acid attack than hydroxyapatite.

[*Answer:* (1) OH^- is a stronger base than F^-, so OH^- reacts more readily with acid; (2) F^- is smaller than OH^-, so F^- experiences stronger interactions with its cationic neighbors and holds the enamel together in a tighter, stronger structure.]

KEY WORDS   Define or explain each term in a written sentence or two.

binary compounds	metallic hydrides	saline hydrides
hydrides	protonation	shift reaction
hydrometallurgical extraction	re-forming reaction	water-splitting reaction
metal-ammonia solution	refractory	

GROUP 13: THE BORON FAMILY

19.8   THE ELEMENTS

KEY CONCEPT   The properties of aluminum and boron

Boron.   Boron is a nonmetal. In almost all of its compounds, it has a +3 oxidation state. It is mined as the minerals borax ($Na_2B_4O_7 \cdot 10H_2O$) and kernite ($Na_2B_4O_7 \cdot 7H_2O$). Extraction of elemental boron involves conversion to boron oxide, followed by reduction with magnesium:

$$B_2O_3(s) + 3Mg(s) \longrightarrow 2B(s) + 3MgO(s)$$

Elemental boron has several allotropes. It is used to harden steels; it is also used to produce fibers to strengthen plastics. Chemically, boron is quite inert.

Aluminum. Aluminum is an amphoteric metal. It reacts with both bases and nonoxidizing acids:

$$2Al(s) + 6H_3O^+(aq) \longrightarrow 2Al^{3+}(aq) + 3H_2(g) + 6H_2O(l)$$

$$2Al(s) + 2OH^-(aq) + 6H_2O(l) \longrightarrow 2[Al(OH)_4]^-(aq) + 3H_2(g)$$

Oxidizing acids lead to the formation of a passivating oxide film on aluminum surfaces that prevents further reaction. Aluminum has a +3 oxidation state in almost all of its compounds. It is mined as bauxite, an impure ore containing Al_2O_3; the aluminum is extracted with the electrolytic **Hall process:**

$$2Al_2O_3(s) + 3C(s) \xrightarrow{\text{electrolysis}} 4Al(s) + 3CO_2(g)$$

The element is a strong, light metal and an excellent conductor. Its low density, wide availability, and resistance to corrosion (because of a passivating oxide film) makes it widely used in structural applications where low density and high strength are required.

EXAMPLE Calculating aluminum and boron extraction

Assume an ore is found that is 8.0% boron. How many grams of Mg are required to extract all the boron from 2.4 kg of the ore?

SOLUTION The reaction is $B_2O_3(s) + 3Mg(s) \rightarrow 2B(s) + 3MgO(s)$. We first calculate the amount of boron that is expected from the percentage in the ore. Because 8.0% translates to a fractional amount of boron in the ore of 0.080 g of boron to 1 g of ore, we get

$$0.080 \frac{\text{g B}}{\text{g ore}} \times (2.4 \times 10^3 \text{ g ore}) = 1.9 \times 10^2 \text{ g B}$$

We now use the coefficients in the chemical equation to calculate the amount of magnesium required to produce 1.9×10^2 g B:

$$1.9 \times 10^2 \text{ g B} \times \frac{1 \text{ mol B}}{10.81 \text{ g B}} \times \frac{3 \text{ mol Mg}}{2 \text{ mol B}} \times \frac{24.31 \text{ g Mg}}{1 \text{ mol Mg}} = 6.4 \times 10^2 \text{ g Mg}$$

EXERCISE How many kilograms of aluminum can be produced electrolytically by a 1.0×10^4 A current running for 24 h? 1 F = 96,485 C.

[*Answer:* 81 kg]

19.9   GROUP 13 OXIDES

KEY CONCEPT   The properties of aluminum and boron

Boron is a nonmetal and, like most nonmetals, has acidic oxides. When B_2O_3 is hydrated, it forms boric acid [H_3BO_3 or $B(OH)_3$]. The boron in boric acid has an incomplete octet, so it can act as a Lewis acid. Boric acid is a mild antiseptic and pesticide; it is also used as a fire retardant in home insulation and clothes. Its main use, however, is as a source of boron oxide; when H_3BO_3 is heated, water is driven off and the anhydride (B_2O_3) is formed:

$$2H_3BO_3(s) \xrightarrow{\Delta} B_2O_3(s) + 3H_2O(l)$$

Boron oxide is used as a flux for soldering or welding and is used in the manufacture of fiberglass and borosilicate glass.

Aluminum oxide (Al_2O_3) is known as **alumina**. Alumina exists in a variety of different crystal structures, and each structure finds uses in science and commerce. One of the less dense forms (γ-alumina) is used as the stationary phase in a specific kind of chromatography. γ-alumina, which is produced by heating aluminum hydroxide, is itself **amphoteric** (that is, it shows both basic and acidic character by reacting with both acids and bases):

$$Al_2O_3(s) + 2OH^-(aq) + 3H_2O(l) \longrightarrow 2Al(OH)_4^-(aq)$$

$$Al_2O_3(s) + 6H_3O^+(aq) + 3H_2O(l) \longrightarrow 2[Al(H_2O)_6]^{3+}(aq)$$

An important aluminum salt is aluminum sulfate, which is prepared by reacting aluminum oxide with sulfuric acid.

$$Al_2O_3(s) + 3H_2SO_4(aq) \longrightarrow Al_2(SO_4)_3(aq) + 3H_2O(l)$$

Aluminum sulfate (papermaker's alum) is used in the preparation of paper. True alums are hydrated sulfates containing two metallic cations (a +1 cation and Al^{3+}); two examples are $KAl(SO_4)_2 \cdot 12H_2O$ and $NH_4Al(SO_4)_2 \cdot 12H_2O$. Sodium aluminate, $NaAl(OH)_4$, is used in water purification (see Case Study 3 in the text).

> **EXAMPLE   Understanding the oxides of boron and aluminum**
>
> Would it be possible for an alum-like compound to have the formula $CaAl(SO_4)_2 \cdot 12H_2O$? If not, explain why.
>
> **SOLUTION** This formula is not possible for an actual compound because the total charge on the compound is not 0, as it must be. The calcium ion (Ca^{2+}) and aluminum ion (Al^{3+}) contribute a total charge of +5 to the formula of the compound; the two sulfate ions (SO_4^{2-}) contribute −4. Thus, the total charge on the proposed formula is +1, which is not permitted for an actual compound.
>
> **EXERCISE** What is the oxidation state of boron in B_2O_3?
>
> [*Answer:* +3]

19.10   CARBIDES, NITRIDES, AND HALIDES

KEY CONCEPT   Aluminum and boron compounds have unusual properties

The properties of the carbides, nitrides, and halides of aluminum and boron are determined, as we might expect, by the underlying structure of each compound.

Boron carbide, $B_{12}C_3$.   When boron is heated strongly with graphite, it forms boron carbide:

$$2B(s) + 3C(s) \longrightarrow B_{12}C_3(s)$$

Boron nitride, BN.   When boron is heated strongly with ammonia, boron nitride is formed:

$$2B(s) + 2NH_3(g) \longrightarrow 2BN(s) + 3H_2(g)$$

Boron halides.   The boron halides are made either by direct reaction of the elements or from the oxide. Two reactions that use the oxide are

$$B_2O_3(s) + 3CaF_2(s) + 3H_2SO_4(l) \longrightarrow 2BF_3(g) + 3CaSO_4(s) + 3H_2O(l)$$

$$B_2O_3(s) + 3C(s) + 3Cl_2(g) \longrightarrow 2BCl_3(g) + 3CO(g)$$

All the halides are electron deficient, with an incomplete octet on boron. The molecules are planar triangular and, because of the incomplete octet, are strong Lewis acids.

Aluminum trichloride, $AlCl_3$.   Aluminum trichloride is formed as an ionic compound by the reaction of chlorine with aluminum or of alumina with chlorine in the presence of carbon:

$$2Al(s) + 3Cl_2(g) \longrightarrow 2AlCl_3(s)$$

$$Al_2O_3(s) + 3Cl_2(g) + 3C(s) \longrightarrow 2AlCl_3(s) + 3CO(g)$$

Heating the ionic $AlCl_3$ to its melting point of 192°C results in a rearrangement to form a molecular liquid of Al_2Cl_6 molecules. Molecular Al_2Cl_6 consists of units in which an unshared pair of electrons on chlorine (a Lewis base) is donated to a vacant orbital on an adjacent aluminum atom (a Lewis acid). Aluminum halides react with water in a highly exothermic reaction. The white solid lithium aluminum hydride, an important reducing agent in organic chemistry, is prepared from aluminum chloride:

$$4LiH + AlCl_3 \longrightarrow LiAlH_4 + 3LiCl$$

EXAMPLE   Understanding the physical properties of boron and aluminum compounds

Explain why the boiling points of the boron halides indicate that London forces are dominant in the intermolecular attractions of these compounds. Why are London forces the dominant forces? The boiling points are

$$BF_3: \ -101°C \qquad BCl_3: \ -107°C \qquad BBr_3: \ 91.3°C \qquad BI_3: \ 210°C$$

SOLUTION Overall, the boiling points increase from BF_3 to BI_3 (even though there is a slight reversal of the trend at BCl_3). Thus, as the compounds get heavier, the boiling points increase. This behavior is typical of situations in which London forces dominate because the strength of London forces increases as the number of electrons increases and a compound becomes more polarizable. The only other forces that might affect the boiling points are dipole-dipole interactions; however, the boron halides are triangular planar molecules, and the individual bond dipoles cancel to give a net dipole moment of 0, so no dipole-dipole attractions exist in these compounds.

EXERCISE Explain why the ionic $AlCl_3$ expands greatly when it changes to molecular Al_2Cl_6.

[*Answer:* The strong ionic bonds in $AlCl_3$ hold the ions together more effectively than the weak London forces hold Al_2Cl_6 molecules together.]

19.11   BORANES AND BOROHYDRIDES

KEY CONCEPT   The properties of boranes and borohydrides

Boranes.   Boranes are compounds of boron with hydrogen, such as B_2H_6 and $B_{10}H_{14}$. Diborane (B_2H_6), which is extremely reactive, is produced by the reaction of sodium borohydride with boron trifluoride. Boranes are electron-deficient compounds with three-center, two-electron B—H—B bonds.

$$4BF_3 + 3BH_4^- \longrightarrow 3BF_4^- + 2B_2H_6(g)$$

Borohydrides, compounds containing BH_4^-.   Sodium borohydride, a common borohydride, is produced by the reaction of sodium hydride on boron trichloride. Borohydride ion is an effective and important reducing agent.

$$4NaH + BCl_3 \longrightarrow NaBH_4 + 3NaCl$$

EXAMPLE   Understanding the oxides of boron and aluminum

The length of the B—H single bond in diborane (see structure 5 in the text on page 746) is 119 pm. Which of the three choices for the three-center B—H bond distance, 102 pm, 119 pm, or 133 pm, is most reasonable?

SOLUTION In the "normal" B—H bond with length 119 pm, one pair of electrons holds two atoms together. In the three-center bond, one pair of delocalized electrons holds three atoms together (not two), and we might expect that the single pair of electrons cannot as effectively bond three atoms as two. Thus, it is most reasonable to assume that the bond length is greater for the three-center bond. 133 pm is the correct answer.

EXERCISE Can sodium borohydride, in principle, be used to reduce Zn^{2+} to Zn? The reduction potential of the Zn^{2+}/Zn couple is -0.76 V.

[*Answer:* Yes]

GROUP 14: THE CARBON FAMILY

19.12 THE ELEMENTS

KEY CONCEPT A large change in metallic character occurs in Group 14

All Group 14 elements have an s^2p^2 valence electron configuration. Carbon and silicon at the top of Group 14 are nonmetals, whereas tin and lead at the bottom are metals. Both tin and lead, however, show amphoteric behavior, indicating that some nonmetal character is present. For instance, tin reacts with both acids and alkalies:

$$Sn(s) + 2HCl(conc, aq) \longrightarrow SnCl_2(aq) + H_2(g)$$

$$Sn(s) + 2OH^-(aq) + 4H_2O(l) \longrightarrow [Sn(OH)_6]^{2-}(aq) + 2H_2(g)$$

Carbon, at the head of Group 14, is unique in the ease with which it forms multiple bonds. Silicon, because of its large size and available empty d-orbitals can act as a Lewis acid by expanding its valence shell; carbon cannot do this. At the head of Group 14, carbon almost always has an oxidation number of $+4$. As we go down the group, the separation in energy of the s- and p- orbitals results in the inert-pair effect (Section 7.18), and the $+2$ oxidation state becomes more stable.

EXAMPLE Predicting the properties of group 14 elements

Based on group trends, what are the stable binary hydrides of carbon (CH_x) and lead (PbH_x)?

SOLUTION Hydrides contain hydrogen in a -1 oxidation state. Because carbon, at the top of Group 14, commonly has a $+4$ oxidation state, its hydride will be CH_4; this is methane, commonly known as natural gas. At the bottom of group 14, the inert pair effect comes into play and the $+2$ oxidation state is the more stable state. The stable hydride of lead is PbH_2.

EXERCISE Would you expect tin or lead to be more metalloidlike in its behavior? (Use a periodic table, but do not look the answer up).

[*Answer:* Tin will be more like a metalloid than lead.]

19.13 THE MANY FACES OF CARBON

KEY CONCEPT Carbon has three allotropic forms

There are three allotropic forms of carbon: graphite, diamond, and the fullerenes. Graphite is the thermodynamically most stable allotropic form.

Graphite. Graphite and activated charcoal are different crystalline forms of graphite; soot and carbon black contain graphite. The carbon in graphite is sp^2 hybridized; structurally, the graphite consists of large sheets of interconnected benzenelike hexagonal units. The delocalized electrons in the sheets give graphite its electrical conductivity. The fact that one large sheet can slip past another when certain impurities are present makes graphite an excellent dry lubricant. Soot and carbon black have commercial uses in rubber and inks. Activated charcoal is an important and versatile purifier; it is used in laboratories, home water filters, and catalytic converters on automobiles to remove unwanted compounds. It does so by adsorption (not absorption) of the unwanted compound onto the microcrystal surface. Graphite is insoluble in liquid solvents.

Diamond.   Diamond is the hardest material known and the best known conductor of heat. Its hardness makes it an excellent abrasive, and synthetic diamonds find wide use as abrasives. The carbon in diamond is sp^3 hybridized, and each carbon atom in a diamond crystal is directly bonded to four other carbon atoms and indirectly interconnected to all of the other carbon atoms in the crystal through C—C single bonds. Diamond is insoluble in liquid solvents.

Fullerenes.   Buckminsterfullerene (C_{60}), a soccer-ball-like molecule (structure 7, page 749 of the text), is the archetype of a family of molecules, containing from 44 to 84 carbon atoms, called fullerenes. You should note that some of the carbon atoms form pentagons and some hexagons in the C_{60} molecules; the fullerenes all have an even number of carbon atoms. The inside of the fullerene molecules are hollow, and chemists are working to discover how to trap different atoms inside the hollow sphere to make useful entities. The fullerenes are soluble in organic solvents.

EXAMPLE   Understanding the allotropic forms of carbon

Using the information provided in this chapter, make a qualitative estimate of the free energy of formation of C_{60}. You cannot do this quantitatively; some intuition is required. The correct answer is not yet available in standard texts.

SOLUTION The free energy of formation of graphite is 0 kJ·mol^{-1} and that of diamond is 2.900 kJ·mol^{-1}. The bonding in C_{60} is closer to that of graphite than to that of diamond and much more extreme conditions are required for the production of diamond than buckminsterfullerene. Both of these lead to the guess that C_{60} will have a free energy closer to that of graphite than of diamond; in addition, graphite is more thermodynamically stable than buckminsterfullerene, so the free energy of formation of buckminsterfullerene should be positive. A guess of about 1 kJ·mol^{-1} seems reasonable.

EXERCISE Use the sketch of C_{60} in your text to determine how many carbon atoms each carbon in buckminsterfullerene is bonded to. Remember there are both hexagonal and pentagonal structures in C_{60}.

[*Answer:* Each carbon is bonded to three others.]

19.14   SILICON, TIN, AND LEAD

KEY CONCEPT   The properties of silicon, tin, and lead

Silicon.   Silicon occurs naturally as silicates (compounds containing SiO_4^{4-}) and as silicon dioxide. Quartz, quartzite, and sand are forms of SiO_2. Pure silicon, which finds wide use in the semiconductor industry, is obtained from quartzite in a three-step process:

$$SiO_2(s) + 2C(s) \longrightarrow Si(s) + 2CO(g) \quad \text{(impure Si formed)}$$

$$Si(s) + 2Cl_2(g) \longrightarrow SiCl_4(l)$$

$$SiCl_4(l) + 2H_2(g) \longrightarrow Si(s) + 4HCl(g) \quad \text{(pure Si formed)}$$

Tin.   Tin is obtained from its ore cassiterite (SnO_2) through reduction with carbon in a process that must be done carefully to avoid contamination with iron. Tin is used mainly for plating.

$$SnO_2(s) + C(s) \longrightarrow Sn(l) + CO_2(g)$$

Lead.   Lead is obtained from its ore galena (PbS) and is very dense and chemically inert. Its high density makes it useful as a radiation shield. Lead forms passivating oxide, chloride, and sulfate films. Because lead is very toxic, its use in plumbing may result in some lead contamination (as Pb^{2+}) of domestic water supplies.

EXAMPLE   Understanding the reactions of the Group 14 elements

Why does the iron in a tin-plated iron can corrode in preference to the tin coat?

SOLUTION Corrosion is an oxidation process. The oxidation of iron is thermodynamically favored over the oxidation of tin:

$$Fe(s) \longrightarrow Fe^{2+}(aq) + 2e^- \qquad E° = +0.44 \text{ V}$$

$$Sn(s) \longrightarrow Sn^{2+}(aq) + 2e^- \qquad E° = +0.14 \text{ V}$$

Thus, if moisture gains access to the iron and permits electron conduction to a cathode reaction, the iron will be oxidized first.

EXERCISE How many liters of H_2 (measured at STP) are produced when 10.0 g Sn is placed in 10.00 mL of hot concentrated HCl (which is approximately 12 M HCl).

[*Answer:* 1.3 L]

19.15   OXIDES OF CARBON

KEY CONCEPT CO_2 and CO are the important oxides of carbon

Carbon dioxide, CO_2.   Carbon dioxide has been discussed earlier. By way of review, we note that it is an acid anhydride, as is expected of a nonmetal oxide; it forms carbonic acid when it is hydrated.

$$CO_2(g) + H_2O(l) \longrightarrow H_2CO_3(aq)$$

Carbon monoxide, CO.   Carbon monoxide is produced when an organic compound or carbon itself is burned in a limited supply of oxygen. Commercially, it is produced as **synthesis gas** (reaction 1). It is produced in the laboratory by dehydrating formic acid (reaction 2).

$$CH_4(g) + H_2O(g) \longrightarrow CO(g) + 3H_2(g) \qquad (1)$$

$$HCOOH(aq) \longrightarrow CO(g) + H_2O(l) \qquad (2)$$

Carbon monoxide is very toxic and a moderately good Lewis base as a result of the presence of lone pairs. For example, in a typical reaction, CO reacts with a *d*-block metal or ion:

$$Ni(s) + 4CO(g) \longrightarrow [Ni(CO)_4](l)$$

The toxicity of CO is partially related to its Lewis base character because it attaches more strongly than oxygen to the iron in hemoglobin and prevents the blood from transporting oxygen. Carbon monoxide is a reducing agent.

EXAMPLE   Understanding the structure and reactions of carbon monoxide

Use standard enthalpies of formation, bond dissociation energies, and the following reaction to estimate the bond dissociation energy of carbon monoxide. Is your result consistent with the Lewis structure of CO?

$$CH_4(g) + H_2O(g) \longrightarrow CO(g) + 3H_2(g)$$

Standard enthalpies of formation are $CH_4(g)$, -74.8 kJ·mol^{-1}; $H_2O(g)$, -241.8 kJ·mol^{-1}; and CO(g), -110.53 kJ·mol^{-1}. Bond dissociation energies are C—H, 435 kJ·mol^{-1}; O—H, 431 kJ·mol^{-1}; and H—H, 436 kJ·mol^{-1}.

SOLUTION We first calculate the reaction enthalpy, using the enthalpies of formation:

$$\Delta H° = [1 \text{ mol} \times \Delta H_f°(CO(g))] - [1 \text{ mol} \times \Delta H_f°(H_2O(g)) + 1 \text{ mol} \times \Delta H_f°(CH_4(g))]$$

$$= [1 \text{ mol} \times -110.53 \text{ kJ·mol}^{-1}] - [(1 \text{ mol} \times -241.8 \text{ kJ·mol}^{-1}) + (1 \text{ mol} \times -74.8 \text{ kJ·mol}^{-1})]$$

$$= +206.1 \text{ kJ}$$

The reaction enthalpy is approximately equal to the energy required to break the bonds in CH_4 and H_2O minus the energy released when bonds are made in CO and H_2:

$$\Delta H^\circ = [(4 \times \text{C—H bond energy}) + (2 \times \text{O—H bond energy})]$$
$$- [\text{CO bond energy} + (3 \times \text{H—H bond energy})]$$

$$= [(4 \times 435 \text{ kJ}) + (2 \times 431 \text{ kJ})] - [(\text{CO bond energy}) + (3 \times 436 \text{ kJ})]$$

We substitute the value of ΔH° and solve for the CO bond energy:

$$206.1 \text{ kJ} = 1294 \text{ kJ} - \text{CO bond energy}$$

$$\text{CO bond energy} = 1088 \text{ kJ}$$

This extremely high bond energy is consistent with the Lewis structure of CO, which indicates that CO is a triple-bonded molecule, $:C\equiv O:$.

EXERCISE Explain, on the basis of its Lewis structure, why CO^- is a Lewis base.

> [*Answer:* The Lewis structure is $:C\equiv O:$. The unshared pairs make it a Lewis base. In most of its reactions as a Lewis base, the unshared pair on carbon is donated.]

19.16 OXIDES OF SILICON: THE SILICATES

KEY CONCEPT   Silicon forms a variety of oxides

Silicon dioxide, SiO_2, *or silica.*   Silica occurs in pure form naturally only as quartz and sand. Silica is very stable, but can be attacked by HF (reaction 1). It is also attacked by the Lewis base OH^- in molten NaOH and by O^{2-} in molten Na_2CO_3 (reaction 2).

$$SiO_2(s) + 6HF(aq) \longrightarrow [SiF_6]^{2-}(aq) + 2H_3O^+(aq) \tag{1}$$

$$SiO_2(s) + 2Na_2CO_3(l) \longrightarrow Na_4SiO_4(l) + 2CO_2(g) \tag{2}$$

Silicates, compounds containing SiO_4^{4-} units.   The compound sodium silicate is formed in reaction 2. **Orthosilicates** are the simplest silicates and contain isolated SiO_4^{4-} ions. Na_4SiO_4 and $ZrSiO_4$ are examples. The **pyroxenes** consist of SiO_4^{4-} chains in which two corner atoms are shared by two neighboring units, so the "average" unit is a SiO_3^{2-} unit. Cations placed along the chain in a more or less random manner provide electrical neutrality. $NaAl(SiO_3)_2$ and $CaMg(SiO_3)_2$ are examples. In **amphiboles,** the chains of silicate ions are linked to form a cross-linked ladderlike structure containing $Si_4O_{11}^{6-}$ units. Tremolite, $Ca_2Mg_5(Si_4O_{11})_2(OH)_2$, is an example. Almost all amphiboles contain hydroxide ions attached to the metal. Sheets containing $Si_2O_5^{2-}$ units, with cations lying between the sheets and linking them together, are also known. $Mg_3(Si_2O_5)_2(OH)_2$ is an example. Other structural types of silicon oxides exist.

Aluminosilicates and ceramics.   Replacing a Si^{4+} ion with an Al^{3+} ion (plus extra cations to make up for the charge difference) results in the formation of **aluminosilicates.** Mica, of which one form is $KMg_3(Si_3AlO_{10})(OH)_2$ is an example. The extra cations hold the sheets of tetrahedra together, which makes this type of material rather hard. Feldspar is a silicate material in which more than half the silicon is replaced by aluminum. Leaching of the cations in a feldspar results in clay. Heating clays to drive out the water trapped between the sheets of aluminosilicate tetrahedra results in formation of **ceramics. Cements** are produced by melting aluminosilicates and then allowing them to solidify. Cement is one of the components used in making concrete.

Silicones.   Silicones are made of long —O—Si—O— chains with the remaining silicon bonding positions occupied by organic groups. They possess both hydrophobic and hydropilic properties, which makes them useful as fabric waterproofers and in biological applications, as surgical and cosmetic implants.

Glasses.   Heating silica (SiO_2) above its melting point and cooling it slowly results in the formation of fused silica, a **glass.** Addition of various metal ions results in what is popularly called "glass." Heating the silica causes rupture of many Si—O bonds, which enables the metal ions to form ionic bonds with some of the oxygen atoms in the silica tetrahedra. Thus, covalent Si—O bonds are replaced by metal-oxygen ionic bonds. The nature of the glass depends on the metal added. Addition of Na^+ and Ca^{2+} (12% Na_2O and 12% CaO) results in **soda-lime glass,** the common glass used for windows and bottles. Reduction of the proportions of Na^+ and Ca^{2+} and addition of 16% B_2O_3 result in **borosilicate glass.** This type of glass, such as Pyrex, is more resistant to heat shock than soda-lime glass. Photochromic sunglasses, which darken in sunlight, contain added Ag^+ and Cu^+ ions. Sunlight causes reduction of the silver ions and darkening of the glass (reaction 3). Removal of the sunlight allows the process to reverse (reaction 4) because the reduction potential (in the glass) of the Ag^+/Ag couple is less than that of the Cu^{2+}/Cu^+ couple:

$$Ag^+(s) + Cu^+(s) \longrightarrow Ag(s) + Cu^{2+}(s) \tag{3}$$

$$Ag(s) + Cu^{2+}(s) \longrightarrow Ag^+(s) + Cu^+(s) \tag{4}$$

EXAMPLE   Understanding the structure of silicon compounds

Use VSEPR theory to confirm that the silicate ion is tetrahedral.

SOLUTION Silicon donates 4 valence electrons and the four oxygens donate 24 valence electrons (4 × 6 = 24). The −4 charge requires 4 additional electrons, so the total number of valence electrons is 32 (4 + 24 + 4 = 32). The Lewis structure is

$$\left[\begin{array}{c} \ddot{\text{O}} \\ | \\ \ddot{\text{O}}\!-\!\text{Si}\!-\!\ddot{\text{O}} \\ | \\ \ddot{\text{O}} \end{array} \right]^{4-}$$

There are four electron pairs surrounding silicon; thus they take on a tetrahedral shape and the ion itself is tetrahedral.

EXERCISE Explain why silicon tetrachloride reacts with water and carbon tetrachloride does not.

$$SiCl_4(l) + 2H_2O(l) \longrightarrow SiO_2(s) + 4HCl(aq)$$

$$CCl_4(l) + 2H_2O(l) \longrightarrow \text{no reaction}$$

[*Answer:* Silicon is larger than carbon and can expand its valence electron shell beyond an octet to accommodate five bonds during reaction as the Lewis base H_2O attacks and Cl^- leaves.]

19.17   CARBIDES

KEY CONCEPT   There are three classes of carbides

Carbides. Carbides are compounds containing the carbide ion (C^{4-}). (Some compounds containing the acetylide ion (C_2^{2-}) are commonly called carbides, even though they are not formally carbides.) Group 1 and 2 metals form **saline carbides** when their oxides are heated with carbon; examples are CaC_2, Li_2C_2, and Be_2C. **Covalent carbides** are carbides formed by metals with electronegativities similar to that of carbon, such as silicon and boron. Silicon carbide, SiC, known as carborundum, is an extremely hard, covalent carbide. It is produced by the reaction of silicon dioxide with carbon:

$$SiO_2(s) + 3C(s) \longrightarrow SiC(s) + 2CO(g)$$

In **interstitial carbides,** carbon atoms lie in holes in close-packed arrays of d-block metal atoms. Bonds between the carbon atoms and metal atoms stabilize the metal lattice and help to hold the metal atoms together. Examples are Cr_3C and Fe_3C.

EXAMPLE   Understanding the carbides

Write and balance the equation that shows the reaction between magnesium carbide (MgC_2) and water.

SOLUTION Magnesium carbide, like calcium carbide, is a saline carbide and reacts with water in the same way that calcium carbide does.

$$MgC_2(s) + 2H_2O(l) \longrightarrow Mg(OH)_2(s) + HC\equiv CH(g)$$

EXERCISE Which of the carbides, BaC_2, Mn_5C_2, or Bi_4C, should have the highest melting point? (It is not necessary to look up the melting points.)

[**Answer:** The interstitial carbide, Mn_5C_2, should have the highest melting point.]

KEY WORDS   Define or explain each term in a written sentence or two.

acid anhydride

chemical plating

covalent carbide

delocalized electron

electron-deficient compound

formal anhydride

Hall process

interstitial carbide

Mond process

saline carbide

synthesis gas

three-center bond

zone refining

SELF-TEST EXERCISES

Hydrogen

1. How many grams of $H_2(g)$ can be formed from 25.0 g of CH_4, using the re-forming reaction (1) followed by the shift reaction (2)?

$$CH_4(g) + H_2O(g) \longrightarrow CO(g) + 3H_2(g) \tag{1}$$

$$CO(g) + H_2O(g) \longrightarrow CO_2(g) + H_2(g) \tag{2}$$

(a) 9.42 g   (b) 3.14 g   (c) 6.28 g   (d) 12.6 g   (e) 25.0 g

2. What is the volume (in liters) of the 3×10^8 kg of H_2 used per year for industrial purposes, if it were all in liquid form (with $d = 0.09 \text{ g} \cdot \text{cm}^{-3}$).

(a) 3×10^9 L   (b) 3×10^8 L   (c) 3×10^{12} L

(d) 3×10^{11} L   (e) 3×10^{15} L

3. What is the volume, at STP, of the 3×10^8 kg H_2 used in the United States each year?

(a) 3×10^8 L   (b) 6×10^8 L   (c) 8×10^{10} L

(d) 3×10^{12} L   (e) 6×10^{12} L

4. Which metal is not active enough to produce H_2 by reaction with water?

(a) Zn   (b) Cu   (c) Na   (d) Mg

5. Which of the isotopes of hydrogen is/are radioactive?

(a) protium   (b) deuterium   (c) tritium

(d) (a) and (b)   (e) all of these

6. How many liters of $H_2(g)$ are required to produce 50.0 L of $NH_3(g)$? Assume all volumes are measured at the same temperature and pressure.

$$N_2(g) + 3H_2(g) \longrightarrow 2NH_3(g)$$

(a) 75.0   (b) 150   (c) 33.3   (d) 50.0   (e) 100

7. In the reaction of the hydride ion with water, H^- acts as all of the following except one. Which is the exception?

$$H^-(aq) + H_2O(l) \longrightarrow OH^-(aq) + H_2(g)$$

(a) reducing agent   (b) electrolyte   (c) Brønsted base

(d) Lewis base   (e) substance oxidized

8. What is the electron configuration of the hydride ion?

(a) $1s^1$ (b) $1s^2$ (c) $1s^22s^1$ (d) $1s^22s^2$ (e) no electrons

9. Which of the following is a saline hydride?

(a) H_2 (b) HCl (c) CuH (d) CaH_2 (e) CH_4

10. What are the products of the following reaction?

$$NaH(s) + H_2O(l) \longrightarrow ?$$

(a) NaO and H_2 (b) $NaOH$ only   (c) $NaOH$ and H_2

(d) Na and H_2O_2 (e) Na, O_2, and H_2

11. Which hydride is a good electrical conductor?

(a) H_2S (b) CH_4 (c) NaH (d) NH_3 (e) CuH

Group 1: The alkali metals

12. All of the following but one describes the alkali metals. Which one does not?

(a) silvery gray   (b) highly reactive   (c) not easily oxidized

(d) soft   (e) low melting point

13. Which of the following reacts with water least violently?

(a) Na   (b) Li   (c) Cs   (d) Rb   (e) K

14. What products are formed when Na reacts with water?

(a) NaH, O_2 (b) NaH, H_2O_2 (c) $NaOH$, H_2

(d) $NaOH$, O_2 (e) $NaOH$, H_2O_2

15. What unusual species is formed when potassium metal is dissolved in liquid ammonia? (am = solvated by ammonia molecules.)

(a) e^- (am)   (b) $H_2(g)$ (c) NH_2^+ (am)

(d) NH_4^+ (am)   (e) $NH_3(g)$

16. How many kilograms of sodium can be produced from the electrolysis of molten NaCl by a 2.50×10^4 A current running for 24.0 h?

(a) 199   (b) 225   (c) 386   (d) 515   (e) 418

17. Only one of the alkali metals forms a nitride when burned in air. Which one?

(a) Na   (b) Cs   (c) Rb   (d) Li   (e) K

18. What is the major product when sodium is burned in air?

(a) Na_2O_2 (b) Na_3N (c) Na_2O (d) NaO_2 (e) $NaNO_3$

19. What is the major product when potassium is burned in air?

(a) K_2O_2 (b) K_3N (c) K_2O (d) KNO_3 (e) KO_2

20. What is the oxidation state of oxygen in the superoxide ion?

(a) -1 (b) -2 (c) $+1$ (d) $-\frac{1}{2}$ (e) 0

21. Which of the following is washing soda?

(a) $NaCl$ (b) $Na_2CO_3 \cdot 10H_2O$ (c) Na_2SO_4

(d) $NaOH$ (e) $NaNO_3$

22. Which of the alkali metals exhibits significant covalency in its compounds?

(a) Li   (b) Na   (c) K   (d) Cs   (e) Rb

23. Which of the following reactions is used commercially to produce Na_2SO_4?

(a) $2Na(s) + K_2SO_4(l) \rightarrow Na_2SO_4(s) + 2K(s)$

(b) $(NH_4)_2SO_4(aq) + 2NaOH(aq) \rightarrow Na_2SO_4(aq) + 2NH_3(g) + 2H_2O(l)$

(c) $H_2SO_4(aq) + 2NaCl(s) \rightarrow Na_2SO_4(aq) + 2HCl(g)$

(d) $K_2SO_4(s) + 2NaCl(aq) \rightarrow Na_2SO_4(aq) + 2KCl(aq)$

24. Which of the following is a common oxidizing agent?

(a) KCl (b) KNO_3 (c) K_2SO_4 (d) K_2CO_3 (e) K

25. Calculate the enthalpy for the half-cell oxidation of Li(s), using the following data:

$$\text{Sublimation energy of Li(s)} = 161 \text{ kJ} \cdot \text{mol}^{-1}$$
$$\text{Ionization energy of Li(g)} = 520.3 \text{ kJ} \cdot \text{mol}^{-1}$$
$$\text{Hydration energy of Li}^+(g) = -520 \text{ kJ} \cdot \text{mol}^{-1}$$

(a) $+1201 \text{ kJ} \cdot \text{mol}^{-1}$ (b) $-161 \text{ kJ} \cdot \text{mol}^{-1}$ (c) $161 \text{ kJ} \cdot \text{mol}^{-1}$

(d) $-1201 \text{ kJ} \cdot \text{mol}^{-1}$ (e) $-879 \text{ kJ} \cdot \text{mol}^{-1}$

Group 2: The alkaline earth metals

26. How many *f*-electrons are there in radium?

(a) 0   (b) 22   (c) 10   (d) 28   (e) 14

27. Which compound results from the compressed deposits of ancient marine organisms?

(a) $MgCO_3$ (b) MgO (c) $BaCO_3$ (d) $CaCO_3$ (e) CaS

28. How many grams of pure magnesium can be produced from 65 kg of $Mg(OH)_2$?

(a) 2.7×10^4 (b) 1.6×10^5 (c) 6.4×10^2

(d) 4.4×10^5 (e) 9.0×10^6

29. Which alkaline earth shares a diagonal relationship with aluminum?

(a) Mg   (b) Ca   (c) Sr   (d) Be   (e) Ba

30. Only one of the alkaline earths forms a significant amount of the nitride when it is burned in air. Which one?

(a) Ba   (b) Sr   (c) Be   (d) Ca   (e) Mg

31. All the alkaline earths but one react with water. Which one doesn't?

(a) Mg   (b) Be   (c) Ca   (d) Sr   (e) Ba

32. What products are formed when barium reacts with water?
(a) $Ba(OH)_2$, H_2 (b) BaH_2, O_2 (c) $Ba(OH)_2$, O_2
(d) BaO, H_2O (e) BaO, H_2

33. What is the oxidizing agent when barium reacts with water?
(a) Ba (b) H_2O (c) Ba^{2+} (d) OH^- (e) O_2

34. Which of the following compounds exhibits substantial covalency?
(a) BaO (b) $CaCl_2$ (c) SrO (d) $Sr(OH)_2$ (e) $BeCl_2$

35. What is the most common structural unit for beryllium?
(a) BeX_4 tetrahedra (b) BeX_6 octahedra
(c) Be—Be dimers (d) BeX dimers

36. Which of the following alkaline earths is most strongly passivated by an oxide coat?
(a) Sr (b) Ca (c) Be (d) Ba (e) Mg

37. Limestone is
(a) $MgCO_3$ (b) $CaCl_2$ (c) CaO (d) $CaCO_3$ (e) CaC_2

38. Hydroxyapatite $[Ca_5(PO_4)_3OH]$ is the structural material in
(a) mortar. (b) tooth enamel. (c) marble. (d) plant stems.

39. At what temperature does the decomposition of $MgCO_3$ into MgO and CO_2 become thermo-dynamically favorable?

	$MgCO_3(s)$	$MgO(s)$	$CO_2(g)$
ΔH°_f, kJ·mol^{-1}	-1095.8	-601.7	-393.51
S°, J(mol·K)$^{-1}$	65.6	26.94	213.74

(a) 575 K (b) 1230 K (c) 2110 K (d) 3610 K (e) 4440 K

Group 13: The boron family

40. Which of the following is an alum?
(a) $CaSO_4 \cdot 2H_2O$ (b) $KAl(SO_4)_2 \cdot 12H_2O$ (c) K_2SO_4
(d) $Na_2CO_3 \cdot 10H_2O$ (e) $Ca_5(PO_4)_3OH$

41. One of the following does not properly describe elemental boron. Which one?
(a) metal (b) acidic oxide (c) several allotropes
(d) hardens steel (e) very inert

42. How many grams of boron are there in 1.00 kg of its ore kernite, $Na_2B_4O_7 \cdot 7H_2O$?
(a) 327 (b) 266 (c) 132 (d) 33.0 (e) 429

43. The common oxidation state of aluminum in its compounds is
(a) +2 (b) +1 (c) +3 (d) 0 (e) −4

44. What is the time (in hours) required to electrolytically reduce 1.0 metric ton of aluminum with a current of 1.0×10^4 A? 1 metric ton = 1000 kg
(a) 3.0×10^2 h (b) 1.0×10^2 h (c) 1.0×10^{-2} h
(d) 1.0×10^6 h (e) 1.0×10^4 h

45. What is the anhydride of boric acid, $B(OH)_3$?

(a) BN   (b) B_2O_3 (c) B_2H_6 (d) $B_{12}C_3$ (e) BF_3

46. The pK_a of boric acid is 9.00. What is the pH of a 0.020 M solution of boric acid?

$$B(OH)_3(aq) + 2H_2O(l) \rightleftharpoons [B(OH)_4]^-(aq) + H_3O^+(aq)$$

(a) 1.78   (b) 7.00   (c) 3.14   (d) 8.65   (e) 5.35

47. Which structure best represents the Al_2Cl_6 molecule?

(a)
```
      Cl  Cl
      |   |
Cl—Al—Al—Cl
      |   |
      Cl  Cl
```

(b)
```
Cl           Cl
  \          |
   Al—Cl—Al—Cl
  /          |
Cl           Cl
```

(c)
```
        Cl—Cl
       /      \
Al—Cl          Cl—Al
       \      /
        Cl—Cl
```

(d)
```
Cl      Cl      Cl
  \    /  \    /
   Al      Al
  /    \  /    \
Cl      Cl      Cl
```

48. The solubility product of $Al(OH)_3$ is approximately 1×10^{-33}. What is the solubility of $Al(OH)_3$ in grams per 100 mL in a solution with pH = 5.44?

(a) 2×10^{-22} (b) 2×10^{-21} (c) 2×10^{-16}

(d) 2×10^{-5} (e) 4×10^{-7}

49. Which of the following is an important industrial catalyst?

(a) CaF_2 (b) BF_3 (c) $AlCl_3$ (d) BN   (e) B_2O_3

50. Which compound possesses a two-electron three-center bond?

(a) NaH   (b) $NaBH_4$ (c) BF_3 (d) B_2H_6 (e) B_2O_3

Group 14: The carbon family

51. Which allotrope of carbon has a hollow core that may be able to trap atoms?

(a) diamond   (b) buckminsterfullerene   (c) graphite   (d) activated carbon

52. Graphite possesses one metallike property. What is the property?

(a) lustrousness   (b) ductility   (c) forms positive ions

(d) electrical conductivity   (e) forms basic oxides

53. It takes extremely high pressures to convert graphite to diamond. Given that the density of graphite is $2.22 \text{ g} \cdot \text{cm}^{-3}$, what is a reasonable density for diamond?

(a) $1.98 \text{ g} \cdot \text{cm}^{-3}$ (b) $2.22 \text{ g} \cdot \text{cm}^{-3}$ (c) $3.51 \text{ g} \cdot \text{cm}^{-3}$

54. What process is used to obtain ultrapure silicon?

(a) fractional distillation   (b) filtration   (c) column chromatography

(d) paper chromatography   (e) zone refining

55. 1.00 kg SiO_2 is processed and results in the production of 126 g of ultrapure silicon. What is the percentage yield?

(a) 26.9%   (b) 12.6%   (c) 18.7%   (d) 44.9%   (e) 35.2%

56. Which carbon species is a reducing agent?

(a) CO   (b) CCl_4 (c) CO_2 (d) CN^- (e) CH_4

57. The boiling points of three carbon compounds are CO_2, $-78°C$; CS_2, $46.2°C$; and CSe_2, $125.1°C$. What is the dominant force that determines the boiling points of these compounds?

(a) ion-dipole   (b) London   (c) dipole-dipole   (d) ion-ion

58. What is the structural unit for silicates?

(a) SiO_2 linear units   (b) SiO_2 angular units   (c) SiO_4 tetrahedra

(d) SiO_4 square planar units   (e) SiF_6 octahedra

59. How many milligrams of silica can be produced by reacting 10.0 mL of 0.200 M HCl with 1.00 g sodium silicate?

$$4HCl(aq) + Na_4SiO_4(s) \longrightarrow SiO_2(s) + 4NaCl(aq) + 2H_2O(l)$$

(a) 266   (b) 41.2   (c) 129   (d) 326   (e) 30.0

60. Which of the following is an interstitial carbide?

(a) Fe_3C (b) CH_4 (c) CaC_2 (d) SiC (e) MgC_2

61. What is the oxidation number of carbon in the acetylide ion?

(a) -2 (b) -1 (c) -4 (d) $+4$ (e) $+2$

CHAPTER 20

THE MAIN-GROUP ELEMENTS:

II THE LAST FOUR FAMILIES

The chemical and physical properties of any element is related to its position in the periodic table. In this chapter we explore the properties of the elements of Groups 15 through 18. As we observed in the previous chapter, we shall see that each element has its own unique properties and there are smooth trends as we go down each group. We shall also correlate and rationalize the properties of the elements with their electron configurations.

GROUP 15: THE NITROGEN FAMILY

20.1   THE ELEMENTS

KEY CONCEPT   Nitrogen and phosphorus

The elements nitrogen and phosphorus at the top of Group 15 are nonmetals, whereas bismuth at the bottom is a metal. Nitrogen differs substantially from other group members in its high electronegativity, small size, ability to form multiply bonds, and lack of usable d-orbitals. Nitrogen displays oxidation numbers from -3 to $+5$. Elemental nitrogen is prepared by fractional distillation of liquid air. The extremely strong triple bond in N_2 makes it very stable. **Nitrogen fixation** is the process in which N_2 is converted to compounds usable by plants. The **Haber-Bosch process** is a major industrial nitrogen-fixation technique. Bacteria found in the root nodules of legumes also fix N_2.

Phosphorus is prepared from phosphate rock such as calcium phosphate:

$$2Ca_3(PO_4)_2(s) + 6SiO_2(s) + 10C(s) \longrightarrow P_4(g) + 6CaSiO_3(l) + 10CO(g)$$

One of the two common allotropic forms of phosphorus, white phosphorus, is a highly reactive, extremely toxic material. It is a molecular solid consisting of P_4 tetrahedra. It bursts into flame spontaneously on contact with air. Red phosphorus, the second common allotrope, likely consists of linked P_4 tetrahedra. It is less reactive than white phosphorus and is used in match heads because it can be ignited by friction.

EXAMPLE   Understanding the uses of nitrogen and phosphorus

Nitrogen-fixation bacteria in the root nodules of legumes convert N_2 to NH_3. How would you classify this reaction (e.g., acid-base, redox, decomposition, precipitation, or other)?

SOLUTION The reaction is a redox reaction because the oxidation state of nitrogen changes from 0 to -3.

EXERCISE How many kilograms of phosphorus are present in 1×10^3 kg $Ca_3(PO_4)_2$?

[*Answer:* 2×10^2 kg]

20.2 COMPOUNDS WITH HYDROGEN AND THE HALOGENS

KEY CONCEPT   Nitrogen has many oxidation states in its compounds

Nitrogen has a rich and complex chemistry, with a wide range of oxidation states, as indicated in the following discussion.

Ammonia, NH_3; oxidation number $= -3$. Ammonia is produced industrially by the Haber-Bosch process. The ammonia produced is the starting point for the production of many nitrogen-containing compounds and is used as a fertilizer. Ammonia is a weak Brønsted base; and when neutralized with acid, it forms ammonium salts. For instance,

$$HCl(aq) + NH_3(aq) \longrightarrow NH_4Cl(aq)$$

The ammonium ion is a weak Brønsted acid:

$$NH_4^+(aq) + H_2O(l) \longrightarrow NH_3(aq) + H_3O^+(aq)$$

Ammonia is a Lewis base, as is evidenced by its reaction with aqueous copper(II) ion:

$$4NH_3(aq) + Cu^{2+}(aq) \longrightarrow [Cu(NH_3)_4]^{2+}(aq)$$

In this reaction, unshared pairs on nitrogen move into unoccupied orbitals on Cu^{2+}. Ammonium salts decompose when heated; if the salt contains a nonoxidizing anion, NH_3 is produced:

$$(NH_4)_2SO_4(s) \longrightarrow 2NH_3(g) + H_2SO_4(l)$$

Hydrazine, N_2H_4; oxidation number $= -2$. Hydrazine is produced by the gentle oxidation of ammonia:

$$2NH_3(aq) + OCl^-(aq) \longrightarrow N_2H_4(aq) + H_2O(l)$$

Hydrazine is a dangerously explosive, oily liquid. A mixture of methylhydrazine (CH_3NHNH_2) and liquid N_2O_4 is used as a rocket fuel because these two liquids ignite on contact and produce a large volume of gas:

$$4CH_3NHNH_2(l) + 5N_2O_4(l) \longrightarrow 9N_2(g) + 12H_2O(g) + 4CO_2(g)$$

Nitrides. Nitrides contain the N^{3-} ion and dissolve in water to produce a hydroxide and ammonia:

$$Zn_3N_2(s) + 6H_2O(l) \longrightarrow 2NH_3(g) + 3Zn(OH)_2(aq)$$

Azides. Azides contain the highly reactive N_3^- ion. Sodium azide is prepared by the reaction of dinitrogen oxide with molten sodium amide ($NaNH_2$):

$$N_2O(g) + 2NaNH_2(l) \longrightarrow NaN_3(l) + NH_3(g) + NaOH(l)$$

Some azides, such as AgN_3, $Cu(N_3)_2$, and $Pb(N_3)_2$ are shock sensitive. The azide ion is a weak base.

Phosphorus compounds vary from the toxic and reactive to important biochemical species. The binary hydride of phosphorus, phosphine (PH_3), is a poisonous gas. Because it does not form hydrogen bonds, it is not very soluble in water. Phosphorus trichloride (PCl_3) and phosphorus pentachloride (PCl_5) are two important halides of phosphorus. They both react with water in a hydrolysis reaction and, without any change in oxidation state, form an acid plus hydrogen chloride gas.

$$PCl_3(l) + 3H_2O(l) \longrightarrow H_3PO_3(s) + 3HCl(g)$$

$$PCl_5(s) + 4H_2O(l) \longrightarrow H_3PO_4(l) + 5HCl(g)$$

EXAMPLE   Understanding the reactions of nitrogen compounds

What is the change in volume (measured at STP) when 100 g of methylhydrazine react with excess dinitrogen tetroxide? Assume the liquids have zero volume.

$$4CH_3NHNH_2(l) + 5N_2O_4(l) \longrightarrow 9N_2(g) + 12H_2O(g) + 4CO_2(g)$$

SOLUTION We first use the molar mass of methylhydrazine ($46.07 \text{ g} \cdot \text{mol}^{-1}$) to calculate the moles of methylhydrazine in 100 g:

$$100 \text{ g} \times \frac{1 \text{ mol CH}_3\text{NHNH}_2}{46.07 \text{ g}} = 2.17 \text{ mol CH}_3\text{NHNH}_2$$

We now use the chemical equation, which tells us that 4 mol of methylhydrazine yield 25 mol of gas (9 + 12 + 4 = 25), to calculate the moles of gas formed:

$$2.17 \text{ mol CH}_3\text{NHNH}_2 \times \frac{25 \text{ mol gas}}{4 \text{ mol CH}_3\text{NHNH}_2} = 13.6 \text{ mol gas}$$

Finally, by using the molar mass of an ideal gas at STP ($V_m = 22.4 \text{ L} \cdot \text{mol}^{-1}$) we determine the

$$13.6 \text{ mol gas} \times 22.4 \text{ L} \cdot \text{mol}^{-1} = 305 \text{ L}$$

EXERCISE In the potentially explosive reaction of ammonium nitrate shown, what is oxidized and what is reduced?

$$2NH_4NO_3(s) \longrightarrow 2N_2(g) + O_2(g) + 4H_2O(g)$$

[*Answer:* N in NH_4^+ is oxidized from oxidation number -3 to 0. N in NO_3^- is reduced from oxidation number $+5$ to 0. Oxygen in NO_3^- is oxidized from oxidation number -2 to 0.]

20.3   NITROGEN OXIDES AND OXOACIDS

KEY CONCEPT   Nitrogen oxides exist with nitrogen oxidation states of +1 to +5

Dinitrogen oxide, N_2O; oxidation number $= +1$. N_2O (laughing gas) is the oxide of nitrogen with the lowest oxidation number for nitrogen in its oxides. It is formed by the gentle heating of ammonium nitrate:

$$NH_4NO_3(s) \longrightarrow N_2O(g) + 2H_2O(g)$$

N_2O is fairly unreactive. It is toxic if inhaled in large amounts.

Nitrogen oxide, NO; oxidation number $= +2$.   NO is manufactured industrially by the catalytic oxidation of NH_3:

$$4NH_3(g) + 5O_2(g) \longrightarrow 4NO(g) + 6H_2O(g)$$

The endothermic formation of NO from the oxidation of N_2 occurs readily at the high temperatures that occur in automobile engines and turbine engine exhausts:

$$N_2(g) + O_2(g) \longrightarrow 2NO(g)$$

NO is oxidized further to NO_2 on exposure to air. Through the formation of NO_2, nitrogen oxide contributes to smog, acid rain, and destruction of the stratospheric ozone layer.

Nitrogen dioxide, NO_2; oxidation number $= +4$. NO_2, like NO, is paramagnetic. It disproportionates in water to form nitric acid:

$$3NO_2(s) + H_2O(l) \longrightarrow 2HNO_3(aq) + NO(g)$$

It is prepared in the laboratory by heating lead(II) nitrate:

$$2Pb(NO_3)_2(s) \longrightarrow 4NO_2(g) + 2PbO(s) + O_2(g)$$

and exists in equilibrium with its dimer, dinitrogen tetroxide:

$$2NO_2(g) \rightleftharpoons N_2O_4(g)$$

Nitrous acid, HNO_2; oxidation number = +3. Nitrous acid can be produced by mixing its anhydride, dinitrogen trioxide, with water:

$$N_2O_3(g) + H_2O(l) \longrightarrow 2HNO_2(aq)$$

Nitrous acid has not been isolated as a pure compound, but it is used in aqueous solution. It is a weak acid (pK_a = 3.4).

Nitric acid, HNO_3; oxidation number = +5. Nitric acid, a widely used industrial and laboratory acid, is produced by the three-step **Ostwald process**:

$$4NH_3(g) + 2O_2(g) \longrightarrow 4NO(g) + 6H_2(g)$$

$$2NO(g) + O_2(g) \longrightarrow 2NO_2(g)$$

$$3NO_2(g) + H_2O(l) \longrightarrow 2HNO_3(aq) + NO(g)$$

The anhydride of nitric acid is dinitrogen pentoxide:

$$N_2O_5(s) + H_2O(l) \longrightarrow 2HNO_3(l)$$

Dinitrogen pentoxide consists of NO_2^+ cations and NO_3^- anions. Nitrogen has its highest oxidation state in nitric acid, and the acid is therefore an excellent oxidizing agent.

EXAMPLE   Understanding the oxides of nitrogen

Explain the meaning of the statement in the text that, because nitrogen has its highest oxidation number (+5) in nitric acid, nitric acid is an oxidizing agent.

SOLUTION Nitrogen has oxidation numbers from +5 to −3. When it's in its +5 oxidation state, it is easy to imagine that nitrogen can gain electrons to change to one of its lower oxidation states. By gaining electrons, it oxidizes something else and, therefore, serves as an oxidizing agent. For example, in the following reaction, nitrogen changes from the +5 to the +2 oxidation state:

$$3H_2S(aq) + 2HNO_3(aq) \longrightarrow S(s) + 2NO(g) + 4H_2O(l)$$

EXERCISE What is the anhydride of HNO_3? (Try to answer this question without using any references.)

[*Answer:* N_2O_5]

20.4   PHOSPHORUS OXIDES AND OXOACIDS

KEY CONCEPT   Phosphorus oxides are based on PO_4 units

Phosphorus(III) oxide (P_4O_6) and phosphorus(V) oxide (P_4O_{10}), are obtained by burning white phosphorus in a limited supply or abundant supply of air, respectively. P_4O_6 can be thought of as originating with PO_3 units (with a tetrahedral phosphorus), whereas P_4O_{10} is built up from interconnected PO_4 tetrahedra (see page 774 of the text for both structures).

Phosphorus(III) oxide, P_4O_6, and *phosphorous acid*, H_3PO_3. Phosphorus(III) oxide is prepared by burning white phosphorus in a limited supply of air. P_4O_6 molecules consist of P_4 tetrahedra, with an oxygen atom lying on each edge of the tetrahedron, between two phosphorus atoms. A tetrahedron has six edges, hence the formula P_4O_6. Phosphorus(III) oxide is the anhydride of phosphorous acid. The structure of phosphorous acid is unusual in that one of the hydrogen atoms is attached to the central phosphorus atom.

Phosphorus(V) oxide, P_4O_{10}, and *phosphoric acid*, H_3PO_4. Phosphorus(V) oxide is prepared by burning phosphorus in excess oxygen. The molecule is the same as that of P_4O_6 (described earlier) but with an additional oxygen attached to each phosphorus at the apices of the tetrahedron. P_4O_{10} is the anhydride of phosphoric acid. Phosphorus(V) oxide traps and reacts with water with great efficiency and is widely used as a drying agent. It must be used carefully, however, because the reaction with water can be violent. Phosphoric acid is a triprotic acid. It is only moderately oxidizing, even though its phosphorus is in a high oxidation state. Phosphoric acid is the parent acid of phosphate salts (salts containing the tetrahedral phosphate ion, PO_4^{3-}). Phosphates are generally not very soluble, which makes them appropriate structural material for bones and teeth.

Polyphosphates. Polyphosphates are compounds that contain linked PO_4^{3-} tetrahedra. The simplest such structure is the pyrophosphate ion ($P_2O_7^{4-}$), in which two phosphate ions are linked by an oxygen atom through —P—O—P— bonds:

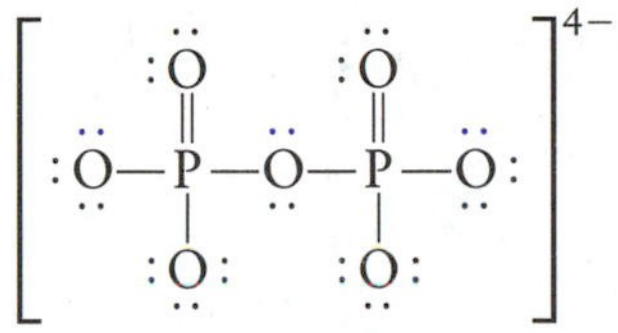

More complicated structures formed by longer chains or rings also exist. The biochemically most important polyphosphate is undoubtedly adenosine triphosphate (ATP), which contains three phosphate tetrahedra linked by —P—O—P— bonds. The hydrolysis of ATP to adenosine diphosphate (ADP) by the rupture of an O—P bond releases energy that is used by cells to drive the biochemical reactions in the cell:

$$\text{ATP} + \text{H}_2\text{O} \longrightarrow \text{ADP} + \text{HPO}_4^{2-} + 41 \text{ kJ}$$

It is not an exaggeration to call ATP the "fuel of life," because it provides the energy for many of the biochemical reactions of cells.

EXAMPLE   Understanding the oxides of phosphorus

Explain why phosphorous acid (H_3PO_3) is a diprotic acid even though it contains three hydrogen atoms.

SOLUTION The Lewis structure of H_3PO_3 is

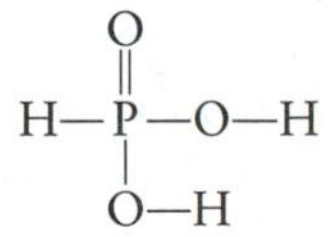

Because of factors discussed earlier in Sections 14.9 and 14.10 of the text, only the hydrogen atoms bonded to oxygen are acidic. The bond strength and bond polarity of the P—H bond do not allow for dissociation of the hydrogen that is bonded to phosphorus. Thus, only two protons ionize, and H_3PO_3 is diprotic.

EXERCISE The structure shown is a reproduction of the P_4O_{10} structure shown on page 774 of the text. How many PO_4 tetrahedra are visible?

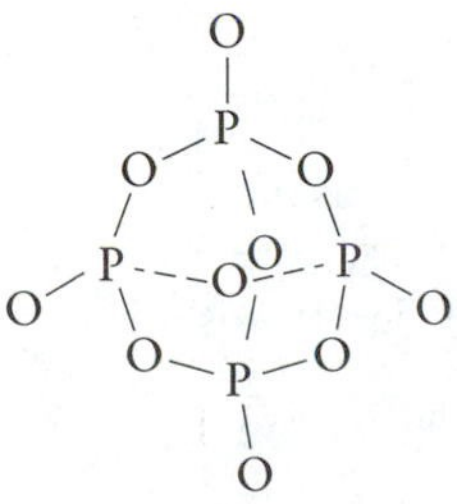

[*Answer:* There are four interconnected PO_4 tetrahedra present, connected by shared "bridging" oxygen atoms. Be certain you can see them!]

Define or explain each term in a written sentence or two.

disproportionation   nitrogen fixation
Haber-Bosch process   Ostwald process
hydrolysis reaction

GROUP 16: THE OXYGEN FAMILY

20.5   THE ELEMENTS

KEY CONCEPT A   Elemental oxygen and sulfur

Oxygen is an important industrial product that is used extensively in the steel industry. It is a colorless, odorless, paramagnetic gas. The paramagnetism of O_2 can be accounted for by an advanced theory of bonding called molecular orbital theory, but a satisfactory Lewis structure for O_2 cannot be drawn. Elemental oxygen exists in two allotropic forms, O_2 and O_3 (ozone). Ozone is produced by an electric discharge through O_2; its pungent odor can be detected near sparking electrical equipment and after lightning strikes. The following table gives some properties of O_2 and O_3.

Property	O_2	O_3
boiling point	$-183°C$	$-112°C$
liquid color	pale blue	dark blue
atmosphere	23% by mass	forms stratospheric ozone layer
magnetism	paramagnetic	diamagnetic

Sulfur is found in ores and as elemental sulfur. The **Frasch process** is used to mine elemental sulfur by melting the sulfur (mp = 165°C) with superheated water and forcing it to the surface with compressed air. The **Claus process** recovers sulfur from H_2S in a process that involves two reactions:

$$2H_2S(g) + 3O_2(g) \longrightarrow 2SO_2(g) + 2H_2O(l)$$

$$2H_2S(g) + SO_2(g) \longrightarrow 3S(s) + 2H_2O(l)$$

Sulfur is used primarily to produce sulfuric acid and to vulcanize rubber. Vulcanization increases the toughness of rubber by introducing cross-links between the natural rubber polymer chains. Elemental sulfur exists in its most stable allotropic form as $S_8(s)$ rings, but for simplicity it is often represented as $S(s)$.

EXAMPLE   Understanding the structure of sulfur and oxygen

Use VSEPR theory to predict the shape of ozone and the hybridization of the central oxygen atom.

SOLUTION The Lewis structure of O_3 (actually, one of two equivalent resonance structures) has 18 valence electrons:

$$:\ddot{O}=\ddot{O}-\ddot{O}:$$

The central atom has three regions of high concentration of electrons around it. Three regions of high concentrations of electrons result in a trigonal orientation for the electron pairs and sp^2 hybridization. Because one of the regions of high concentration of electrons is a lone pair, the molecule has a bent or angular shape.

EXERCISE In the second reaction associated with the Frasch process, which atom is oxidized and which atom is reduced?

$$2H_2S(g) + SO_2(g) \longrightarrow 3S(s) + 2H_2O(l)$$

[*Answer:* S in H_2S is oxidized, and S in SO_2 is reduced.]

KEY CONCEPT B   Trends and differences in Group 16 elements

There is a gradual change from nonmetallic behavior at the top of Group 16 to metallic behavior at the bottom, as in Groups 14 and 15. Other changes occur in going down the group. Sulfur, selenium, and tellurium are able to form relatively long chains (**catenate**), whereas oxygen is not. Sulfur and oxygen are both quite reactive and combine with most of the other elements.

EXAMPLE   Understanding trends in Group 16

What trend in boiling points would you expect for the diatomic molecules X_2 for the first four Group 16 elements.

SOLUTION For a homonuclear diatomic molecule, the only intermolecular forces are London forces. As the molecular masses of the molecules increase, the London forces increase and boiling points increase.

EXERCISE The boiling points of the Group 16 elements increase from oxygen to tellurium, but the boiling point of polonium is lower than that of tellurium. Suggest a reason for the reversal in the trend.

[*Answer:* The type of force that determines the boiling point changes from London forces (for O through Te) to metallic bonding (for Po).]

20.6   COMPOUNDS WITH HYDROGEN

KEY CONCEPT A   Water

Because of its unique set of physical and chemical properties, water is one of the most remarkable compounds known. Water for domestic water supplies must undergo many types of purification. High-purity water for special applications is obtained by distillation or ion exchange. Water has a higher boiling point than expected because of the extensive hydrogen bonding between H_2O molecules. Hydrogen bonding also causes a more open structure and, therefore, a lower density for the solid than for the liquid. With its high polarity, water is an excellent solvent for ionic compounds. Chemically, it can both donate and accept protons (it is amphiprotic); so it is both a Brønsted acid and a Brønsted base. It can act as an oxidizing agent and a reducing agent. With its unshared electron pairs on oxygen, water can also act as a Lewis base. Thus, it can form complexes with transition metals and can hydrolyze certain substances (such as PCl_5). In its hydrolysis reactions, a bond between oxygen and another element is formed. Hydrolysis reactions can occur with or without a change in oxidation number. Oxides of metals are bases because the oxide ion is a stronger base than OH^-, and oxide ion immediately reacts with water to form OH^-. Some of the reactivities of water are

As a Brønsted base:	$CH_3COOH(aq) + H_2O(l) \longrightarrow CH_3CO_2^-(aq) + H_3O^+(aq)$
As a Brønsted acid:	$H_2O(l) + NH_3(aq) \longrightarrow NH_4^+(aq) + OH^-(aq)$
As an oxidizing agent:	$2Na(s) + 2H_2O(l) \longrightarrow 2NaOH(aq) + H_2(g)$
As a reducing agent:	$2H_2O(l) + 2F_2(g) \longrightarrow 4HF(aq) + O_2(g)$
Hydrolysis without a change in oxidation state:	$PCl_5(s) + 4H_2O(l) \longrightarrow H_3PO_4(aq) + 5HCl(aq)$
Hydrolysis with a change in oxidation state (of Cl, in this case):	$Cl_2(g) + H_2O(l) \longrightarrow ClOH(aq) + HCl(aq)$

EXAMPLE   Understanding the reactions of water

Describe how water can be considered a Lewis base in its reaction with PCl_5. What is the Lewis acid?

SOLUTION The reaction is

$$PCl_5(s) + 4H_2O(l) \longrightarrow H_3PO_4(aq) + 5HCl(aq)$$

In this reaction, the unshared electron pairs on the oxygen move into empty orbitals on the phosphorus atoms to form new P—O bonds as P—Cl bonds are broken. Thus, water acts is the Lewis base, and PCl_5 is the Lewis acid.

EXERCISE What is the reduction potential at 25°C for the following half-reaction at pH = 11.0? $E° = -0.83\,V$. Assume that the partial pressure of H_2 is 1 atm exactly.

$$2H_2O(l) + 2e^- \longrightarrow 2OH^-(aq) + H_2(g)$$

[*Answer:* -0.65 V]

KEY CONCEPT B   Hydrogen peroxide (H_2O_2)

H_2O_2 is a highly reactive, pale blue liquid. Many of its physical properties are similar to those of water; one exception is its substantially higher density ($1.44\ \mathrm{g \cdot mL^{-1}}$). Chemically, hydrogen peroxide is more acidic than water ($pK_a = 11.75$). It is a strong oxidizing agent, but it can also function as a reducing agent.

As an oxidizing agent: $2Fe^{2+}(aq) + H_2O_2(aq) + 2H^+(aq) \longrightarrow 2Fe^{3+}(aq) + 2H_2O(l)$

As a reducing agent: $Cl_2(g) + H_2O_2(aq) + 2OH^-(aq) \longrightarrow 2Cl^-(aq) + 2H_2O(l) + O_2(g)$

The O—O bond in H_2O_2 is very weak. Oxygen has a -1 oxidation state in H_2O_2.

EXAMPLE   Predicting the physical properties of hydrogen peroxide

The boiling point of hydrogen peroxide is very different from that of water. Predict whether it is higher or lower than water's, and give the reason for your prediction.

SOLUTION H_2O_2 has two oxygens, so it seems reasonable to expect that it may be more extensively hydrogen bonded than H_2O, which has only one oxygen. The more extensive hydrogen bonding should lead to a higher boiling point; thus, we predict that H_2O_2 has a higher boiling point than water. This prediction is correct; the boiling point of H_2O_2 is 152°C.

EXERCISE Explain why H_2O_2 is more dense than H_2O.

[*Answer:* More extensive hydrogen bonding holds the molecules closer together.]

20.7   SULFUR OXIDES AND OXOACIDS

KEY CONCEPT A   Sulfur dioxide and sulfites

Sulfur burns in air to form SO_2, which is a poisonous gas. Volcanic activity, the combustion of fuels contaminated with sulfur, and the oxidation of H_2S are the major sources of the SO_2 found in the atmosphere. The equation for the oxidation of H_2S is

$$2H_2S(g) + 3O_2(g) \longrightarrow 2SO_2(g) + 2H_2(g)$$

SO_2 is an acidic oxide and forms sulfurous acid when it reacts with water:

$$SO_2(g) + H_2O(l) \longrightarrow H_2SO_3(aq)$$

In aqueous solutions, H_2SO_3 exists as a mixture of $SO_2 \cdot 7H_2O$ in equilibrium with H_3O^+ and SO_3^{2-}, with few or no H_2SO_3 molecules present. In $SO_2 \cdot 7H_2O$, each SO_2 molecule is located in a cage formed by surrounding water molecules in a structure called a **clathrate.** SO_2 and sulfite ion (SO_3^{2-}) can act as oxidizing or reducing agents:

SO_3^{2-} as an oxidizing agent: $4HSO_3^-(aq) + 2HS^-(aq) \longrightarrow 3S_2O_3^{2-}(aq) + 3H_2O(l)$

SO_3^{2-} as a reducing agent: $SO_3^{2-}(aq) + S(s) \longrightarrow S_2O_3^{2-}(aq)$

EXAMPLE   Understanding the structure and reactions of SO_2

What is $\Delta G°$ for the reaction

$$2H_2S(g) + 3O_2(g) \longrightarrow 2SO_2(g) + 2H_2O(g)$$

Would you expect the H_2S that escapes into the atmosphere to stay as H_2S or to form SO_2?

SOLUTION For $\Delta G°$, we use the standard free energies of formation of the reactants and products:

$$\Delta G° = [\{2 \text{ mol} \times \Delta G_f°(H_2O(g))\} + \{2 \text{ mol} \times (\Delta G_f°(SO_2))\}]$$
$$- [\{2 \text{ mol} \times \Delta G_f°(H_2S(g))\} + \{3 \text{ mol} \times \Delta G_f°(O_2(g))\}]$$

$$= [\{2 \text{ mol} \times (-228.57 \text{ kJ} \cdot \text{mol}^{-1})\} + \{2 \text{ mol} \times (-300.19 \text{ kJ} \cdot \text{mol}^{-1})\}]$$
$$- [\{2 \text{ mol} \times (-33.56 \text{ kJ} \cdot \text{mol}^{-1})\} + \{3 \text{ mol} \times (0 \text{ kJ} \cdot \text{mol}^{-1})\}]$$

$$= [-1057.52 \text{ kJ}] - [-67.12 \text{ kJ}]$$

$$= -990.40 \text{ kJ}$$

The large negative free energy of reaction indicates that the products are thermodynamically strongly favored, so we would expect H_2S to form SO_2.

EXERCISE Use VSEPR theory to predict the shape of the SO_3^{2-} ion.

[*Answer:* Trigonal pyramidal]

KEY CONCEPT B   Sulfur trioxide and the sulfates

Sulfur dioxide can be oxidized directly to sulfur trioxide by reaction with O_2:

$$2SO_2(g) + O_2(g) \longrightarrow 2SO_3(g)$$

The uncatalyzed reaction is slow in the atmosphere, but it can be catalyzed by the presence of metal cations in droplets of water or by certain surfaces. Indirect pathways also exist in the atmosphere for the conversion of SO_2 to SO_3. SO_3 is a trigonal planar molecule and forms S_3O_9 trimers in the solid. The **contact process** for the production of H_2SO_4 involves the production of SO_3, which, by dissolving in "oleum" (98% H_2SO_4) and reacting with water, is converted to H_2SO_4.

$$S(g) + O_2(g) \xrightarrow{1000°C} SO_2(g)$$

$$2SO_2(g) + O_2(g) \xrightarrow{500°C, V_2O_5} 2SO_3(g)$$

Sulfuric acid is a strong Brønsted acid, a powerful dehydrating agent, and a mild oxidizing agent. In water, its first ionization is complete:

$$H_2SO_4(aq) + H_2O(l) \longrightarrow HSO_4^-(aq) + H_3O^+(aq)$$

HSO_4^- is a weak acid, with $pK_a = 1.92$:

$$HSO_4^-(aq) + H_2O(l) \rightleftharpoons SO_4^{2-}(aq) + H_3O^+(aq)$$

As a dehydrating agent, sulfuric acid is able to extract water from a variety of compounds, including sucrose ($C_{12}H_{22}O_{11}$) and formic acid (HCOOH):

$$C_{12}H_{22}O_{11}(s) \xrightarrow{\text{Conc. } H_2SO_4} 12C(s) + 11H_2O(l)$$

$$HCOOH(l) \xrightarrow{\text{Conc. } H_2SO_4} CO(g) + H_2O(l)$$

As a result of the large negative enthalpy of solution involved, mixing sulfuric acid with water can cause violent splashing; so for reasons of safety, sulfuric acid is always carefully added to water, not the water to acid. The sulfate ion is a mild oxidizing agent:

$$4H^+(aq) + 2e^- + SO_4^{2-}(aq) \longrightarrow SO_2(aq) + 2H_2O(l) \qquad \Delta E^\circ = 0.17 \text{ V}$$

EXAMPLE Understanding the reactions of sulfuric acid

Mixing sulfuric acid with solid sodium chloride results in the formation of dry hydrogen chloride. Write the equation to the reaction, state what type of reaction it is, and suggest why the hydrogen chloride is dry.

SOLUTION Because HCl is one of the products formed by the reaction of H_2SO_4 with NaCl, it is reasonable to assume that the Na^+ and SO_4^{2-} that remain form sodium sulfate. The equation is

$$H_2SO_4(l) + 2NaCl(s) \longrightarrow 2HCl(g) + Na_2SO_4(s)$$

The reaction is a Lewis acid-base reaction in which the Lewis base Cl^- reacts with the Lewis acid H^+ to form HCl, and the salt Na_2SO_4 forms from the remaining ions. The HCl is dry because sulfuric acid is a strong dehydrating agent and removes any stray water present.

EXERCISE Pure sulfuric acid shows very high conductivity. Explain why.

[*Answer:* It undergoes the self-ionization $2H_2SO_4(l) \rightleftharpoons HSO_4^- + H_3SO_4^+$; the ions formed conduct an electric current.]

20.8   SULFUR HALIDES

KEY CONCEPT   Sulfur hexafluoride and the sulfur chlorides

Sulfur has oxidation number +6 (its highest oxidation state) in SF_6, but the large number of strongly electronegative fluorine atoms around the sulfur protect the compound from attack. It is a good gas-phase electrical insulator. S_2Cl_2 and SCl_2 both have bad odors. S_2Cl_2 is used for the vulcanization of rubber; it can react with more chlorine to form SCl_2.

EXAMPLE   Understanding sulfur hexafluoride and the sulfur chlorides

What is the shape of SF_6?

SOLUTION The Lewis structure of SF_6 is

There are six regions of high electron concentration (the single bonds) surrounding the central sulfur atom, each with a fluorine atom attached; this arrangement results in an octahedral shape for the molecule.

EXERCISE Which of the three sulfur halide compounds discussed, SF_6, S_2Cl_2, or SCl_2 has the lowest vaporization temperature? (Do not look this up; reason the answer through.)

[*Answer:* SF_6, because it is nonpolar.]

KEY WORDS   Define or explain each term in a written sentence or two.

catenate	clathrate	Frasch process
chalcogens	Claus process	ion exchange
chemiluminescence	contact process	paramagnetic

GROUP 17: THE HALOGENS

20.9 THE ELEMENTS

KEY CONCEPT A   The properties of the Group 17 elements

The Group 17 elements F, Cl, Br, I, and As are called the halogens.

Fluorine.   Fluorine is notable for being the most electronegative element. Elemental fluorine is obtained by the electrolysis of an anhydrous, molten KF–HF mixture. Fluorine is highly reactive and highly oxidizing; it is used in the production of SF_6, Teflon, and UF_6 (for isotopic enrichment).

Chlorine.   Chlorine is obtained by the electrolysis of molten or aqueous NaCl. It is a pale green-yellow gas that is used in a number of industrial processes.

Bromine.   Bromine, a corrosive, reddish brown liquid of Br_2 molecules, is produced by the oxidation of Br^- by Cl_2. Bromine has a number of industrial uses, including organic chemical synthesis and photographic processes.

$$Cl_2(g) + 2Br^-(aq) \longrightarrow Br_2(l) + 2Cl^-(aq)$$

Iodine.   Iodine occurs as I^- in seawater and as an impurity in Chile saltpeter. The element is produced by the oxidation of I^- with Cl_2. It is a blue-black, lustrous solid of I_2 molecules. I_2 is slightly soluble in water, but it dissolves well in iodide solutions because it reacts with aqueous I^- to form a complex:

$$I_2(aq) + I^-(aq) \longrightarrow I_3^-(aq)$$

Because iodine is an essential trace element in human nutrition, iodides are often added to table salt. This "iodized salt" can prevent iodine deficiency, which leads to goiter (enlargement of the thyroid gland).

EXAMPLE   Understanding reactions and properties of the halogens

Comment on why it is possible to use Cl_2 to produce Br_2 from Br^-. What is $E°$ for the reaction?

SOLUTION The reaction is an oxidation-reduction reaction. Using the reduction potentials for the two half-reactions that are involved allows us to calculate $E°$ for the reaction:

$$
\begin{array}{ll}
Cl_2(g) + 2e^- \longrightarrow 2Cl^-(aq) & E°_{red} = 1.36 \text{ V} \\
\underline{Br^-(aq) \longrightarrow Br_2(l) + 2e^-} & \underline{E°_{ox} = -1.07 \text{ V}} \\
Cl_2(g) + 2Br^-(aq) \longrightarrow Br_2(l) + 2Cl^-(aq) & E° = +0.29 \text{ V}
\end{array}
$$

Because the standard reaction potential is positive, the reaction is thermodynamically favored and can be used to produce Br_2.

EXERCISE Is the lustrous appearance of I_2 consistent with its position in Group 17?

> [*Answer:* Yes. A lustrous appearance is a metallic property; going down a group results in an increase in metallic properties. I_2 is a nonmetal, but it has a metallike appearance.]

KEY CONCEPT B   Trends and differences

The halogens show smooth variations in the properties of the elements and their compounds, with some exceptions for fluorine. Many of their physical properties are determined by London forces. Fluorine displays only a -1 oxidation number in its compounds, whereas the other halogens typically show -1, $+1$, $+3$, $+5$, and $+7$ in their compounds. Because of fluorine's small size, high electronegativity, and lack of usable *d*-orbitals, fluorine compounds are somewhat unusual. For instance, fluorine tends to bring out high oxidation numbers in elements to which it is bonded (e.g., $+7$ for I in IF_7). The small size of the fluoride ion results in high lattice enthalpies for fluorides, a property that makes them less soluble than other halides. The low bond energy of F_2 and high strength of fluorine's bonds with other elements result in highly exothermic reactions when F_2 reacts. For instance, the reaction of F_2 with CH_4 is more exothermic than the reaction of Cl_2 with CH_4:

$$F_2(g) + CH_4(g) \longrightarrow CH_3F(g) + HF(g) \qquad \Delta H° = -434 \text{ kJ}$$

$$Cl_2(g) + CH_4 \longrightarrow CH_3Cl(g) + HCl(g) \qquad \Delta H° = -98 \text{ kJ}$$

The strong bonds formed by fluorine make many of its compounds relatively inert. Fluorine's ability to form hydrogen bonds results in relatively high melting points, boiling points, and enthalpies of vaporization for many of its compounds.

EXAMPLE   Understanding the properties of covalent halogens

HCl, HBr, and HI are all strong acids, whereas HF is a weak acid. Explain.

SOLUTION The HF bond is exceptionally strong (fluorine forms strong bonds in its compounds), so it does not ionize nearly as easily as HCl, HBr, and HI.

EXERCISE Fluoridation of domestic water supplies can be accomplished by adding NaF to the water to get about 1 ppm (part per million) by mass of F^- ion. In other words, for every million grams of water, 1 g of fluoride ion is present. How many grams of NaF must be added per gallon of water to get to 1 ppm of F^-? 1 qt = 0.9463 L.

[*Answer:* 0.008 g]

20.10   COMPOUNDS OF THE HALOGENS

KEY CONCEPT A   The interhalogens

The interhalogens are compounds in which a heavier halogen atom at the center of the compound bonds to an odd number of smaller halogens on the periphery (I_2Cl_6 does not exactly follow this pattern). They are typically formed by the direct reaction of the elements.

$$I_2(g) + F_2(g) \xrightarrow{275°C} IF_7(g)$$

As the central atom gets larger, the bond enthalpies in the compound increase, resulting in a generally (but not invariably) lower reactivity for the compounds as the size of the central atom increases. Table 20.6 of the text is a compilation of the known interhalogens.

EXAMPLE   Understanding the interhalogens

Explain the trend in bond enthalpies for the X—F bond in the following interhalogens: ClF_5, 154 kJ·mol^{-1}; BrF_5, 187 kJ·mol^{-1}; IF_5 269 kJ·mol^{-1}.

SOLUTION All these compounds have similar Lewis structures and the same shape; the differences must originate with some other factor. That factor is the size of the central atom. As the central atom gets smaller, the fluorine atoms on the periphery get more crowded and the compound less stable. That is, if we try to break an X—F bond, the crowding helps to "push" fluorine atoms away from the central atom, thus lowering the bond enthalpy.

EXERCISE The I—F bond enthalpy in IF_7 is one of the following: 232 kJ·mol^{-1}, 269 kJ·mol^{-1}, or 306 kJ·mol^{-1}. Which one? (Do not look this up; use the information in the preceding example.)

[*Answer:* 232 kJ·mol^{-1}]

KEY CONCEPT B   The hydrogen halides

The hydrogen halides (HX, X = halogen) can be prepared by the direct reaction of the elements or by the action of a nonvolatile acid on a metal halide:

Direct reaction of elements: $$Cl_2(g) + H_2(g) \longrightarrow 2HCl(g)$$

Acid + metal halide: $$CaF_2(s) + 2H_2SO_4(aq, conc) \longrightarrow Ca(HSO_4)_2(aq) + 2HF(g)$$

$$KI(s) + H_3PO_4(aq) \longrightarrow KH_2PO_4(aq) + HI(g)$$

All hydrogen halides are colorless, pungent gases above 19°C. HF has the unusual property of attacking and dissolving silica glasses and is used for glass etching. In this reaction, F^- acts as a Lewis base and replaces oxygen atoms in the silicate structure:

$$6HF(aq) + SiO_2(s) \longrightarrow H_2SiF_6(aq) + 2H_2O(l)$$

In the etching process, the disruption of the glass structure at the surface of the glass causes the glass to become opaque.

EXAMPLE Explaining the physical and chemical properties of the hydrogen halides

Predict the trend in enthalpies of vaporization for the hydrogen halides.

SOLUTION Enthalpies of vaporization are determined largely by intermolecular attractions. There are three intermolecular attractions to consider: hydrogen bonding, dipole-dipole attractions, and London forces. The presence of hydrogen bonding in HF makes its enthalpy of vaporization the highest among the halogen halides. Dipole-dipole attractions decrease slightly in going down the group because the electronegatives of the halogens decrease slightly from 3.0 for Cl to 2.5 for I. However, London forces increase substantially because there is a significant increase in the number of electrons in going from HCl to HI. The increase in London forces should result in an increase in enthalpies of vaporization in going from HCl to HI.

EXERCISE Why doesn't HCl dissolve glass like HF does?

[**Answer:** The Si—F bond that forms when HF reacts with SiO_2 is very strong, whereas the Si—Cl bond that would form with HCl is not.]

KEY CONCEPT C   Halogen oxides and oxoacids

Hypohalous acid, HOX, halogen oxidation number $= +1$.   The hypohalous acids are prepared by the direct reaction of a halogen with water, and the corresponding hypohalite salts, are prepared by the reaction of a halogen with aqueous alkali:

Hypochorous acid: $$Cl_2(g) + H_2O(l) \longrightarrow HOCl(aq) + HCl(aq)$$

Sodium hypochlorite: $$Cl_2(g) + NaOH(aq) \longrightarrow NaOCl(aq) + HCl(aq)$$

Hypochlorites.   Hypochlorites oxidize organic material in water by producing oxygen in aqueous solution; the O_2 readily oxidizes organic materials. The O_2 production occurs in a two-step process, the net result of which is

$$2OCl^-(aq) \longrightarrow 2Cl^-(aq) + O_2(g)$$

Chlorates, ClO_3^-, chlorine oxidation number $= +5$.   Chlorates are prepared by the reaction of chlorine with hot aqueous alkali:

$$3Cl_2(g) + 6OH^-(aq, hot) \longrightarrow ClO_3^-(aq) + 5Cl^-(aq) + 3H_2O(l)$$

Chlorates decompose when heated, with the identity of the final product depending on the presence of a catalyst:

$$4KClO_3(l) \xrightarrow{\Delta} 3KClO_4(s) + KCl(s)$$

$$2KClO_3(l) \xrightarrow{\Delta, MnO_2} 2KCl(s) + 3O_2(g)$$

Chlorates are good oxidizing agents. They are also used to produce the important compound ClO_2, frequently through reduction with SO_2:

$$2NaClO_3(aq) + SO_2(g) + H_2SO_4(aq) \longrightarrow 2NaHSO_4(aq) + 2ClO_2(g)$$

Chlorine dioxide, ClO_2, chlorine oxidation number = +4. Chlorine dioxide is an odd-electron, paramagnetic, yellow-green gas that is used to bleach paper pulp. It is highly reactive and may explode violently if the circumstances are right.

Perchlorates, salts containing ClO_4^-, chlorine oxidation number = +7. Perchlorates are prepared by the electrolysis of aqueous chlorates. The half-reaction involved is

$$ClO_3^-(aq) + H_2O(l) \longrightarrow ClO_4^-(aq) + 2H^+(aq) + 2e^-$$

Perchlorates and perchloric acid ($HClO_4$) are powerful oxidizing agents; perchloric acid in contact with small amounts of organic matter may explode.

EXAMPLE **Understanding the reactions and structure of halogen oxides**

The hypohalous ions ClO^-, BrO^-, and IO^- all undergo the disproportion reaction

$$3XO^-(aq) \longrightarrow XO_3^-(aq) + 2X^-(aq)$$

The reaction is slowest with hypochlorite, faster with hypobromate, and fastest with hypoiodate. Suggest a reason for this order of reactivities.

SOLUTION The reaction is a Lewis acid-base reaction in which oxygen atoms form new bonds with the halogen atoms and the old halogen-oxygen bonds break. As the size of the central halogen atom increases, their is more room for the electrons on the oxygen atom to approach the halogen and form a bond, so the reaction becomes faster.

EXERCISE Predict the relative strengths of the following acids in water: HOCl, HOBr, and HOI.

[*Answer:* Strongest HOCl > HOBr > HOI weakest]

KEY WORDS Define or explain each term in a written sentence or two.

interhalogens
iodized salt

GROUP 18: THE NOBLE GASES

20.11 THE ELEMENTS

KEY CONCEPT The noble gases are monatomic, unreactive gases

The Group 18 elements are He, Ne, Ar, Xe, Kr, and Ra. Until 1962 no compounds of these so-called inert gases had been synthesized. Now that synthesis of some compounds has been accomplished, the elements in this group are called the noble gases. All have close-shell configurations and are chemically quite unreactive. All occur as monatomic gases.

Helium. Helium is the second most abundant element in the universe. It is formed in the Earth through the emission of alpha particles (helium nuclei), which capture two electrons to become helium atoms. Its low density and lack of flammability (unlike hydrogen) make it an ideal gas for lighter-than-air aircraft. Helium has the lowest boiling point of any known substance (4.2 K) and finds extensive use as a **cryogenic** coolant. (Cryogenics is the science of low temperatures.) It cannot be solidified at normal pressure. Below 1 K, it becomes a **superfluid,** a fluid that has zero viscosity and, therefore, no resistance to flow.

Neon, argon, krypton, xenon. These noble gases find specialized uses as cryogenic fluids and in lighting applications.

Radon. Radon is highly radioactive. It is formed as a natural product of radioactive decay in the Earth and seeps out of the ground in some areas of the United States (and elsewhere). In those areas, there is the possibility that it may find its way into houses, where it may pose a health threat because of its radioactivity.

EXAMPLE   Explaining the physical properties of the noble gases

The phase diagram of helium follows. At what temperature and pressure are the liquid, solid, and vapor at equilibrium?

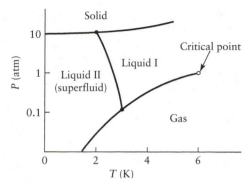

SOLUTION One of the many unique features of helium is its lack of a standard triple point. There is no temperature and pressure at which the liquid, solid, and vapor coexist at equilibrium.

EXERCISE Explain why helium does not form a solid at normal pressures.

[*Answer:* The intermolecular London forces are so weak that they cannot bring the atoms together to form the solid.]

20.12   COMPOUNDS OF THE NOBLE GASES

KEY CONCEPT   There are very few noble gas compounds

Helium, neon, and argon form no stable compounds. Krypton forms one stable neutral compound, KrF_2. Xenon displays a rich chemistry involving, for the most part, Xe—O and Xe—F bonds. Because of radon's radioactivity, its chemistry is largely unexplored. Compounds with relatively unstable Xe—N, Xe—C, and Kr—N bonds have been reported. Xenon displays oxidation states of +2, +4, +6, and +8. Direct reaction of fluorine with xenon at high temperature results in the formation of XeF_2 (Xe oxidation number = +2), XeF_4 (Xe oxidation number = +4), and XeF_6 (Xe oxidation number = +6). All three compounds are covalent molecules in the gas phase. They are powerful **fluorinating agents** (reagents that attach fluorine to other substances). Xenon trioxide can be synthesized by the hydrolysis of xenon tetrafluoride:

$$6XeF_4(s) + 12H_2O(l) \longrightarrow 2XeO_3(aq) + 4Xe(g) + 3O_2(g) + 24HF(aq)$$

The trioxide is the anhydride of xenic acid (H_2XeO_4); in aqueous alkali, the acid forms $HXeO_4^-$, which further disproportionates to the perxenate ion (XeO_6^{4-}, Xe oxidation state = +8).

$$2HXeO_4^-(aq) + 2OH^-(aq) \longrightarrow XeO_6^{4-}(aq) + Xe(g) + O_2(g) + 2H_2O(l)$$

EXAMPLE   Predicting the oxidizing ability of noble gas compounds

Would you expect the perxenate ion to be a good oxidizing agent? Why?

SOLUTION We would predict it to be a good oxidizing agent because the xenon in perxenate has a very high oxidation number (+8) and a strong tendency to gain electrons in order to assume a lower oxidation number.

EXERCISE Suggest why XeF_4 is such a good fluorinating agent in reactions such as

$$\text{Compound} + XeF_4(s) \longrightarrow \text{compound—F} + Xe(g)$$

[*Answer:* When XeF_4 gives up its fluorine atoms, it forms the very stable Xe(g). The formation of a stable product helps to make a reaction occur.]

Define or explain each term in a written sentence or two.

cryogenics
interhalogen
superfluidity

SELF-TEST EXERCISES

Group 15: The nitrogen family

1. One of the following does not describe nitrogen. Which one?

(a) high electronegativity (b) highly reactive (c) forms multiple bonds
(d) small size (e) no usable d-orbitals

2. On a mole basis, air is 78% N_2. What is the partial pressure of N_2 in a 8.5-L flask filled with air at 325 K? There is a total of 0.544 mol of air in the flask. $R = 0.0821 \text{ L} \cdot \text{atm} \cdot \text{mol}^{-1} \cdot \text{K}^{-1}$.

(a) 3.1 atm (b) 1.3 atm (c) 2.4 atm (d) 1.7 atm (e) 3.6 atm

3. What is the P—P—P bond angle in P_4?

(a) 90° (b) 120° (c) 45° (d) 60° (e) 180°

4. What partial pressure (in Torr) of P_4 develops when 10.0 g $Ca_3(PO_4)_2$ are completely reacted as shown at 800°C in a 5.0-L closed vessel?

$$2Ca_3(PO_4)_2(s) + 6SiO_2(s) + 10C(s) \longrightarrow P_4(g) + 6CaSiO_3(s) + 10CO(g)$$

(a) 444 Torr (b) 677 Torr (c) 326 Torr
(d) 918 Torr (e) 216 Torr

5. What is the pH of a 0.15 M solution of nitrous acid? $pK_a = 3.47$.

$$HNO_2(aq) + H_2O(l) \rightleftharpoons H_3O^+(aq) + NO_2^-(aq)$$

(a) 3.40 (b) 2.15 (c) 1.45 (d) 1.71 (e) 4.38

6. What is the oxidation number of phosphorus in the pyrophosphate ion, $P_2O_7^{4-}$?

(a) +3 (b) +9 (c) +5 (d) −6 (e) −2

Group 16: The oxygen family

7. Use that fact that O_2 is 23% by mass of the atmosphere to estimate its partial pressure when atmospheric pressue is 1.0 atm. Assue the remaining atmospheric gas is N_2.

(a) 0.16 atm (b) 0.21 atm (c) 1.0 atm
(d) 0.23 atm (e) 0.32 atm

8. From the boiling points of O_2 (−183°C) and O_3 (−112°C), predict the boiling point of S_2. (Because of chemical reactions, this boiling point has not yet been measured.)

(a) −254°C (b) −41°C (c) −183°C (d) −112°C (e) −148°C

9. How many kilograms of solid sulfur can be produced from 1.0 kg of H_2S?

(a) 1.4 kg (b) 0.82 kg (c) 0.65 kg (d) 0.94 kg (e) 0.79 kg

10. Which of the Group 16 elements is best described as a nonmetal with some metallic characteristics?

(a) O   (b) S   (c) Se   (d) Te   (e) Po

11. Which of the Group 16 elements has the least ability to catenate?

(a) O   (b) Se   (c) Po   (d) Te   (e) S

12. What is the role of water in the following reaction?

$$PCl_5(s) + 4H_2O(l) \longrightarrow H_3PO_4(aq) + 5HCl(aq)$$

(a) Brønsted base   (b) reducing agent   (c) oxidizing agent

(d) Lewis base   (e) solvent

13. What is the pH of a 3.0% by mass solution of H_2O_2? Assume the solution has a density of $1.0 \text{ g} \cdot \text{mL}^{-1}$. The pK_a of H_2O_2 is 11.75.

(a) 11.86   (b) 5.90   (c) 8.43   (d) 2.35   (e) 7.00

14. What is a clathrate?

(a) a compound in which molecules are trapped in a cage

(b) an oxygen-containing compound that is amphiprotic

(c) a Lewis base that is also an oxidizing agent

(d) a strongly hydrated covalent compound

15. Sulfur trioxide is the anhydride of which acid?

(a) H_2SO_3 (b) $H_2S_2O_6$ (c) H_2S (d) $H_2S_2O_3$ (e) H_2SO_4

16. The high boiling point of H_2SO_4 is largely due to

(a) London forces   (b) dipole-dipole attractions

(c) self-ionization and ion-ion attractions   (d) hydrogen bonding

17. The ionization energy of SF_6 is high because of the

(a) low electronegativity of sulfur   (b) highly symmetrical octahedral shape

(c) high electronegativity of fluorine   (d) large number of unshared electron pairs

(e) expanded octet of sulfur

Group 17: The halogens

18. Which of the halogens is found in nature in its elemental form?

(a) fluorine   (b) chlorine   (c) bromine

(d) iodine   (e) none of these

19. What is the hybridization of the central iodine atom in I_3^-?

(a) sp^2 (b) sp^3 (c) sp (d) sp^3d^2 (e) sp^3d

20. From the trends in Group 17, predict which of the following does not describe astatine. Because of its short half-life, the properties of At are not well known.)

(a) lustrous   (b) colorless   (c) solid

(d) exists as an At_2 molecule   (e) forms -1 ion

21. Which compound has the highest lattice enthalpy?

(a) NaCl   (b) NaF   (c) NaI   (d) NaBr

22. How many liters of HF(g) measured at STP can be produced from 25.0 g of CaF_2?

(a) 22.4 L   (b) 28.6 L   (c) 14.3 L   (d) 7.15 L   (e) 11.2 L

23. Which hydrogen halide has the highest enthalpy of vaporization?

(a) HI   (b) HBr   (c) HCl   (d) HF

24. Chlorine has oxidation numbers ranging from -1 to $+7$. With what oxidation state is it most likely to participate in disproportionation reactions?

(a) $+1$ (b) -1 (c) $+7$ (d) $+3$

25. What products are formed when $KClO_3$ is heated with no catalyst present?

(a) $KClO_4$, KCl   (b) $KClO_2$, KCl   (c) KCl, O_2

(d) KCl, ClO_2 (e) no reaction occurs

26. Perchloric acid is prepared by the aqueous electrolytic oxidation of

(a) ClF_3 (b) ClO_2 (c) Cl_2 (d) ClO_3^- (e) HCl

Group 18: The noble gases

27. Which noble gas has the highest boiling point?

(a) Kr   (b) Ne   (c) Ar   (d) Xe   (e) Rn

28. A superfluid is a liquid

(a) that has zero viscosity

(b) with unusually strong hydrogen bonds

(c) that will not vaporize or solidify

(d) that can dissolve any known solid

(e) that reacts with almost all known substances

29. Three noble gases have been observed to form stable, neutral compounds. Which noble gas is least likely to form such compounds?

(a) He   (b) Ne   (c) Ar   (d) Xe   (e) Kr

30. Xenon fluorides are prepared by heating Xe with

(a) HF   (b) NaF   (c) IF_7 (d) F_2 (e) PtF_4

31. Xenon has been found to form stable bonds with only one element besides fluorine. What element is it?

(a) chlorine   (b) carbon   (c) oxygen   (d) sodium   (e) hydrogen

32. All of the following but one are known xenon species. Which one has not been synthesized?

(a) XeO_6^{4-} (b) $HXeO_4^-$ (c) XeO_2 (d) XeO_3

CHAPTER 21

THE *d*-BLOCK: METALS IN TRANSITION

Many of the familiar metals we use as structural metals and in our everyday technology are transition metals, as are the precious metals silver, gold, and platinum. Mercury, used in thermometers, and copper, used as wire, are also part of the *d*-block. It is not surprising that the text calls the transition metals the "workhorse" elements.

THE *d*-BLOCK ELEMENTS AND THEIR COMPOUNDS

21.1 TRENDS IN PHYSICAL PROPERTIES

KEY CONCEPT   Orbital shapes affect many transition metal properties

The members of the *d*-block of the periodic table are called the **transition metals.** An important aspect of their chemistry is that their *d*-electrons are more strongly bound than their valence *s*-electrons. The elements in the last row of the *d*-block (Period 6, from Lu through Hg) also have chemical and physical properties that are strongly affected by the presence of underlying *f*-electrons. The lack of shielding from *d*-electrons causes the metallic radii to decrease slightly in going from left to right and the first ionization energies to increase slightly. The trend is not uniform, however. Iron has the smallest radius and highest ionization energy; electron-electron repulsions in doubly occupied *d*-orbitals increase the atomic radii and decrease the ionization energies after iron. The trends are approximately the same for the second and third rows of the *d*-block. However, the **lanthanide contraction** decreases the metallic radii and increases the ionization energies of the Period 6 elements, so they are similar to the values of the Period 5 elements. The lanthanide contraction also results in high densities for the third row (Period 6) of the *d*-block because the relatively heavy elements have small radii and correspondingly small volumes.

EXAMPLE   Predicting the radius of a *d*-block element

The metallic radius of Ni is 125 pm and that of the element directly below it, Pd, is 137 pm. Predict the radius of Pt, the element below Pd.

SOLUTION Owing to the lanthanide contraction, the metallic radius of each of the elements in the third row of the *d*-block is very close to that of the element directly above. In fact, from tantalum through mercury, the metallic radius is from 0 pm to 3 pm greater than the element above. Lanthanum and hafnium are still only 9 pm to 7 pm greater, respectively, than the elements directly above. We thus predict that the radius of Pt is between 137 and 140 pm, no more than 3 pm greater than that of Pd. (The actual radius is 139 pm.)

EXERCISE The metallic radius of Cu is 128 pm and that of the element directly below it, Ag, is 144 pm. Predict the radius of Au, the element below Ag.

[*Answer:* Predicted between 144 and 147 pm; actual 144 pm]

21.2   TRENDS IN CHEMICAL PROPERTIES

KEY CONCEPT   The transition metals display a large variety of oxidation states

The d-block elements display a wide range of oxidation numbers. Most of the elements have only one or two important oxidation numbers—that is, the oxidation numbers that appear in most of their compounds. Compounds in which elements have oxidation numbers high in their range tend to be good oxidizing agents; for instance, MnO_4^-, in which Mn has a +7 oxidation number, is a common and important oxidizing agent. Compounds with elements that have oxidation numbers low in their range tend to be good reducing agents; for example, Fe^{2+} is a commonly used reducing agent. Most d-metal oxides are basic, but there is a shift toward acidic oxides as the metal oxidation number increases. The elements on the left of the d-block resemble their s-block neighbors, whereas those on the right resemble their p-block neighbors.

EXAMPLE   Predicting the chemical properties of the d-block elements

Predict whether V_2O_5 (an orange solid) is a good oxidizing agent, a good reducing agent, or neither. Will it dissolve best in acid or base?

SOLUTION Vanadium has a +5 oxidation number in V_2O_5. According to Figure 21.6 of the text, this is the highest oxidation number that vanadium attains. We thus predict it will be a good oxidizing agent and that it will tend to be an acidic oxide. Because it is an acidic oxide, it will dissolve best in base.

EXERCISE Only one of the following transition metal compounds is acidic: Tc_2O_7, Fe_2O_3, NiO. Which one?

[*Answer:* Tc_2O_7]

21.3   SCANDIUM THROUGH NICKEL

KEY CONCEPT A   The elements scandium through manganese

Scandium.   Scandium is highly reactive and has few commercial uses. It exists only with a +3 oxidation number, and its aqueous ion $[Sc(H_2O)_6]^{3+}$ is fairly acidic.

Titanium.   Titanium is a light, strong metal that is passivated by an oxide coat. It appears in its ore with its highest oxidation number (+4) and, as might be expected, is difficult to remove from its ore. The +4 oxidation number is the most stable oxidation number for titanium and appears in its most important compound, TiO_2. This nontoxic compound is used as a pigment in paints. Titanium(IV) forms a series of compounds containing the titanate ion, TiO_3^{2-}. Barium titanate ($BaTiO_3$) is **piezoelectric,** which means that it develops an electric potential when it is mechanically stressed.

Vanadium.   Vanadium is a soft, silver-gray metal used in the production of iron alloys. Its most important compound is V_2O_5, which is used as a catalyst in the contact process for sulfuric acid. It forms many different colored compounds and thus finds use in glazes in the ceramics industry.

Chromium.   Chromium is a bright corrosion-resistant metal that forms compounds with many different colors. It is used in the production of stainless steel, as a coating for magnetic tapes (as CrO_2), and for chromium plating. The yellow solid sodium chromate (Na_2CrO_4), in which chromium has a +6 oxidation number, is one of chromium's most useful compounds. Placing the chromate ion in acid solution converts it to the dichromate ion ($Cr_2O_7^{2-}$) with no change in oxidation number:

$$2CrO_4^{2-}(aq) + 2H^+(aq) \longrightarrow Cr_2O_7^{2-}(aq) + H_2O(l)$$

Dichromates are common laboratory oxidizing agents; they are also corrosive and toxic, so they must be handled with care.

Manganese.   Manganese is a gray metal. Because of its lack of resistance to corrosion, it is rarely used alone. It is an important component of many alloys, however. In steel, it removes traces of sulfur by forming a sulfide that is not incorporated into the final steel produced. It is also used to make alloys of nonferrous metals. Manganese displays the oxidation numbers +1 through +7 in its compounds. The most stable oxidation number is +2, but +4 and +7 are also important. MnO_2 takes part in the cathode reaction in the Lelanche dry cell. MnO_2 is also the starting point for the production of many manganese compounds. Potassium permanganate, which contains Mn(VII) in the permanganate ion (MnO_4^-) is an extremely versatile laboratory oxidizing agent in acid solution. (It decomposes in basic solution.) Permanganate ion is reduced to Mn^{2+} when it is used as an oxidizing agent in acid solution:

$$MnO_4^-(aq) + 8H^+(aq) + 5e^- \longrightarrow Mn^{2+}(aq) + 4H_2O(l)$$

EXAMPLE   Predicting the reactions and properties of some *d*-block elements

Is an aqueous solution of Cr^{3+} likely to be acidic or neutral?

SOLUTION Cr^{3+} is a small, highly charged ion. As with Al^{3+} and Ti^{3+}, we would expect Cr^{3+} to form a highly hydrated ion in which electrons in the solvating water molecules are polarized, thus making the ion acidic. The equilibrium follows; in fact, Cr^{3+} has a pK_a of 4:

$$[Cr(H_2O)_6]^{3+}(aq) + H_2O(l) \rightleftharpoons [Cr(H_2O)_5(OH)]^{2+}(aq) + H_3O^+(aq)$$

EXERCISE $TiCl_4$ melts at $-23°C$ and boils at $136°C$. Is its bonding covalent or ionic?

[*Answer:* Covalent]

KEY CONCEPT B   The elements iron through nickel

Iron.   Iron is an important structural metal. It is extracted from its ore in a blast furnace. Calcium oxide (from the decomposition of limestone in the furnace charge) is used to remove silicon, aluminum, and phosphorus impurities from the ore through the reactions

$$CaO(s) + SiO_2(s) \longrightarrow CaSiO_3(l)$$

$$CaO(s) + Al_2O_3(s) \longrightarrow Ca(AlO_2)_2(l)$$

$$6CaO(s) + P_4O_{10}(s) \longrightarrow 2Ca_3(PO_4)_2(l)$$

The first stage in making steel (an alloy of iron) is to lower the carbon content of the pig iron that comes from a blast furnace and to remove the remaining impurities. This is done in a modern version of the **Bessemer converter,** using the **basic oxygen process,** in which oxygen and powdered limestone are forced through the molten metal. Iron is not a particularly hard metal. Alloying it makes it harder and improves its resistance to corrosion. Iron, its oxides, and many of its alloys display the property of **ferromagnetism,** the ability to be permanently magnetized. This is accomplished by forcing the magnetic fields of many of the electrons in the iron atoms to become aligned in the same direction, resulting in a large overall magnetic field. Iron is an essential dietary element because it is part of the hemoglobin complex that is responsible for oxygen transport.

Cobalt.   Cobalt is used mainly in iron alloys. However, as a component of vitamin B_{12}, it also is an essential dietary element.

Nickel.   Nickel is a hard, silver-white metal that is used mainly in the production of alloys; its most stable oxidation number is +2. Nickel is found mostly as a sulfide ore (NiS). Roasting this ore converts it to NiO. In the **Mond process,** NiO is reduced to Ni with hydrogen; the impure product thus produced is converted to $Ni(CO)_4$ (nickel carbonyl), and the impurities are removed. Pure nickel is then recovered from the nickel carbonyl. Nickel(III) is reduced to nickel(II) in the operation of nickel-cadmium (nicad) batteries.

EXAMPLE   Explaining the properties of iron and nickel

Explain why iron pentacarbonyl, $Fe(CO)_5$, has such a low boiling point (103°C).

SOLUTION $Fe(CO)_5$ is made up of neutral CO molecules bonded to a neutral Fe atom. No ions are involved. Thus, the intermolecular interactions in this compound do not involve ion-ion interactions. We have not learned how to predict the shape of this compound, but its low boiling point is an indication that it has the highly symmetric trigonal-bipyramidal shape and, therefore, no net dipole moment. This means that the only form of intermolecular attractions present are weak London forces, and the boiling point of the compound is correspondingly low.

EXERCISE How many liters of CO(g) measured at 50°C and 1.00 atm pressure are required to convert 125 g Ni to $Ni(CO)_4$ in the Mond process?

$$Ni(s) + 4CO(g) \longrightarrow Ni(CO)_4(g)$$

[*Answer:* 226 L]

21.4   GROUPS 11 AND 12

KEY CONCEPT   The copper and zinc families

The elements Cu, Ag, and Au (with the configuration $d^{10}s^1$) and Zn, Cd, and Hg (with the configuration $d^{10}s^2$) are all relatively unreactive. Cu, Ag, and Au are known as the **coinage metals.**

Copper.   Copper ore is separated from unwanted rock by using **froth flotation.** The metal is then extracted from its ore by using either the **pyrometallurgical process** or the **hydrometallurgical process.** Copper is an excellent electrical conductor and forms useful alloys such as bronze (when alloyed with tin) and brass (when alloyed with zinc). Copper metal does not displace H^+ from acids, but it can be oxidized by acids (such as HNO_3) that have anions that are good oxidizing agents. Copper corrodes in moist air to form $Cu_2(OH)_2CO_3(s)$, a green compound that is apparent on aged decorative copper trim:

$$2Cu(s) + H_2O(l) + O_2(g) + CO_2(g) \longrightarrow Cu_2(OH)_2CO_3(s)$$

Copper forms compounds with an oxidation number of +1, but in water Cu(I) tends to disproportionate to Cu(s) and Cu(II).

Silver.   Silver lies above H^+ in the electrochemical series, so it does not produce H_2 in acid solutions. It is oxidized by nitric acid in a reaction that is reminiscent of copper's reaction with nitric acid:

$$3Ag(s) + 4H^+(aq) + NO_3^-(aq) \longrightarrow 3Ag^+(aq) + NO(g) + 2H_2O(l)$$

$$3Cu(s) + 8H^+(aq) + 2NO_3^-(aq) \longrightarrow 3Cu^{2+}(aq) + 2NO(g) + 4H_2O(l)$$

Aqueous Ag(I), unlike aqueous Cu(I), does not disproportionate, and almost all silver compounds have silver with a +1 oxidation number. AgF and $AgNO_3$ are the only common soluble silver salts. Silver is used extensively in the photographic industry.

Gold.   Gold is extremely inert. Its most common oxidation number in compounds is +3. Aqueous Au(I) tends to disproportionate into Au(s) and Au(III).

Zinc.   Zinc occurs mainly as ZnS. It is obtained from its ore by froth flotation followed by smelting with coke. It is a silvery, reactive amphoteric metal. The oxidation number of zinc in most of its compounds is +2. It reacts with acids to form Zn^{2+} and with alkalis to form $[Zn(OH)_4]^{2-}$:

$$Zn(s) + 2H^+(aq) \longrightarrow Zn^{2+}(aq) + H_2(g)$$

$$Zn(s) + 2OH^-(aq) + 2H_2O(l) \longrightarrow [Zn(OH)_4]^{2-} + H_2(g)$$

Cadmium.   Cadmium is similar to zinc but more metallic, and its oxides are more basic than zinc's. Cadmium salts are extremely toxic. The oxidation number of cadmium in most of its compounds is +2.

Mercury. Mercury occurs as HgS (cinnabar). It is obtained from its ore by froth flotation followed by roasting in air. The volatile metal distills off the roasting mixture as Hg(g). It is then condensed to Hg(l). Mercury lies above hydrogen in the electrochemical series, so it does not form H_2 when placed in acids. It can be oxidized by nitric acid, in a reaction that is very similar to those we saw earlier for copper and silver:

$$3Hg(l) + 8H^+(aq) + 2NO_3^-(aq) \longrightarrow 3Hg^{2+}(aq) + 2NO(g) + 4H_2O(l)$$

Mercury is found in its compounds with oxidation numbers +1 and +2. The +1 ion is unusual; it is a covalently bonded, diatomic cation $Hg-Hg^{2+}$, written as Hg_2^{2+}. In a reaction that is often used to test chemically for the presence of mercury, Hg_2Cl_2 disproportionates when mixed with aqueous ammonia:

$$Hg_2^{2+}(aq) \longrightarrow Hg(l) + Hg^{2+}(aq)$$

Mercury vapors and salts are both toxic.

EXAMPLE   Understanding the coinage metals

Explain why gold is less reactive than silver.

SOLUTION Because of the lanthanide contraction, gold and silver have the same metallic radii (144 pm). However, gold has 32 more protons in the nucleus (which result in an additional +32 nuclear charge), so its outer electrons are much more tightly bound than silver's. Gold's valence electrons are, therefore, much less easily lost than silver's, so it is much less reactive.

EXERCISE Why don't Zn, Cd, and Hg form compounds with a +3 oxidation number?

[*Answer:* The high, poorly shielded nuclear charge holds the *d*-electrons too tightly.]

KEY WORDS   Define or explain each term in a written sentence or two.

basic oxygen process	hydrometallurgical process	smelted
domain	lanthanide contraction	transition metal
ferromagnetism	piezoelectric	
froth flotation	pyrometallurgical process	

COMPLEXES OF THE *d*-BLOCK ELEMENTS

21.5   THE STRUCTURES OF COMPLEXES

KEY CONCEPT A   Complexes are formed by a Lewis acid-base reaction

The Lewis acid in a complex is always a metal cation or metal atom and is frequently referred to as the *central metal ion* or *central metal atom*. The Lewis base, which bonds to the central metal ion or atom, is called a **ligand**. The term **coordination compound** refers to a complex or compound that contains a complex. The ligands that are directly bonded to the central ion are collectively referred to as the **coordination sphere**. Two common and important types of complexes are those that have six ligands in their coordination sphere (a *six-coordinate* complex) and those with four ligands in their coordination sphere (a *four-coordinate* complex). Six-coordinate complexes are usually octahedral, whereas four coordinate complexes are either tetrahedral or square planar. Because water is a Lewis base, most aqueous ions are complexed by the water (unless a stronger complexing agent is present). Some ligands, such as ethylenediamine and ethylenediaminetetraacetate ion

(EDTA$^-$), contain more than one atom that can bond to the central metal ion; such ligands are called **chelates**.

EXAMPLE **Predicting features of complexes**

What is the likely shape of the complex [PtCl$_4$]$^{2-}$?

SOLUTION There are four chloride ions complexed to the central metal ion, so we expect the complex to be either tetrahedral or square planar. To decide which structure is more likely, we must determine the electron configuration of the central metal ion. Because there are four Cl$^-$ ions and the total charge on the complex is -2, the platinum must have a $+2$ charge. Pt has a [Xe]$4f^{14}5d^96s^1$ electron configuration; thus, the Pt^{2+} ion has an [Xe]$4f^{14}5d^8$ configuration. Ions with a d^8 configuration tend to form square-planar complexes and we conclude (correctly) that this is the shape of [PtCl$_4$]$^{2-}$.

EXERCISE What is the coordination number of an octahedral complex?

[*Answer:* Six]

KEY CONCEPT B   Naming complexes

A complex is named by naming the ligands in alphabetical order, with a prefix to indicate the number of each ligand, followed by the name of the central metal, including its oxidation state in roman numerals in parentheses. If the whole complex is neutral or positively charged, the metal name used is the same as the element name. If the complex is negatively charged, the metal is named by adding the suffix -*ate* to the stem name of the metal. Complicated ligand names are placed in parentheses, with the prefix *bis-* (rather than *di-*) to indicate two of them are present or *tris-* (instead of *tri-*) to indicate three are present. Finally, an ionic coordination compound is named like any other ionic compound, with the name of the cation first, followed by the name of the anion.

EXAMPLE **Naming complexes**

(a) Name the complex [TiCl$_2$F$_4$]$^{2-}$, and (b) write the formula of tetraaquaethylenediamminechromium (IV). State the shape of both.

SOLUTION (a) This complex has four fluoride ions and two chloride ions as ligands. Each ligand has a -1 charge, so the ligands contribute a charge of -6 to the complex. Because the total charge of the complex is -2, the titanium must have a $+4$ charge and a $+4$ oxidation number. Because the complex is negatively charged, it has an -*ate* suffix and is named dichlorotetrafluorotitanate(IV) ion. Because there are six ligands, it is a six-coordinate complex and is probably octahedral. (b) This complex has four water molecules (*tetraaqua*) and one ethylenediamine (en) molecule as ligands. All the ligands are neutral, so the total charge on the complex is the same as that of the central metal ion, $+4$. The formula is [Cr(en)(H$_2$O)$_4$]$^{4+}$. Because ethylenediamine is a bidentate ligand, the complex is a six-coordinate complex and has an octahedral shape.

EXERCISE (a) Name the complex [Ni(H$_2$O)$_2$(NH$_3$)$_4$]$^{2+}$ and (b) write the structural formula of monocarbonylpentacyanomolybdate(IV)

[*Answer:* (a) Tetraamminediaquanickel(II); (b) [Mo(CN)$_5$(CO)]$^-$]

21.6   ISOMERS

KEY CONCEPT   Isomers are different compounds with the same molecular formula

Isomers are different compounds that have the same molecular formula. There are two broad classes of isomers. **Structural isomers** are isomers in which one or more atoms are linked to different neighbors in different isomers. We will consider four different types of structural isomers.

Ionization isomers.   In a pair of ionization isomers, a ligand and anion in one isomer switch positions to form the other isomer. Thus, when the compound ionizes, different species form the ions. Symbolically, $[MA_4B_2]C_2$ and $[MA_4C_2]B_2$ are ionization isomers. An example of this is found in the compounds $[PtCl_2(NH_3)_2]Br_2$ and $[PtBr_2(NH_3)_2]Cl_2$.

Hydrate isomers. Hydrate isomers are actually ionization isomers in which water is one of the ligands whose position is different in different isomers. Symbolically, $[M(H_2O)_6]B_3$ and the hydrated salt $[MB(H_2O)_5]B_2 \cdot H_2O$ would be hydrate isomers. An example of this is found in the hydrate isomers $[Cr(H_2O)_6]Cl_3$ and $[CrCl(H_2O)_5]Cl_2 \cdot H_2O$.

Linkage isomers. Linkage isomerism can occur when the ligand(s) in the complex have more than one atom that can bond to the central metal ion. Two isomers differ with respect to which of the atoms in the isomer is actually bonded to the central metal ion. For instance, in $[Co(NH_3)_5NCS]^{2+}$, the nitrogen in the thiocyantae ion (SCN^-) is bonded to Co^{3+}. In $[Co(NH_3)_5SCN]^{2+}$, it is the sulfur that bonds to Co^{3+}.

Coordination isomerism. Coordination isomerism is reminiscent of ionization isomerism. It occurs in situations in which both the anion and cation of a compound are complexes and a ligand in the anionic complex in one isomer is exchanged with a ligand in the cationic complex to form the second isomer. Thus, $[MA_6][NB_6]$ and $[MB_6][NA_6]$, in which M and N are metals, are coordination isomers. $[Cr(NH_3)_6][Fe(CN)_6]$ and $[Cr(CN)_6][Fe(NH_3)_6]$ are coordination isomers.

Stereoisomers are isomers in which all the atoms are linked to the same neighbors in different isomers, but the arrangement of atoms in space is different for different isomers. We will consider two different types of stereoisomers.

Geometrical isomerism. In geometrical isomers, the ligands are arranged such that neighboring ligands are different for the different isomers. The cis-trans isomerism that occurs for octahedral complexes of the form MX_4Y_2 is a type of geometrical isomerism. In the trans isomer, the two Y ligands are across from each other; in the cis isomer, the two Y ligands are next to each other:

cis   trans

Optical isomerism. In optical isomerism, a molecule or ion and its mirror image are not the same. To understand the relationship of a mirror image to the object making the image when optical isomerism occurs, place your left hand next to your right hand, palm to palm. One hand appears as a mirror image of the other. But are your left and right hands the same? Try putting a left-hand glove on your right hand; clearly, they are not the same. The relationship of a right hand to a left hand is the same as the relationship between two optical isomers. They are mirror images of each other, but different. Optical isomers are not geometrical isomers. A molecule that is different from its mirror image is said to be **chiral**. We should be aware that in many cases a molecule is the same as its mirror image and is, therefore, not chiral. Each of the two types of a chiral molecule (the molecule and its mirror image) is called an **enantiomer**. Enantiomers have identical properties except that they rotate the plane of **plane-polarized light** in opposite but equal directions. A molecule or ion that rotates the plane of plane-polarized light is said to be **optically active**. The (+)-enantiomer rotates the plane clockwise (as seen by an observer the light beam is aimed at) and the (−)-**enantiomer** rotates the plane counterclockwise by exactly the same amount.

EXAMPLE 1   Predicting properties of isomers

What ions form when $[PtCl_2(NH_3)_2]Br_2$ ionizes, and what ions form when its ionization isomer $[PtBr_2(NH_3)_2]Cl_2$ ionizes?

SOLUTION We must understand the structures of these isomers to answer the question. $[PtCl_2(NH_3)_2]Br_2$ is made up of the $[PtCl_2(NH_3)_2]^{2+}$ cation (in which Pt exists in the +4 oxidation state) and two Br^- anions.

When it ionizes, it forms the $[PtCl_2(NH_3)_2]^{2+}$ cation and two Br^- anions. Using a similar analysis for $[PtBr_2(NH_3)_2]Cl_2$ (in which Pt also exists in the +4 oxidation state), $[PtBr_2(NH_3)_2]^{2+}$ cation and two Cl^- anions are formed upon ionization.

EXERCISE The cis form of a square-planar complex is represented below. Draw the structure of the trans form.

B—A
M
B—A

Answer:

A—B
M
B—A

EXAMPLE 2 Predicting the chirality of complexes

Is the following octahedral complex chiral?

A
B
B——A
B
C

SOLUTION If the mirror image of the complex is different from the complex, the complex is chiral. Otherwise, it is not chiral. To find out the relationship between the complex and its mirror image, we draw the complex (a) and its mirror image (b) in the following figure. Now, we reorient the mirror image (on paper) to see if it is identical to the original complex. In this case, a 90° clockwise rotation (looking down from the top of the complex) results in (c). A point-by-point comparison indicates that (a) and (c) are identical. Thus, the complex and its mirror image are identical and the complex is not chiral.

(a) Mirror (b) → (c)

EXERCISE Is the following octahedral complex chiral?

[*Answer:* Yes]

KEY WORDS   Define or explain each term in a written sentence or two.

chelate
coordinate
coordination compounds
coordination number
coordination sphere

isomerism (ionization, hydrate, linkage, coordination, optical)
isomers
ligand
octahedral complex

square-planar complex
substitution reaction
tetrahedral complex

21.7   THE EFFECTS OF LIGANDS ON d-ELECTRONS

KEY CONCEPT   Ligands affect central metal d-orbital energies

The starting point of **crystal field theory** is the assumption that the ligands in a complex have bonded to and oriented around the central metal atom. We then ask what effects the ligands have on the central metal d-electrons. The ligands are often negatively charged or, at the very least, have a large electron cloud directed at the central atom. In an octahedral complex, two of the five d-orbitals point directly toward one or more ligands, and three of the five point between the ligands. Electrons in the two orbitals that point directly at the ligands experience repulsions from the electrons on the ligands and have a higher energy than they would if they were uniformly dispersed around the nucleus. These higher energy orbitals are called **e-orbitals.** Electrons in the three orbitals that point between the ligands have a lower energy than they would if they were uniformly dispersed around the nucleus. These lower energy orbitals are called **t-orbitals.** This behavior results in the following orbital diagram.

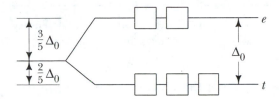

The difference in energy between the two sets of orbitals is called the **ligand field splitting,** symbolized Δ_0.

EXAMPLE   Predicting the effect of ligands on d-orbital energies

Water bonds more strongly to Cr^{3+} in $[Cr(H_2O)_6]^{3+}$ than does chloride ion in $[CrCl_6]^{3-}$. Which of these two complexes will have the highest ligand field splitting (largest Δ_0)?

SOLUTION We assume that the stronger bonding of H_2O means that it is closer to the central Cr^{3+} ion than the Cl^- is to the central metal Cr^{3+} in their respective complexes. So, the H_2O interacts more strongly with the Cr^{3+} d-orbitals than does the Cl^-; this, in turn, makes the ligand field splitting larger in $[Cr(H_2O)_6]^{3+}$ than in $[CrCl_6]^{3-}$.

EXERCISE Assume we have an imaginary tetrahedral complex involving a central berylium atom with sp^3 hybrid orbitals interacting with the four ligands. (Such a complex does not exist.) Comment, qualitatively, on the size of the ligand field splitting in such a complex.

[*Answer:* It is zero. There would be no ligand field splitting in such a complex.]

21.8   THE EFFECTS OF LIGANDS ON COLOR

KEY CONCEPT   Electrons in d-orbitals can absorb energy

Electrons in the lower energy d-orbitals of a complex can absorb photons of energy $E = \Delta_0$, an absorption that often results in a colored complex. Using the relationship $E_{photon} = h\nu$, we find that the wavelength λ of the photon is related to Δ_0 through the relationship

$$\Delta_0 = hc/\lambda$$

Some ligands, called **weak-field ligands,** produce a small Δ_0 when they form a complex with a metal. Others, called **strong-field ligands,** result in a large Δ_0. The **spectrochemical series** is a list of ligands ordered according to the value of Δ_0 they typically cause in a complex. The color of complexes is caused by the absorption of specific wavelengths of light corresponding to Δ_0. Let's assume that white light, a mixture of photons of all colors with wavelengths ranging from about 420 nm (violet) to about 700 nm (red), impinges on a complex. If the complex absorbs photons of only one color, the transmitted light will no longer be white but will manifest the fact that one of the colors of light has been absorbed. The following table (and the color wheel in Fig. 21.41 of the text) indicate the relationship between the light absorbed and the color of the complex (the light transmitted). The wavelengths of absorbed light are approximate.

Color of Light Absorbed	Color of Complex
violet (420 nm)	yellow
blue (460 nm)	orange
blue-green (490 nm)	red
green (520 nm)	purple
yellow-green (560 nm)	red-violet
yellow (570 nm)	violet
orange (600 nm)	blue
red (700 nm)	green

The corresponding colors in the two columns are called **complementary colors**. However, we must be careful in interpreting the colors of complexes because complexes usually absorb a wide range of colors, not a single wavelength.

EXAMPLE   Calculating Δ_0 from the absorption of light by complexes

$[Co(NH_3)_6]^{3+}$ displays an absorption maximum at 437 nm. What is Δ_0 (in $kJ \cdot mol^{-1}$) for the complex? What color is the complex?

SOLUTION We use the relationship $\Delta_0 = hc/\lambda$ to calculate Δ_0. Substituting the values $h = 6.63 \times 10^{-34}$ J·s, $c = 3.00 \times 10^8$ m·s^{-1} and $\lambda = 437$ nm (437×10^{-9} m) gives

$$\Delta_0 = \frac{(6.63 \times 10^{-34}\ \text{J} \cdot \text{s})(3.00 \times 10^8\ \text{m} \cdot \text{s}^{-1})}{437 \times 10^{-9}\ \text{m}}$$

$$= 4.55 \times 10^{-19}\ \text{J}$$

This is the energy absorbed by a single complex that absorbs a single photon. For a mole of the complex, we must multiply this by Avogadro's number, 6.02×10^{23}:

$$\Delta_0\ (\text{kJ} \cdot \text{mol}^{-1}) = (4.55 \times 10^{-19}\ \text{J})(6.02 \times 10^{23})$$

$$= 2.74 \times 10^5\ \text{J} \cdot \text{mol}^{-1}$$

$$= 274\ \text{kJ} \cdot \text{mol}^{-1}$$

From the preceding table, a complex that absorbs light at 437 nm is yellow-orange.

EXERCISE $[CrF_6]^{3-}$ displays an absorption maximum at 658 nm. What is Δ_0 (in $kJ \cdot mol^{-1}$) for the complex? What color is the complex?

[*Answer:* 182 kJ·mol^{-1}; blue-green]

21.9   THE ELECTRONIC STRUCTURES OF MANY-ELECTRON COMPLEXES

KEY CONCEPT   The electronic configuration of complexes

To build up the valence-electron configuration of the central metal electrons in the t- and e-orbitals of a complex, we must add electrons to the orbitals so as to achieve the lowest possible energy.

Each orbital can hold a maximum of two electrons, and the two electrons in an orbital must have paired spins. Hund's rule applies, but it must be applied properly: When Δ_0 is large, the low-energy, t-orbitals fill before any electrons enter the e-orbitals. Hund's rule is applied to the lower t-orbitals first and then, after they fill, to the upper e-orbitals. When Δ_0 is small, electrons enter all five orbitals (following Hund's rule) before any pairing of electrons begins. The filling of orbitals for octahedral complexes is as follows:

	Large Δ_o		Small Δ_o	
	t-orbitals	e-orbitals	t-orbitals	e-orbitals
d^1	↑ □ □	□ □	↑ □ □	□ □
d^2	↑ ↑ □	□ □	↑ ↑ □	□ □
d^3	↑ ↑ ↑	□ □	↑ ↑ ↑	□ □
d^4	↑↓ ↑ □	□ □	↑ ↑ ↑	↑ □
d^5	↑↓ ↑↓ ↑	□ □	↑ ↑ ↑	↑ ↑
d^6	↑↓ ↑↓ ↑↓	□ □	↑↓ ↑ ↑	↑ ↑
d^7	↑↓ ↑↓ ↑↓	↑ □	↑↓ ↑↓ ↑	↑ ↑
d^8	↑↓ ↑↓ ↑↓	↑ ↑	↑↓ ↑↓ ↑↓	↑ ↑
d^9	↑↓ ↑↓ ↑↓	↑↓ ↑	↑↓ ↑↓ ↑↓	↑↓ ↑
d^{10}	↑↓ ↑↓ ↑↓	↑↓ ↑↓	↑↓ ↑↓ ↑↓	↑↓ ↑↓

Note that for the d^1, d^2, d^3, d^8, d^9, and d^{10} configurations, the size of Δ_0 does not affect the electron configuration. But for the d^4, d^5, d^6, and d^7 configurations, different electron configurations are obtained for the different sizes of Δ_0. For octahedral d^4 through d^7 complexes, the configurations obtained with a large Δ_0 minimize the number of unpaired electrons and are called **low-spin complexes**. Conversely, octahedral d^4 through d^7 complexes with a small Δ_0 have a maximum of unpaired electrons and are called **high-spin complexes**.

EXAMPLE Predicting the electronic configuration of a complex

What is the electron configuration of $[Fe(CN)_6]^{4-}$? Is it high spin or low spin?

SOLUTION $[Fe(CN)_6]^{4-}$ is an octahedral complex in which the d-orbitals are split into t-orbitals and e-orbitals. From page 835 of the text, we note than CN^- is a strong-field ligand, so we expect the ligand field splitting Δ_0 to be relatively large. The Fe^{2+} ion has six d-electrons; in a strong field, these six electrons stay in the lower energy t-orbitals. The orbital diagram is shown below; the electron configuration is t^6 and it is low spin.

e — □ □
t — ↑↓ ↑↓ ↑↓

EXERCISE What is the electron configuration of $[Fe(H_2O)_6]^{2+}$? Is it high spin or low spin?

Answer:

$$e \; \boxed{\uparrow} \; \boxed{\uparrow}$$
$$t \; \boxed{\uparrow\downarrow} \; \boxed{\uparrow} \; \boxed{\uparrow}$$

$t^4 e^2$, high spin

21.10   THE MAGNETIC PROPERTIES OF COMPLEXES

KEY CONCEPT   Magnetic properties are determined by unpaired electrons

The presence of unpaired electrons makes a species **paramagnetic**. We can see that many of the configurations presented in the previous section result in unpaired electrons; low-spin d^1–d^5 complexes, low-spin d^7–d^9 complexes, and all high-spin complexes are paramagnetic. On the other hand, low-spin d^6 and all d^{10} complexes have no unpaired electrons and are **diamagnetic**.

EXAMPLE   Predicting the properties of a complex

Predict whether $[Co(CN)_6]^{3-}$ is a high-spin or low-spin complex. How many unpaired electrons does it possess? Is it paramagnetic or diamagnetic?

SOLUTION We must first determine how many d-electrons are present in the complex. Because each CN^- contributes a -1 charge to the complex, the cobalt must have a $+3$ charge. A neutral cobalt atom has an $[Ar]3d^74s^2$ configuration, so Co^{3+} possesses a d^6 configuration. CN^- is very high in the spectrochemical series; thus, we expect a very large Δ_0. This, in turn, leads to a low-spin complex. A low-spin d^6 complex has six electrons in the three low-energy t-orbitals; so all the electrons are paired and the complex is diamagnetic.

EXERCISE Predict whether $[Fe(SCN)_6]^{3-}$ is a high-spin or low-spin complex. How many electrons does it possess? Is it paramagnetic or diamagnetic?

[*Answer:* High-spin d^5 complex; five unpaired electrons; paramagnetic]

KEY WORDS   Define or explain each term in a written sentence or two.

complementary color	high-spin complex	spectrochemical series
crystal field theory	ligand field splitting	strong-field ligand
diamagnetism	low-spin complex	t-orbital
e-orbital	paramagnetism	weak-field ligand

SELF-TEST EXERCISES

The d-block elements and their compounds

1. What electron configuration should lead to the largest metallic radius?
(a) $[Ar]3d^14s^2$ (b) $[Kr]4d^15s^2$ (c) $[Kr]4d^55s^1$
(d) $[Ar]3d^{10}4s^2$ (e) $[Kr]4d^75s^1$

2. Which element is expected to have a metallic radius nearest to that of Pd?
(a) Ni   (b) Pt   (c) Ag   (d) Ru   (e) Co

3. Five transition metal species are shown along with the range of oxidation numbers the metal displays in its compounds. Which species is likely to be the best oxidizing agent?

(a) Fe^{2+} ($+2 \longrightarrow +6$)　　　(b) MnO_2 ($+1 \longrightarrow +6$)　　　(c) V_2O_5 ($+1 \longrightarrow +5$)

(d) NiO ($+2 \longrightarrow +4$)　　　(e) Sc^{3+} ($+3$ only)

4. Which oxide of vanadium is most acidic?

(a) V_2O_3　　　(b) VO　　　(c) VO_2　　　(d) V_2O_5

5. Judging by electron configuration and position in the periodic table, which of the following would you expect to be, overall, the most reactive?

(a) Sc　　　(b) Ti　　　(c) V　　　(d) Mn　　　(e) Fe

6. The Sc^{3+} ion is

(a) basic.　　　(b) neutral.　　　(c) acidic.

7. Which compound of chromium is a common laboratory oxidizing agent?

(a) $CrCl_3$　　　(b) $K_4[Cr(CN)_6]$　　　(c) Cr_2O_3

(d) $KCrOF_4$　　　(e) $K_2Cr_2O_7$

8. What is the oxidation state of manganese in the hypomanganate ion, MnO_4^{3-}?

(a) $+5$　　　(b) $+7$　　　(c) $+3$　　　(d) $+9$　　　(e) $+8$

9. A ferromagnetic material is one that

(a) can permanently magnetize iron.　　　(b) can be permanently magnetized.

(c) can be chemically reduced by Fe^{2+}.　　　(d) can be chemically oxidized by Fe^{3+}.

(e) has no unpaired electrons.

10. Which element is biologically important in oxygen transport in mammals?

(a) Zn　　　(b) Fe　　　(c) Se　　　(d) Cr　　　(e) Au

11. What products are obtained when NiS is roasted in air?

(a) Ni, SO_2　　　(b) Ni, H_2S　　　(c) $NiSO_4$

(d) NiO_2, SO_3　　　(e) NiO, SO_2

12. What is the oxidation state of nickel in $Ni(CO)_4$?

(a) 1　　　(b) 3　　　(c) 0　　　(d) 4　　　(e) 2

13. The low reactivity of the coinage metals is due to

(a) the high shielding ability of s-electrons.　　　(b) their large number of oxidation states.

(c) the poor shielding ability d-electrons.　　　(d) their high electrical conductivity.

(e) the lanthanide contraction.

14. How many milliliters of H_2 (measured at STP) are required to reduce the Cu^{2+} found in 100 mL of 0.250 M $CuCl_2$?

$$Cu^{2+}(aq) + H_2(g) \longrightarrow Cu(s) + 2H^+(aq)$$

(a) 422 mL　　　(b) 220 mL　　　(c) 900 mL

(d) 1.23×10^4 mL　　　(e) 560 mL

15. All of the statements that follow except one describe gold. Which one does not describe gold?

(a) reduces H^+ to form H_2　　　(b) excellent conductor　　　(c) yellow color

(d) dissolves in aqua regia　　　(e) forms complexes

16. What is the formula of mercury(I) chloride?

(a) HgCl   (b) Hg_2Cl_2 (c) Hg_2Cl (d) $HgCl_2$

17. What is the anhydride of the acid H_2CrO_4?

(a) Cr_2O_3 (b) CrO_2 (c) CrO   (d) CrO_3 (e) Cr_2O_5

18. What is the reducing agent in the following reaction, which occurs in a blast furnace?

$$Fe_2O_3(s) + 3CO(g) \longrightarrow 2Fe(s) + 3CO_2(g)$$

(a) Fe_2O_3 (b) CO   (c) CO_2 (d) Fe

Complexes of the *d*-block elements

19. What is the ligand in the compound $K_4[Fe(CN)_6]$?

(a) Fe^{2+} (b) K^+ (c) C   (d) CN^- (e) N

20. What is the coordination number of the central metal ion (or atom) in an octahedral complex?

(a) 2   (b) 5   (c) 8   (d) 4   (e) 6

21. All of the following but one describe complexes in general. Which one does not?

(a) colored   (b) used in analysis

(c) exist as isomers   (d) biologically irrelevant

(e) undergo substitution reactions

22. Which of the following can act as a ligand?

(a) H_2O (b) Fe   (c) Ca^{2+} (d) C_2H_2 (e) CH_4

23. What is the name of $[CoCl(NH_3)_5]^{2+}$?

(a) pentaammoniamonochlorocobaltate   (b) pentaammoniamonochlorocobalt(II)

(c) pentamminechlorocobalt(II)   (d) pentamminechlorocobalt(III)

24. What is the formula (including the charge) of aquatribromoplatinate(II) ion?

(a) $[PtBr_3(H_2O)]^-$ (b) $[Pt(Br_2)_3(H_2O)]^{2+}$ (c) $[PtBr_2(H_2O)]^0$

(d) $[PtBr_3(H_2O)_3]^-$ (e) $[PtBr_3(H_2O)]^{2+}$

25. Which of the following represents a cis isomer?

(a) 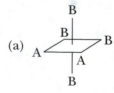   (b) 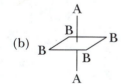   (c)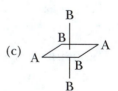

26. Which of the following is chiral?

(a) 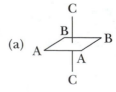   (b) 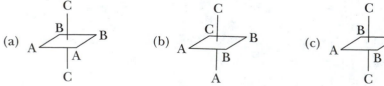   (c)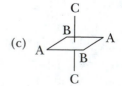

27. Which of the following are examples of ionization isomers?

(a)
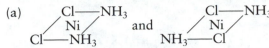
and

(b) $[Co(NH_3)NCS]^{2-}$ and $[Co(NH_3)SCN]^{2-}$

(c) $[FeCl(H_2O)_5]F$ and $[FeF(H_2O)_5]Cl$

(d) $[Cr(NH_3)_6][Fe(CN)_6]$ and $[Cr(CN)_6][Fe(NH_3)_6]$

Crystal field theory

28. What is Δ_0 (in $kJ \cdot mol^{-1}$) for a complex that absorbs light at 425 nm?

(a) $321 \ kJ \cdot mol^{-1}$ (b) $4.68 \times 10^{-9} \ kJ \cdot mol^{-1}$ (c) $3.11 \times 10^{-19} \ kJ \cdot mol^{-1}$

(d) $355 \ kJ \cdot mol^{-1}$ (e) $282 \ kJ \cdot mol^{-1}$

29. What color would you expect for a complex with $\Delta_0 = 222 \ kJ \cdot mol^{-1}$?

(a) red   (b) green-yellow   (c) red-violet   (d) green   (e) red-orange

30. Which of the following has the largest Δ_0?

(a) $[Fe(H_2O)_6]^{3+}$ (b) $[Fe(SCN)_6]^{4-}$ (c) $[Co(CN)_6]^{3-}$

(d) $[AuCl_4]^-$ (e) $[MnF_6]^{2-}$

31. Which ligand is most likely to result in the formation of a high-spin complex?

(a) SCN^- (b) CN^- (c) CO (d) H_2O (e) NH_3

32. How many unpaired electrons are present in $[Fe(CN)_6]^{3-}$?

(a) 0   (b) 1   (c) 3   (d) 4   (e) 5

33. Which type of complex is diamagnetic?

(a) d^6 (high spin)   (b) d^5 (high spin)   (c) d^3

(d) d^6 (low spin)   (e) d^4 (low spin)

34. Which of the following is the correct orbital diagram for six electrons in an octahedral complex with weak-field ligands?

(a)
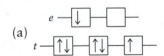

(b)
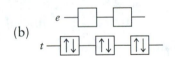

(c)
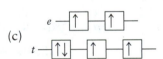

(d)
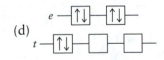

(e)
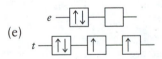

CHAPTER 22

NUCLEAR CHEMISTRY

Most of the reactions we have studied in our chemistry course up to now have involved changes in electronic structure; in the field of nuclear chemistry, we consider processes that involve changes in the nucleus. The energies involved in nuclear processes are considerably larger than those involved in chemical processes. Nuclear processes are involved in a variety of important technological applications, from cancer therapy to production of electricity.

NUCLEAR STABILITY

22.1 NUCLEAR REACTIONS

KEY CONCEPT A   Recognizing isotopes and nuclides

Isotopes are atoms with the same atomic number but different mass numbers. For instance, carbon-12 and carbon-14 both have an atomic number of 6, but they have different mass numbers, so both are isotopes of carbon. A **nuclide** is a specific isotope of an element: carbon-12 is a nuclide, carbon-14 is a nuclide, and vanadium-54 is a nuclide. Approximately 850 nuclides are known. A common way of representing a nuclide is with the symbol $^A_Z E$ where E is the element symbol of the nuclide, A is the mass number, and Z is the atomic number. (The symbolism is also used for fundamental particles, in which case Z is the charge on the particle.) Because the mass number equals the total number of neutrons and protons (the total number of **nucleons**), $A - Z$ equals the number of neutrons in the nucleus of the nuclide. The symbol $^A E$ is also used to represent a nuclide because writing the atomic number is redundant; once the element symbol is known, the atomic number is known.

> **EXAMPLE   Determining the number of nucleons in a nuclide**
>
> How many neutrons, protons, and nucleons are there in an atom of sodium-25. What is the symbol for this nuclide?
>
> **SOLUTION** The number of protons is determined by the identity of the element. Sodium has an atomic number of 11, so it has 11 protons ($Z = 11$). The mass number of this nuclide is 25. This equals the sum of the number of neutrons plus protons, so the number of neutrons is $25 - 11$ or 14. The protons and neutrons are the nucleons in the atom, so there are 25 nucleons. The symbol for the nuclide is $^{25}_{11}\text{Na}$.
>
> **EXERCISE** How many neutrons, protons, and nucleons are there in an atom of rubidium-96? What is the symbol for the nuclide?
>
> [*Answer:* 59 neutrons, 37 protons, 96 nucleons; $^{96}_{37}\text{Rb}$]

KEY CONCEPT B   Nuclear reactions are different from chemical reactions

By the term "nuclear reaction" we mean any transformation that the nucleus of an atom undergoes. This may involve, for example, the emission of radioactivity, nuclear fission, or nuclear fusion. Nuclear reactions differ from chemical reactions in three important aspects:

1. Isotopes of an element often have vastly different nuclear properties; in contrast, isotopes have substantially similar chemical properties.
2. Nuclear reactions usually involve formation of a new element from the old one. The goal of alchemists to transform lead to gold would have involved a nuclear reaction!
3. Nuclear reactions involve enormous energies relative to chemical reactions. This is evident when we consider, for instance, the difference in the destructive power of a nuclear weapon relative to a conventional bomb.

EXAMPLE Comparing nuclear and chemical reactions

Suggest why the nuclear reactions different isotopes of an element undergo are generally different whereas the chemical reactions are very similar.

SOLUTION The electronic structures of different isotopes are the same; because chemical reactions involve the electrons, we fully expect chemical reactions to be the same for different isotopes. However, because there are a different number of neutrons present in the nuclei of different isotopes, the structures of the isotopic nuclei are very different. These different structures lead to different nuclear reactions for the different isotopes.

EXERCISE After the Chernobyl disaster, in which a nuclear reactor exploded, a significant problem involving the spread of radioactive iodine throughout parts of eastern Europe developed. Does this mean that the original reactor fuel contained large amounts of iodine?

[*Answer:* No. The iodine formed from nuclear reactions.]

22.2   NUCLEAR STRUCTURE AND NUCLEAR RADIATION

KEY CONCEPT   Radiation from nuclei

The term *radiation* refers to particles or photons that are emitted from the nuclei of atoms. (X-rays are an exception, because the extranuclear electrons are involved in their production.) There are four major types of radiation. Three of them (alpha, beta, and positron) occur through the emission of a particle from the nucleus. The last (gamma) corresponds to the emission of a high-energy photon from a nucleus in an excited state. The different radiations were first characterized by their behavior as they travel between electrically charged plates of opposite charge. Positively charged particles are attracted by the negative plate and repelled by the positive plate, whereas negatively charged particles are attracted by the positive plate and repelled by the negative one. Table 22.1 of the text lists some properties of various forms of radiation. It should be noted that physicists often use the term *electron* and the symbol β^- for what we call beta emission.

EXAMPLE   Distinguishing the different types of radiative emissions

Comment on the relative direction and extent of deflection experienced by α particles and β particles as they travel between electrically charged plates of opposite charge.

SOLUTION Alpha particles are relatively massive (mass number = 4) and have a +2 charge. They are attracted by the negative plate and repelled by the positive plate, but they are not deflected much because of their large mass. Beta particles have very little mass (mass number = 0) and a negative charge. They are attracted by the positive plate and repelled by the negative plate; they undergo large deflections because of their low mass.

EXERCISE Comment on the direction and extent of deflection of a γ-ray photon as it passes between electrically charged plates of opposite charge.

[*Answer:* No deflection occurs.]

22.3 NUCLEAR DECAY

KEY CONCEPT   Nuclear decay causes a nuclear transmutation

The emission of a particle (other than a neutron) from the nucleus of an atom results in a change in the atomic number of the nuclide involved. The new nuclide that results is called the **daughter nucleus**, or daughter nuclide, and the conversion of one element into another is called **nuclear transmutation**. The daughter nucleus is usually formed in an excited state and emits one or more γ-ray photons as it relaxes to its ground state.

EXAMPLE 1   Recognizing a nuclear transmutation

Does the equation that follows represent a nuclear transmutation? Explain your answer.

$$^{239}_{93}\text{Np} \longrightarrow ^{239}_{94}\text{Pu} + ^{0}_{-1}\beta$$

SOLUTION The equation shows the conversion of a neptunium-239 nucleus into a plutonium-239 nucleus through the emission of a β particle (or 1 mol neptunium-239 into 1 mol plutonium-239 through the emission of 1 mol of β particles). Because the equation represents the conversion of one element into another, it is a nuclear transmutation.

EXERCISE What is the daughter nuclide in the transmutation equation given in the example?

[*Answer:* Plutonium-239]

EXAMPLE 2   Understanding and calculating the wavelength of γ radiation

The nuclear disintegration in the preceding example includes the emission of a γ-ray photon of energy 8.88×10^{-14} J. What is the origin of the γ radiation?

SOLUTION The plutonium nucleus is formed in an excited state (represented in the following equation by an asterisk). When it relaxes to the ground state, it emits a γ-ray photon:

$$^{239}_{94}\text{Pu}^* \longrightarrow ^{239}_{94}\text{Pu} + ^{0}_{0}\gamma$$

EXERCISE What is the wavelength, in picometers, of the γ-ray photon emitted? The energy of the photon is given in this example; $h = 6.63 \times 10^{-34}$ J·s; $c = 3.00 \times 10^{8}$ m·s^{-1}.

[*Answer:* 2.24 pm]

Toolbox 22.1   HOW TO IDENTIFY THE PRODUCTS OF A NUCLEAR REACTION

KEY CONCEPT   Alpha and beta disintegration, electron capture, and positron emission

The three radioactive particle emissions and electron capture result in different types of changes in the nucleus that undergoes transmutation. These are summarized in the table.

Event	Change in Number of Protons	Change in Number of Neutrons	Change in Mass Number
α emission	decrease by 2	decrease by 2	decrease by 4
β emission	increase by 1	decrease by 1	no change
positron emission	decrease by 1	increase by 1	no change
electron capture	decrease by 1	increase by 1	no change

When the original nuclide and the type of reaction are known, the correct product of a nuclear reaction can always be determined by writing a balanced nuclear equation. A balanced nuclear equation is one in which the sum of the mass numbers on the left equals the sum of the mass numbers on the right, and the sum of charges on the left equals the sum of the charges on the right. Practically, this means that the sum of the superscripts must be the same on the left as on the right and that the sum of the subscripts must be the same on the left as on the right. For instance, for the equation

$$^{133}_{53}I \longrightarrow ^{0}_{-1}e + ^{133}_{54}Xe$$

Mass balance: $133 = 133 + 0$ Charge balance: $53 = -1 + 54$

It should be recognized that the electron emitted as a β particle is emitted from the nucleus; it is not one of the extranuclear electrons. However, electron capture involves the capture of one of the atom's extranuclear s-electrons by the nucleus.

EXAMPLE 1   Identifying the element formed by α decay

What element is formed when polonium-210 undergoes α decay?

SOLUTION We first formulate the problem by recognizing that, for α ($^{4}_{2}$He) decay, an α particle is one of the products of the nuclear reaction. We thus have

$$^{210}_{84}Po \longrightarrow ^{A}_{Z}E + ^{4}_{2}He$$

To achieve mass balance, the sum of the superscripts on the left side of the equation must equal the sum of the superscripts on the right side. Thus, $A = 206$ ($210 = 206 + 4$). Similarly, for charge balance, the sum of the subscripts on the left side of the equation must equal the sum of the subscripts on the right. Thus, $Z = 82$ ($84 = 82 + 2$). The daughter nuclide formed is lead-206:

$$^{210}_{84}Po \longrightarrow ^{206}_{82}Pb + ^{4}_{2}He$$

EXERCISE What element is formed when neptunium-237 undergoes α decay?

[*Answer:* Protactinium-233, $^{233}_{91}$Pa]

EXAMPLE 2   Identifying the element formed by β decay

What daughter nuclide is formed when actinium-227 undergoes β decay?

SOLUTION We formulate the problem by recognizing that a β particle ($^{0}_{-1}$e) is one of the products of the nuclear reaction:

$$^{227}_{89}Ac \longrightarrow ^{A}_{Z}E + ^{0}_{-1}e$$

Mass balance requires that $A = 227$ and charge balance requires that $Z = 90$ ($89 = 90 - 1$). Thus, the daughter nuclide is thorium-227:

$$^{227}_{89}Ac \longrightarrow ^{227}_{90}Th + ^{0}_{-1}e$$

EXERCISE What daughter nuclide is formed when tantalum-186 undergoes β decay?

[*Answer:* Tungsten-186, $^{186}_{74}$W]

EXAMPLE 3   Identifying the element formed by electron capture

What daughter nuclide is formed when iridium-186 undergoes electron capture?

SOLUTION We formulate the problem by recognizing that for electron capture, an electron is one of the reactants. We must determine E in the equation

$$^{186}_{77}Ir + ^{0}_{-1}e \longrightarrow ^{A}_{Z}E$$

Mass balance requires that $186 + 0 = A$, so $A = 186$. Charge balance requires that $77 + (-1) = Z$, so $Z = 76$. The daughter nuclide is osmium-186:

$$^{186}_{77}Ir + ^{0}_{-1}e \longrightarrow ^{186}_{76}Os$$

EXERCISE What daughter nuclide results from electron capture by xenon-116?

[*Answer:* Iodine-116, $^{116}_{53}$I]

EXAMPLE 4   Identifying the nuclide formed by positron emission

What daughter nuclide is formed when bismuth-192 undergoes positron emission?

SOLUTION We formulate the problem by recognizing that a positron ($^{0}_{1}$e) must be one of the products of the nuclear reaction when positron emission occurs. The problem is to determine E in the equation

$$^{192}_{83}\text{Bi} \longrightarrow\ ^{A}_{Z}\text{E} +\ ^{0}_{1}\text{e}$$

Mass balance requires that $192 = A + 0$, so $A = 192$. Charge balance requires that $83 = Z + 1$, so $Z = 82$. The daughter nuclide is lead-192:

$$^{192}_{83}\text{Bi} \longrightarrow\ ^{192}_{82}\text{Pb} +\ ^{0}_{1}\text{e}$$

EXERCISE What daughter nuclide results from positron emission from tin-111?

[*Answer:* Indium-111, $^{111}_{49}$In]

EXAMPLE 5   Identifying the type of transmutation that occurs

What process occurs when platinum-197 forms the daughter nuclide gold-197?

SOLUTION In this case we know the original nuclide and the daughter nuclide and must determine E in the equation

$$^{197}_{78}\text{Pt} \longrightarrow\ ^{A}_{Z}\text{E} +\ ^{197}_{79}\text{Au}$$

Mass balance requires that $197 = A + 197$, so $A = 0$. Charge balance requires that $78 = Z + 79$, so $Z = -1$. E is a β particle, so the process that occurs is β emission. Another approach is to realize that the process in which the number of protons increases by 1 and the mass number stays the same is β emission.

EXERCISE What process transmitters strontium-83 to rubidium-83?

[*Answer:* Electron capture or positron decay]

22.4   THE PATTERN OF NUCLEAR STABILITY

KEY CONCEPT   The band of stability and sea of instability

The underlying reasons that explain why some nuclides are stable and others are radioactive are not well understood; however, an important empirical observation allows us to organize our thinking. If we plot all the stable (nonradioactive) nuclei, with atomic number on the x-axis and mass number of the y-axis, we find that the stable nuclei lie in a narrow band (the **band of stability**) that ends at $Z = 83$. The inference that can be made from the existence of this band is that the relative number of protons and neutrons helps to determine the stability of nuclei. A nucleus may be unstable because it has too many protons (is proton rich) or too many neutrons (is neutron rich). The fact that the band of stability ends at $Z = 83$ informs us that nuclei with more than 83 protons are unstable, suggesting that to be stable a nucleus cannot be too massive. The nuclides that lie outside the band of stability are said to be in the **sea of instability**. In Fig. 22.12 of the text, nuclides to the upper left of the band of stability are neutron rich, those to the lower right are proton rich, and those to the right of $Z = 83$ are too massive to be stable. Nuclides that are too massive *tend* to decay by α emission; those that are neutron rich, by β emission; and those that are proton rich, by electron capture or positron emission.

Type of Nuclide	Most Common Type of Decay	Result of Decay
too massive	α decay	mass number decreases by 4, Z decreases by 2
neutron rich	β decay	Z increases, number of neutrons decreases
proton rich	electron capture or positron emission	Z decreases, number of neutrons increases

EXAMPLE 1   Predicting the mode of decay

What is a likely mode of decay for $^{236}_{92}U$?

SOLUTION For nuclides with Z > 83, α decay is a common and likely mode of decay. Without information to the contrary, we assume that uranium-236 undergoes α decay.

EXERCISE Which of the following is most likely to undergo α decay?

$$^{242}_{96}Pu \quad ^{188}_{79}Au \quad ^{165}_{69}Tm$$

[*Answer:* Plutonium-242]

EXAMPLE 2   Predicting the mode of decay

What is a likely mode of decay for $^{176}_{79}Au$? (Use Figure 22.12 of the text for help.)

SOLUTION Plotting this nuclide on Figure 22.12 of the text shows that it lies below the band of stability. Nuclides in this area of the plot have too many protons relative to the number of neutrons present; that is, if the number of protons is decreased, the point for the nuclide moves toward the band of stability. (Check this for yourself!) Nuclides that are proton rich may decay by either electron capture or positron emission. In a sample of gold-176, some nuclei will decay by positron emission, some by electron capture.

EXERCISE What is a likely mode of decay for $^{201}_{78}Pt$?

[*Answer:* β emission]

PITFALL   Do all nuclides with Z > 83 undergo α emission?

Not all nuclides with Z > 83 undergo α emission as their mode of decay. Even though α emission is quite common for massive nuclides, it is not the only mode of decay. For example, $^{245}_{94}Pu$ and $^{237}_{92}U$ both decay entirely by β emission.

22.5   NUCLEOSYNTHESIS

KEY CONCEPT   Neutron- and charged-particle-induced transmutation

Nucleosynthesis, as the word implies, is the formation (synthesis) of elements. Because neutrons have zero charge, they can approach and enter a positively charged nucleus with relative ease. A neutron entering a nucleus does not result in a new element but forms a new isotope of the original element. However, the nuclide formed frequently undergoes radioactive decay that results in a new element. As an example, bombarding iron-54 with a neutron results in the emission of a proton and the formation of manganese-54:

$$^{54}_{26}Fe + ^{1}_{0}n \longrightarrow ^{54}_{25}Mn + ^{1}_{1}H$$

High-energy charged particles can also be used to accomplish nucleosynthesis reactions.

EXAMPLE 1   Neutron-induced transmutation

When $^{238}_{92}U$ is bombarded with a neutron, the product formed undergoes β decay. What is the final product?

SOLUTION   The reactants in this nuclear reaction are uranium-238 and a neutron. The products are a β particle and the unknown product E:

$$^{238}_{92}U + ^{1}_{0}n \longrightarrow ^{0}_{-1}\beta + ^{A}_{Z}E$$

Mass number balance requires that $238 + 1 = 0 + A$, so $A = 239$. Charge balance requires that $92 + 0 = -1 + Z$, so $Z = 93$. E is a neptunium-239, $^{239}_{93}Np$. The starting nuclide is uranium-238, the bombarding particle is a neutron, the particle emitted is a β particle, and the nuclide formed is neptunium-239.

EXERCISE   What nuclide is produced in the atmosphere through the bombardment of nitrogen-14 by high-energy neutrons? A proton ($^{1}_{1}H$) is emitted as part of the overall nuclear reaction.

[*Answer:* $^{14}_{6}C$]

EXAMPLE 2   Charged–particle-induced transmutation

What is E in the following nucleosynthesis?

$$^{27}_{13}Al + ^{4}_{2}He \longrightarrow ^{1}_{0}n + ^{A}_{Z}E$$

SOLUTION   Mass number balance requires that $27 + 4 = 1 + A$, so $A = 30$. Charge balance requires that $13 + 2 = 0 + Z$, so $Z = 15$. Therefore, E is phosphorus-30.

EXERCISE   What is the bombarding particle in the following nucleosynthesis?

$$^{246}_{96}Cm + ^{A}_{Z}E \longrightarrow 5^{1}_{0}n + ^{254}_{102}No$$

[*Answer:* $^{13}_{6}C$]

KEY WORDS   Define or explain each term in a written sentence or two.

α particle	magic number	radioactivity
β particle	nuclear equation	sea of instability
γ radiation	nuclear transmutation	spontaneous neutron emission
band of stability	nucleosynthesis	transuranium elements
daughter nucleus	radioactive series	

RADIOACTIVITY

22.6   THE EFFECTS OF RADIATION

Alpha, beta, and gamma radiation are each a form of **ionizing radiation.** Ionizing radiation is able to cause ejection of electrons from atoms and molecules, thus disrupting chemical bonds. The actual damage done to an organism by radiation is dependent on the extent to which the radiation can reach sensitive tissue as well as on the amount of radiation that does so. The radiation **dose** is a quantitative measure of how much energy is deposited in a sample; some measures of dose take into account the intrinsic ability of a particular radiation to do damage. The radiation dose is the energy deposited in a sample per unit of mass. One standard unit for measuring radiation dose is the rad (*r*adiation *a*bsorbed *d*ose). One **rad** corresponds to exactly 10^{-2} J of energy deposited in a kilogram of sample:

$$1 \text{ rad} = 10^{-2} \text{ J} \cdot \text{kg}^{-1}$$

A dose of 500 rad from γ radiation is enough to cause death in 50% of the humans that receive this dose. This is equivalent to only 5 J of energy per kilogram of tissue. Alpha radiation is approximately 20 times as effective per rad as β and γ radiation in causing biological damage. To account for the difference in biological damage that different types of radiation cause, each type of radiation is assigned a **relative biological effectiveness, Q.** Beta and gamma radiation both have $Q = 1$, whereas alpha radiation has $Q = 20$. (These numbers are approximate because Q also depends on the tissue being irradiated.) The dose in rads times Q is a measure of the expected biological effect of any radiation and is called the **dose equivalent.** The unit for the dose equivalent is the **rem** (*r*adiation *e*quivalent *m*an).

$$\text{rem} = \text{rad} \times Q$$

EXAMPLE   Doses from radiation

What is the dose equivalent (in rem) when 2.5 J of energy is delivered to 0.85 kg of tissue by α radiation?

SOLUTION We first calculate the energy delivered per kilogram of tissue, by dividing the total energy delivered by the mass of the tissue in kilograms:

$$\text{Energy per kg} = 2.5 \text{ J}/0.85 \text{ kg}$$

$$= 2.9 \text{ J} \cdot \text{kg}^{-1}$$

Now we use the fact that 1 rad $= 10^{-2}$ J$\cdot$kg^{-1} to calculate the dose in rads:

$$\text{Dose in rad} = 2.9 \text{ J} \cdot \text{kg}^{-1} \times \frac{1 \text{ rad}}{10^{-2} \text{ J} \cdot \text{kg}^{-1}}$$

$$= 2.9 \times 10^2 \text{ rad}$$

Finally, we use $Q = 20$ for α radiation to calculate the equivalent dose in rems:

$$\text{Rem} = \text{rad} \times Q$$

$$= 2.9 \times 10^2 \times 20 \text{ rem}$$

$$= 5.8 \times 10^3 \text{ rem}$$

EXERCISE A 58-kg person is bathed in β radiation. Because β radiation is only moderately penetrating, only 2.0 kg of the person's tissue is irradiated, with a total dose received of 23 rem. What is the total energy absorbed by the tissue?

[*Answer:* 0.46 J]

22.7   MEASURING RADIOACTIVITY

KEY CONCEPT   The activity of radioactive sources

The **activity** of a radioactive source is measured by the number of nuclear disintegrations that occur per second. The number of disintegrations per second can be measured by a **Geiger counter** or **scintillation counter.** A source that undergoes 3.7×10^{10} nuclear disintegrations per second is called a **1-curie** (Ci) source. In the SI system, the unit of activity is the **becquerel** (Bq); one becquerel is equivalent to one disintegration per second (dis$\cdot$s^{-1}).

$$1 \text{ Ci} = 3.7 \times 10^{10} \text{ dis} \cdot \text{s}^{-1}$$

$$1 \text{ Bq} = 1 \text{ dis} \cdot \text{s}^{-1}$$

EXAMPLE   Calculating the activity of a source

How many disintegrations per second occur in a 40.0 Ci cobalt-60 radioactive source?

SOLUTION A 1-Ci source undergoes 3.7×10^{10} dis$\cdot$s^{-1}. We can use this definition as a conversion factor to answer the problem:

$$40.0 \ \cancel{Ci} \times \frac{3.7 \times 10^{10} \ dis \cdot s^{-1}}{1 \ \cancel{Ci}} = 1.5 \times 10^{12} \ dis \cdot s^{-1}$$

EXERCISE A sample of a radioactive source undergoes $3.3 \times 10^{11} \ dis \cdot s^{-1}$. What is the activity of the source, in curies and in becquerels?

[*Answer:* 8.9 Ci; 3.3×10^{11} Bq]

22.8   THE LAW OF RADIOACTIVE DECAY

KEY CONCEPT A   Radioactive decay follows a first-order rate law

The radioactive decay of a radioactive nuclide follows a first-order rate law. If N equals the number of radioactive nuclei present at any time, the rate of decay is proportional to N:

$$\text{Rate of decay} = kN$$

In addition, if N_0 is the number of radioactive nuclei at time $= 0$, and N is the number of radioactive nuclei at time $= t$, then the familiar first-order integrated rate law is obtained:

$$\ln\left(\frac{N_0}{N}\right) = kt$$

where k is the first-order rate constant associated with the decay and is called the **decay constant.** Like any first-order rate constant, it has units of reciprocal time (1/time, or time^{-1}). The **half-life** of a radioactive element is defined as the time it takes for one-half of a sample to undergo radioactive decay. The following bar graph shows the decay of a radioactive sample through three half-lives. Note that the mass of the radioactive sample (shaded bars) decreases by half for every half-life. At the same time, the mass of the daughter nuclides increases, as shown by the unshaded bars. On the scale of this experiment, mass is conserved. The sum of the masses of the radioactive sample and the daughter nuclides is always 8 g.

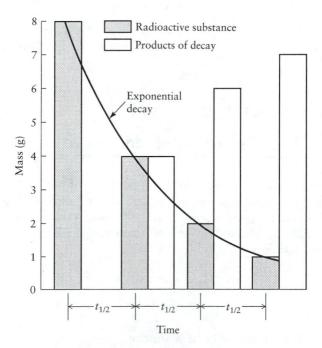

The half-life of an element is related to the decay constant through the relation

$$t_{1/2} = \frac{0.693}{k}$$

EXAMPLE 1   First-order decay kinetics

The decay constant for the decay of rhodium-99 is 0.043 day^{-1}. How many grams of rhodium-99 remain if a 15-g sample is allowed to decay for exactly 10 days?

SOLUTION Because the mass of a sample is proportional to the number of nuclei present, we can write the integrated rate law as

$$\ln\left(\frac{m_0}{m}\right) = kt$$

where m_0 is the mass of the radioactive sample present at time = 0 and m is the mass present at time = t. Substituting $k = 0.043$ day^{-1} and $t = 10$ day into the equation gives

$$\ln\left(\frac{m_0}{m}\right) = 0.043 \; \text{day}^{-1} \times 10 \; \text{day}$$

$$= 0.43$$

We next take the exponential of both sides of the equation. Recalling that $e^{\ln y} = y$ and calculating $e^{0.43} = 1.54$ gives

$$\frac{m_0}{m} = 1.54$$

We finally use the fact that $m_0 = 15$ g to calculate the desired result:

$$\frac{15 \text{ g}}{m} = 1.54$$

$$m = 9.7 \text{ g}$$

EXERCISE The decay constant for the decay of lead-209 is 0.21 h^{-1}. How many grams of lead-209 remain if a 2.6-kg sample is allowed to decay for 3.0×10^2 min?

[*Answer:* 0.91 kg]

EXAMPLE 2   Calculating half-lifes

What is the half-life of an element with $k = 0.387$ (ms)$^{-1}$?

SOLUTION The relationship between the half-life of an element and the decay constant for its decay is

$$t_{1/2} = 0.693/k$$

Substituting in $k = 0.387$ (ms)$^{-1}$ gives

$$t_{1/2} = \frac{0.693}{0.387 \text{ (ms)}^{-1}}$$

$$= 1.79 \text{ ms}$$

EXERCISE What is the half-life of an element that has a first-order decay constant of 0.020 s^{-1}?

[*Answer:* 35 s]

EXAMPLE 3   Using the first-order integrated rate law for radioactive decay

A 26-mg sample of thorium-215 is allowed to decay until 20 mg remain. How long was the sample allowed to decay? The half-life of thorium-215 is 1.2 s.

SOLUTION The integrated rate law for radioactive decay using the mass of a sample is

$$\ln\left(\frac{m_0}{m}\right) = kt$$

We first calculate k, using the relation

$$k = \frac{0.693}{t_{1/2}}$$

$$= \frac{0.693}{1.2 \text{ s}}$$

$$= 0.58 \text{ s}^{-1}$$

Substituting $m = 20$ mg, $m_0 = 26$ mg, and $k = 0.58$ s^{-1} into the integrated rate law gives

$$\ln\left(\frac{26 \text{ mg}}{20 \text{ mg}}\right) = 0.58 \text{ s}^{-1} \times t$$

$$\ln(1.3) = 0.58 \text{ s}^{-1} \times t$$

$$0.26 = 0.58 \text{ s}^{-1} \times t$$

$$t = \frac{0.26}{0.58 \text{ s}^{-1}}$$

$$= 0.45 \text{ s}$$

EXERCISE How much time is required for the decay of a sample of rhenium-192 to 0.20% of its original mass? $t_{1/2} = 16$ s.

[*Answer:* 143 s]

KEY CONCEPT B   Radiocarbon dating

A small percentage of the carbon atoms in atmospheric CO_2 is radioactive carbon-14, which has a half-life of 5730 years(y). Plants and animals incorporate carbon-14 into their tissue, plants through photosynthesis and animals by being part of the food web. The carbon-14 present in any living organism decays at its normal rate, but the organism takes in fresh carbon-14 as long as it lives. Thus a type of dynamic equilibrium is reached for a live organism, in which the relative amount of carbon-14 in the organism is the same as that in atmospheric CO_2. However, when the organism dies, the incorporated carbon-14 continues to decay, but it is no longer replenished with fresh carbon-14. The relative amount of carbon-14 found in a sample of the dead organism's tissue is a measure of the time since the organism died. For instance, if a sample has exactly one-half of the radioactive carbon it originally carried when it was alive, we would be safe in concluding that the organism died 5730 y (one half-life) previously. The technique of determining the "age" (actually, the time since the organism died) of a formerly living sample by the amount of carbon-14 it contains is called **radiocarbon dating.**

EXAMPLE   Interpreting radiocarbon dating

A sample of charcoal from a hearth discovered in an archaeological dig undergoes 6100 carbon-14 disintegrations per gram of carbon in a 24-h period. Carbon from a modern source undergoes 920 disintegrations per hour (dis·h^{-1}) per gram of carbon. How old is the piece of charcoal; that is, how long ago did the tree from which the charcoal was formed die?

SOLUTION We assume that the modern carbon source is an accurate reflection of the carbon-14 present at the time the tree was alive. This assumption permits the calculation of the number of disintegrations per gram of carbon that would have occurred in a 24-h period when the tree was alive:

$$920 \frac{\text{dis}}{\text{h}} \times 24 \text{ h} = 2.2 \times 10^4 \text{ dis} \quad \text{(in 24 h)}$$

Because the number of disintegrations is proportional to the number of carbon-14 nuclei present, we can write the first-order integrated rate law as

$$\ln\left(\frac{dis_0}{dis}\right) = kt$$

where dis = the number of disintegrations in a certain time interval at time $= t$ and dis_0 = the number of disintegrations in the same time interval at time $= 0$. Also, $k = 0.693/t_{1/2} = 1.21 \times 10^{-4}$ y^{-1}. Substituting $dis = 6100$ and $dis_0 = 2.2 \times 10^4$ into the integrated rate equation along with the value of k just calculated gives

$$\ln\left(\frac{2.2 \times 10^4 \ \cancel{dis}}{6100 \ \cancel{dis}}\right) = 1.21 \times 10^{-4} \ y^{-1} \times t$$

$$\ln(3.6) = 1.21 \times 10^{-4} \ y^{-1} \times t$$

$$1.3 = 1.21 \times 10^{-4} \ y^{-1} \times t$$

$$t = \frac{1.3}{1.21 \times 10^{-4} \ y^{-1}}$$

$$= 1.1 \times 10^4 \ y$$

The tree died approximately 11,000 years ago.

EXERCISE A sample of wood from an ancient boat undergoes 2800 carbon-14 disintegrations per gram carbon in a 30-h period. How old is the piece of wood?

[*Answer:* 1.9×10^4y]

KEY WORDS   Define or explain each term in a written sentence or two.

activity	Geiger counter	rad
absorbed dose	ionizing radiation	relative biological effectiveness (Q)
decay constant	isotopic dating	rem
dose equivalent	law of radioactive decay	scintillation counter

NUCLEAR ENERGY

22.9   MASS-ENERGY CONVERSION

KEY CONCEPT   Nuclear binding energy

The nuclear binding energy of a nuclide is the energy released when the neutrons and protons in the nucleus combine to form the nucleus. The binding energy is a measure of the stability of the nucleus. For instance, for sulfur-35, which contains 16 protons and 19 neutrons,

16 protons + 19 neutrons $\longrightarrow$ sulfur-35 nucleus   energy released = binding energy

A nuclide with a high binding energy is more stable than one with a low binding energy. The nuclear binding energy E_{bind} is given by

$$E_{bind} = \Delta m \times c^2$$

where c is the speed of light (3.00×10^8 $m \cdot s^{-1}$) and Δm is the difference in mass between the nucleus and the separated neutrons and protons that make up the nucleus.

$$\Delta m = \text{(mass of separated neutrons + protons)} - \text{(mass of nucleus)}$$

EXAMPLE   Calculating the binding energy of a nuclide

What is the binding energy (in $kJ \cdot mol^{-1}$) of nitrogen-14, which has an atomic mass of 14.0031 u? The mass of a proton is 1.0078 u, and that of a neutron 1.0087 u.

SOLUTION   The binding energy is given by

$$E_{bind} = \Delta m \times c^2$$

Nitrogen has atomic number 7, so it contains 7 protons. The isotope of mass number 14 contains 7 neutrons $(14 - 7 = 7)$. Thus,

$$\Delta m = [(7 \times \text{mass of proton}) + (7 \times \text{mass of neutron})] - (\text{mass of nitrogen-14 nucleus})$$

$$= [(7 \times 1.0078 \text{ u}) + (7 \times 1.0087 \text{ u})] - (14.0031 \text{ u})$$

$$= 0.1124 \text{ u}$$

We convert this mass difference from atomic mass units to kilograms so that the calculated energy will be expressed in joules. The conversion factor is $1 \text{ u} = 1.661 \times 10^{-27} \text{ kg}$.

$$0.1124 \text{ u} \times \frac{1.661 \times 10^{-27} \text{ kg}}{1 \text{ u}} = 1.867 \times 10^{-28} \text{ kg}$$

Substituting this value into the original equation gives the binding energy for one atom:

$$E_{bind} = \Delta m \times c^2$$

$$= (1.867 \times 10^{-28} \text{ kg})(3.00 \times 10^8 \text{ m} \cdot \text{s}^{-1})^2$$

$$= (1.867 \times 10^{-28} \text{ kg})(9.00 \times 10^{16} \text{ m}^2 \cdot \text{s}^{-2})$$

$$= 1.681 \times 10^{-11} \text{ J}$$

$$= 1.681 \times 10^{-14} \text{ kJ}$$

Avogadro's number is used to convert to $kJ \cdot mol^{-1}$; the binding energy of nitrogen-14 is $1.012 \times 10^{10} \text{ kJ} \cdot mol^{-1}$:

$$1.681 \times 10^{-14} \frac{\text{kJ}}{\text{atom}} \times 6.022 \times 10^{23} \frac{\text{atom}}{\text{mol}} = 1.012 \times 10^{10} \text{ kJ} \cdot mol^{-1}$$

EXERCISE   What is the binding energy, in $kJ \cdot mol^{-1}$, of copper-62, which has an atomic mass of 61.9326 u?

[*Answer:* 5.228×10^{10} $kJ \cdot mol^{-1}$]

22.10   NUCLEAR FISSION

KEY CONCEPT A   Types of nuclear fission

Fission is a process in which a nucleus breaks up into smaller nuclei. The breakup is accompanied by the release of a large amount of energy. **Spontaneous nuclear fission** occurs when a nucleus breaks up without first absorbing a particle. When bombarding a nucleus with a neutron causes the nucleus to break up into smaller nuclei, **induced nuclear fission** is said to occur, as shown by the following reaction:

$$^1_0 n + ^{235}_{92}U \longrightarrow ^{131}_{50}Sn + ^{103}_{42}Mo + 2^1_0 n$$

A nuclide that can undergo induced nuclear fission is called a **fissionable** nuclide. A nuclide that can undergo fission when it is bombarded by a slow-moving neutron is called a **fissile nuclide**. Uranium-235, plutonium-239, and uranium-233 are examples of fissile nuclides. A self-sustaining fission reaction can occur in a sample if the fission reaction produces neutrons that are captured by nuclei in the sample, which then undergo fission, producing more neutrons, and so on. If the sample is too small or is inappropriately shaped, many neutrons will escape instead of hitting nuclei and a sustained fission reaction does not occur. For a large sample of the proper shape, few neutrons escape and a sustained nuclear reaction can occur. A sample of the proper shape and size to sustain fission is said to possess a **critical mass.** A sample with a **supercritical mass** is large enough not only to sustain a fission reaction but also to explode. In a nuclear reactor, fission is assisted by the **moderator,** a substance that slows neutrons so that they can collide more effectively with nuclei. The fission is controlled both by the geometry of the reactor fuel and by **control rods** made of a substance that absorbs neutrons and prevents a runaway nuclear reaction. The difference between a fission explosion and controlled fission is schematically indicated in the figure. Each × represents a fission event. The neutrons from the fission are shown but the products are not.

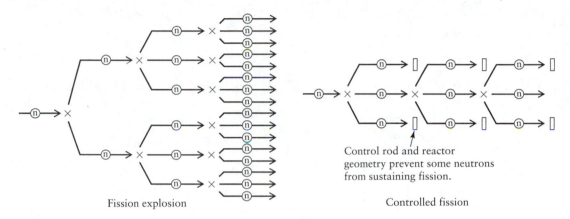

Fission explosion

Control rod and reactor geometry prevent some neutrons from sustaining fission.

Controlled fission

EXAMPLE   Writing a nuclear fission reaction

What is the missing product E in the following fission reaction?

$$\,^{1}_{0}n + \,^{235}_{92}U \longrightarrow \,^{A}_{Z}E + \,^{139}_{56}Ba + 3\,^{1}_{0}n$$

SOLUTION For mass number balance, the sum of the superscripts on the left must equal the sum of the superscripts on the right. Thus, $1 + 235 = A + 139 + 3$, and $A = 94$. For charge balance, the sum of the subscripts on the left must equal the sum of the subscripts on the right; thus, $0 + 92 = Z + 56 + 0$, so $Z = 36$. The missing product is krypton-94, $\,^{94}_{36}Kr$.

EXERCISE What is the missing product in the following nuclear fission reaction?

$$\,^{1}_{0}n + \,^{235}_{92}U \longrightarrow \,^{A}_{Z}E + \,^{137}_{52}Te + 2\,^{1}_{0}n$$

[*Answer:* Zirconium-97, $\,^{97}_{40}Zr$]

KEY CONCEPT B   The energy released during fission

The energy released during a fission reaction is reflected in the difference in the masses of the reactants and products through the equation $E = c^2 \Delta m$. For this situation, E is the energy released, and Δm is the difference in the masses of the products and reactants,

$$\Delta m = (\text{mass of reactants}) - (\text{mass of products})$$

EXAMPLE   Calculating the energy released during fission

What is the energy released, in kJ·mol, uranium-235^{-1}, in the fission reaction

$$\,^{1}_{0}n + \,^{235}_{92}U \longrightarrow \,^{97}_{40}Zr + \,^{137}_{52}Te + 2\,^{1}_{0}n$$

The masses of the various species are neutron, 1.0087 u; uranium-235, 235.0439 u; zirconium-97, 96.9110 u; and tellurium-137, 136.9254 u.

SOLUTION The energy released is given by

$$E = c^2 \, \Delta m$$

$$\Delta m = (m_{neutron} + m_{uranium\text{-}235}) - (m_{zirconium\text{-}97} + m_{tellurium\text{-}137} + 2m_{neutron})$$

$$= (1.0087 \text{ u} + 235.0439 \text{ u}) - (96.9110 \text{ u} + 136.9254 \text{ u} + 2 \times 1.0087 \text{ u})$$

$$= (236.0526 \text{ u}) - (235.8538 \text{ u})$$

$$= 0.1988 \text{ u}$$

We convert the mass to kilograms and at the same time calculate the energy:

$$E = (0.1988 \text{ u} \times 1.661 \times 10^{-27} \text{ kg} \cdot \text{u}^{-1}) \times (3.00 \times 10^8 \text{ m} \cdot \text{s}^{-1})^2$$

$$= 2.97 \times 10^{-11} \text{ J}$$

$$= 2.97 \times 10^{-14} \text{ kJ}$$

This is the energy released when one atom of uranium-235 undergoes fission. For a mole of uranium-235, we get

$$E \text{ (per mole)} = 2.97 \times 10^{-14} \, \frac{\text{kJ}}{\text{atom}} \times 6.02 \times 10^{23} \, \frac{\text{atom}}{\text{mol}}$$

$$= 1.79 \times 10^{10} \text{ kJ} \cdot \text{mol}^{-1}$$

EXERCISE What is the energy released (in $kJ \cdot mol^{-1}$ of uranium-235) during the following fission:

$$^{1}_{0}n + \,^{235}_{92}U \longrightarrow \,^{90}_{38}Sr + \,^{143}_{54}Xe + 3^{1}_{0}n$$

The masses of the nuclides involved are strontium-90, 90.9102 u, and xenon-143, 142.9296 u. The preceding example problem has additional required masses.

[**Answer:** 1.68×10^{10} kJ·mol^{-1}]

22.11   NUCLEAR FUSION

KEY CONCEPT   The energy released during fusion

In the nuclear fission process, a heavy nucleus disintegrates into lighter nuclei, with the release of energy. In the **nuclear fusion** process, lighter nuclei are fused together into a heavier nucleus, also with the release of energy. In both cases, the binding energy of the products is more than that of the reactants, so there is a release of energy when the process occurs. The energy released during fusion is reflected in the difference in the masses of the reactants and products and is calculated in the same way as for fission.

EXAMPLE   Calculating the energy released during fusion

Calculate the energy released when 1.00 g ^2H (deuterium) reacts in the fusion reaction

$$^{2}_{1}H + \,^{2}_{1}H \longrightarrow \,^{3}_{2}He + \,^{1}_{0}n$$

The masses of the species involved are $^{2}_{1}H$, 2.0140 u; $^{3}_{2}He$, 3.01603 u; and $^{1}_{0}n$, 1.0087 u.

SOLUTION The energy released is given by the relationship

$$E = c^2 \, \Delta m$$

$$\Delta m = (\text{mass of reactants}) - (\text{mass of products})$$

For our problem,

$$\Delta m = (2 \times \text{mass deuterium}) - (\text{mass helium-3} + \text{mass neutron})$$

$$= (2 \times 2.0140 \text{ u}) - (3.01603 \text{ u} + 1.0087 \text{ u})$$

$$= 0.0033 \text{ u}$$

We convert this to kilograms and solve for E simultaneously:

$$E = (3.00 \times 10^8 \text{ m} \cdot \text{s}^{-1})^2 \times 0.0033 \text{ u} \times 1.661 \times 10^{-27} \text{ kg} \cdot \text{u}^{-1}$$

$$= 4.9 \times 10^{-13} \text{ J}$$

$$= 4.9 \times 10^{-16} \text{ kJ}$$

This result is for the fusion of two atoms of deuterium. The energy released per atom is half this amount, or 2.5×10^{-16} kJ. For 1.00 g of deuterium, we get

$$1.00 \text{ g }^2\text{H} \times \frac{1 \text{ mol }^2\text{H}}{2.0140 \text{ g}} \times 6.02 \times 10^{23} \frac{\text{atom}}{\text{mol}} = 2.99 \times 10^{23} \text{ atom}$$

This is the number of deuterium atoms that undergo fusion. Multiplying this number by the energy released per atom gives us the answer to the problem:

$$2.99 \times 10^{23} \text{ atom} \times 2.5 \times 10^{-16} \frac{\text{kJ}}{\text{atom}} = 7.5 \times 10^7 \text{ kJ}$$

EXERCISE Calculate the energy released when 1 mol ^{2}H fuses with 1 mol ^{3}H (tritium) in the fusion reaction

$$^2_1\text{H} + {}^3_1\text{H} \longrightarrow {}^4_2\text{He} + {}^1_0\text{n}$$

The masses of the species not given in the example are $^3_1$H, 3.01605 u, and $^4_2$He, 4.00260 u

[*Answer:* 1.69×10^9 kJ]

22.12 THE CHEMISTRY OF NUCLEAR POWER

KEY CONCEPT   Nuclear fuel and nuclear waste

Uranium is mined as the ore **pitchblende** (UO_2), which contains the desired fissile uranium-235 as 0.7% of the uranium present. Extraction is the process of converting the uranium in the ore to a desired chemical form. A common extraction process involves converting the UO_2 to UF_6. The 0.7% concentration of uranium-235 is not high enough to be useful as a fuel, so the percentage of uranium-235 must be increased. The process of increasing the percentage of uranium-235 in a sample is called **isotopic enrichment,** and the reason for producing UF_6 is to use it in the enrichment process. The enrichment process uses the difference in masses of $^{235}UF_6$ (mass = 349.03 u) and $^{238}UF_6$ (mass = 352.04 u). In the gas diffusion process, the difference in masses leads to a difference in rates of diffusion and the possibility of enrichment.

$$\frac{\text{Rate of diffusion of }^{235}UF_6}{\text{Rate of diffusion of }^{238}UF_6} = \sqrt{\frac{\text{mass of }^{238}UF_6}{\text{mass of }^{235}UF_6}}$$

$$= \sqrt{\frac{352.04 \text{ u}}{349.03 \text{ u}}}$$

$$= 1.0043$$

The enriched UF_6 is converted back to UO_2 or to uranium metal for use as a fuel. Fission reactors result in the production of a large amount of radioactive products. It is hoped that a way can be discovered to store the unwanted, highly radioactive products for many thousands of years until they are no longer dangerously radioactive. One idea for long-term storage that has been extensively studied is underground storage in deep mines. One of the problems with underground storage is

leaching, the process by which underground water slowly dissolves the stored wastes and disperses it to deep underground water sources. One way to avoid leaching is to encapsulate the highly radioactive fission (HRF) products into a glass. A crack in an underground storage encapsulating material might allow ground water to come into contact with the HRF products; glasses are preferred because they do not form cracks as easily as do crystalline materials.

EXAMPLE   Understanding fission fuel and waste

Why is it necessary to enrich the U-235 content from the 0.7% found in ores to the 3% needed for some fission reactors?

SOLUTION In order for fission to be self-sustaining in a reactor, a certain fraction of the neutrons released during the fission event must collide with a uranium-235 nucleus. If the concentration of the U-235 fuel is not high enough, the released neutrons will not find a U-235 nucleus to collide with.

EXERCISE Why does storage of HRF products in a glass rather than, say, storage as a liquid in drums reduce the possibility of a chemical explosion?

[*Answer:* When stored in a glass, the HRF products cannot diffuse toward each other, which they must do in order to react.]

KEY WORDS   Define or explain each term in a written sentence or two.

breeder reactor	fissionable	nuclear fusion
critical mass	induced nuclear fission	nuclear reactor
enrich	moderator	supercritical mass
fissile nuclide	nuclear binding energy	thermonuclear explosion

DESCRIPTIVE CHEMISTRY TO REMEMBER

- The three natural **radioactive series** are

  Uranium-238 series: starts at uranium-238 and ends at lead-206

  Uranium-235 series: starts at uranium-235 and ends at lead-207

  Thorium-232 series: starts at thorium-232 and ends at lead-208

- The typical annual dose of radiation received by a person in the United States is **0.2 rem.** The actual figure depends on where the individual lives.
- In a **breeder reactor,** neutrons are not moderated (slowed down). Some the high-energy neutrons are used to convert ^{238}U, which is not fissile, to ^{239}Pu, which is fissile.
- ^{56}Fe has the highest **binding energy** of any nuclide and is, therefore, the energetically most stable nuclide.
- The three common **fissile nuclides** are uranium-235, uranium-233, and plutonium-239.
- The **critical mass** of normal-density plutonium in the shape of a sphere is 15 kg (about the size of a grapefruit).
- Gaseous UF_6 is used in the gaseous diffusion method of uranium enrichment.
- Thermonuclear bombs use **lithium-6** as a fuel.
- The principal ore used in mining extraction is **pitchblende,** UO_2.
- **Enriching** uranium ore for use in some reactors changes the U-235 concentration from 0.7% (the natural abundance) to 3%.
- A **moderator** (in a reactor) slows down neutrons so they have a greater probability of colliding with a nucleus. Water and graphite are good moderators.
- **Control rods** in a reactor absorb neutrons so that the reactor does not run out of control. Boron and cadmium are useful moderators.

NUCLEAR EQUATIONS TO KNOW

- **Carbon-14** (detected as the basis for carbon dating) emits a beta particle to form nitrogen-14:

$$^{14}_{6}\text{C} \longrightarrow {}^{14}_{7}\text{N} + {}^{0}_{-1}\text{e} + {}^{0}_{0}\gamma$$

- Radium-226 is an **alpha emitter;** it forms radon-222 when it decays:

$$^{226}_{88}\text{Ra} \longrightarrow {}^{222}_{86}\text{Rn} + {}^{4}_{2}\text{He}$$

- In a multistep **neutron-induced transmutation,** iron-58 absorbs two neutrons to form cobalt-60 and a beta particle. The overall process only is shown:

$$^{58}_{26}\text{Fe} + 2{}^{1}_{0}\text{n} \longrightarrow {}^{60}_{27}\text{Co} + {}^{0}_{-1}\text{e}$$

- In a typical **fission reaction,** uranium-235 absorbs a neutron and breaks up into barium-142 and krypton-92, with release of two neutrons:

$$^{235}_{92}\text{U} + {}^{1}_{0}\text{n} \longrightarrow {}^{142}_{56}\text{Ba} + {}^{92}_{36}\text{Kr} + 2{}^{1}_{0}\text{n}$$

- In a typical multistep **fusion reaction,** six deuteriums form helium, protons, and neutrons. The overall process is

$$6{}^{2}_{1}\text{H} \longrightarrow 2{}^{4}_{2}\text{He} + 2{}^{1}_{1}\text{p} + 2{}^{1}_{0}\text{n}$$

MATHEMATICAL EQUATIONS TO KNOW AND UNDERSTAND

$\text{Rem} = \text{rad} \times Q$	definition of rem
$\text{Rate of decay} = k \times N$	rate of radioactive decay
$\ln\left(\dfrac{N_0}{N}\right) = kt$	integrated rate law for radioactive decay
$t_{1/2} = 0.693/k$	relation of half-life to decay constant
$E = c^2 \Delta m$	mass-energy relation

SELF-TEST EXERCISES

Nuclear stability

1. What particle is deflected most as it moves between two oppositely charged plates?
(a) neutron   (b) $^{4}_{2}\text{He}$ (c) γ (d) $^{12}_{6}\text{C}$ (e) β^{+}

2. What is the energy of a γ-ray photon with $\lambda = 2.65$ pm? $h = 6.63 \times 10^{-34}$ J·s; $c = 3.00 \times 10^{8}$ m·s^{-1}.
(a) 6.54×10^{-8} J   (b) 7.51×10^{-14} J   (c) 1.76×10^{-45} J
(d) 7.95×10^{-4} J   (e) 1.16×10^{-12} J

3. What product is formed when ^{47}K undergoes beta decay?
(a) ^{47}Ar (b) ^{47}Ca (c) ^{46}Ca (d) ^{46}K (e) ^{48}K

4. What product is formed when ^{72}As undergoes positron emission?
(a) ^{72}Ge (b) ^{73}As (c) ^{74}As (d) ^{73}Ge (e) ^{72}Se

5. What product is formed when ^{226}Th undergoes alpha decay?

(a) ^{224}Rn   (b) ^{224}Ra   (c) ^{222}Ra   (d) ^{222}Rn   (e) ^{230}Th

6. What product is formed when ^{160}Er undergoes electron capture?

(a) ^{159}Ho   (b) ^{159}Er   (c) ^{160}Ho   (d) ^{161}Yb   (e) ^{160}Tm

7. What is a likely mode of decay for ^{241}Am?

(a) positron emission   (b) electron capture

(c) beta emission   (d) alpha emission

8. What is a likely mode of decay for ^{184}Tl?

(a) electron capture   (b) positron emission

(c) alpha emission   (d) gamma emission

(e) both electron capture and positron emission

9. What is a likely mode of decay for a light element that is neutron rich?

(a) positron emission   (b) electron capture

(c) gamma emission   (d) beta emission

(e) both positron emission and electron capture

10. What is the product, E, of the nucleosynthesis, $^{239}_{94}\text{Pu} + ^{4}_{2}\text{He} \longrightarrow ^{1}_{0}\text{n} + ^{A}_{Z}\text{E}$

(a) $^{242}_{96}$Cm   (b) $^{241}_{96}$Cm   (c) $^{243}_{96}$Cm   (d) $^{235}_{92}$U   (e) $^{236}_{92}$U

11. Bombarding $^{239}_{94}$Pu with a single neutron results in beta emission and formation of

(a) $^{238}_{95}$Am   (b) $^{239}_{93}$Np   (c) $^{240}_{93}$Np   (d) $^{240}_{94}$Pu   (e) $^{240}_{95}$Am

Radioactivity

12. Which of the following types of radioactivity has the highest penetrating power?

(a) alpha   (b) beta   (c) gamma   (d) positron

13. What is the activity of a radioactive source that undergoes 5.5×10^{13} dis $\cdot$ s^{-1}?

(a) 1.5 kCi   (b) 6.7 mCi   (c) 20 Ci

(d) 49 mCi   (e) 7.9×10^2 Ci

14. What is the dose in rem when 70 kg of tissue receive the equivalent of 6.2 J of alpha radiation? 1 rad $= 10^{-2}$ J $\cdot$ kg^{-1}; $Q = 20$ for alpha radiation.

(a) 18 rem   (b) 87 rem   (c) 1.2 rem

(d) 1.8×10^2 rem   (e) 5.6 rem

15. How many grams of potassium-44 remain if a 25-g sample is allowed to decay for 1.0 h? The half-life of potassium-44 is 22.4 min.

(a) 0.44 g   (b) 1.6×10^2 g   (c) 3.1 g   (d) 3.9 g   (e) 2.6 g

16. If 13.2 g of a radioactive sample remain after the sample ages for 12.0 min, how much sample was originally present? The half-life of the substance is 5.40 min.

(a) 122 g   (b) 2.82 g   (c) 1.43 g   (d) 61.6 g   (e) 166 g

17. A sample of wood from an ancient boat is found to undergo 5200 carbon-14 disintegrations per gram of carbon in an 18-h period. How old is the boat? Current samples of carbon decay at the rate of 920 dis $\cdot$ h^{-1}. The half-life of carbon-14 is 5730 y.

(a) 2.6×10^4 y   (b) 1.8×10^3 y   (c) 2.2×10^3 y

(d) 4.6×10^3 y   (e) 9.6×10^3 y

Nuclear energy

18. What is the missing product in the following fusion reaction?

$$\,^2_1H + \,^3_1H \longrightarrow \,^4_2He + ?$$

(a) $\,^1_0n$ (b) $\,^1_1p$ (c) $\,^0_{-1}e$ (d) $\,^0_1e$ (e) $\,^2_2He$

19. What is the missing product in the following nuclear fission reaction?

$$\,^1_0n + \,^{235}_{92}U \longrightarrow ? + \,^{89}_{37}Rb + 3\,^1_0n$$

(a) $\,^{144}_{55}Cs$ (b) $\,^{146}_{55}Cs$ (c) $\,^{143}_{55}Cs$ (d) $\,^{144}_{54}Xe$ (e) $\,^{146}_{54}Xe$

20. What is the binding energy, in $kJ \cdot mol^{-1}$, of $\,^{34}_{15}Si$? The mass of silicon-34 is 33.9764 u, that of a proton is 1.0078 u, and that of a neutron is 1.0087 u.

(a) 3.18×10^{11} (b) 5.17×10^{11} (c) 4.59×10^{-11}

(d) 6.54×10^{10} (e) 2.76×10^{10}

21. How much energy (in $kJ \cdot mol^{-1}$ of uranium-235) is released in the following fission reaction? The masses of the various species involved are neutron, 1.0087 u; uranium-235, 235.0439 u; lanthanum-147, 146.9281 u; and bromine-87, 86.9207 u.

$$\,^1_0n + \,^{235}_{92}U \longrightarrow \,^{147}_{57}La + \,^{87}_{35}Br + 2\,^1_0n$$

(a) 2.57×10^{10} (b) 8.31×10^{10} (c) 1.68×10^{10}

(d) 7.40×10^{10} (e) 5.45×10^{10}

22. How much energy (in kilojoules) is released when 1.00 g $\,^2H$ undergoes fusion? The masses of the species involved are $\,^2H$, 2.0140 u; $\,^3H$, 3.01603 u; $\,^4He$, 4.00260 u; and n, 1.0087 u. The reaction is $\,^2_1H + \,^3_1H \rightarrow \,^4_2He + \,^1_0n$.

(a) 1.69×10^9 (b) 8.36×10^8 (c) 4.28×10^{10}

(d) 8.36×10^{11} (e) 1.69×10^{12}

23. UO_3 is converted to UF_6 in a four-step process. How many grams of UF_6 can be produced from 100 g UO_3? It is not necessary to do a calculation for each step.

(a) 81.3 g   (b) 66.8 g   (c) 91.4 g   (d) 100 g   (e) 123 g

Descriptive chemistry

24. What compound is used in the enrichment of uranium-235 by the gas diffusion method?

(a) UO_2 (b) UF_6 (c) UF_4 (d) UO_3 (e) U

25. Which of the following is a fissile nuclide?

(a) uranium-238   (b) iron-56   (c) plutonium-239

(d) lithium-6   (e) deuterium

26. One of the following nuclides is not at the head of a natural radioactive series. Which one?

(a) uranium-238   (b) uranium-235   (c) plutonium-239   (d) thorium-232

27. What is the annual dose of radiation received by an average U.S. citizen?

(a) 2 rem   (b) 0.2 rem   (c) 500 rem   (d) 5 krem

28. which of the following is the most stable nuclide?

(a) uranium-235   (b) helium-4   (c) nitrogen-14

(d) iron-56   (e) carbon-14

29. What percentage of uranium-235 is required in the fuel for a typical light water reactor in the United States?

(a) 3%   (b) 0.7%   (c) 8%   (d) 100%   (e) 95%

30. What is the function of a moderator in a fission reactor?

(a) keeps the reactor at a moderate temperature   (b) absorbs neutrons

(c) absorbs electrons   (d) slows down neutrons

(e) emits protons

31. What is the function of the control rods in a fission reactor?

(a) keeps the reactor at a moderate temperature   (b) absorbs neutrons

(c) absorbs electrons   (d) slows down neutrons

(e) emits protons

32. Which of the following is a good reactor moderator?

(a) Li-6   (b) U-238   (c) Na   (d) Fe   (e) C(graphite)

33. Which of the following is a good material for control rods?

(a) B   (b) H_2O (c) Ca   (d) He   (e) Li

34. What is the formula for pitchblende, the principle uranium ore?

(a) UF_6 (b) UO_2 (c) UO_3 (d) U   (e) UF_2

ANSWERS TO SELF-TEST EXERCISES

Chapter 1: 1e, 2c, 3c, 4a, 5c, 6c, 7e, 8e, 9c, 10c, 11e, 12e, 13c, 14b, 15b, 16c, 17a, 18d, 19c, 20a, 21c, 22e, 23b, 24e, 25b, 26d, 27b, 28b, 29d, 30c, 31c, 32a, 33b, 34e, 35b, 36b, 37b, 38a, 39d, 40c, 41b, 42a, 43c, 44c, 45e, 46e, 47b, 48b, 49b, 50c, 51d, 52b, 53d, 54e, 55d, 56a, 57a, 58c, 59b

Chapter 2: 1c, 2d, 3a, 4c, 5e, 6c, 7e, 8d, 9b, 10b, 11e, 12d, 13d, 14b, 15a, 16e, 17c, 18b, 19d, 20a, 21a, 22a, 23a, 24e, 25e, 26e, 27a, 28d, 29a, 30e, 31c, 32c, 33a, 34b, 35b, 36c, 37d, 38c, 39b, 40a, 41e, 42b, 43b, 44d, 45a, 46c, 47c, 48c, 49b, 50c, 51d, 52e, 53b, 54c, 55a, 56e, 57c, 58d, 59b, 60e

Chapter 3: 1c, 2b, 3c, 4a, 5d, 6c, 7b, 8b, 9b, 10b, 11a, 12e, 13e, 14a, 15c, 16d, 17a, 18c, 19e, 20b, 21a, 22d, 23b, 24d, 25e, 26b, 27c, 28a, 29c, 30a, 31b, 32d, 33d, 34e, 35b, 36b, 37a, 38c, 39c, 40a, 41d, 42b, 43a, 44b, 45d, 46c, 47b, 48b, 49a, 50d, 51e, 52b

Chapter 4: 1c, 2e, 3a, 4b, 5b, 6c, 7a, 8d, 9e, 10b, 11a, 12d, 13a, 14e, 15a, 16d, 17a, 18b, 19c, 20e, 21d, 22c, 23c, 24d, 25a, 26b, 27e, 28d, 29a

Chapter 5: 1c, 2a, 3d, 4b, 5d, 6a, 7e, 8c, 9a, 10c, 11b, 12a, 13b, 14a, 15e, 16d, 17e, 18a, 19c, 20c, 21b, 22b, 23b, 24b, 25e, 26d, 27c, 28a, 29b, 30c, 31d, 32b, 33b, 34b, 35e, 36c, 37a, 38c, 39b, 40b, 41e, 42d, 43c, 44a, 45c, 46e, 47d, 48d, 49a, 50c, 51a, 52c, 53e, 54a

Chapter 6: 1a, 2d, 3a, 4a, 5c, 6b, 7e, 8e, 9c, 10b, 11c, 12a, 13c, 14b, 15d, 16a, 17d, 18a, 19e, 20c, 21a, 22a, 23b, 24d, 25e, 26a, 27e, 28c, 29d, 30c, 31c, 32e, 33c, 34b, 35c, 36b, 37c, 38c, 39a, 40a, 41b, 42b, 43d, 44c, 45b, 46e, 47e, 48a, 49a, 50c, 51a, 52b, 53d

Chapter 7: 1c, 2c, 3d, 4b, 5a, 6a, 7d, 8e, 9c, 10c, 11b, 12e, 13e, 14d, 15a, 16b, 17a, 18a, 19a, 20e, 21b, 22b, 23d, 24a, 25e, 26c, 27c, 28e, 29e, 30a, 31a, 32c, 33b, 34d, 35b, 36a, 37c, 38d, 39a, 40a, 41c, 42e, 43b, 44e, 45d, 46b, 47a, 48b, 49c, 50d, 51d, 52b, 53b, 54e, 55b

Chapter 8: 1d, 2c, 3e, 4a, 5b, 6c, 7e, 8a, 9d, 10b, 11c, 12d, 13d, 14b, 15e, 16c, 17d, 18c, 19a, 20d, 21e, 22c, 23b, 24a, 25a, 26d, 27d, 28c, 29a, 30b, 31e, 32a, 33c, 34c, 35c, 36a, 37d, 38e, 39b, 40e, 41e, 42a, 43d, 44e, 45a, 46b, 47b, 48e, 49d, 50e, 51a, 52a, 53a, 54a

Chapter 9: 1b, 2a, 3d, 4e, 5c, 6a, 7a, 8e, 9e, 10c, 11b, 12b, 13e, 14c, 15b, 16b, 17d, 18e, 19e, 20b, 21e, 22c, 23d, 24b, 25a, 26d, 27e, 28a, 29c, 30b, 31e, 32b, 33c, 34d, 35a, 36e, 37b, 38c, 39b, 40b, 41e, 42d, 43c, 44b, 45b, 46c, 47b

Chapter 10: 1d, 2c, 3a, 4b, 5c, 6b, 7b, 8c, 9e, 10b, 11b, 12a, 13d, 14a, 15b, 16a, 17b, 18d, 19a, 20e, 21d, 22a, 23a, 24b, 25c, 26e, 27a, 28d, 29e, 30d, 31b, 32e, 33c, 34a, 35d, 36a, 37d, 38a, 39b, 40a, 41b, 42b, 43d, 44b, 45c, 46e, 47b, 48a, 49d, 50e, 51c, 52e, 53d, 54e, 55a, 56c, 57d

Chapter 11: 1b, 2c, 3a, 4e, 5d, 6b, 7e, 8c, 9e, 10d, 11d, 12b, 13d, 14c, 15b, 16c, 17a, 18a, 19c, 20d, 21e, 22d, 23b, 24d, 25a, 26b, 27a, 28c, 29a, 30c, 31e, 32a, 33c, 34b, 35d, 36a, 37a, 38d, 39e, 40b, 41a, 42b, 43e, 44b, 45e, 46a, 47b, 48a, 49e, 50c, 51d, 52d, 53a, 54d, 55b

Chapter 12: 1b, 2b, 3d, 4b, 5e, 6c, 7c, 8a, 9a, 10c, 11d, 12d, 13e, 14a, 15d, 16d, 17d, 18d, 19b, 20d, 21a, 22b, 23a, 24c, 25d, 26d, 27e, 28e, 29b, 30c, 31e, 32e, 33d, 34a, 35e, 36d, 37e, 38e, 39a, 40a, 41a, 42c, 43c, 44b, 45e, 46c, 47a, 48e, 49b, 50c, 51b, 52a, 53d

Chapter 13: 1c, 2c, 3a, 4b, 5d, 6c, 7d, 8e, 9a, 10b, 11b, 12b, 13e, 14d, 15e, 16d, 17c, 18c, 19e, 20e, 21a, 22a, 23b, 24d, 25b, 26b, 27a, 28d, 29c, 30c, 31b, 32a, 33c, 34b, 35a, 36c, 37c, 38a, 39d, 40a, 41b, 42e

Chapter 14: 1d, 2a, 3a, 4a, 5c, 6c, 7b, 8a, 9b, 10d, 11e, 12c, 13d, 14b, 15c, 16c, 17d, 18c, 19a, 20a, 21b, 22e, 23d, 24a, 25d, 26c, 27c, 28b, 29d, 30d, 31e, 32d, 33c, 34d, 35e, 36e, 37a, 38a, 39c, 40a, 41d, 42b, 43b, 44c, 45e, 46d, 47d, 48b, 49d, 50d, 51a, 52d, 53b, 54a, 55d, 56d, 57c, 58c

Chapter 15: 1e, 2b, 3a, 4a, 5a, 6c, 7b, 8a, 9c, 10e, 11c, 12d, 13b, 14a, 15c, 16b, 17c, 18b, 19d, 20a, 21b, 22c, 23b, 24a, 25d, 26e, 27c, 28a, 29d, 30d, 31a, 32c, 33d, 34e, 35a, 36a, 37a, 38b, 39d, 40e, 41b, 42d, 43a, 44b, 45b, 46c, 47e, 48e, 49b, 50c, 51a, 52c, 53e, 54b, 55e, 56d, 57d, 58a, 59b, 60a, 61d

Chapter 16: 1a, 2c, 3b, 4a, 5d, 6d, 7e, 8b, 9d, 10e, 11c, 12b, 13b, 14d, 15c, 16d, 17e, 18c, 19a, 20c, 21e, 22e, 23c, 24a, 25b, 26d, 27b, 28a, 29c, 30b, 31c, 32a, 33b, 34d, 35b, 36b, 37a, 38b, 39b, 40a, 41c, 42d, 43c, 44a

Chapter 17: 1a, 2a, 3d, 4c, 5b, 6d, 7d, 8b, 9c, 10e, 11a, 12c, 13c, 14a, 15e, 16a, 17d, 18e, 19b, 20c, 21c, 22c, 23a, 24a, 25b, 26b, 27a, 28a, 29c, 30a, 31d, 32b, 33e, 34d, 35d, 36e, 37b, 38a, 39c, 40e, 41b, 42d, 43d

Chapter 18: 1a, 2e, 3c, 4b, 5b, 6d, 7b, 8c, 9b, 10e, 11a, 12a, 13d, 14a, 15e, 16e, 17e, 18b, 19d, 20c, 21b, 22a, 23a, 24b, 25b, 26d, 27d, 28b, 29d, 30c, 31c, 32a, 33a, 34d, 35c, 36d, 37b, 38c, 39e, 40d, 41d, 42c

Chapter 19: 1d, 2a, 3d, 4b, 5c, 6a, 7b, 8b, 9d, 10c, 11e, 12c, 13b, 14c, 15a, 16d, 17d, 18a, 19e, 20d, 21b, 22a, 23c, 24b, 25c, 26e, 27d, 28a, 29d, 30e, 31b, 32a, 33b, 34e, 35a, 36c, 37d, 38b, 39a, 40b, 41a, 42c, 43c, 44a, 45b, 46e, 47d, 48e, 49b, 50d, 51b, 52d, 53c, 54e, 55a, 56a, 57b, 58c, 59e, 60a, 61b

Chapter 20: 1b, 2b, 3d, 4e, 5b, 6c, 7b, 8b, 9d, 10d, 11a, 12d, 13b, 14a, 15e, 16d, 17c, 18e, 19e, 20b, 21b, 22c, 23d, 24a, 25a, 26d, 27e, 28a, 29a, 30d, 31c, 32c

Chapter 21: 1b, 2b, 3c, 4d, 5a, 6c, 7e, 8a, 9b, 10b, 11e, 12c, 13c, 14e, 15a, 16b, 17d, 18b, 19d, 20e, 21d, 22a, 23d, 24d, 25a, 26b, 27c, 28e, 29c, 30c, 31a, 32c, 33d, 34c

Chapter 22: 1e, 2b, 3b, 4a, 5c, 6c, 7d, 8e, 9d, 10a, 11e, 12c, 13a, 14d, 15d, 16d, 17e, 18a, 19a, 20e, 21c, 22b, 23e, 24d, 25c, 26c, 27b, 28d, 29a, 30d, 31b, 32c, 33a, 34b